# Fundamentals of Electron Emission Physics

*Kevin L. Jensen*

*Institute for Research in Electronics and Applied Physics*
*University of Maryland*
*College Park, MD, USA*

WILEY

Published by John Wiley & Sons, Inc., Hoboken, New Jersey.
Published simultaneously in Canada.



*Library of Congress Cataloging-in-Publication Data:*

**Hardback**: 9781394199556
**ePDF**: 9781394199600
**epub**: 9781394199617

Cover Design: Wiley
Cover Image: © Saha Entertainment/Getty Images

*To those gone but whose mark remains:*
*William D. Jensen, Noel J. Baker, Jerome K. Percus*

*In appreciation for love and shared passions:*
*Barbara L. Jensen*

*With love for her support and patience with my odyssey*
*(and devotion and plucky comic relief):*
*Janet Jensen*

# Contents

## Preface

*Our doubts are traitors,*
*And makes us lose the good we oft might win,*
*By fearing to attempt.*

W. Shakespeare, *Measure for Measure* I.4

A party trick of rather doubtful repute is the ability to hold one's breath until the discourse gets quiet or the engagement wanes after confessing to mirthful revelers that one is a physicist who studies processes of fundamental interest with great importance to a broad swathe of critical technologies relying on electron beams. It is not entirely unexpected. Even at the start of the careers of those who view with some trepidation what is required (and they only perceive a sliver of it) but who come to relish that physics, the journey appears laborious and constantly beset with unanticipated but necessary side quests. So it was at the start of my odyssey. Research, like odysseys in general, does not come with a map. And yet, a good travel log can help, and those who write them return a benefit to future generations of physicists and engineers, paid for by past generations and the public at large.

For decades after the tectonic onset of quantum mechanics and its ability to make sense of seemingly incomprehensible phenomena associated with thermal, field, and photo-induced electron emission, electron beam devices (particularly those behind militarily essential technologies like radar) used thermal emission cathodes. By the late 1950s, momentum built for investigating whether the other mechanisms could provide benefit or opportunity. Microfabrication methods in the late 1960s and early 1970s made clear that opportunity was afoot, even as thermal cathode development itself steadily improved. The topics were very new, as were their closely related disciplines like resonant tunneling that earned Leo Esaki the Nobel Prize in 1973: his delightful Nobel talk began with a model of field emission. In the 1980's, through DOD, NASA, DOE, and DARPA funding in the United States, paralleled by similar investments abroad, electron source programs took root in the fertile soils of research laboratories, industrial centers, and academic institutions. With great serendipity, that was when my career began. Not many maps were to be had.

After a decade of benefitting from governmental support, signs increasingly appeared suggesting that the investments in me be repaid forward. There were the article contributions and updated encyclopedia shorts, but then the book chapters [147, 164, 245] began to be solicited. As authorship was in my DNA, it did not take long for my whetted desires to try to make a book-level account of what I learned and was taught, resulting in Electron Emission Physics [149], a tome outside the reach of most students. Travel logs, to be useful though, have to be more accessible, and so I made a more instructional account with *Introduction to the Physics of Electron Emission* [154], which sought to explain a much wider range of the basics implicit in the prior work. That book explained, but did not necessarily show how to do, and in the years since its publication, it became apparent that a tome teaching how to perform the calculations behind the prior books would be welcome to those who, like me decades before, pined for a travel log. The present work is just that. It is hoped that this contribution will help reveal what is needed to calculate, and so appreciate what came in the book before it, just as that book sought to explain ideas utilized in the book before it.

The present work enlarges on a method of the prior work: before, some code was given for the interested reader, but mostly those algorithms were in FORTRAN, a hugely powerful but underappreciated programming language that seemingly replicates in scientific circles (in my experience) the same reactions as the party trick on which this Preface began. Programming languages evolve, and it was felt that to provide true pedagogical benefit, a widely used platform replete with online resources and excellent graphics available to students as well as academic and research groups might prove more beneficial. To that end, the code that generates many of the figures used in the body of the text are given in Appendix B to help students learn the HOW as well as the WHAT. As Albert Einstein sagely advised,

> *If you want to find out anything from the theoretical physicists about*
> *the methods they use, I advise you to stick closely to one principle:*
> *don't listen to their words, fix your attention on their deeds.*
>
> Albert Einstein[1]

Travel logs should be enjoyable, not merely a chronicle of tempests, anthropophagi, and misfortunes in strange lands. I have taken liberties to make that so. May it encourage an adventurous spirit. And to my investors and many, many shipmates (named and unnamed), thank you.

College Park, MD, USA
July 2025

*K. L. Jensen*

1 A. Einstein. The Herbert Spencer lecture, in *Ideas and Opinions*. Crown Publishers. p. 270, 1954.

## Acknowledgments

... let your reason serve
To make the truth appear, where it seems hid,
And hide the false seems true.

W. Shakespeare, Measure for Measure V.1

I thank J. J. Petillo (Leidos); N. M. Riga, D. A. Shiffler (AFRL); J. L. Lebowitz (Rutgers); P. O'Shea, D. Feldman (UMD); N. A. Moody, D. Dimitrov, V. Pavlenko (LANL); M. Osofsky, F. A. Buot, M. McDonald, A. Shabaev, J. E. Yater, E. Rittersdorf, J. Shaw, A. Shih, B. Pate, A. K. Ganguly, P. Mukhopadhyay-Phillips, M. A. Kodis, R. K. Parker, T. Reinecke, E. Zaidman (NRL); D. Finkenstadt (USNA); R. Seviour (U. Huddersfield); Y. Y. Lau, B. Gilchrist (U. Michigan); O. Chubenko (NIU); M. Cahay (U. Cincinnati); R. Forbes (U. Surrey); J. W. Luginsland (AFOSR); A. I. Akinwande (MIT); P. Zhang (Michigan State); E. Montgomery (Euclid); C. Holland, P. Schwoebel (SRI); D. Dowell, J. Smedley (SLAC); S. T. Purcell (U. Lyon), J. P. Xanthakis (NTUA); A. Kyritsakis (U. Tartu); W. Mackie (Linfield); P. Hawkes (CEMES-CNRS) – many who shared their insights over decades. Great appreciation to L. Phillips for decades of casual but strangely useful insights and puncheons of coffee. The expertise required to bring this work into existence was made possible by substantial funding by the Naval Research Laboratory (NRL), the Office of Naval Research (ONR), the Department of Energy (DOE), the Air Force Research Laboratory (AFRL), the Air Force Office of Scientific Research (AFOSR), the Joint Technology Office (JTO), and the University of Maryland (UMD-IREAP): I am honored by their investments in my research. I remember and acknowledge influential colleagues who have passed away: J. K. Percus (NYU), A. Rokhlenko (Rutgers), C. A. Spindt (SRI), F. M. Charbonnier (Linfield), P. J. Price (IBM – T. J. Watson), M. Virgo (ANL), V. T. Binh (U. Lyon), I. Brodie (UC Davis), C. K. Chu (Columbia), and C. L. Bohn (NIU).

*Kevin L. Jensen*

## Acronyms

| | |
|---|---|
| **AK** | Anode–Cathode |
| **AL** | Accumulation Layer |
| **CBM** | Conduction Band Minimum |
| **CL** | Child–Langmuir |
| **FD** | Fowler–DuBridge (photoemission) or<br>Fermi–Dirac (electron statistics) or<br>Forbes–Deane (elliptical integral function) |
| **FN** | Fowler–Nordheim |
| **HPM** | High power microwave |
| **LDR** | Lorentz–Drude–Resonant |
| **MD** | Molecular dynamics |
| **MM** | Moments model |
| **MTE** | Mean Transverse Energy |
| **Nd:YAG** | Neodymium-doped yttrium aluminum garnet |
| **NEA** | Negative Electron Affinity |
| **oGTFP** | Original General Thermal-Field-Photoemission |
| **PIC** | Particle-in-Cell |
| **rGTFP** | Revised General Thermal-Field-Photoemission |
| **RLD** | Richardson–Laue–Dushman |
| **SCL** | Space charge limited |
| **SEY** | Secondary electron yield |
| **SMM** | Simple Moments Model |
| **SN** | Schottky–Nordheim |
| **TMA** | Transfer Matrix Approach |
| **TWT** | Traveling Wave Tube |
| **UEM** | Ultrafast electron microscopy |
| **VBM** | Valence Band Minimum |
| **VNE** | Vacuum Nano-Electronics |
| **WKB** | Wentzel–Kramers–Brillouin, sometimes appearing as JWKB for<br>Jeffreys–Wentzel–Kramers–Brillouin |

# Part I

# Foundations

# 1

# Historical Background

*For 'tis your thoughts that now must deck our kings,*
*Carry them here and there, jumping o'er times,*
*Turning th' accomplishment of many years*
*Into an hour-glass; for the which supply,*
*Admit me Chorus to this history ...*

W. Shakespeare, *Henry V* I.Pro

As the Renaissance came to a close, previously heretical notions gained support in what came to be known as the Scientific Revolution. The organization of the heavens, in which Earth was no longer the center of the Universe (the Copernican model), championed by Galileo Galilei (Italian) is perhaps the better-known instance, but the atomic hypothesis beautifully articulated by the Roman poet Lucretius and strenuously advocated by Giordano Bruno (Italian) and supported by Galileo and others like Thomas Harriot (English) and Johannes Kepler (German), was comparable in importance but perhaps not as widely appreciated by modern audiences who lean towards big spectacle. Shortly before Shakespeare wrote *Hamlet*, Bruno was burnt at the stake in 1600 by the Inquisition, and a little more than 20 years after Shakespeare wrote *The Tempest*, Galileo was under house arrest in 1633, having been threatened with torture by the Inquisition if he did not recant.

The atomic hypothesis was too explanatory to suppress for long. Even the brilliant Descartes (French), who was not an atomist because he opposed the idea of vacuum and thought particles could be divided into smaller particles, incorporated ideas of atomism into his influential mechanistic philosophy. John Dalton (English) was able to explain chemical reactions by the union of molecules, and by 1890, Amedeo Avogadro (Italian) wrote about his intuition that two equal volumes of different gases kept at the same temperature and pressure would nevertheless contain the same number of molecules. By 1839, Michael Faraday (English) would write "On the Absolute Quantity of Electricity Associated with the Particles or Atoms of Matter" in *Experimental Researches in Electricity*, introducing the idea that charge, like particles of matter, came in units in equal proportion to the number of electrolytes. In his treatise on electricity and magnetism, James C. Maxwell (Scottish) remarked that the idea of a "molecule of electricity" was "gross... and out of harmony" with his treatise, and predicted that when an understanding of electrolysis materialized, it would not include the notion of "molecular charges." Nevertheless, by 1874, George Johnstone Stoney (Irish) made a crude estimate of the basic unit of charge, which he termed "$e$" (but which herein shall be $q$) and christened it as "electron" in 1891.

*Fundamentals of Electron Emission Physics*, First Edition. Kevin L. Jensen.

In between the estimate of $q$ and the naming of the electron, Edwin J. Houston (American) presented in October 1884 at the first *American Institute of Electrical Engineers* meeting in Philadelphia an inexplicable high vacuum phenomena observed by the inventor Thomas A. Edison (American). Edison had taken his incandescent lamp, in which a carbon fiber was enclosed in a glass vessel, the ends of which were connected to a battery, and inserted a strip of platinum connected to a galvanometer. Under ordinary operation, light was produced with no unusual effects, but as the current levels rose so that the luminosity of the lamp increased markedly (a factor of 3 to 12), the needle of the galvanometer was strongly deflected by current passing through its coils. For the physical theories of the time, and as Houston reported to the audience, the origin of the current was a mystery. Even more mysterious, when the leads were switched so that the galvanometer was connected to the negative terminal, then the current flowed in the opposite direction, but at about 1/40 of the level. Houston was baffled and told his audience

> *The question is, what is the origin of this current? How is it produced? Since we have within the globe a nearly perfect vacuum, we cannot conceive the current as flowing across vacuous space, as this is not in accordance with our pre-conceived ideas connected with higher vacua... I have no theory to propound as to the origin of these phenomena.*
>
> E. J. Houston [132]

The phenomena became known as the "Edison Effect" and the current as "Cathode Rays." Up until 1890s, very different conceptions about their nature were debated. One explanation, identified with the German physicists, was that cathode rays were a process occurring in the aether because their motion in magnetic fields was in circular paths rather than straight lines. A competing explanation argued that the cathode rays were material and were the paths of particles of negative charge. In 1895, Jean Baptiste Perrin (French) demonstrated the particle nature of the cathode rays, and then in 1897, Joseph J. Thomson (English) settled the matter definitively by showing the deflection of the rays passing between a capacitor plate [340], proving that all of the particles had the same charge-to-mass ratio that was independent of the potential between the electrodes, and indifferent to either the residual gases within the tube or the chemical nature of the materials from which the rays were produced.

Physicists rapidly overcame their skepticism about the existence of elementary particles of negative charge [70]. The quantum theory pioneered by Erwin Schrödinger (Austrian), Werner Heisenberg (German), and brought to the United States by J. Robert Oppenheimer (American), was used by Niels H. D. Bohr (Danish) to explain the spectral lines of hydrogen and Arnold Sommerfeld (German) to put forth a model of metals. By the 1930s, through the efforts of Owen W. Richardson (English), Laue (German), and Dushman (Russian/American) the canonical equation for thermionic emission was developed bearing their names; Ralph H. Fowler (English) and Lothar W. Nordheim (German/American) did the same for field emission, and Fowler and DuBridge (American) for photoemission.

The development of vacuum electron tubes and radio communications was a watershed technology reliant on electron emission [98], but given that they were superseded by transistors in the 1960's, and cathode ray tubes (CRT's) by flat panel displays by 2007, focusing on a different technology here is more useful. The realization of how to manipulate electron beams inaugurated an era of relentless technological innovation, the most important of which may very well have been radar, a technology that has not ceded its empire to solid-state, especially for high-power devices. Given its importance in directing the outcome of the Second World War, radar deserves far more attention than what is often accorded to it.

*There were many factors that allowed the Allies to defeat the Axis forces in World War II, but the introduction of radar was one of the most important. There is no way to know whether the Allies would have lost the war if they did not have radar, but it is quite clear that there would have been more losses and a longer time needed to win the war if there were no radar.*

M. Skolnik [316] / IEEE

Atomic weapons, credited in the popular imagination as having outsized importance in WWII, were not a part of bringing the war in Europe to a close in 1944. Both Truman and his advisors understood that the ending of the war with Japan in 1945 had options other than the atomic bomb [351].

Other applications of electron sources rapidly materialized. Microwave ovens are perhaps the most obvious vacuum device, using a magnetron familiar from early radar devices. The most familiar devices to older generations are CRTs, which were the basis of all televisions, computer monitor displays, and oscilloscopes for decades. The heart of the CRT was a thermionic cathode that generated a beam of electrons that were accelerated towards a phosphor screen such that the beam scanned back and forth ("raster"). Before being overtaken by LCD and plasma displays, flat-panel field emission displays (FEDs) were developed in the 1990s by using small clusters of microfabricated field emitters to individually address pixel elements [330, 339].

Field emission microscopy [72], flash X-ray sources [247], and microwave devices using metallic wires [41] received intense investigation beginning in the late 1950s as a possible alternate to thermionic cathodes (the ability of field emission to be strongly modulated being the driver). Microwave amplifiers and high-power microwave (HPM) devices had strong military applications [110, 309]. Although solid-state electronics are justifiably venerated with solid-state devices dominating the lower-power, lower-frequency regimes, it is still the case that high average power and frequency need the "tubes" [29]. Vacuum electronics [99, 112, 274], and after the 1980s vacuum microelectronics [379] and vacuum nanoelectronics [77] led to much cathode development for microwave amplifiers and "tubes" [107] like klystrons, TWTs, magnetrons, crossed field amplifiers, gyrotrons, particle accelerators, rf injectors [229], free electron lasers (FELs) [30, 275], energy recovery linacsI (ERLs) [266, 287, 313], and X-ray FELs [344] use various cathode technologies.

Particular military and commercial interests in addition to radar include communications, electronic countermeasures, directed energy, long-range surveillance, long-range weather detection, airborne weather, missile tracking and guidance, marine radar, air and ballistic missile defense, high-resolution mapping and satellite altimetry, and airport surveillance [316]. Electron beam lithography [162], an analog of photolithography using particles instead of light, is a method to create very small structured circuits and nanotechnology by using an electron beam to expose a resist, which can then be selectively removed. Similarly, field emission scanning electron microscopy (FESEM) images secondary electrons that are emitted from the surface of a sample that are created by primaries emitted from a field emission source and accelerated through a high-gradient potential (generally in the kV range) [276], a development over the typically thermionic tungsten source that required long scan times. A higher brightness field emission source enables a modern FESEM to achieve resolution of feature sizes on the order of 4 nm. Other uses include charge neutralization cathodes for ion and Hall thrusters used for satellite propulsion [246], and thermionic and non-thermionic cathode alternatives for electrodynamic tethers [192] for propellantless propulsion for satellites passing through the Earth's magnetic field [106, 260].

Field and secondary emission are not always opportunities: they can cause dark current on the surface of multialkali antimonide photocathodes used in photoinjectors [270] or lead to breakdown on metal surfaces subject to high fields. Copper surfaces of particle accelerators (for example, the Stanford Linear Accelerator (SLAC)) exhibit regions of melting and protrusion formation that ultimately give rise to undesirable emission off of the copper cavity walls, leading in turn to breakdown [267, 361, 362]: breakdown phenomena, in general, are intimately associated with the emission of electrons [32, 199, 305, 354].

Behind these myriad applications, opportunities, and problems lurk four mechanisms for extracting electrons from materials and one process that limits how many of them can be emitted. They are thermal, field, photo, and secondary electron emission, and how much current associated with each that can be drawn from a cathode (or a breakdown site) is constrained by space charge. They have characteristic behaviors and result in characteristic equations. It is the aim of the present work to instill a sense of familiarity in the calculation of quantities associated with these processes. Knowledge is borne of observation, but understanding is revealed in the ability to manipulate to a chosen end.

# 2

# Basic Methods

*We fortify in paper and in figures,*
*Using the names of men instead of men ...*

W. Shakespeare, Henry 4.2 I.3

## 2.1 Units

### 2.1.1 SI Units

The great lengths to which theoretical physicists often go to render their equations in dimensionless form belie a faint chauvinism and evidence of an unwise distaste for units. That is unkind. Units are an asset to calculations foresworn at peril. The *International System of Units* (SI) is the basis for the metric system, and is made up of seven base units of which six are important here. The ignored unit is a candela (a measure of perceived brightness of light). The remaining six, and their SI units, are

1. Length in meters (m)
2. Time in seconds (s)
3. Mass in kilograms (kg)
4. Electric current in ampere (A)
5. Temperature in kelvin (K)
6. Amount in mole (mole)

The last two (kelvin and mole) will be retained in the units to be used in later sections, leaving the first four of meters, kilograms, seconds, and amperes (or MKSA as previously known) to be examined. The SI units are the units of engineering on a human scale as in Table 2.1, reflected in the fact that in looking at the third column, one sees dead people. Electron emission occurs at an atomic scale, and therefore units that are comparable to the quantities associated with the hydrogen atom tend to be easier to work with, easier still if one sticks to the basic six units. Part of the difficulty of working with large exponents is mitigated by relying instead on the prefixed SI units of nanometers (nm) and femtoseconds (fs). If, instead of units of [V/m] for electric field $\mathcal{E}$, gigavolts per meter, or [GV/m], are used, then fortuitously, $q\mathcal{E}$ has the same numerical value in MKSA as $F$ does in [efnq], and will make discussions especially for field emission more transparent.

The hydrogen atom recommends a few more changes.

*Fundamentals of Electron Emission Physics*, First Edition. Kevin L. Jensen.

**Table 2.1** Units conversions: [MKSA] = factor × [efnq]. The number $u = 6.241509074$ is the mantissa of the factor where shown.

| Quantity | MKSA symbol | MKSA unit | Factor | efnq unit |
|---|---|---|---|---|
| Length | m | Meter | $10^9$ | nm |
| Time | s | Second | $10^{15}$ | fs |
| Energy | J | Joule | $u \times 10^{18}$ | eV |
| Charge | C | Coulomb | $u \times 10^{18}$ | $q$ |
| Mass | kg | Kilogram | $u \times 10^{30}$ | eV (fs/nm)$^2$ |
| Force | N | Newton | $u \times 10^9$ | eV/nm |
| Potential | V | Volt | 1 | eV/$q$ |
| Current | A | Ampere | $u \times 10^3$ | $q$/fs |
| Current Density | J | A/cm$^2$ | $u \times 10^{-11}$ | $q$/(fs-nm$^2$) |
| Field | $\mathcal{E}$ | V/m | $10^{-9}$ | eV/($q$-nm) |
| Power | W | Watt | $u \times 10^3$ | eV/fs |
| Capacitance | F | Farad | $u \times 10^{18}$ | $q^2$/eV |
| Resistance | Ω | Ohm | $1/u$ | eV fs/$q^2$ |

### 2.1.2 Hydrogen Atom

The hydrogen atom is the simplest arrangement of interacting charged particles, and so units based on its physics are uncommonly useful in dealing with emission. A quantum mechanical description of the hydrogen atom suggests two alterations to the focus of the [MKSA] units: although length and time are still useful base units, electron transitions in the hydrogen atom are more easily expressed in units of electron volts [eV], and the charge of the proton and electron is more readily set to one unit of charge and its negative, henceforth called [q] here. The electron charge is $-q$ in these preferred units. As a result, the recommended units are eV-fs-nm-q or [efnq] hereafter. The correspondence between [MKSA] and [efnq] is shown in Table 2.1. Included in that table are some legacy units common in the electron emission literature, in particular, [Amp/cm$^2$] for current density, and units such as [GV/m = $10^9$ V/m], [MV/m = $10^6$ V/m], or [kV/m = $10^3$ V/m] for field strength are often encountered. Some passing deference is paid to them because, face it, if one is typing on a `qwerty` keyboard, one is living in a glass house.

The evaluation of formulae in [efnq] units deals with quantities of order unity in electron emission phenomena, and is therefore simpler to manipulate. A plus is that there are fewer units to keep track of. As an example, consider the following two evaluations of a quantity known as the fine structure constant $\alpha$ that assembles a multitude of units into a dimensionless quantity that is arguably one of the most important constants. In MKSA, it is evaluated by[1]

$$\begin{aligned} \alpha &= \frac{q^2}{4\pi\varepsilon_0 \hbar c} \\ &= \frac{(1.602 \times 10^{-19}\ \mathrm{C})^2}{4\pi(8.854 \times 10^{-12}\ \mathrm{F/m})(6.626 \times 10^{-34}\ \mathrm{Js})(2.998 \times 10^8\ \mathrm{m/s})} \\ &= 0.00729735 \end{aligned} \tag{2.1}$$

1 The formula uses 4-digit precision in representation for compactness, but the actual calculation uses the full 8 digits of accuracy of Table 2.2.

compare that traffic jam to

$$\alpha = \frac{4Q}{\hbar c} = \frac{1.43996 \text{ eV nm}}{(0.658212 \text{ eV fs})(299.792 \text{ nm/fs})} = 0.00729735 \tag{2.2}$$

which lends itself to a fast approximation of $\alpha \approx (7/5)/[(2/3)300] = 7/1000$. The parameter $Q$ introduced in Table 2.2 shall emerge as one of the most often used parameters in the emission equations due to its relationship to barriers associated with metals. As a second example, for order of magnitude estimates, $\hbar \approx \pi/5$ and $c \approx 100\pi$ in [efnq], and with approximating $\pi^2 \approx 10$, then $\hbar c \approx (100/5)10 = 200$ eV-nm compared 197.327 eV-nm as would be found using values in Table 2.2. As a historical curiosity, Heisenberg mischievously offered $\alpha = \pi/2^4 3^3$ at a time when the constants of nature were thought to have a mathematical genesis by some eminent physicists [16].

Both energy and angular momentum in the hydrogen atom are quantized, and so units derived from the energy levels or the unit of angular momentum $\hbar$, respectively, are suggested. Conjoining the magnitude of the electron charge, or $q$, with a potential $\varphi$ allows $q\varphi \equiv V$ to be expressed in [eV], the natural unit to express work function $\Phi$ in. The ionization energy of the electron in the ground state of hydrogen is the Rydberg energy $R_y = 13.6057$ eV. The angular momentum of the electron is quantized ($mvr = n\hbar$ for $n$ an integer) for which the units are of Planck's constant $\hbar$. Calculation then shows that the velocity of the electron in the ground state of the hydrogen atom is $v \equiv \alpha c$ because the kinetic energy is $(1/2)mv^2$ and

$$\alpha = \sqrt{\frac{2R_y}{mc^2}} = \sqrt{\frac{27.2114 \text{ eV}}{510999 \text{ eV}}} = \frac{1}{137.036} \tag{2.3}$$

The planetary model of Bohr for the hydrogen atom suggests a length scale: the electromagnetic force of attraction counters the centrifugal force, or

$$\frac{mv^2}{r} = \frac{q^2}{4\pi\varepsilon_0 r^2} \Leftrightarrow \frac{m(\alpha c)^2}{r} = \frac{\alpha\hbar c}{r^2} \tag{2.4}$$

**Table 2.2** Fundamental constants[a] in units of eV, nm, fs, q, and K to nine significant digits.

| Symbol | Definition | Value | Unit |
|---|---|---|---|
| $-q$ | Electron charge | $-1$ | q |
| $c$ | Speed of light | 299.792458 | nm/fs |
| $k_B$ | Boltzmann's constant | 1/11604.5181 | eV/K |
| $\hbar$ | Planck's constant/$2\pi$ | 0.658211957 | eV fs |
| $\hbar c$ | Planck × light speed/$2\pi$ | 197.326981 | eV nm |
| $\alpha$ | Fine structure constant | 1/137.035999 | - |
| $mc^2$ | Electron rest energy | 510998.950 | eV |
| $a_o$ | Bohr radius $\hbar/m\alpha c$ | 0.0529177211 | nm |
| $Q$ | $\alpha\hbar c/4$ | 0.359991137 | eV nm |
| $R_y$ | $2Q/a_o$ | 13.6057 | eV |
| $N_A$ | Avogadro's number | $6.02214076 \times 10^{23}$ | - |

a) NIST Database https://physics.nist.gov/cuu/Constants/Public Domain

The particular value of $r$ for which this holds is the Bohr radius $a_o$, and so

$$a_o = \frac{\hbar}{\alpha mc} = 0.0529177 \text{ nm} \tag{2.5}$$

and the Rydberg energy is seen to be equivalent to

$$R_y = \frac{\alpha \hbar c}{2a_o} = 13.6057 \text{ eV} \tag{2.6}$$

It will prove convenient to write this relationship instead as

$$R_y = \frac{2Q}{a_o} \tag{2.7}$$

which serves as the first definition of the parameter $Q = \alpha\hbar c/4$. A characteristic time for the hydrogen atom is how long an electron with a velocity $\alpha c$ takes to cross the distance of the Bohr radius, or $\tau = a_o/\alpha c = \hbar a_o/4Q$. A characteristic current is, therefore $q/\tau$, related to the Alfvén–Lawson current $I_o$ encountered in the beam envelope equation [293] that will arise when electron beams are analyzed and given by

$$I_o \equiv \frac{q}{\alpha^3 \tau} = \frac{qc}{\alpha^2 a_o} = \frac{5665.26}{\alpha^2} \text{ [q/fs]} \tag{2.8}$$

It is more commonly encountered as $I_o = 4\pi\varepsilon_0 mc^3/q = 17045.1$ A in treatments using SI units of amperes.

### 2.1.3 Emission Units

The use of $Q$ instead of $R_y$ is motivated by the treatment of the Schottky–Nordheim (SN) emission barrier associated with metals. In the simplest approach, a cube of metal of sides $L$ has eigenstates $\psi_n(x) \propto \cos(k_n x)$ where $k_n = 2\pi n/L$, and similarly for the $y$- and $z$-coordinates, characterized by $k_m$ and $k_l$ respectively. Electrons put into the cube fill the lowest available eigenstate two at a time. The last electron into the cube occupies the highest filled eigenstate, the energy $E_F$ of which is designated the Fermi energy: in statistical mechanics, it is referred to as the "chemical potential" $\mu$ at $T = 0$. For an electron to escape the cube, as Einstein characterized the problem [76], an electron needs to acquire an additional amount of energy called the "work function" $\Phi$. If the zero of energy is taken as the bottom of the conduction band of the metal, then the height of the barrier to emission is $V_o = \mu + \Phi$. If an electric field is applied to the metal in the $x$-direction and the sides of the cube are so large that the problem can be treated as a one-dimensional problem, then the energy barrier as a function of the distance $x$ from the surface declines linearly with distance. From classical electrostatics, a mathematically elegant method to enforce boundary conditions is using the method of images [78]. A departing electron appears to see its image charge, equidistant from the surface but on the opposite side. Thus, the distance to the image charge is $2x$. In all, then, the barrier is given by (first in the more familiar SI units, second in a form on which subsequent theory is based)

$$\begin{aligned} V_{SN}(x) &= \mu + \Phi - q\mathcal{E}x - \frac{q^2}{16\pi\varepsilon_0 x} \\ &\equiv V_o - Fx - \frac{Q}{x} \end{aligned} \tag{2.9}$$

and is called the Schottky–Nordheim (SN) barrier of electron emission. The canonical equations of electron emission presume it. The quantity $Q$ therefore is the simplest compilation of terms making up the image charge contribution to the SN barrier. Because it is the interaction of two equal and oppositely charged particles, it shares terms with the more familiar Rydberg energy via $R_y = 2Q/a_o$.

In what follows, the subscript "SN" shall be suppressed for visual simplicity, but it is understood that $V(x)$ presently refers to the SN barrier.

The maximum of $V(x)$ occurs where $dV/dx = 0$, and here is called $x_o$: it is easily shown to be

$$\frac{dV}{dx} = -F + \frac{Q}{x^2} \Rightarrow x_o \equiv \sqrt{\frac{Q}{F}} \tag{2.10}$$

and is independent of both the Fermi level $\mu$ and work function $\Phi$. An electron with energy $E$ will pass over the barrier ("fly-over") if $E > V(x_o)$ and tunnel under it if $E < V(x_o)$ where

$$V(x_o) = \mu + \Phi - \sqrt{4QF} \equiv \mu + \phi \tag{2.11}$$

where the effective or reduced work function $\phi = \Phi - \sqrt{4QF}$ is introduced. Some dimensionless quantities are inordinately useful. An important one is $y(E)$ which measures how far a barrier $V(x)$ is reduced by the application of an electric field compared to $\mu + \Phi - E$ or

$$y(E) = \frac{\sqrt{4QF}}{\mu + \Phi - E} \tag{2.12}$$

for the SN barrier explicitly, but will be generalized in Eq. (9.2), where additional properties of the SN barrier will be found in Section 9.

Emission current density $J$ is defined as the number of electrons in unit time from unit area that pass from the metal into vacuum over or through a barrier. In the emission literature, the SN barrier is often implicitly used, for which $y(\mu)$ is a particular quantity related to the utilization of elliptical integral functions $v(y)$ and $t(y)$ introduced by Fowler and Nordheim, and used to develop the most common form of the Fowler–Nordheim (FN) equation by Murphy and Good [264]. In the works of R. Forbes [87, 90, 91], who provides an alternate but equivalent formulation of the FN equation to the approach taken below, $y(\mu)^2 \equiv F/F_R \equiv f$ is designated the "scaled field" and is the ratio between the applied $F$ and a reference $F_R = \Phi^2/4Q$ that pulls the SN barrier down to the Fermi level $\mu$. As a result, where $y(\mu) = 1$ functions as a boundary to when tunneling by an electron at the Fermi level is occurring, as tunneling occurs only for $y < 1$ at zero temperature.

A proper evaluation of current density past the SN barrier requires a quantum mechanical treatment, but statistical mechanics can be quite gratifying. The current density $J$ and the number density $\rho$ are integrals over a distribution function [209]. When applied to particles like atoms of a gas, the distribution function $f(\vec{r}, \vec{k})$ is a six-dimensional phase space of position $\vec{r}$ and momentum $\hbar\vec{k}$, but if the current flow is axisymmetric and the transverse directions are uniform (e.g. no electric fields in the transverse direction), then the problem can be narrowed to position $x$ and momentum $\hbar k_x$ in only one direction, so that

$$f(x, k_x) = \frac{1}{(2\pi)^2} \iint dy\, dz\, dk_y\, dk_z\, f(\vec{r}, \vec{k}) \tag{2.13}$$

is more simply considered. The moments of the distribution function

$$\begin{aligned} \rho(x,t) &= \frac{1}{2\pi} \int_{-\infty}^{\infty} f(x, k_x, t)\, dk_x = \langle k_x^0 \rangle \\ J(x,t) &= \frac{q}{2\pi} \int_{-\infty}^{\infty} \frac{\hbar k_x}{m} f(x, k_x, t)\, dk_x = \frac{\hbar}{m} \langle k_x^1 \rangle \end{aligned} \tag{2.14}$$

identify $\rho(x, t)$ as a number density; the inclusion of $q$ in $J(x, t)$ makes it a charge current density. Properly, moments [202] are generally defined around the average $\langle k \rangle$ so that the $n^{th}$ moment is $\langle (k - \langle k \rangle)^n \rangle$, but that is fastidious for present needs: here and below it is easier to speak of $\langle k^n \rangle$ as the "moment," realizing that the conventional moment can be obtained from it and the lower moments if needed. A simple analysis neglects the spatial variation of the distribution, which only matters near the barrier, and assumes the distribution is spatially uniform up to the surface (or barrier). Thus, the distribution collapses to $f(k)$, sometimes called the "supply function," from which quantities like average velocity $\langle v_x \rangle = \hbar \langle k_x \rangle / m$ and kinetic energy $\hbar^2 \langle k_x^2 \rangle / 2m$ are evaluated. Two iconic examples illustrate the point. For the first, consider a spatially uniform environment with no forces. Were this a classical problem like a non-interacting gas of atoms, then a Maxwell–Boltzmann distribution $f_{MB}[E(k)]$ would be in play, but instead, consider a quantum mechanical gas of electrons governed by a Fermi–Dirac distribution $f_{FD}[E(k)]$). For the second, consider a one-dimensional current of charge.

In the first example, for the electron gas, the energy $E$ depends only on the magnitude of a three-dimensional momentum $\hbar \vec{k}$, not direction, and so $E(k) = \hbar^2 k^2 / 2m$. Thus, instead of cube of material as considered above, it is more practical to consider spherical coordinates for which the angular coordinates may be integrated over to give $4\pi$. For no spatial variation and steady state conditions, the density $\rho$ is simply a constant. Electrons, being fermions, do not have overlapping quantum numbers, one of which is energy: the last added electron to the energy "box" is at the Fermi level $E = \mu$. Its momentum is therefore $\hbar k_F \equiv \sqrt{2m\mu}$. At zero temperature, a Fermi–Dirac distribution is unity up to $E = \mu$ and zero thereafter (*aka* a step function), and so using $g = 2$ for spin and $d^3k = 4\pi k^2 dk$

$$\rho = \frac{4\pi g}{(2\pi)^3} \int_0^{k_F} k^2 dk = \frac{k_F^3}{3\pi^2} \tag{2.15}$$

a well-known result in statistical mechanics [209].

In the second example, Poisson's equation in one dimension relates the density of charge $\rho(x)$ to the variation of $V(x)$. In [efnq] units, it is

$$\Delta F = 16\pi Q \rho(x) \Delta x = 16\pi Q \Delta \sigma(x) \tag{2.16}$$

Unlike a single charge, a sheet of charge produces an $F$ which does not decline with distance (assumed to be known here, but proven in Section 15.1). Hence, the total $F$ across a gap is just its value at the cathode, $F(0)$ plus a summation over all the incremental forces due to the incremental sheets of charge. It follows [78, 87, 170]

$$F(x) = F_o + 16\pi Q \int_0^x \sigma(x')dx' \tag{2.17}$$

but rather than try to solve this by integration again, consider a simpler approach: A slab of charge is produced by a current density acting over a time increment, or $\sigma = J\tau$, where $\tau$ is the transit time of a sheet of charge across the AK gap. As a thought experiment, imagine that all of the charge in the gap was emitted in one tumultuous infinitesimal instant, the sheet of charge being the same as if a constant charge density was emitted over $\tau$. If the charge density of that sheet is such as to eliminate the field behind it at the cathode, then that charge density is by definition the Child–Langmuir limit, or $J_{CL}$. Although this does not appear to help, because $\tau$ is unknown, it is clear that $\tau$ has to be larger than, but of a comparable size to, the transit time for a single charge in a constant background field of $F_o = V_a/D$, or $\tau > \tau_o = \sqrt{2mD/F_o} = D\sqrt{2m/V_a}$. So, what happens if $\tau_o$ is used in place of $\tau$? It follows $V_a \approx 16\pi Q J_{CL} \tau$, or, putting in the pieces,

$$J_{CL}(V_a, D) \lesssim \frac{q}{16\pi Q \sqrt{2m}} \frac{V_a^{3/2}}{D^2} \tag{2.18}$$

Such a simple analysis gets across the essential finding that $J_{CL} \propto V_a^{3/2}/D^2$, while acknowledging the fact that the actual transit time for sheets moving in a gap where charge is present and fields reduced takes longer, enough that the 16 in the denominator should be 18 instead (Eq. (2.33) below).

The focus on a moments-based approach to looking at electron emission has therefore, by necessity, drawn attention to the first two moments in particular – the number density $\rho$ and the current density $J$. In time, they will be further related by the continuity equation of Eq. (4.17), as also follows from a distribution function approach, but a quantum mechanical version of the same equation will arise when considering the Wigner distribution function (WDF).

## 2.2 Elementary Emission Models

Electrons emitted from a metal allow the distribution of electrons to be thought of as an electron gas (the Sommerfeld model used by [93, 268] in one of the early successes of the then-new quantum theory). The meaning of $f(k)$ will change when quantum methods are considered, but for now, treat it heuristically. Inside the metal, $f(k)$ is the distribution of electrons, but for calculations in the vacuum outside a metal surface, $f(k)$ becomes the distribution of electrons that have escaped over or through the surface barrier. Such a conception allows for a quick introduction to electron emission and gives crude representations for what will come to be known as the "canonical" equations.

The canonical equations refer to four equations of electron emission plus one law. Those who first developed them have bequeathed their surnames to those equations. They are (ordered to correlate with Figure 2.1):

- **Field emission** for electron emission by the application of high electric fields that enable quantum-mechanical tunneling, leading to the Fowler–Nordheim equation [95];
- **Thermal emission** for electron emission by heating the cathode so that a thermal tail of the distribution function has sufficient energy to overcome the surface barrier, leading to the Richardson–Laue–Dushman equation [70];
- **Photoemission** for electron emission by the absorption of photons such that the electrons have sufficient energy to pass over the barrier, leading to the Fowler–DuBridge equation [69];
- **Secondary emission** for electron emission by bombardment with high-energy primary electrons that transfer energy to secondary electrons such that they have sufficient energy to pass over the surface barrier, leading to the Baroody equation [15]; and
- **Space charge** limited emission restricts what can be emitted due to the presence of charge in the AK gap that suppresses the field at the surface of the cathode or constrains the amount of charge able to cross the gap, leading to the Child–Langmuir law [220].

Equations describing photo, thermal, and field emission existed by the 1930s, and secondary by the 1950s; all have been generalized, improved, and extended since. The canonical forms are nevertheless pervasively used in the literature. Their popularity derives from their qualitative account of the processes they describe and the simplicity of how they do so. To understand them, a good deal of physics must first be considered, as it will in due course. A stripped-down version of each process, though, gets the point across and gives a glimpse of their commonality.

### 2.2.1 Thermal Emission

The earliest method to liberate electrons from the filament of a lightbulb [132] used intense heating, although the intent was to generate light. The electrons that escape their bondage to the metal are in

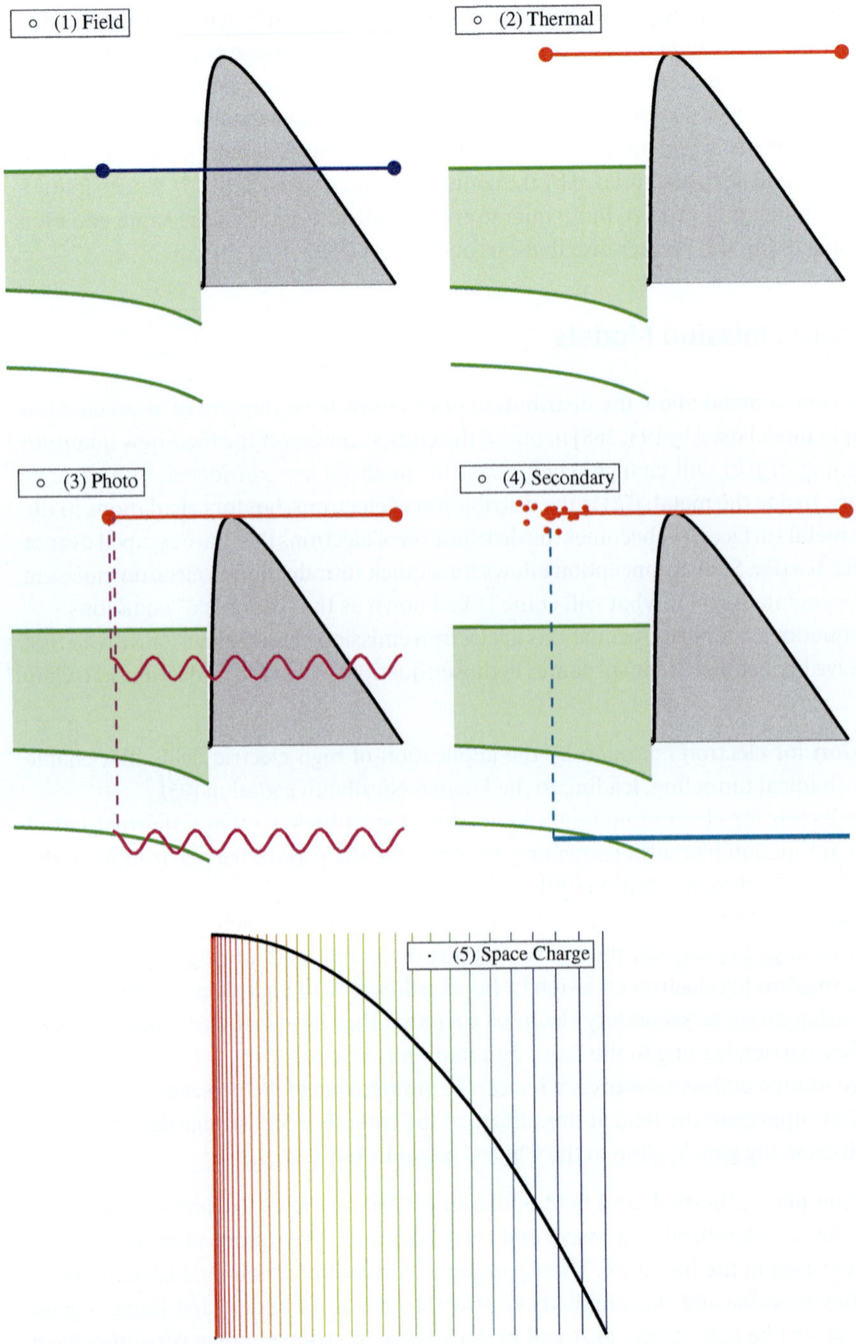


**Figure 2.1** Electron emission schema. (1–4) Material (green) is to the left, vacuum region (gray) is to the right; (5) Cathode to the left, anode to the right, vacuum in anode–cathode (AK) gap. Photoemission shows excitation from both conduction band (metal-like) and valence band (semiconductor). Field emission relies on tunneling through the surface barrier, the others on emission over the barrier. Space charge is not emission but current limitation: black line is $F(x)$.

the thermal tail of the Fermi–Dirac distribution. The function $f(k)$ is therefore crudely obtained by assuming that 1/6 of the electrons travel with an average velocity in any given Cartesian direction (a cube has six faces). The proportion of electrons that then surmount the surface barrier $V_o = \mu + \Phi - \sqrt{4QF}$ is (1/6) the total number of electrons with $k > k_o$, both of which can be estimated using the density formulation of Eq. (2.14) extended to three dimensions, or letting $\phi = \Phi - \sqrt{4QF}$, then

$$p \approx \frac{1}{6}\frac{\int_{k_o}^{\infty} f_{FD}(k)4\pi k^2 dk}{\int_{0}^{\infty} f_{FD}(k)4\pi k^2 dk} \approx \frac{mk_o}{2\beta\hbar^2 k_F^3} e^{-\beta\phi} \tag{2.19}$$

The average velocity of an electron in the thermal tail is, by the equipartition theorem, $(1/2)m\langle v\rangle^2 = k_B T/2$ for each direction. Assuming $T = 1273$ K, $d_o = 102000$ kg/m$^3$, $M_W = 0.18384$ kg/mole, then $k_o = 14.2928$ [1/nm], and $\langle v\rangle = 0.1389$ nm/fs. If each tungsten atom gives up but one electron, then $\mu = 9.1862$ eV, and $p = 4.1188 \times 10^{-11}$, giving

$$J_T \approx qp\left(\frac{d_o}{M_W}\right)\langle v\rangle = 5.7951 \text{ A/cm}^2 \tag{2.20}$$

A flat thermionic cathode with a radius of 1/4 cm, operating at that temperature, is then crudely expected to produce a little over 1 amp of current.

The history of the canonical thermal emission equation is as follows. O. W. Richardson in 1901 predicted that $\ln(J/\sqrt{T})$ was a linear function of $1/T$. Motivated by the difficulties faced by that theory in the explanation of conductivity and specific heat, Richardson (in 1912) and M. von Laue (in 1918) ascertained the correct $T^2$ dependence of the coefficient. Richardson wrote

> *Taken together, these experiments prove that the emission of electrons does not arise from any interaction between the hot filament and surrounding gases or vapors, nor from any process involving consumption of the material of the filament. It thus follows that the emission of electrons from hot tungsten, which there is no reason for not regarding as exhibiting this phenomenon in a typical form, is not a chemical but a physical process.*
>
> O. Richardson [296]

In 1923, Dushman's quantum mechanical analysis related the coefficient $A_{RLD}$ of Richardson's equation to fundamental constants (see Eq. (8.4)). R. H. Fowler in 1929 modified $A_{RLD}$ to include a factor of 2 for electron spin. By the 1930s, the so-called "Edison effect" was dubbed "thermionic emission," and, given its preeminence in vacuum tubes used in radio communications, experiments were directed to exploring low work function coatings such as barium, cesium, and thorium could improve electron emission at lower temperatures. It rapidly became clear that the surface work functions were not uniform, nor characterized by a single work function $\Phi$; rather, different crystal faces of polycrystalline refractory metal surfaces (often tungsten) exhibited different values of $\Phi$. In fact, most early thermionic cathodes were tungsten because they were dependable, easy to clean by heating, and hard to kill: by comparison, more exotic cathodes that relied on other materials were poisoned by unknown processes. The high work function, however, was a limitation to achieving high current density, and so thorium (which was unfortunately radioactive) and submonolayer coverages of alkali (e.g. cesium) and alkali earth (e.g. barium) metals were investigated and found to substantially reduce the work function and greatly improve thermal emission current.

In one of the more emblematic instances of the impact of good luck on the fortunes of science, Arthur Wehnelt, in 1904 in Germany, was working on cathode ray tubes using thermionic emitters made of platinum which, like other metals, had a fairly high work function. Vacuum technique being what it was in those days, stopcock grease was used in the vacuum systems, and some of it had contaminated the cathode surface, curiously leading to much higher emitted current. Wehnelt, who was fluent in the technology behind vacuum tubes, electron guns, and emission from metal oxides, surmised that the metal oxides were behind the superior emission, and thus undertook an exploration of what the best metal oxide was for thermal emission. He concluded that it was barium oxide (BaO). Thus, was born the metal oxide thermionic cathode, used in radio, radar, TV, displays, microwave ovens, and all manner of electron beam devices. As A.H. Sommer observed

> *The BaO cathode is a truly amazing case of a researcher interpreting an accidental observation and putting it to practical use. It is ironic that this observation could not have been made in a modern high vacuum system!*
>
> A. H. Sommer [323]

The need for radar in World War II drove the development of semiconductor, oxide, and matrix emitter thermionic cathodes. A pivotal improvement was the invention of "dispenser cathodes" that continually rejuvenated the low-work function barium coatings on the surface by diffusion through a porous sintered tungsten matrix, to replace the barium lost to evaporation [119]. Increasing the temperature of a dispenser cathode increases the current density but does so at the cost of increasing evaporation rates. As barium exhaustion signals the lifetime of the dispenser cathode has been reached, keeping the current density low is accommodated by using beam compression to get the needed current densities in the beam tunnel.

### 2.2.2 Field Emission

Tunneling is a consequence of the wave nature of the electron postulated by de Broglie [34]. That the electron manifests wave behavior entails that treating an electron only as a particle that emerges from a metal surface with zero velocity is insufficient: quantum mechanically, the average velocity of the emitted electrons, which tend to be the most energetic ones and therefore from near the Fermi level, is $v_x \sim \hbar k_F/m$. Within the barrier, the oscillatory behavior of the wave function $\psi_k(x)$ switches over to attenuation. Were the barrier rectangular, then the attenuation would go as $\kappa(k)x$, where $x$ is the penetration into the barrier and $\kappa(k) = \sqrt{k_0^2 - k^2}$. The density, being proportional to $|\psi_k(x)^2|$, is therefore reduced by the exponential of a factor $\theta(k) = 2\kappa L$, where $L$ is the thickness of the barrier, and $\theta(k)$ will come to be known as the Gamow factor. However, an electric field at the surface of a metal gives rise to a barrier of the form $V(x) = V_o - Fx$ (neglecting the image charge), and so $\theta(k) \to 2\sigma\kappa(k)L(k)$ for a triangular-like barrier, where $\sigma$ is a factor of order unity (and presciently set equal to 2/3 here, see Section 8.3). The WKB approximation is dominated by $e^{-\theta}$, and so suggests that the probability of tunneling should be comparable to $p \approx D(k_F) \sim \exp[-\theta(k_F)]$ if the electrons most likely to tunnel are those near the Fermi level. All together,

$$J_F \approx q\left(\frac{\hbar k_F}{m}\right) p \left(\frac{k_F^3}{3\pi^2}\right) = \frac{4qm\mu^2}{3\hbar^3\pi^2} \exp\left[-\frac{4}{3\hbar F}\sqrt{2m\Phi^3}\right] \tag{2.21}$$

Consider typical metal parameters of $\mu = 7$ eV, $\Phi = 4.5$ eV, and $F = 7.5$ eV/nm. The calculated values are then $\hbar k_F/m = 1.56919$ nm/fs, $L(k_F) = \Phi/F = 0.6$ nm, $p = \exp(-8.69430)$, and $\rho = 84.1088$ nm$^{-3}$, for which $J \approx 131.983 \exp(-8.69430)$ [q/fs nm$^2$] $= 3.54274 \times 10^8$ [A/cm$^2$]. A single conical field emitter with a tip having an emission area of 10 $nm^2$ will then produce a current of 35.4274 $\mu$A. Although including image charge effects will increase this estimate, the notional emission area for actual conical metallic field emitters is smaller, making 1–10 $\mu$A per emitter a typical value.

The history of the canonical field emission equation is as follows. In 1897, R. W. Wood found that subjecting surfaces to very strong electric fields results in the emission of electrons [119]. Later in Germany, Julius E. Lilienfield (University of Leipzig) was laboring to improve his X-ray tubes in the early 1900s, when he found that the measured current between two metals in vacuum was larger than he had anticipated, with a fluorescence that was observed in the "anti-cathode" region of his tube. Lilienfeld described the phenomena by

> *Es kann an jeder Röntgenröhre beobachtet werden, dass die Fläche, in welcher die Kathodenstrahlen auf die Antikathode auftreffen, ein Licht aussendet, dessen Helligkeit reichlich genügt, um eine ganz bestimmte Farbenempfindung hervorzurufen, und zwar diejenige eines bläulichen Graus. Diese sichtbare Strahlung hat mit der Wärmestrahlung des Antikathodenspiegels nichts Gemeinsames. (trans): On each X-ray tube, an area where the cathode rays are incident on the anticathode can be observed to emit a light whose brightness is enough to cause a bluish gray coloration. This visible radiation has no relation to radiated heat from the anticathode mirror.*
>
> J. E. Lilienfeld [231]

and later, in 1922, coined the phrase "Autoelektronische Entladung" (or "auto-electron discharge," which became "field emission"), attributing it to electrons being liberated from his metal tips [204]. In 1923, Schottky gave a non-quantum mechanical account of both thermal and field emission [307] by proposing a lowering of the emission barrier whereby $\phi = \Phi - \sqrt{4QF}$, and speculated that small bumps on larger bumps resulted in multiplicative field enhancements that lead to the fields required for tunneling. In January 1928, J. R. Oppenheimer predicted that atoms subjected to fields would eventually lose their electron in a process unaccounted for by classical physics. He conjectured [268] that the "pulling of electrons out of a metal" by extremely high electric fields explained the data of Millikan and Eyring [253]. Oppenheimer predicted that the current $J$ observed by Millikan and Eyring was such that $\ln(J/F^{1/4})$ was linear in $1/F$. On May 1 of the same year, Fowler and Nordheim [95] argued instead that $\ln(J/F^2)$ was linear in $1/F$: as a practical matter, Millikan's data supported both formulations fairly well given the uncertainty attending the measurements of those times. Fowler and Nordheim's theory was amended by Nordheim afterwards to include the Schottky factor for barrier lowering that led to $\Phi \to \phi = \Phi - \sqrt{4QF}$ in the exponential argument.

### 2.2.3 Photo Emission

An electron near the Fermi level that absorbs an incident photon of energy $\hbar\omega$ will travel in a random direction specified by the polar angle $\theta$ relative to the surface normal,[2] with an energy $E = \mu + \hbar\omega$. If the "normal" energy $E_x = E\cos^2\theta = \hbar^2 k_x^2/2m$ exceeds the surface barrier $\mu + \phi$, then the electron is photoemitted (even the generally high fields in photo injectors do not approach the sizes needed for appreciable tunneling). The excited electron need not be at the Fermi level; if $\theta = 0$, then the initial energy can be at $\phi - \hbar\omega$ below the Fermi level, and photoemission can still occur, in keeping with Einstein's identification of the work function $\Phi$ being the minimum energy a photon must possess to cause electron ejection [76] in the absence of an electric field at

2 Be mindful that here, $\theta$ is the polar angle, not the Gamow factor of the previous section.

the surface. Let $I_\omega$ be the laser intensity [MW/cm$^2$], and $[1-R(\omega)]I_\omega$ the amount absorbed, where $R(\omega)$ is the reflectivity. The number of photoexcited electrons is then $J_\omega A\Delta t/q$, where $\Delta t$ is the duration of the laser pulse and $A$ is the illumination area, whereas the number of photons absorbed is $[1-R]I_\omega A\Delta t/\hbar\omega$. Quantum efficiency $QE$ is a measure of that ratio, and so, if the duration of the current pulse is equal to that of the laser pulse, then

$$\begin{aligned} QE &= \left(\frac{J_\omega A\Delta t}{q}\right)\left(\frac{\hbar\omega}{I_\omega A\Delta t}\right) = \frac{\hbar\omega}{q}\frac{J_\omega}{I_\omega} \\ &\equiv [1-R(\omega)]\, F_\lambda \Omega_p \end{aligned} \tag{2.22}$$

where $F_\lambda$ measures losses associated with electrons migrating to the surface, generally because of scattering events, and $\Omega_p$ is the fraction of a sphere (a cap) of electrons in energy space that have a normal energy component in excess of $\mu+\pi$. A crude estimate of $\Omega_p$ is to let it be the ratio of the volume of the Fermi–Dirac sphere with $\mu \geq E \geq \mu - (\hbar\omega - \phi)$, such that the forward energy is in excess of the barrier height compared to the total area of the sphere. To that end, introduce the quantities $\Delta = (\hbar\omega - \phi)/\mu$ and $k_w = k_F\sqrt{1-\Delta}$, so that $\Omega_p$ is approximated by

$$\begin{aligned} \Omega_p &\approx \frac{2\int_{k_w}^{k_F} dk_x \int_0^{\sqrt{k_F^2-k_x^2}} 2\pi k_\perp dk_\perp}{\int_0^{k_F} 4\pi k^2 dk} \\ &= 1 - \frac{1}{2}(2+\Delta)\sqrt{1-\Delta} \approx \frac{3}{8}\Delta^2 \end{aligned} \tag{2.23}$$

to leading order in $\Delta$. Observe that in the integrals, the differential volume element of the numerator is $d^3k = 2\pi\, dk_x\, k_\perp\, dk_\perp$ (cylindrical coordinates) and the denominator is $d^3k = 4\pi k^2 dk$ (spherical coordinates) for ease in evaluating the integrals. From the definition of $\Delta$, then to leading order

$$QE \propto (\hbar\omega - \phi)^2 \tag{2.24}$$

For metals like copper under laboratory conditions, $\mu = 7$, $\Phi = 4.5$ eV, $F \approx 0.001$ eV/nm, and a commonly used laser wavelength of $\lambda = 266$ nm imply $\hbar\omega = 4.66106$ eV. Therefore, $\Delta = 0.0284296$ and $\Omega_p \approx 3.03090 \times 10^{-4}$. Measured QEs are lower by about a factor of 20, giving an idea of the magnitude of $(1-R)F_\lambda$ (reflectivity and losses due to scattering) for metals.

The history of the canonical photoemission equation is as follows. In 1887, Heinrich Hertz was conducting experiments to demonstrate the existence of electromagnetic waves, in which he used a Tesla transformer to produce sparks in a primary gap that caused sparking in a secondary gap a distance away. His investigations led him to conclude that premature sparking at the secondary gap was a consequence of the ultraviolet light from the primary spark [28]. Hertz's assistant, Philipp Lenard, continued the investigations and observed that light striking certain metals induced the emission of electrons, for which Lenard received the Nobel Prize in 1905; the muted enthusiasm shown to him by posterity is a measure of his firm embrace of Nazi ideology and antisemitism. The difficulties in explaining both the promptness of photoemission and the insensitivity of the energy of photoemitted electrons to the intensity of light (particularly, red light of any intensity did not lead to photoemission) paved the way for Einstein's explanation using Planck's corpuscular theory of light, in which Einstein boldly proposed

*The usual conception that the energy of light is continuously distributed over the space through which it propagates, encounters very serious difficulties when one attempts to explain the photoelectric phenomena, as has been pointed out in Herr Lenard's pioneering paper. According to the concept that the incident light consists of energy quanta... one can conceive of the ejection of electrons by light in the following way. Energy quanta penetrate into the surface layer of the body, and their energy is transformed, at least in part, into kinetic energy of electrons. The simplest way to imagine this is that a light quantum delivers its entire energy to a single electron: we shall assume that this is what happens.*

A. Einstein [76]

By 1928, Fowler [92] was able to exploit Sommerfeld's electron theory of metals in much the same way as he did in his treatment of field emission to show that the photoelectric threshold frequency was sharp (at a particular frequency $\omega_o$), and that the corresponding energy was equal to the thermionic work function, or $\hbar\omega_o = \Phi$. DuBridge and Hergenrother [68] modified Fowler's formulation and consequently accounted for the effects of temperature on the energy distribution of the electrons, as well as the total yield, the metric of which is the QE. Photoemission from semiconductors, particularly from compound photoemitters, was important in the development of semiconductor band structure theory.

### 2.2.4 Secondary Emission

A beam of high-energy electrons (primaries) striking a material will result in the subsequent emission of electrons from the surface, but there are different kinds [154]. The primaries may collide with electrons in the material and be redirected back to the surface with roughly half their incident energy (back-scattered primaries), or the primaries may undergo repeated scatterings and yet still have enough energy to escape (intermediate-energy primaries), or the electrons with which the primaries scatter have sufficient energy ($\lesssim$ 20 eV) to be emitted themselves (low-energy secondaries). It is the last group of interest here for $E_o \lesssim 5$ keV.

As the energy $E_o$ of the primary electron beam increases, the secondary electron yield (SEY) $\delta$ from solids initially increases, peaks, and then decreases. The maximum yield $\delta_m$ occurs at $E_m$. In general, for the metals, the values of the maximum yield typically range between $0.6 < \delta_m < 1.6$. Semiconductors generally have higher SEYs. As primaries penetrate a material, they lose energy after collisions with other electrons, for which a useful approximation used by Baroody [15] and others [155, 244, 311, 372] is

$$\frac{dE}{dx} \approx -\frac{A}{E^{n-1}} \tag{2.25}$$

where $A$ is a proportionality constant and $n = 2$ is known as Whiddington's law. The decline in energy $E(x)$ is then

$$\int_0^x dx' = \int_{E_0}^{E(x)} \frac{dx}{dE}\, dE \to E(x) = [E_o^n - nAx]^{1/n} \tag{2.26}$$

where $E_o$ is the initial energy of the primary electron. The range $R(E_o)$ is when the primary has lost all of its energy ($E(R) = 0$), and so is found to be $R(E_o) = E_o^n/nA$. If the loss of primaries as a function of $x$ due to various mechanisms is parametrically represented as $N(x) = N_o \exp(-\gamma x)$, where $N(x)$ is the number of primaries at $x$, $N_o$ is the initial number of primaries, and $\gamma$ is a decay depth factor, then

$$\begin{aligned}\delta(E_o) &= -B \int_0^{R(E_0)} \frac{dE}{dx} e^{-\gamma x} dx \\ &= BE_o G_n[\gamma R(E_o)] \\ G_n(\lambda) &\equiv e^{-\lambda} \int_0^1 \exp(\lambda s^n) ds\end{aligned} \tag{2.27}$$

where $\lambda \equiv \gamma R(E_o) = \gamma E_o^n / nA$ (here, it is not a wavelength), and $B = D/\Delta E$ is taken as the ratio of a surface emission probability $D$ with the average energy $\Delta E$ required to free a core electron from one of the atoms making up the secondary emission material. The factor $e^{-\gamma x}$ is the probability that a secondary electron generated at a depth $x$ into the material can reach the surface without losing energy due to scattering which renders it emission-ineligible. For small $\lambda$, then $G_n(\lambda) \approx 1$, so that $\delta(E_o) \propto E_o$ and so grows, but when $\lambda$ is large, then $G_n(\lambda) \approx 1/\lambda n$ so that $\delta(E_o) \propto 1/E_o^{n-1}$ and so declines. It follows that in-between, a maximum in $\delta(E_o)$ occurs at $E_o = E_m$. For many materials, plots of $\delta(E_o)/\delta(E_m)$ as a function of $E_o/E_m$ overlap, that is, secondary emission exhibits a universal curve, although instead of $n = 2$ as Baroody considered, a value of $n \approx 1.35$ better represents many metals, as will be seen in Section 14.1.

The history of the canonical secondary emission equation is as follows. Secondary electron emission was discovered in 1902 by Austin and Starke, but its theoretical description was complicated by the fact that the energy-loss mechanism of the primaries and the generation of the secondaries demanded a quantum mechanical treatment that was difficult [55], one of the first being by Wooldridge [363]. Baroody lamented this fact, observing

> *There would seem to be qualitative value, however, in a much simpler theory in which the primary is considered to interact in a classical manner with a free electron gas, the possibility of momentum transfer to the lattice being introduced by assuming a finite mean free path for elastic collision ... Although the use of classical principles greatly simplifies the calculations, the problem remains complex, and it is not clear that all the mathematical difficulties have been adequately met. The final results are therefore probably more limited in significance than would have been judged from the starting assumptions. It is partly because of this circumstance, and partly because it is felt that more attention should be given the (sic) Sommerfeld model, that the present paper is offered.*
>
> E. M. Baroody [15]

And so unlike the other canonical equations, Baroody's formulation came after the madness of the Second World War rather than before it. His formulation makes use of concepts concerning penetration into the metal, generation of the excited electrons, and their transport back to and emission from the surface, for which analogs exist in Spicer's semi-empirical three-step model of photoemission. Secondary emission is utilized in photomultipliers, magnetrons, crossed-field amplifiers [107], and diamond current amplifiers [198, 317, 368], but secondary emission processes also figure in damage and degradation, as when some of a high-energy beam in a particle accelerator strikes cavity walls; a rough distinction between the secondaries utilized by electron sources and those contributing to wall damage is that the secondary yield of cathodes is desired to be high, but the secondary yield of walls and cavities is purposely suppressed by exploiting low SEY materials.

### 2.2.5 Space Charge

Consider a one-dimensional current flow of charge. Let there be $\Delta N$ particles, each of which has a charge $q$, passing through a unit area $\Delta A = (\Delta x)^2$ in a time $\Delta t$. Current density $J$ is then defined as

$$J = q\frac{\Delta N}{\Delta x^2 \Delta t} = q\left(\frac{\Delta N}{\Delta V}\right)\frac{\Delta x}{\Delta t} \to q\rho(x)v_x \tag{2.28}$$

where $\Delta V = \Delta A\ \Delta x$. $\Delta A$ need not be related to $\Delta x$, but it is convenient to suppose $\Delta A = \Delta x^2$. In the limit of an infinitesimal time increment $\Delta t \to dt$, then $\Delta x \to dx = v_x dt$, and likewise for infinitesimal spatial increment, $\Delta N/\Delta x^3 \to \rho(x)$ is the number charge density at the location $x$. Let the electrons have a velocity $v_x$ in the $\hat{x}$–direction. Observe that the sign of $J$ denotes the direction of current flow, but given the miserable convention that current flow and electron flow are in opposite directions, that tidbit will not be used unless needed. The form of Eq. (2.28) makes possible the consideration of charge injected to a gap between the cathode (the source of the electrons) and the anode (the collector of electrons) separated by $D$ (the AK gap). Electrons within the gap impede the flow of electrons emitted at a later time by pushing on them with a force $F = dV/dx$ that, from simple Gaussian pillbox arguments [78] applied to a sheet of charge $\sigma$, sets $F = q^2\sigma/\varepsilon_0$, constant and independent of distance from the sheet. If the sheet is one of many, then $F \to dF$, and if the charge per unit area is from a thin slice of density or $\sigma = \rho(x)dx$, then Poisson's equation pleasingly becomes

$$dF = \frac{q^2}{\varepsilon_0}\rho(x)dx = 4\pi Q\rho(x)dx \tag{2.29}$$

Adding up all the small forces $dF$ from all the sheets of charge $\rho(x)dx$ then gives the field at the surface of the cathode $F$ in contrast to the field across the AK gap when no charge is present $F_o = V_a/D$, or

$$V_a = FD + 16\pi Q\int_0^D \rho(x')dx' \tag{2.30}$$

Space charge limited current density $J_{CL}$ is said to occur when $F$ at the surface of the cathode is driven to zero. The amount of charge per unit area in the gap is the integral in Eq. (2.30). But charge per unit area is also the product of a current density with a time, and so that integral could be replaced with $J_{CL}\tau$ where $\tau$ is a characteristic time. Like what? Near the cathode surface, the fields are reduced, so the emitted electrons are not traveling that fast, but out near the center of the gap and to the anode, the forces are comparable to $V_a/D$ (but smaller), and that suggests that the factor $\tau$ is simply how long it takes an electron to get from the cathode to the anode. Were no charge present in the AK gap, then $\tau$ would then simply be the time of flight of a free electron subject to a constant force $F_o$, or [170, 345, 376]

$$D = \frac{1}{2}\frac{F_o}{m}\tau_o^2 \to \tau_o = \sqrt{\frac{2mD^2}{V_a}} \tag{2.31}$$

and so, the smallest $\tau$ could be is $\tau_o$. Therefore, the largest $J_{CL}$ could be is given by $V_a = 16\pi Q J_{CL}\tau_o$, or

$$J_{CL} \lesssim \frac{q}{16\pi Q\sqrt{2m}}\frac{V_a^{3/2}}{D^2} \tag{2.32}$$

It will be shown in Section 15.1, Eq. (15.9), that the equality holds when 16 in the denominator is replaced by 18, but that is not the point; what is the point is that the characteristic behavior of space charge limited current, namely, that $J_{CL} \propto V_a^{3/2}/D^2$, comes out of such a simple analysis. To avoid leaving a misleading equation about, the correct form of the Child–Langmuir equation for space charge limited current density is

$$J_{CL}(V_a, D) = \frac{q}{18\pi Q\sqrt{2m}}\frac{V_a^{3/2}}{D^2} = \frac{4\varepsilon_0}{9D^2}\sqrt{\frac{2q}{m}}\varphi_a^{3/2} \tag{2.33}$$

when the [efnq] units are replaced by [MKSA] via $V_a = q\varphi$ for potential [25, 377]. The simplicity of the units behind Eq. (2.33) in [efnq] units allows for quick comparisons using $q/(16\pi Q\sqrt{2m}) =$ 0.0163883 [q/eV$^{3/2}$ fs] $= 2.62570 \times 10^{-6}$ A/eV$^{3/2}$. By way of example, an AK gap of $D = 1$ cm and $J_{CL} = 1$ A/cm$^2$ requires a potential difference of 5.254 kV ($V_a = 5254$ eV).

The history of the canonical law of space charge limited flow is as follows. Space charge limited emission theory is often associated with thermal emission: Irving Langmuir's ground-breaking treatment observed that an understanding of space charge was crucial in showing that current due to thermal emission was not, as argued in Germany at the time, due to chemical reactions. Because current from a thermal cathode rose as the temperature rose, departures were observed from Richardson's equation for a given anode voltage. Langmuir wrote

> *This apparent discrepancy between the results of calculation by Richardson's equation and the facts observed with a tungsten lamp seemed at first to confirm the growing opinion that in a very high vacuum the thermionic current is very small, if not entirely absent. Experiments on the Edison (sic) effect in tungsten lamps, made some time ago by the writer, throw a great deal of light on the cause of the apparent failure of Richardson's equation at high temperatures. The observations therefore seem of sufficient interest to warrant their publication.*
>
> I. Langmuir [220]

Langmuir mentioned in a footnote that Clement D. Child found the same relation for the maximum current pulled across a planar gap for positive ions in 1911. For electrons, the relation therefore goes by the name of the Child–Langmuir equation (or as *Child's Law*), and it has since become

> *... one of the best-known limits in the fields of non-neutral plasma physics, accelerator physics, sheath physics, vacuum electronics, and high power microwaves ...*
>
> J. Luginsland et al. [238]

Space charge is inextricably linked with geometry: 10 years after Langmuir's first paper, he and Blodgett [221] found space charge limited current between coaxial cylinders, and a year later, between concentric spheres [222]. Later refinements include initial velocities, voltage differences of such a magnitude that electrons traversing the gap became relativistic [25], and ultimately, quantum mechanical effects for very small AK gaps [4]. Space charge effects additionally appear for electron transport in semiconductors: when scattering dominates, the Child–Langmuir law transforms into the Mott–Gurney law [243]. Within thin films, space charge effects during transport can lead to self-quenching behavior as shown in simulations by Mumford and Cahay [261] of Metal/CdS/LaS cold cathodes. Because the origins of space charge effects reside in the interaction between charged particles, space charge physics is a central concern to recent and technologically cutting-edge applications like electron beam wake field accelerators, ultrafast electron microscopy (UEM), kinetics and transients in materials science and biology, the operation of quantum dots, plasmonic devices, photodiodes [377], and nanoscale vacuum channel transistors [122, 123].

## 2.3 Exercises

**Exercise 2.1** Assume a simple one-dimensional normalized distribution $f(k_z)$ in wave number such that

$$\langle k_z \rangle = \frac{1}{2\pi} \int_{-\infty}^{\infty} k_z f(k_z)\, dk_z$$

Show that

$$\langle (k_z - \langle k_z \rangle)^2 \rangle = \langle k_z^2 \rangle - \langle k_z \rangle^2$$

thereby showing the connection between "moments" as used herein and their more conventional definition. Demonstrate the relation explicitly assuming

$$f(k_z) \propto \int k_\perp dk_\perp f_{FD}[E(k)]$$

in the limit of zero temperature where $f_{FD}(E) \propto \Theta(\mu - E)$ (Eq. 7.14) and $E(k)$ is parabolic in $k = |\vec{k}|$.

**Solution** $\hbar^2 \langle k_z^2 \rangle / 2m = \mu/3$.

**Exercise 2.2** Using the information of Tables 2.1 and 2.2, calculate the coefficient $A_{RLD}$ of the Richardson–Lau–Dushman equation in both [MKSA] and [efnq] units.

**Solution** Eq. (8.4).

**Exercise 2.3** As used in Eq. (2.21), show

$$\frac{4}{3}\kappa(k_F)L(k_F) = \frac{4}{3\hbar F}\sqrt{2m\Phi^3}$$

for the triangular barrier and an electron at the Fermi level $\mu$.

**Exercise 2.4** Typical current densities $J$ from the surface of a thermionic emitter are 1 A/cm$^2$. For a crystal of tungsten with an area of 6 $\mu$m$^2$, how many electrons per nanosecond are emitted (round to lowest integer)?

**Solution** 374.

**Exercise 2.5** Typical currents $J$ from a single Spindt-type molybdenum field emitter are 1 $\mu$A. How many electrons per picosecond are emitted (round to lowest integer)?

**Solution** 6.

**Exercise 2.6** Calculate what field causes the height of a Schottky–Nordheim (SN) barrier $V(x_o)$ to be lowered to the Fermi energy $\mu$. Assume the work function is $\Phi = 4.5$ eV.

**Solution** $F = 14.0628$ eV/nm.

**Exercise 2.7** How many hydrogen atoms would fit between the surface $x = 0$ of a metal and the maximum point $x_o$ of $V(x)$ for an SN barrier if $F = 5$ eV/nm (typical field emission conditions)? Assume the work function is $\Phi = 4.5$ eV and $F = 8.1$ eV/nm?

**Solution** Two hydrogen atoms.

**Exercise 2.8** Drive laser frequencies are typically doubled, tripled, and even quadrupled ($\omega_n/\omega_1 = 2, 3, 4$) for photoemission experiments. An Nd:YAG laser has a primary wavelength of $\lambda_1 = 1064$ nm. Calculate the photon energies $\hbar\omega_n$ for those cases.

**Solution** $\hbar\omega_n = (1.165, 2.331, 3.496, 4.661)$ eV for $n \in (1, 2, 3, 4)$.

**Exercise 2.9** In secondary emission from a semiconductor like diamond (bandgap $E_g = 5.5$ eV), the number of secondaries generated internally (not emitted) is approximately the ratio of the primary energy with the energy loss per collision $\Delta$ transferred to a daughter electron. Assume $\Delta$ is $2.5 \times$ the bandgap of the semiconductor. For a 3 keV primary, how many daughter electrons are created?

**Solution** 218.

**Exercise 2.10** At what anode potential will the space charge limit $J_{CL}$ for a parallel plate diode with a separation of $D = 0.05$ cm be equal to 33.213 A/cm$^2$?

**Solution** $\varphi_a = V_a/q = 1$ kV.

# Part II

# A Review of Methods

# 3

# Electrostatics

*I flam'd amazement. Sometime I'ld divide,*
*And burn in many places; on the topmast,*
*The yards, and boresprit, would I flame distinctly,*
*Then meet and join. Jove's lightning, the precursors*
*O' th' dreadful thunder-claps...*

W. Shakespeare, *Tempest* I.2

## 3.1 Method of Images

Insofar as the Schottky–Nordheim (SN) barrier describes the barrier to emission, all of the canonical equations depend on $F$. Inside a metal, $F$ vanishes (i.e. $V(x < 0) = 0$) because mobile electrons shield out external fields. On a curved metal surface, therefore, all field lines are normal to the local plane of the surface. Because $F$ at the surface of a not necessarily planar conductor needs to be known to understand its impact on the barrier, some consideration of the variety of methods available to enable the calculation of $F$ must be examined. The Uniqueness theorem aids that examination: if two solutions to Laplace's equation $\nabla^2 V(\vec{r}) = 0$ agree on the boundary, then the two solutions are identical [78]. That means that a clever placement of charges can solve a problem by virtue of superposition: charge and charge alone produce a potential, meaning that if an arrangement of charges replicates equipotential lines that match the conducting surfaces, then the problem of the potential generated by Poisson's equation has been solved. A second method relies on finding orthogonal coordinate systems that allow the multidimensional Laplace's equation to be separated into independent equations of one variable alone ("separation of variables"). If a conducting surface matches one of the "equipotential lines" (a line where the potential does not vary for changes in that coordinate) then again, the problem of the potential is solved.

The method of images is considered first. Something like it has been encountered already in the Schottky–Nordheim barrier by the addition of the image charge term. All electric fields on a metallic conducting surface are normal to the surface where they terminate. Therefore, a flat, planar conductor with a charge above it has all its field lines bending in the proximity to the surface until they meet the surface at right angles. The surface equidistant between a charged particle and its equal and opposite charged partner is flat, and therefore is the solution for a charge outside a planar conductor as in Figure 3.1 (see Algorithm B.2.3): as that arrangement of charges satisfies

*Fundamentals of Electron Emission Physics*, First Edition. Kevin L. Jensen.

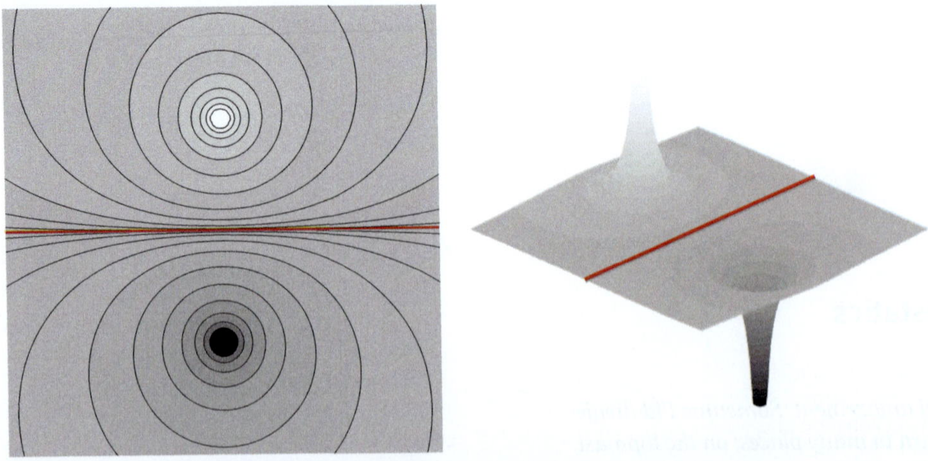

**Figure 3.1** Two representations of a charge (white) and its image (black): (*left*) contour plot of equipotentials, for which electric field lines are normal to the equipotential lines. (*right*) three-dimensional representation. In both, the straight red line represents the surface of a metallic conductor.

the boundary conditions that the potential vanishes on the surface, then the problem for $V(\vec{r})$ is solved.[1]

The force between an external electron $(-q)$ and its image $(+q)$ inside the metal, both a distance $x$ from the surface, is given by

$$F(x) = -\frac{q^2}{4\pi\varepsilon_0(2x)^2} = -\frac{Q}{x^2} \tag{3.1}$$

where the sign is negative because the force is attractive: the image charge is of opposite sign. The potential energy is then determined by integrating over the force from the location of an emitted charge out to infinity (where the potential energy is taken as zero), or

$$V_{image}(x) = -Q\int_x^\infty \frac{dx'}{(2x')^2} = -\frac{Q}{x} \tag{3.2}$$

Hence, the SN barrier of Eq. (2.9) is the sum of the triangular barrier $V_o - Fx$ and an image charge term $-Q/x$. The image charge contribution is the consequence of an electron in vacuum interacting with a charge redistribution that forms to keep the metal surface at an equipotential as the electron moves away, a redistribution that is mathematically equivalent to an equal and opposite charge the same distance away from the surface but inside the metal.

## 3.2 Point Charge Models

More complex conducting surfaces can be approximated by collections of point charges. Instead of a fictitious image charge inside the surface of a metal, consider instead a line of charges of one sign held separate from an equivalent line of charges but of different sign. By superposition,

1 Remember that $V(\vec{r})$ is a potential energy, so that the usual negative sign that accompanies Poisson's equation in the standard accounts is affected by choosing the electron as the charge in that potential.

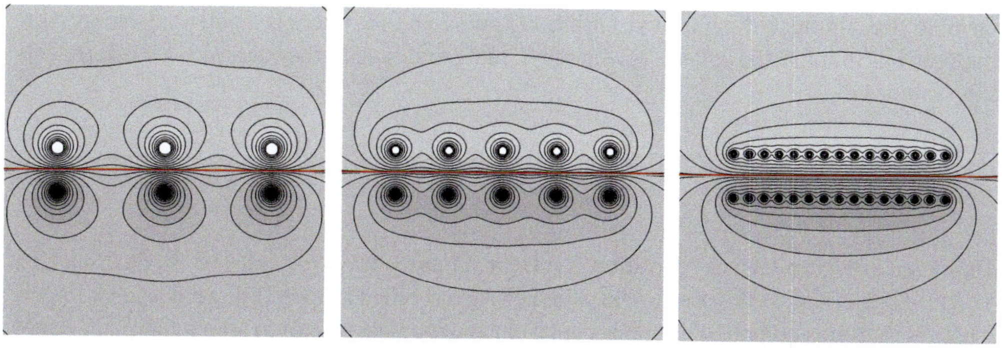

**Figure 3.2** (*left* to *right*) The equipotential lines for a line of $N = (3, 5, 15)$ charges (white) at $z = D/2$ and their oppositely charged counterparts at $z = -D/2$, whereas the number of charges increases, the total charge is held constant and equally distributed amongst them. As the number of charges increases, the arrangement takes on the appearance of a classic capacitor plate. The red line is the midpoint.

the equipotential lines are found by adding up all the contributions of the individual charges $q_j$

$$V(\vec{r}) = V(x, y, z) = -\sum_{j=1}^{N} \frac{qq_j}{\sqrt{(x - x_j)^2 + (y - y_j)^2 + (z - z_j)^2}} \tag{3.3}$$

with $j \in (1, \cdots, N)$ denoting the $N$ charges located at $\vec{r}_j = [x_j, y_j, z_j]$. Consider two simple cases, the first leading to a model of a capacitor plate, and the second leading to a model of a dipole.

The capacitor plate is normally two flat metal sheets of opposite sign a short distance apart, but consider a variant that is more mathematically useful as a first introduction. Let there be $N$ charges in a line at $z = D/2$, with an additional $N$ oppositely charged counterparts at $z = -D/2$ but otherwise sharing the same equi-spaced $x$ and $y$ coordinates. The plates are therefore long, but the width is very thin, but that idealization is enough to make a point. The behavior of $V(x, y, z)$ along the plane $y = 0$ is shown for increasingly larger values of $N$ in Figure 3.2 (see Algorithm B.2.3). As the number of charges increases with the total amount of charge on each plate held fixed, the equipotential lines become increasingly flat near the center of the capacitor. The electric field is a measure of the number of lines in a unit area, and so the electric field becomes uniform and in the $\hat{z}$-direction. Because the equipotential lines are equally spaced to a good approximation for large $N$, the electric field approaches a constant value and is uniform, converging to $F = V_a/D$ where $V_a$ is the potential energy difference between the two plates. To recover the more familiar configuration, where the plates of the capacitor are areas ("sheets"), the single line of $N$ charges is replicated into $N$ parallel lines of $N$ charges apiece, and a very similar calculation shows that the field between the anode and cathode sheets is not surprisingly $F = V_a/D$.

The dipole represents a conceptually opposite limit, namely $N = 2$ equal and opposite charges ($\pm q_o$) a distance $\delta$ apart, and in a background field directed along the $\hat{z}$-direction. Now $V(\vec{r})$ is

$$\begin{aligned} V(x, y, z) &= Fz + \frac{C}{R_+(x, y, z)} - \frac{C}{\sqrt{x^2 + y^2 + (z - \delta)^2}} \\ R_\pm(x, y, z) &= \sqrt{x^2 + y^2 + (z \pm \delta)^2} \end{aligned} \tag{3.4}$$

where $C$ is to be determined by boundary conditions, namely, where the metal surface is at zero potential. Clearly, the equipotential lines are going to be like a bump (given that equipotential surfaces around single charges are spherical), and far from that protrusion, the potential will be flat (remember that although the narrative speaks of "potential," $V(\vec{r})$ here is a potential energy). Let the height of the bump to be $a$. Require, therefore, that on the $\hat{z}$-axis, the value at $z = a$ is zero. This sets the value of $C$

$$\begin{aligned} V(0,0,a) &= F_o a + \frac{C}{a+\delta} - \frac{C}{a-\delta} = 0 \\ C &= \frac{F_o a}{2\delta}(a^2 - \delta^2) \end{aligned} \tag{3.5}$$

and is shown in Figure 3.3 (see Algorithm B.2.3) for the cases $\delta/a = 0.4$ and $0.04$. In the limit that $\delta \to 0$, the contribution to $V(\vec{r})$ due to the paired charges is called a dipole, but what is more important here is that the equipotential line becomes a hemisphere (red curve) atop a flat plane (cyan line). The bending of the equipotential lines in the vicinity of the hemisphere (or boss) causes the electric field near the hemisphere to change, and therefore, $\vec{F} = -\nabla V$ to change. Taking the gradient of Eq. (3.4) is straightforward, but a shortcut for calculating the field at the apex of the hemisphere makes use of the following limits as $\delta \to 0$ by rewriting $V(\vec{r})$ as

$$V(x,y,z) = Fz + \frac{C(R_- - R_+)}{R_+ R_-} = Fz - \frac{C(R_+^2 - R_-^2)}{R_+ R_-(R_- + R_+)} \tag{3.6}$$

Several limiting relations are useful (Exercise 3.3)

$$\begin{aligned} &\lim_{\delta \to 0} \quad C[R_+(x,y,z)^2 - R_-(x,y,z)^2] = 2za^3 \\ &\lim_{\delta \to 0} \quad R_+(x,y,z)R_-(x,y,z) = r^2 \\ &\lim_{\delta \to 0} \quad R_+(x,y,z) + R_-(x,y,z) = 2r \end{aligned} \tag{3.7}$$

where $r^2 = x^2 + y^2 + z^2$. As a result, in the $\delta \to 0$ limit,

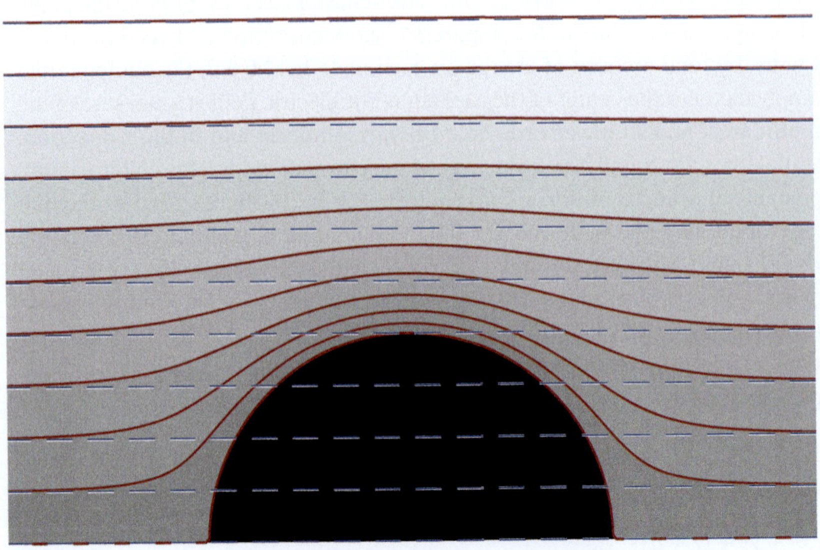

**Figure 3.3** A conducting hemisphere on a plane subject to a background field $\vec{F} = F_o\hat{z}$. Both the hemisphere and the plane are grounded. Red lines are equipotentials for the hemisphere; blue dashed lines are equipotentials when the hemisphere is absent.

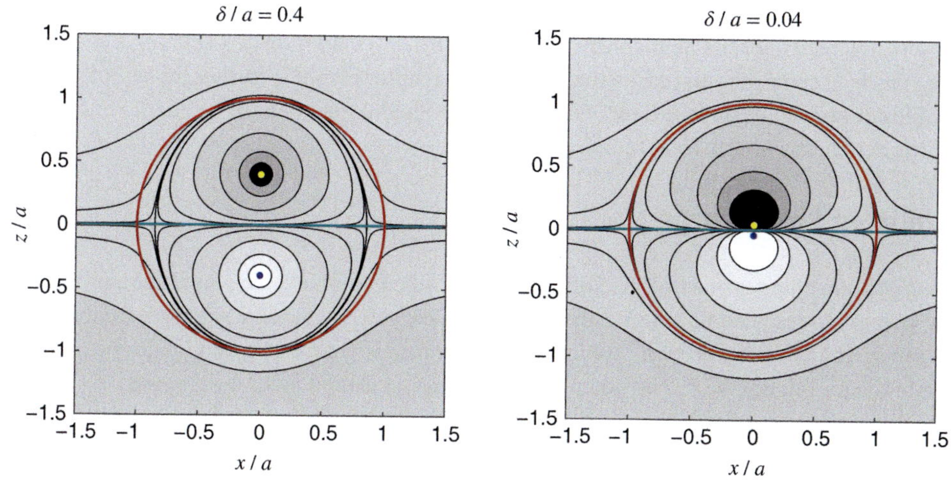


**Figure 3.4** Potential energy $V(x, y, z)$ for $y = 0$ for two closely spaced charges of equal magnitude and opposite sign in a background field $F_o$ such that at $r = a$, $V = 0$. The charges are located at $\delta/a = \pm 0.4$ (left) and $\pm 0.04$ (right). The black lines are equipotential lines, spaced to show detail but at the same height in each case. The curve of a hemisphere (red) is overlaid. The cyan equipotential is at 0. Yellow dot is negative charge, blue dot is positive charge. $F_o$ is along $\hat{z}$-axis.

$$V(x, y, z) = F_0 z \left[1 - \left(\frac{a}{r}\right)^3\right] = F_o r \cos\theta \left[1 - \left(\frac{a}{r}\right)^3\right] \tag{3.8}$$

where the last form is in spherical coordinates, for which the "dipole" contribution is $V_d = F_o a^3 \cos\theta/r^2$. Lastly, the value of $F$ along the surface of the hemisphere is radially outward with the value $F(a, \cos\theta) = 3F_o \cos\theta$ in spherical coordinates, which is maximum on axis ($\theta = 0$): the field enhancement factor, defined as the ratio of the apex field to the background field, for a hemisphere is therefore $F(a, 0)/F_o = 3$, as indicated by Figure 3.4 (see Algorithm B.2.4): the field is a measure of the number of equipotential lines in a unit length, and at the apex of the hemisphere, more lines are squeezed into a small space than further away from the hemisphere.

Bumps have a radius of curvature which is revealed by the behavior of the equipotential lines for $V(\vec{r}) = V(\rho, z)$ in cylindrical coordinates for which $\rho^2 = x^2 + y^2$. Let $z_s(\rho)$ be the equation of an equipotential coinciding with the surface of the metal such that $V(\rho, z_s(\rho)) = 0$. If the equipotential is a hemisphere as considered above, then

$$z_s(\rho) = \sqrt{a^2 - \rho^2} \approx a - \frac{\rho^2}{2a} \tag{3.9}$$

Compare this to a Taylor expansion of $V(\rho, z)$

$$V(\rho, z) = V(0, a) + (z - a)\partial_z V(0, a) + \frac{1}{2}\rho^2 \partial_\rho^2 V(0, a) \tag{3.10}$$

where $V(0, a) = 0$ and where $\partial_z V(0, a) = \partial_z V(\rho, z)|_{\rho=0,z=a}$ (and similarly for $\partial_\rho^2 V(0, a)$). A comparison of the two indicates that

$$a \equiv \frac{\partial_z V(0, a)}{\partial_\rho^2 V(0, a)} \tag{3.11}$$

defines the local radius of curvature. Although the very important relationship expressed by Eq. (3.11) is trivially satisfied for the hemisphere using Eq. (3.8), the evaluation of local radius of curvature will be far more consequential when the shape of the emitter is cones, ellipses, hyperbolas, and more complex entities to be encountered, which will change the field enhancement factor from 3 to something depending on the local radius of curvature.

## 3.3 Constant Line Charge

A collection of point charges in a line is a straightforward albeit cumbersome summation problem. When the point charges become infinitesimal and the locations are side by side, the summation becomes a more tractable integration problem. Let the constant charge per unit length be $q\lambda$ and the length of the line charge be $L$. The potential energy $V(\rho, z)$ is then given by [78, 148].

$$V_L(\rho, z) = 4Q\lambda \int_{-L}^{L} \frac{dz'}{\sqrt{\rho^2 + (z - z')^2}} = 4Q\lambda\, Q_0\left[\frac{L}{u(\rho, z)}\right] \tag{3.12}$$

in cylindrical coordinates, where $Q_0(x)$ is a Legendre function of the second kind (Appendix A.4) and should not be confused with $Q = \alpha\hbar c/4$ in the coefficient. The ellipsoidal functions $u$ and $w$ are defined by

$$u(\rho, z) \pm w(\rho, z) \equiv \sqrt{\rho^2 + (z \pm L)^2} \tag{3.13}$$

$$u(\rho, z)w(\rho, z) = Lz \tag{3.14}$$

In analogy with the dipole model in Section 3.2, the equipotential line describing the metal surface is such that at the apex $z_o$, then $V(0, z_o) = 0$. The potential everywhere for a line of charge in a constant background field is then

$$V(\rho, z) = F_o z_o \left[-\frac{z}{z_o} + \frac{Q_0\left[L/u(\rho, z)\right]}{Q_0\left[L/u(0, z_0)\right]}\right] \tag{3.15}$$

The factor $F_o z_o$ sets the energy scale. Introduce

$$z_o \equiv L(1 + \delta) \tag{3.16}$$

where $\delta$ acts like the fraction of the emitter above the extent of $L$ and therefore is expected to bear a very simple relation to the apex radius $a$ when $\delta$ is small. In terms of $\delta$, the easier relation

$$Q_0\left[\frac{L}{u(0, z_0)}\right] = \frac{1}{2}\ln\left(\frac{2 + \delta}{\delta}\right) \tag{3.17}$$

is shown by substitution of $z_0$ into Eq. (A.23).

The evaluation of the field enhancement factor $\beta = F/F_o$ to relate field at the apex of the line charge to the background field, requires $\partial_z u(\rho, z)$ evaluated at $(\rho, z) = (0, z_o)$. Yes, $\beta$ is an annoyingly common symbol to use for many different parameters in physics tomes in general and electron emission literature in particular, but continuity with language and notations established in that literature unfortunately requires forbearance and vigilance about what the symbol means: here, the "beta factor" is the dimensionless geometrical factor showing how much a field increases near a curved conducting surface. Now, because $u(0, z > L) = z$, it is found $\partial_z u(0, z) = 1$. Consequently, for small $\delta \ll 1$, $F(\rho, z)$ at the apex is

$$
\begin{aligned}
F(0, z_0) &= -F_o z_0\, \partial_z \left[ -\frac{z}{z_0} + \frac{Q_0(L/z)}{Q_0(L/z_o)} \right] \Bigg|_{z=z_0} \\
&= F_o + \frac{F_o L z_0}{(z_0^2 - L^2) Q_0(L/z_o)} \\
&\approx F_o \left[ 1 + \frac{2(1+\delta)}{\delta(2+\delta) \ln\left(\frac{2+\delta}{\delta}\right)} \right]
\end{aligned} \tag{3.18}
$$

Thus, an elongated line charge has a higher field enhancement than a hemisphere as may have been expected.

Finding the apex radius requires $\partial_\rho^2 u(\rho, z)$, and is best evaluated in stages. Let the dimensionless function $U_c(\rho, z)$ be defined as the term in the square brackets $[\cdots]$ in Eq. (3.15) such that $V(\rho, z) = F_o z_0\, U_c(\rho, z)$. Then

$$
a = \frac{\partial_z U_c(\rho, z)}{\partial_\rho^2 U_c(\rho, z)} \Bigg|_{\rho=0, z=z_0} \tag{3.19}
$$

The numerator was considered in Eq. (3.18). The denominator requires the relations where in the second line, use has been made of the following relations, valid when $\rho \to 0$ and $z \to z_o$ after the differentiations:

$$
\begin{aligned}
\partial_\rho u(\rho, z)|_{z=z_0, \rho=0} &= 0 \\
\partial_\rho^2 u(\rho, z)|_{z=z_0, \rho=0} &= \frac{z_0}{z_0^2 - L^2}
\end{aligned} \tag{3.20}
$$

to find

$$
\begin{aligned}
\partial_z U_c(\rho, z)|_{z=z_0, \rho=0} &= -\frac{1}{z_0} - \frac{L}{(z_0^2 - L^2) Q_0(L/z_o)} \\
\partial_\rho^2 U_c(\rho, z)|_{z=z_0, \rho=0} &= -\frac{L z_0}{(z^2 - L^2)^2 Q_0(L/z_o)}
\end{aligned} \tag{3.21}
$$

Lastly, using Eq. (3.16), $a$ is then

$$
\frac{a}{L} = \frac{\delta(2+\delta)}{1+\delta} + \left[ \frac{\delta(2+\delta)}{1+\delta} \right]^2 Q_0\left( \frac{1}{1+\delta} \right) \tag{3.22}
$$

The behavior of the equipotential lines is shown in Figure 3.5, and the behavior of $a/L$ and $\beta \times \delta$ as a function of $\delta$ in Figure 3.6 (see Algorithms B.2.5).

## 3.4 Tapered Line Charge

Instead of a constant charge per unit length, now allow the charge to vary, and to be of opposite sign on each side of the $x - y$ plane. This is accomplished by replacing $\lambda \to \lambda_o z/L$, so that the charge density varies linearly over the length of the wire (a "tapered dipole" model). For it, $V(\rho, z)$ is

$$
V(\rho, z) = \frac{4Q\lambda_o}{L} \int_{-L}^{L} \frac{z' dz'}{\sqrt{\rho^2 + (z - z')^2}} - F_o z \tag{3.23}
$$

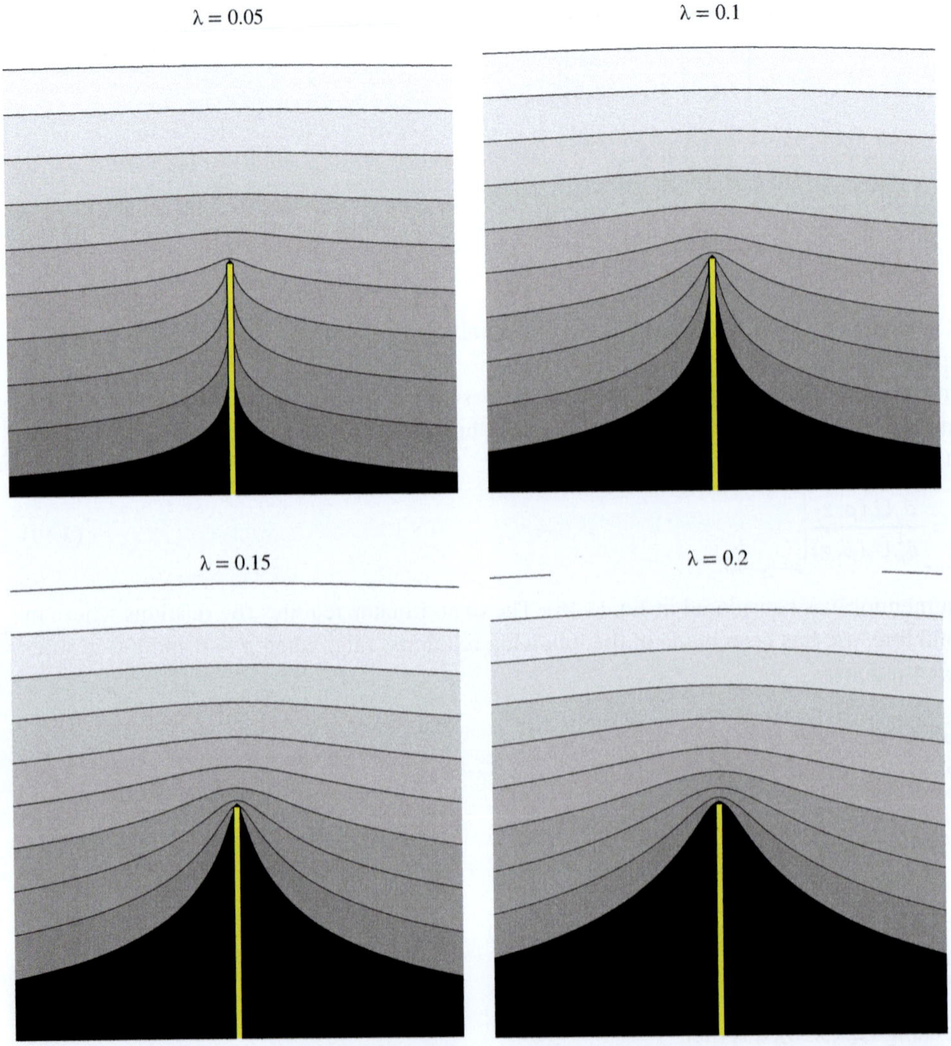


**Figure 3.5** Equipotential surfaces of the constant line charge model for various line charge densities $\lambda$. Size of view region is $2L \times 2L$. $x$-axis is along base, $z$-axis is center vertical. Yellow center line is location and extent of constant line charge.

where now $V(\rho, z_o) = 0$ identifies the apex and defines $z_o$. In terms of the $u(\rho, z)$ and $w(\rho, z)$ functions of Eq. (3.13), it becomes

$$V(\rho, z) = F_o z_o \left[ -\frac{z}{z_o} + \frac{zQ_0\,(L/u(\rho, z)) - w(\rho, z)}{z_o Q_0\,(L/z_o) - L} \right] \tag{3.24}$$

As before, the factor $F_o z_o$ sets the energy scale. Again let $z_o = (1 + \delta)L$ to find

$$\beta = \frac{F(0, z_o)}{F_o} = \frac{1}{\delta(2 + \delta)\left[(\delta + 1)Q_0\left(\frac{1}{1+\delta}\right) - 1\right]} \tag{3.25}$$

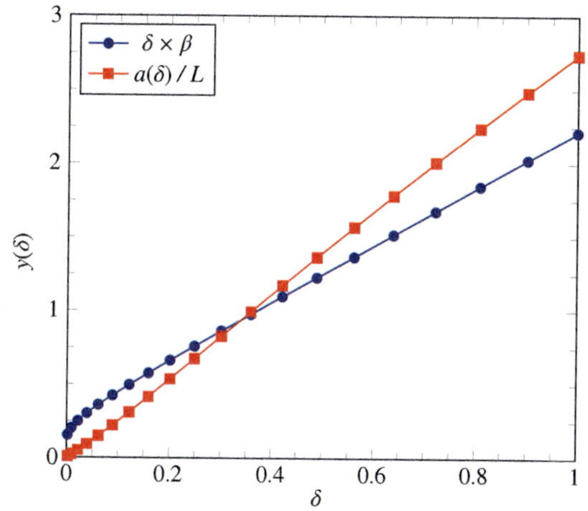


**Figure 3.6** Behavior of the field enhancement factor $\beta$ and apex radius $a$ as a function of $\delta = (z_o - L)/L$ for a line charge with constant charge per unit length $\lambda$.

Similarly, for apex radius, the derivatives of Eq. (3.19) are harder, but give

$$a \to L\delta\left(\frac{2+\delta}{1+\delta}\right) \tag{3.26}$$

which is seen to be the first term in the constant line charge radius of Eq. (3.22). Because of the taper, the width of the zero equipotential ellipse-like structures is much narrower than for the constant line charge shapes: as a result, larger values of $\lambda_o$ are used in Figure 3.7 (see Algorithm B.2.5).

## 3.5 Orthogonal Polynomials

Orthogonal polynomial expansions can be used to recreate shapes in a controlled and often useful manner. Legendre polynomials (Appendix A.4) are good for various point charge configurations and particular surfaces with rotational symmetry. The hemisphere on a plane of Figure 3.4 has such rotational symmetry about the $\hat{z}$-axis, so in spherical coordinates, the solution everywhere can be expressed as [78]

$$V(r,\theta) = \sum_{l=0}^{\infty}\left(A_l r^l + \frac{B_l}{r^{l+1}}\right)P_l(\cos\theta) \tag{3.27}$$

where the coefficients $A_l$ an $B_l$ are dictated by boundary conditions:

- Far away from the hemisphere, the field is constant, and so $V(r,\theta) \to -F_o z = -F_o r\cos\theta$. As a result, all $A_l$ vanish, save for $A_1 = -F_o$ because all higher $P_l$ involve higher powers of $\cos\theta$.
- The hemisphere is held at ground, and so $V(a,\theta) = 0$. Because all $A_l$ for $l \neq 1$ are zero so too must $B_l$ be zero for all $l \neq 1$ because otherwise there is nothing to cancel them at $r = a$.
- And finally, for $l = 1$ and $r = a$, it follows that $B_1 = -F_o a^3$ to cancel the $A_1$ term.

These boundary conditions determine that only $P_1(\cos\theta) = \cos\theta$ survives, and so the final form is

$$V(r,\theta) = -F_o r\cos\theta\left\{1 - \left(\frac{a}{r}\right)^3\right\} \tag{3.28}$$

which matches Eq. (3.8), but obtained with less complication.

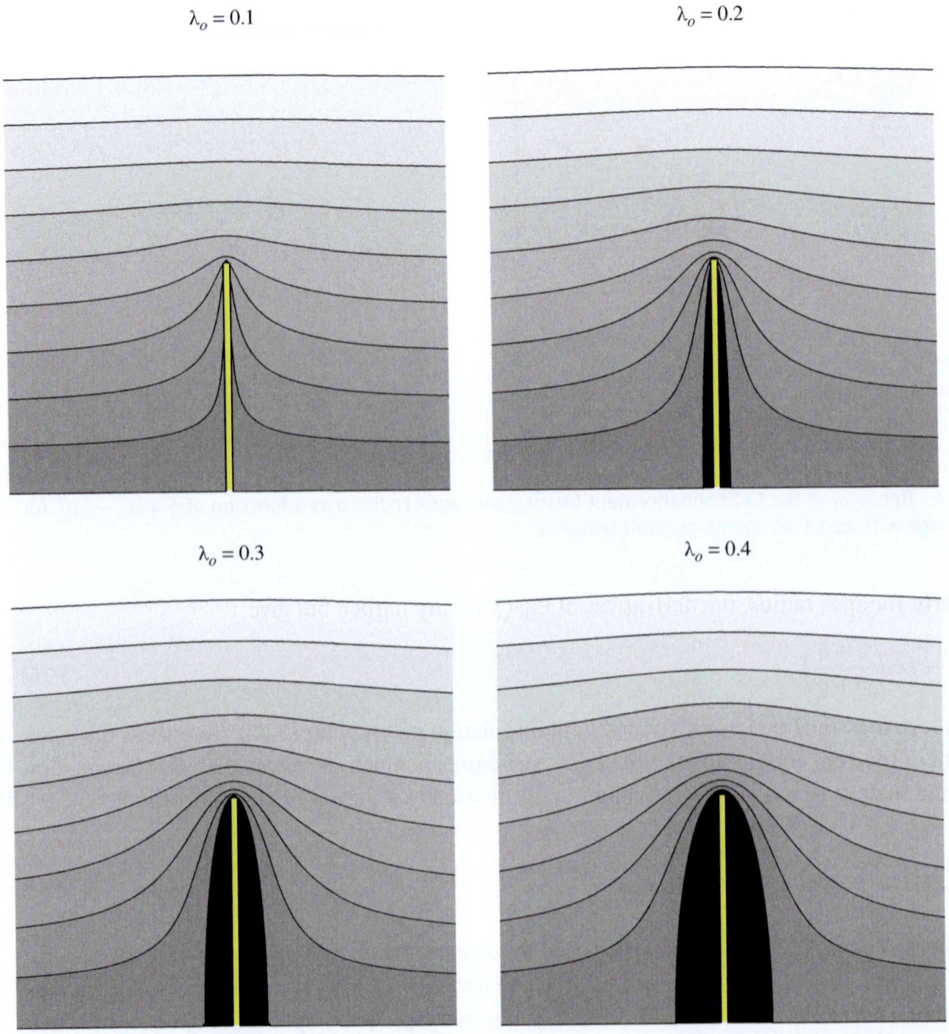


**Figure 3.7** Equipotential surfaces of the tapered line charge model for various line charge densities $\lambda_o$. Size of the view region is $2L \times 2L$. Yellow center line is location and extent of tapered line charge $\lambda(z) = \lambda_o z/L$.

## 3.6 Prolate Spheroidal Models

The prior methods all sought to match the boundary of the metal cathode to solvable mathematical models. The final analytical method to consider, then, is when the mathematics dictates which boundary is even to be considered. The solution of Laplace's equation is best done in a coordinate system that allows for a separation of variables, such that equipotential lines are along one coordinate, and field lines along another. There are many to choose from [6], but the ones of paramount interest here are the prolate spheroidal coordinates that can model two-dimensional ellipsoids and hyperboloids, or their revolutions about the $\hat{z}$-axis for three-dimensional emitters. In these coordinate systems, the metal surface is taken to lie along one of the orthogonal planes.

The advantages of the prolate spheroidal approach to modeling field emission structures are well-recognized [126, 147, 207, 383] in particular, but as a means of determining local fields at the surface

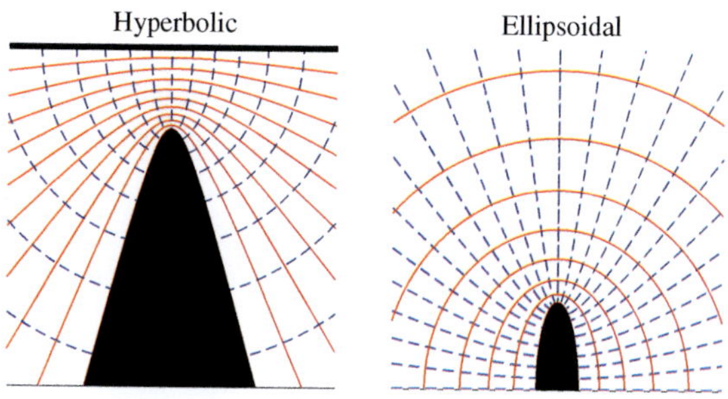


**Figure 3.8** The prolate coordinate system: (*left*) the hyperbolic system, where equipotential lines (red solid) are for constant $v$ and where blue dashed lines are field lines, and the anode (black line) is on top; (*right*) the ellipsoidal system, where equipotential lines (red solid) are for constant $\eta$ and where blue dashed lines are the field lines.

of structures, the methods have wider application. Of the various ways to represent the prolate spheroidal coordinate system, specifying the orthogonal coordinates using angles $(\eta, v)$ is preferred. Their relation to the cylindrical coordinates $(\rho, z)$ is defined by

$$\rho = L\sinh(\eta)\sin(v) \tag{3.29}$$

$$z = L\cosh(\eta)\cos(v) \tag{3.30}$$

where $L$ is a length scale (the more general case in Appendix A.6 includes a third angle $\gamma$ where $x = \rho\cos\gamma$ and $y = \rho\sin\gamma$). Hyperbolic surfaces are defined by constant $v$, and ellipsoidal by constant $\eta$, both of which are shown in Figure 3.8 (see Algorithm B.2.4). The hyperbolic representation allows for a close-spaced anode along $z = 0 \leftrightarrow v = \pi/2$, whereas the ellipsoidal representation allows putting in a constant background field $F_o$, but this apparent freedom is rather inflexible: the hyperbolic model is exact only for a diode geometry with the anode a fixed distance away and related to the cone angle, whereas elliptical models are exact for an ellipsoidal emitter all by itself in the presence of a uniform field. The prolate spheroidal treatment, therefore, does not describe arrays, but it does reveal how the fields along the surface $F(\eta, v)$, and particularly the apex fields, depend on geometry. Importantly for prolate spheroidal coordinate systems, the lines denoting fields and equipotentials are reversed in the hyperbolic versus the ellipsoidal representations: in Figure 3.8, the red lines and blue dashes of the hyperbolic representation are the blue dashes and red lines, respectively, of the ellipsoidal representation, but upside down.

Separation of variables recasts Laplace's equation into two equations. Letting $V(\eta, v) = U(\eta)W(v)$ gives

$$\frac{1}{V(\eta, v)}\nabla^2 V(\eta, v) = \frac{1}{U(\eta)}\nabla_\eta^2 U(\eta) + \frac{1}{W(v)}\nabla_v^2 W(v) = 0 \tag{3.31}$$

The Laplacian components of Eqs. (A.35) and (A.36) give solutions that can describe hyperbolic and ellipsoidal emitters. Using the $\nabla^2$ from Eq. (A.33) in Appendix A.6, Eq. (3.31) can be rewritten as

$$\frac{1}{U}\left[\frac{1}{\sinh\eta}\frac{\partial}{\partial\eta}\left(\sinh\eta\,\frac{\partial U}{\partial\eta}\right)\right] = -\frac{1}{W}\left[\frac{1}{\sin v}\frac{\partial}{\partial v}\left(\sin v\,\frac{\partial W}{\partial v}\right)\right] \tag{3.32}$$

Since the left-hand side and the right-hand side depend on different coordinates, then they both must be equal to a constant, conventionally set to $n(n+1)$ for integer $n$, as can be shown by ladder operation methods to be considered in Chapter 6. Which Legendre polynomial satisfies the equation is then labeled by $n$, as in $W_n(\eta)$.

### 3.6.1 Hyperbolic Geometry

A hyperbolic cone along the equipotential defined by $\upsilon_o$ with a planar anode at $\upsilon = \pi/2$ is modeled by setting each side of Eq. (3.32) to the same constant, then expanding out the derivatives and simplifying. In the case of $W_0$, that equation is

$$\left(\sin \upsilon \frac{\partial^2}{\partial \upsilon^2} + \cos \upsilon \frac{\partial}{\partial \upsilon}\right) W_0(\upsilon) = 0 \tag{3.33}$$

for which the solution is

$$W_0(\upsilon) = V_o\left(1 - \frac{Q_0(\cos \upsilon)}{Q_0(\cos \upsilon_o)}\right) \tag{3.34}$$

The gradient along its surface $\hat{\upsilon} \cdot \vec{F} \equiv F_\eta(\eta, \upsilon)$, using Eq. (A.32), results in

$$F_\upsilon(\eta, \upsilon) = -\frac{1}{L\left(\sinh^2\eta + \sin^2\upsilon\right)}\partial_\upsilon W_0(\upsilon) \tag{3.35}$$

which can be recast in terms of its value at the apex $F(0, \upsilon_o)$ as

$$\begin{aligned} F(\eta, \upsilon_o) &= \frac{\sin \upsilon_o}{\sqrt{\sin^2 \upsilon_o + \sinh^2 \eta}} F(0, \upsilon_o) \\ F(0, \upsilon_o) &= \frac{V_o}{LQ_0(\cos \upsilon_o)\sin^2 \upsilon_o} \end{aligned} \tag{3.36}$$

Lastly, the surface radius of curvature $a$ is found from Eq (3.19) to be

$$a = L\frac{\sin^2\upsilon_o}{\cos \upsilon_o} = D\tan^2\upsilon_o \tag{3.37}$$

where the tip-to-anode distance $D = L\cos \upsilon_o$ is introduced.

### 3.6.2 Ellipsoidal Geometry

An ellipsoidal geometry in a background field $F_o$ is defined by $\eta = \eta_0$. As with the hemispherical case, far from the ellipsoid, the potential goes to

$$\lim_{\eta \to \infty} V(\eta, \upsilon) = -F_o z = -F_o L \cosh \eta \cos \upsilon \tag{3.38}$$

From the asymptotic behavior, $W = \cos \upsilon$, so that from the behavior of the $\eta$ part

$$-\frac{1}{\cos \upsilon}\left\{\frac{1}{\sin \upsilon}\frac{\partial}{\partial \upsilon}\left(\sin \upsilon \frac{\partial}{\partial \upsilon} \cos \upsilon\right)\right\} = 2 = n(n+1) \tag{3.39}$$

or $n = 1$. As a result, $U(\eta)$ is expected to be of the form

$$U(\eta) = -\cosh \eta + A_1 Q_1(\cosh \eta) \tag{3.40}$$

where $Q_1(x)$ is defined in Eq. (A.24) but for $x > 1$, or

$$Q_1(\cosh \eta) = \frac{\cosh \eta}{2}\ln\left(\frac{\cosh \eta + 1}{\cosh \eta - 1}\right) - 1 \tag{3.41}$$

The coefficient $A_1$ is specified by boundary conditions. The metal surface is defined by $(\eta = \eta_o)$ such that $V(\eta_o, \upsilon) = 0$, from which

$$U(\eta) = -\cosh\eta \left[1 - \frac{\cosh\eta_o\, Q_1(\cosh\eta)}{\cosh\eta\, Q_1(\cosh\eta_o)}\right] \tag{3.42}$$

As with the hemispherical problem contained in Eq. (3.28) of Section 3.5, a similar line of argument gives

$$V(\eta, \upsilon) = -F_o L \cosh\eta_o \cos\upsilon \left(\frac{\cosh\eta}{\cosh\eta_o} - \frac{Q_1(\cosh\eta)}{Q_1(\cosh\eta_o)}\right) \tag{3.43}$$

The hemisphere is a special case of an ellipsoid for which the major axis is equal to the minor axis, and so the form of Eq. (3.28) is obtained in the large $\eta$ limit

$$\begin{aligned} V(\eta \gg 1, \upsilon) &= -F_o L \cosh\eta \cos\upsilon + F_o L \cos\upsilon \frac{\cosh^3\eta_o}{\cosh^2\eta} \\ &= F_o r \cos\upsilon \left[1 - \left(\frac{a}{r}\right)^3\right] \end{aligned} \tag{3.44}$$

where $a = L\cosh\eta_o$.

In the case of the ellipsoid, the field along the surface is obtained using Eq. (A.32) to find

$$F(\eta, \upsilon) = \frac{\sinh\eta_o \cos\upsilon}{\sqrt{\sin^2\upsilon + \sinh^2\eta_o}} F(\eta_o, 0) \tag{3.45}$$

$$F(\eta_o, 0) = \frac{F_o}{Q_1(\cosh\eta_o)\sinh^2\eta_o} \tag{3.46}$$

The apex radius can be found using Eq. (3.19), but another simple method suggests itself. From the definitions of $z$ and $\rho$ in Eqs. (3.29) and (3.30), the surface defined by $z = z_s(\rho)$ satisfies

$$1 = \frac{\rho^2}{L^2\sinh^2\eta_o} + \frac{z_s(\rho)^2}{L^2\cosh^2\eta_o} \tag{3.47}$$

Isolating $z_s(\rho)$ and comparing it to parabolic relation $z_s(\rho) \approx z_o - (\rho^2/a)$ in the small $\rho$ limit allows for the identification

$$a = L\frac{\sinh^2\eta_o}{\cosh\eta_o} = z_o \tanh^2\eta_o \tag{3.48}$$

## 3.7 Exercises

**Exercise 3.1** What is the force exerted on an electron by its image charge when the electron just passes over the highest point of an SN barrier $V(x)$? Assume copper-like parameters and field emission conditions of $F = 6$ eV/nm.

**Solution** 6 eV/nm.

**Exercise 3.2** What is the width $L$ of the SN barrier for an electron tunneling at the Fermi level $\mu$? Assume the work function is $\Phi = 4.5$ eV and $F = 6.57$ eV/nm. What are the values of $(x_o, x_\pm(\mu))$?

**Solution** $L = 0.5$ nm, $(x_o, x_\pm) = (0.234, 0.342 \pm 0.250)$ nm.

**Exercise 3.3** Show the limiting relations of Eq. (3.6) and use them to show Eq. (3.8).

**Exercise 3.4** For a hemispherical surface of radius $a$, how much does the field on axis drop compared to its apex value $F(a, \theta = 0)$ when the distance from the apex is 1/8 of the hemisphere radius?

**Solution** 80.2%.

**Exercise 3.5** A "monopole" is a single point charge at the origin, but it can be thought of as the limit $L \to 0$ and $\delta \to \infty$ of Eq. (3.18) such that $z_0$ stays finite. Find the field enhancement factor of a monopole.

**Solution** $\beta = 2$.

**Exercise 3.6** (Hard) Assume an ellipsoidal emitter such that that $R$ is the ratio of the semi-major axis $L \cosh \eta_o \cos 0$ to the semi-minor axis $L \sinh \eta_o \sin(\pi/2)$ or $R = \coth \eta_o$. Find the field enhancement factor $\beta(\eta_o)$ in the limits of a sphere ($\eta_o \to \infty$) and a needle ($\eta_o \to 0$)

- Spherical Limit: $\beta(\eta_o \to \infty) = 15/(7 - 2R^2) \to 3$
- Needle limit: $\beta(\eta_o \to 0) = (2R^2 - 3)/[2(\ln(2R) - 1)] \to R^2/\ln(2R)$.

**Exercise 3.7** Moving away from the apex of a hyperbolic or ellipsoidal emitter, $V(x)$ no longer follows an SN barrier where $x$ is the distance to the apex, but close to the emitter apex, an effective SN barrier $V(x) \approx \mu + \Phi_a - F_a x - Q_a/x$ can still be constructed. If $x_o$ is near the maximum of $V(x)$, then if three values $V(x_o)$ and $V(x_o \pm \delta)$ are known (numerically or otherwise), find the effective parameters $(\Phi_a, F_a, Q_a)$.

**Solution** Define $\xi(x_j) = (V(x_j) - \mu)x_j$ for $j \in (x_o - \delta, x_o, x_o + \delta)$ to give $(\xi_m, \xi_o, \xi_p)$, respectively. The effective parameters are

$$\begin{aligned}
\Phi_a &= \frac{(\xi_p - \xi_m)\,\delta - 2x_o\,(\xi_p - 2\xi_o + \xi_m)}{2\delta^2} \\
F_a &= \frac{(\xi_p - 2\xi_o + \xi_m)x_o^2 - (\xi_p - \xi_m)x_o\delta + 2\xi_o\delta^2}{2\delta^2} \\
Q_a &= \frac{-\xi_p + 2\xi_o - \xi_m}{2\delta^2}
\end{aligned} \tag{3.49}$$

# 4

# Statistical Mechanics

*So distribution should undo excess,*
*And each man have enough.*

W. Shakespeare, *King Lear* IV.1

## 4.1 Maxwell–Boltzmann Distribution

As late as 1651, a list of prohibited doctrines assembled by the Jesuits (*Ordinatio pro studiis superioribus*) included propositions that the continuum could not be made up of indivisible, or atomic, parts [3]. Nevertheless, the atomic hypothesis gained ascendancy when John Dalton (1766–1844) introduced it into chemistry, and the experiments of Robert Brown (1773–1858) showed that small particles suspended in a liquid were buffeted about and resulted in rapid and zigzag motion, explainable only by the assumption that small bodies (water molecules) were colliding with the grains (cytoplasmic granules shed by pollen, but not pollen itself as is often assumed). The zigzags require a microscope to discern [290], but when one does, then the random paths taken by the grains can be mapped out, as in Figure 4.1, where the $(x, y)$ positions of the grains can be recorded every $\Delta t = 30$ seconds using a grid with spacings of about 3 $\mu$m [278].

Atomic collisions are extraordinarily rapid, far more often than the short time interval over which grains change direction. The calculation of an average velocity in each coordinate direction can be had by forming the ratio $\Delta x/\Delta t$ (in the $x$-direction), and similarly in the $y$-direction, and that it has a Gaussian shape is a consequence of after each collision, the grain has a new random velocity: the average of those random velocities is, by the Central Limit Theorem [8], normally distributed. Put another way, if $u_j$ is a random number, then the average of $N$ of them can be related to the average of a bunch of smaller averages: Let $N = m \times n$, where there are $m$ smaller groupings of size $n$ of random numbers. Then

$$\langle u \rangle = \frac{1}{N}\sum_{j=1}^{N} u_j = \frac{1}{m}\sum_{i=1}^{n}\frac{1}{n}\sum_{k=1}^{m} u_{(m-1)n+k} = \frac{1}{m}\sum_{k=1}^{m}\langle u_m \rangle \tag{4.1}$$

or the sum over the averages of the $m$ smaller groupings. The average of averages of random numbers is normally distributed.

The kinetic theory of gasses posits that particles have an energy of motion (KE or $E$), the probability of having that energy ($\mathcal{P}(E)$) monotonically decreases as $E$ increases, and that the energy of

*Fundamentals of Electron Emission Physics*, First Edition. Kevin L. Jensen.

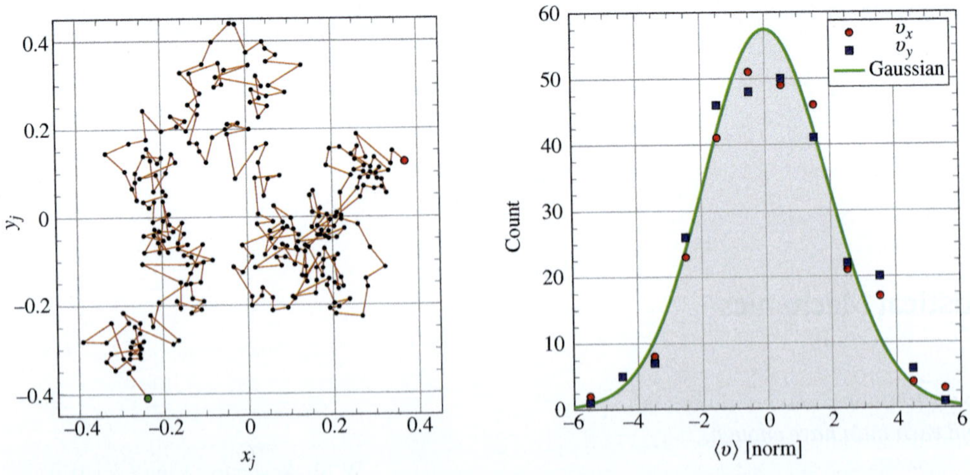


**Figure 4.1** (*left*) Coordinates of a grain every 30 seconds (data digitally extracted from Figure 9 of Ref. [278] / Jean Perrin): the view area was comparable to 100 $\mu$m on a side. Red and green dots are the (presumed) beginning/end points. (*right*) Velocity histogram where $\langle v \rangle$ is the average velocity in each of 12 bins of the histogram, normalized to the size of the bin, and overlaid on a Gaussian distribution.

a level in that ordering decreases such that each incremental increase $\Delta E$ in that energy results in the same proportionate decrease in $\mathcal{P}(E)$. The probability $\mathcal{P}(E)$ that an atom has a kinetic energy $E$ therefore does not depend on where the energy reference point is, and so

$$\frac{\mathcal{P}(E_1 + \Delta E)}{\mathcal{P}(E_1)} = \frac{\mathcal{P}(E_2 + \Delta E)}{\mathcal{P}(E_2)} \tag{4.2}$$

Rewriting using logarithms, and putting a $\Delta E$ in the denominator gives

$$\frac{\ln[\mathcal{P}(E_1 + \Delta E)] - \ln[\mathcal{P}(E_1)]}{\Delta E} = \frac{\ln[\mathcal{P}(E_2 + \Delta E)] - \ln[\mathcal{P}(E_2)]}{\Delta E} \tag{4.3}$$

But in the limit $\Delta E \to 0$, either side of the equation becomes a derivative. But that then means that the derivative of $\ln[\mathcal{P}(E)]$ with respect to $E$ is a constant independent of $E$, or $(d/dE)\ln[\mathcal{P}(E)] = -\beta$, with the minus sign meaning that as the energy goes up, the probability goes down, and so

$$\mathcal{P}(E) = \frac{e^{-\beta E}}{\int_0^\infty e^{-\beta E} dE} = \beta e^{-\beta E} \tag{4.4}$$

The factor $\beta$ is related to the idea of temperature $T$: as temperature increases, more hot molecules exist, and so $\mathcal{P}(E)$ increases. To make $\beta$ have the units of inverse energy so that $\beta E$ is dimensionless, then $\beta = 1/k_B T$.

The energy so far is just the kinetic energy, or $E = (1/2)Mv^2$ for a classical gas, and it depends only on velocity. The potential energy is generally a function of position. To specify what is known about the gas molecule, then, its three velocity components and three position components have to be known. If they are plotted on a six-dimensional graph of positions and velocities, the distribution of dots becomes a phase space description. Consider the case where the potential energy is so weak that changes in it matter little compared to the kinetic energy, a situation that occurs for gasses at sea level. Because $v^2 = |\vec{v}|^2$, only the magnitude of the velocity matters, for which $v^2 = v_x^2 + v_y^2 + v_z^2$. Let $a = M/2k_BT$ for simplicity. Then the manner in which $v$ is distributed shall be called $f(v)$. Because of the form of kinetic energy and the behavior of Gaussian functions,

$$\begin{aligned} f(v) \propto e^{-av^2} &= e^{-a(v_x^2+v_y^2+v_z^2)} \\ &= e^{-av_x^2}e^{-av_y^2}e^{-av_z^2} \propto f(v_x)f(v_y)f(v_z) \end{aligned} \tag{4.5}$$

or a product of Boltzmann distributions for each spatial coordinate. The product of the distribution function with some little volume in phase space is then the probability that a gas atom has those phase space coordinates. The phase space volume element contains a cube $dx\,dy\,dz \equiv d^3r$ for position, and a similar cube $dv_x\,dv_y\,dv_z = d^3v$ for velocity. And so, the probability that the element of (a six-dimensional) phase space at $\vec{r}$ and $\vec{v}$ contained in a volume $d^3rd^3v$ is just $\mathcal{P}(E) \leftrightarrow f(\vec{r}, \vec{v})d^3rd^3v$. The sum over all the phase space points, that is, the sum over all the probabilities, is just the probability that the atom is somewhere, anywhere, and that is unity. As a consequence of Eq. (4.5)

$$1 = \int f(v)d^3rd^3v = \iiint e^{-av^2}d^3v \iiint d^3r = V\left[\int_{-\infty}^{\infty} e^{-as^2}ds\right]^3 \tag{4.6}$$

where $V = \int d^3r$ is spatial volume and $s$ is a dummy integration variable. The Maxwell–Boltzmann (MB) distribution is then

$$f(v)d^3v = \frac{e^{-av^2}d^3v}{V\displaystyle\int e^{-av^2}d^3v} \equiv \frac{1}{\mathcal{Z}\hbar^3}e^{-Mv^2/2k_BT}d^3v \tag{4.7}$$

where $\mathcal{Z}(T)$ is called the partition function [290]. The term $\hbar k/M$ takes over the role of $v$ and is preferable to what follows, in terms of which

$$f(k)d^3k = \frac{1}{\mathcal{Z}\hbar^3}e^{-M\hbar^2k^2/2k_BT}d^3k \tag{4.8}$$

From the known solution to Gaussian integrals $\int_{-\infty}^{\infty} e^{-as^2}ds = \sqrt{\pi/a}$,

$$\mathcal{Z}(T) = \frac{V}{\hbar^3}\left(\frac{Mk_BT}{2\pi}\right)^{3/2} \equiv \frac{V}{\lambda_T^3} \tag{4.9}$$

where $\lambda_T$ is called the "thermal wavelength" for a single molecule.

## 4.2 Entropy

A box of molecules of volume $V$ has $N$ such molecules. The energy $E$ for one of them is now replaced by a sum over energies $\sum_{j=1}^{N} E_j$ for all of them. But because the distribution functions of all the molecules are Maxwell–Boltzmann, that sum is a product of $N$ distribution functions – that is, $\exp[\sum_j E_j] = e^{E_1}e^{E_2}\cdots e^{E_N}$. And so, for $N$ particles, $\mathcal{Z}(T) \to [\mathcal{Z}(T)]^N$. The energy of the molecules is now expressed in terms of their free energy $\mathcal{F}$, the energy in a system that is available to do work [291], defined by $\mathcal{Z}^N = \exp(-\beta\mathcal{F})$. It follows

$$\mathcal{F} = -Nk_BT\ln(\mathcal{Z}) \tag{4.10}$$

When a side of the box is acted on to expand (work is done by the gas) or contract (work is done on the gas), the free energy changes. Work is a force $F$ applied over a distance $x$, and pressure is the force $F$ over an area $A$, such that the volume is $\mathcal{V} = Ax$. The change in free energy at constant $T$ as a result of a change in volume $\mathcal{V}$ then causes a change in pressure $\mathcal{P}$ by

$$\mathcal{P} = -\left(\frac{d\mathcal{F}}{d\mathcal{V}}\right)\Big|_T = Nk_B T \frac{d}{d\mathcal{V}} \ln\left(\frac{\mathcal{V}}{\lambda_T^3}\right) = \frac{Nk_B T}{\mathcal{V}} = \rho k_B T \tag{4.11}$$

where $\rho = N/\mathcal{V}$ is the density of the gas. The equation is more commonly encountered as $\mathcal{P}\mathcal{V} = Nk_B T$ (the ideal gas equation), but here more compactly written as $\beta\mathcal{P} = \rho$. It follows from the behavior of the Maxwell–Boltzmann (MB) distribution of Eq. (4.8), which was position-independent: when a potential energy term is reinserted (the ideal gas is in a potential gradient such as gravity), the analysis can be generalized [154, 225] so that the distribution function acquires a position dependence.

In addition to pressure $\mathcal{P}$, the entropy $\mathcal{S}$ is also found from the free energy. Using Eq. (4.10) then results in

$$\mathcal{S} = -\left(\frac{d\mathcal{F}}{dT}\right)\Big|_{\mathcal{V}} = Nk_B\left[\ln\left(\frac{\mathcal{V}}{\lambda_T^3}\right) + \frac{3}{2}\right] \tag{4.12}$$

This form relies on the particles being distinguishable, but that assumption leads to the Gibbs Paradox: like volume and particle number, entropy should be additive, so that if a partition separating two sides of a box enclosing a gas at the same temperature is removed, then not only does the new volume become the sum of the two partition volumes, and the number of particles become the sum of the two partition number of particles, but so too must entropy be the sum of the entropies of the two partitions. That does not happen with Eq. (4.12). If the two partitions are of the same size and particle number (to make things easy), then $\mathcal{V} \to 2\mathcal{V}$ and $N \to 2N$ do not result in $\mathcal{S} \to 2\mathcal{S}$, even though the pressure $\mathcal{P} = (2Nk_B T)/(2\mathcal{V}) = (Nk_B T)/\mathcal{V}$ does not change (as it should not). The prescription $\mathcal{Z}(T) \to [\mathcal{Z}(T)]^N$ is to blame: if the particles are indistinguishable, then their ordering does not matter. By way of example, the number of distinguishable ways to arrange the five coins (penny, nickel, dime, quarter, half-dollar) is $5! = 120$, but the number of distinguishable ways to arrange five coins (2 pennies, 3 dimes) is $5!/2!3! = 10$ because the order of the pennies and dimes does not matter. In the same way, going from one particle to $N$ particles demands that $\mathcal{Z}(T) \to \mathcal{Z}(T)^N/N!$ in Eq. (4.10) to obtain

$$\mathcal{F} = -k_B T \ln\left(\frac{\mathcal{Z}^N}{N!}\right) \approx -Nk_B T \ln\left(\frac{e\mathcal{V}}{N\lambda_T^3}\right) \tag{4.13}$$

where Stirling's approximation $\ln(N!) \approx \ln\left(N^N e^{-N}\sqrt{2\pi N}\right) \approx N\ln(N/e)$ is used for when $N$ is large (say, comparable to Avogadro's number or the number of conduction electrons in a small piece of metal). The inclusion of $N!$ is termed "correct Boltzmann counting" and insures that the MB distribution is the limit of the quantum distributions to be encountered below. When the proper $\mathcal{F}$ of Eq. (4.13) is used in Eq. (4.12), then the correct result is

$$\mathcal{S} = -\left(\frac{d\mathcal{F}}{dT}\right)\Big|_{\mathcal{V}} = Nk_B\left[\ln\left(\frac{\mathcal{V}}{N\lambda_T^3}\right) + \frac{5}{2}\right] \tag{4.14}$$

where $5/2 = \ln(e) + 3/2$ has been used. Now the change $\mathcal{V} \to 2\mathcal{V}$ and $N \to 2N$ results in $\mathcal{S} \to 2\mathcal{S}$ as it should.

## 4.3 Phase Space

The canonical equations of electron emission are one-dimensional, and so to describe and understand them, only one position coordinate $x$ and one wave number coordinate $k_x \to k$ are needed, with the recognition that the results can be generalized back to three dimensions when required. Such is the approach taken here. Thus, consider a distribution $f(x, k_x) \equiv f(x, k)$. Imagine that $f(x, k)$ is made up of a bunch of phase space points at the location $(x_j, k_l)$ for $(j, l) \in (0, \cdots, n)$, where $n^2$ is the number of points (particles), where, for simplicity, the trapezoid is chosen so that the number of points in each dimension is the same. Assume that the particles are evolving under a force that causes a constant acceleration g (meant to be suggestive of the gravitational analog). Place a Cartesian grid over phase space where each square element on that grid is specified by the indices $(l, k)$ such that the grid elements for the initial condition are specified by $0 \leq x_j \leq L$ and $0 \leq k_l \leq k_n$. Then, as a function of time, the evolution of a phase space point initially in the $(j, l)$ grid element proceeds by the usual ballistic equations

$$\begin{aligned} x(t) &= x_j + k_l t + \frac{1}{2} g t^2 \\ k(t) &= k_l + g t \end{aligned} \tag{4.15}$$

two examples for different g of which are shown in Figure 4.2. As time evolves, the phase space points act like ballistic particles, but remain inside a trapezoid, the corners of which evolve ballistically as well. The area of a trapezoid is the product of half the average of the top and bottom with the height. For a given $k_l$, the length of the trapezoid is

$$\begin{aligned} x(t) - x(t) &= (x_n + k_l t + g t^2/2) - (x_0 + k_l t + g t^2/2) \\ &= x_n - x_0 = L \end{aligned}$$

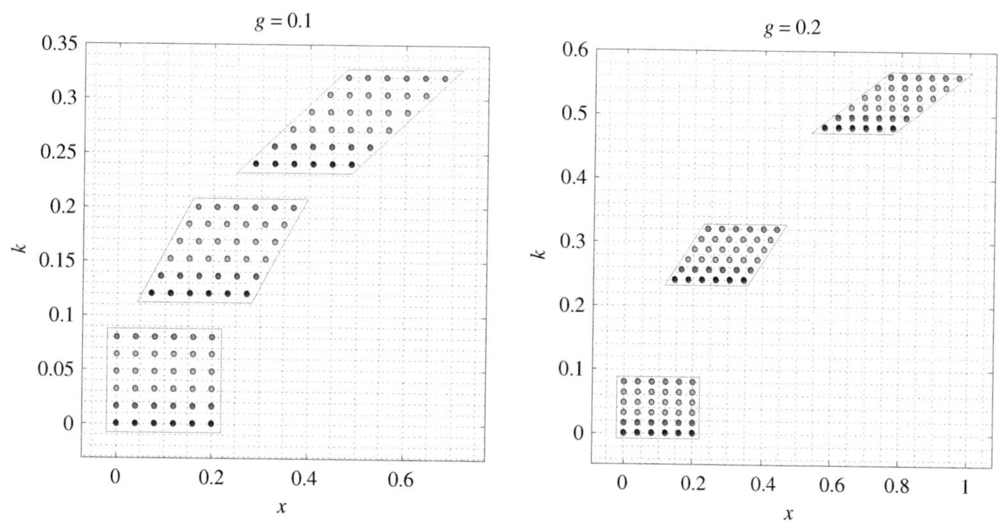


**Figure 4.2** Evolution of 64 phase space points $[x(t), k(t)]$ where $k(t) = k_o + gt$ and $x(t) = x(0) + k(0)t + (g/2)t^2$, where $x(0)$ and $k(0)$ are uniformly distributed between $0 \leq x(0) \leq L = 0.2$ and $0 \leq k(0) \leq= k_n = 0.08$, and $g = (0.1, 0.2)$. The three gray trapezoids for $t = [0, 1.2, 2.4]$ (lower left to upper right) enclose all the phase space points as they evolve (the corners also follow the ballistic equations).

and so, the top and bottom are of equal length. The height is always $(k_n + gt) - (k_0 + gt) = k_n$ since $k_0 = 0$. And so, the area of all three trapezoids shown is the same and equal to $Lk_n$. That means that the number of points in the trapezoidal area is unchanging as the trapezoid deforms, a conclusion known as Liouville's Theorem when $L \to dx$ and $k_n \to dk$ in the infinitesimal limit [154]. It is equivalent to asserting the points in the trapezoids are neither created nor destroyed (number of particles is conserved).

If the mass (or charge) of the point is smeared out over the grid element $dxdk$, then $f(x, k)$ acts as a local density and $f(x, k)\, dx\, dk$ takes over representing the mass (or charge). $f(x, k)$ is then a distribution function, generally normalized so that an integral of it over all position and momentum is unity. Moments of the distribution (i.e. integrals over the product of $f(x, k)$ with a power of $k$) give number density and current density via Eq. (2.14). Although $\rho(x, t)$ is a number density, the inclusion of $q$ in the definition of $J(x, t)$ renders it a charge current density as is more convenient. Because particle number is conserved, then

$$0 = \frac{df}{dt} = \partial_t f + \dot{x}\partial_x f + \dot{k}\partial_k f \tag{4.16}$$

where the Newtonian over-dot notation refers to the time derivative of $x$ and $k$. Eq. (4.16) is Boltzmann's transport equation [154, 243, 382]. Observe that the velocity term is $\dot{x} = \hbar k/m$. If $V(x)$ is independent of time, then $-\partial_x V(x) = F = -\hbar\dot{k}$ is the force term. Assuming $F$ is constant, performing an integration over all $k$, and requiring $f(x, \pm\infty) \to 0$ results in the continuity equation

$$q\partial_t \rho(x, t) + \partial_x J(x, t) = 0 \tag{4.17}$$

The relationship between the time rate of change of density $\rho(x, t)$ and the spatial variation in current density $J(x, t)$ is, in this derivation, based on a phase space composed of moving point particles, be they atoms of an ideal gas (in which case, the $q$ is absent) or even electrons and ions (in which case the $q$ is present). The same equation will be seen to pass over to quantum mechanical quantities having a wave nature, but the meaning of the terms will be subtly changed.

## 4.4 Diffusion

A group of atoms in thermal equilibrium with each other can be treated to a first approximation as a group of hard spheres that interact through scattering, which serves to redistribute energy and randomize direction. Let the number of atoms per unit volume (the density) at a point $x$ and time $t$ be $\rho(x, t)$, so that the number of particles in a slab of area $A$ and thickness $\Delta x$ is $N = \rho(x, t) A\Delta x$. The area $A$ appears on both sides of all subsequent equations, and so factors out (equivalently, work in units where $A = 1$ below). To understand what happens to a group of atoms in that slab, consider a simple one-dimensional model of their motion where, after a given time increment $\Delta t$, a particular atom has a probability $p$ of jumping to the right or left, and a probability of $(1 - p)$ of staying where it is. Let position points be demarcated by $x_j = j\Delta x$ and time events by $t_n = n\Delta t$, so that a useful notation is $\rho(x_j, t_n) \equiv \rho_j^n$. Then, if the number of atoms in the slab at time $t_n$ is known, the predicted number at the next time step is affected by the *net* number of atoms that jump into the slab from the right $(p/2)(\rho_{j+1}^n - \rho_j^n)$, and the net number that jumps into the slab from the left and proportional to $(p/2)(\rho_j^n - \rho_{j-1}^n)$, where that net number can be negative if more depart than arrive. It follows that

$$\rho_j^{n+1} = \rho_j^n + \frac{p}{2}\left(\rho_{j+1}^n - \rho_j^n\right) + \frac{p}{2}\left(\rho_j^n - \rho_{j-1}^n\right) \tag{4.18}$$

Expanding and regrouping gives

$$\rho_j^{n+1} - \rho_j^n = \frac{p}{2}\left(\rho_{j+1}^n - 2\rho_j^n - \rho_{j-1}^n\right) \tag{4.19}$$

The probability $p$ is dimensionless. A diffusion constant $D \equiv p\Delta x^2/2\Delta t$ can be introduced, in terms of which expanding and regrouping gives

$$\frac{\rho_j^{n+1} - \rho_j^n}{\Delta t} = D\,\frac{\rho_{j+1}^n - 2\rho_j^n - \rho_{j-1}^n}{2\Delta x^2} \tag{4.20}$$

The finite-difference forms of first and second derivatives [320] in the limit $(\Delta t, \Delta x) \to 0$ allow the equation to become

$$\frac{d}{dt}\rho(x,t) - D\frac{d^2}{dx^2}\rho(x,t) = 0 \tag{4.21}$$

a form easily extended to three dimensions (3D) through the replacement $d^2/dx^2 \to \nabla^2$. The resulting equation is then known as Fick's law of diffusion [154, 290].

The shape of $\rho(x, t)$ can be anticipated by a simple simulation: take $N$ points and place them at the origin at $t = 0$ and then at each time step $t_n$ for $n \in (1, \cdots, N_t)$, allow a fraction $p$ of them to jump, half to the left and half to the right, as per Eq. (4.18), and then plot a histogram of their positions. Because the histogram will spread out over time, it is useful to calculate $\sigma$, where the variance is defined by $\sigma^2 = \langle (x_j^n - \langle x_j^n \rangle)^2 \rangle$, and then consider the shape of the histogram as a function of $(x/\sigma)$ instead. If $p = 1/2$, then this can be easily done by $x_j^{n+1} = x_j^n +$ `round`$)((2u-1)\Delta x$ where `round(x)` function rounds $(2u - 1)$ to the nearest integer and $u$ is a uniformly distributed random number over a unit interval. Letting $\Delta x = 1$ unit and $N = 2^9 = 512$, then the distribution after $N_t$ time steps is shown in Figure 4.3 (see Algorithm B.2.6) along with a Gaussian proportional to $\exp[-(x/\sigma)^2/2]$

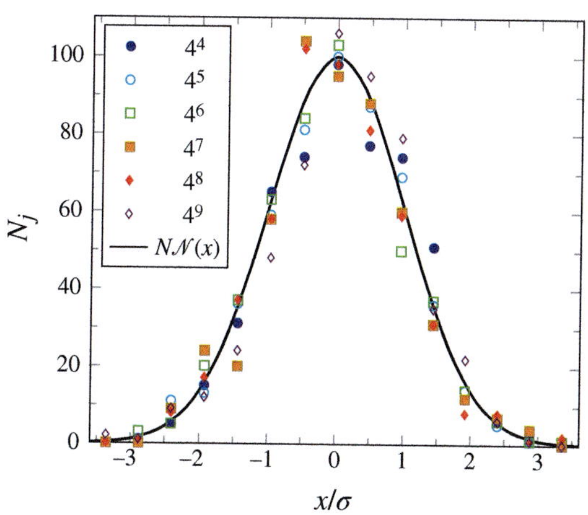


**Figure 4.3** The distribution of 512 particles in 16 bins (the number the $j^{th}$ bin is $N_j$) after $4^n$ time steps where the particles jump with a probability $p = 0.5$. The variance $\sigma^2$ for each $N_t = 4^n$ gives a measure of how far the particles have migrated from their starting position at $x = 0$, and $\langle x_j \rangle \approx 0$. The curve $\exp[-(x/\sigma)^2/2]$ is scaled to match the height. The distribution over $N_j$ mimics the Normal distribution $\mathcal{N}(x)$ scaled so an integral over it is $N = 512$.

for $N_t = 4^k$ and $k \in (4, \cdots, 9)$. Such behavior suggests that $\rho(x,t) = A\,\exp(-Bx^2)$, where $A(t)$ and $B(t)$ may depend on time $t$. When that ansatz is put into Eq. (4.21), then

$$D\frac{d^2}{dx^2}\rho(x,t) = -2DBA(1-2Bx^2)\rho(x,t)$$
$$\frac{d}{dt}\rho(x,t) = \left[\frac{1}{A}\frac{dA}{dt} - x^2\frac{dB}{dt}\right]\rho(x,t) \tag{4.22}$$

Equating them and rearranging

$$\frac{1}{A}\frac{dA}{dt} + 2DB = x^2\left(\frac{dB}{dt} + 4DB^2\right) \tag{4.23}$$

which must hold for arbitrary $x$, and so each side is separately equal to zero. On the right-hand side, the ansatz $B = B_o t^\nu$ results in

$$0 = \frac{dB}{dt} + 4DB^2 = \nu B_o t^{\nu-1} + 4B_o^2 D t^{2\nu} \tag{4.24}$$

which suggests $t^{-\nu-1} = -4DB_o/\nu$, which only holds for all $t$ if $(\nu = -1)$. Consequently, $B = 1/(4Dt)$. Setting the rest of Eq. (4.23) to zero gives

$$\frac{A}{2t} + \frac{dA}{dt} = 0 \Rightarrow A = \frac{A_o}{\sqrt{t}} \tag{4.25}$$

The integral over all $x$ of $\rho(x,t)$ at any time should give $N$, the number of particles. Therefore,

$$\frac{\rho(x,t)}{N} = \frac{1}{\sqrt{4\pi Dt}}\exp\left(-\frac{x^2}{4Dt}\right) \equiv \mathcal{N}(x) \tag{4.26}$$

where $\mathcal{N}(x)$ is the normal distribution [8] for zero mean ($\langle x\rangle = 0$) with $\sigma = \sqrt{2Dt}$. The 1D result generalizes to 3D by letting $x^2 \to x^2 + y^2 + z^2$ in the exponent. Observe that the diffusion term satisfies the principle of superposition: if two (or more) distributions $\rho_1(x,t)$ and $\rho_2(x,t)$ separately satisfy Eq. (4.21), then so does their sum.

## 4.5 Quantum Distributions

The quantum version of the partition function $\mathcal{Z}$ takes into account the number of particles $n_l$ that can occupy an energy state $l$: the discussion is somewhat hampered by the fact that the notion of a quantum mechanical eigenstate characterized by an energy $E_l$ has yet to be introduced (as it will be in Section 6.4), so for now, a willingness to accept that the number of particles in a state $n_l$ can vary depending on the statistics of the particles involved, and one can wait on the quantum notions that help make sense of that. The calculation of $\mathcal{Z}$ then sums over all those occupation numbers for a given energy $E_l$ and must be performed for each state $l$, and then the product of terms for all those states must be taken. In the case of particles following Bose–Einstein statistics (bosons), the number $n_l$ is not constrained, and so [133, 154, 209, 289]

$$\mathcal{Z}_{BE} \equiv \prod_{l=1}^{\infty}\sum_{n_l=0}^{\infty} \exp\left[\beta n_l(\mu - E_l)\right] \tag{4.27}$$

where $\mu$ is the chemical potential. The summation over $n_l$ is analytic, and given by $1/(1 - \exp[\beta(\mu - E_l)]$ so that the grand potential $\Omega$ is

$$\Omega_{BE} = \frac{1}{\beta} \ln[\mathcal{Z}_{BE}] = k_B T \sum_l \ln\{1 - \exp[\beta(\mu - E_l)]\} \tag{4.28}$$

The density $\rho$ is the ratio of the average number of particles $\langle N \rangle$ over the volume $\mathcal{V}$, but

$$\langle N \rangle = -\left.\frac{\partial \Omega_{BE}}{\partial \mu}\right|_{T,V} = \sum_l \frac{1}{e^{\beta(E_l - \mu)} - 1} \equiv \sum_l \langle n_l \rangle \tag{4.29}$$

Thus, the probable number of bosons as a function of temperature (recall $\beta = 1/k_B T$) characterized by an energy $E_l$ is given by the Bose–Einstein distribution $f_{BE}(E_l) = \langle n_l \rangle$ or

$$f_{BE}(E) \equiv \frac{1}{e^{\beta(E-\mu)} - 1} \tag{4.30}$$

as it is conventionally represented.

Particles satisfying Fermi–Dirac distributions (fermions) differ, in that only one particle or none can occupy a given state characterized by $E_l$, or

$$\mathcal{Z}_{FD} \equiv \prod_{l=1}^{\infty} \sum_{n_l=0}^{1} \exp[\beta n_l(\mu - E_l)] \tag{4.31}$$

The summation over $n_l$ now consists of only two terms, 0 or 1, but if spin is taken into account, then actually there are two $n_l$, one for spin up (call it $n_l^+$) and one for spin down (call it $n_l^-$), and so the $\sum_{n_l}[\cdots] \to \sum_{n_l^+}[\cdots]\sum_{n_l^-}[\cdots]$. As a result,

$$\mathcal{Z}_{FD} = \prod_{l=1}^{\infty} (1 + \exp[\beta(\mu - E_l)])^g \tag{4.32}$$

where $g = 2$ for the spin of electrons. The same operations as for BE now yield

$$\Omega_{FD} = \frac{1}{\beta} \ln[\mathcal{Z}_{FD}] = k_B T g \sum_l \ln\left[1 + e^{\beta(\mu - E_l)}\right] \tag{4.33}$$

$$\langle N \rangle = \sum_l \frac{1}{1 + e^{\beta(E_l - \mu)}} \tag{4.34}$$

from which the Fermi–Dirac distribution is defined as

$$f_{FD}(E) \equiv \frac{g}{1 + e^{\beta(E-\mu)}} \tag{4.35}$$

In going to the continuum, the summation over states $l$ becomes an integration over momentum and so density becomes

$$\rho(T) = \frac{\langle N \rangle}{\mathcal{V}} = \frac{g}{(2\pi)^3} \int f_{FD}[E(k)]\, d^3k = \frac{g}{(2\pi)^3} \int \frac{d^3k}{1 + e^{\beta[E(k) - \mu]}} \tag{4.36}$$

At zero temperature, $f_{FD}(E) = g\Theta(\mu - E)$, and so $d^3k \to 4\pi k^2 dk$ and the upper limit of the integral is $k_F = \sqrt{2m\mu}/\hbar$, and so

$$\rho(T = 0) = \frac{1}{\pi^2} \int_0^{k_F} k^2 dk = \frac{k_F^3}{3\pi^2} \tag{4.37}$$

as found previously in Eq. (2.15).

## 4.6 Temperature

Temperature is a purely statistical concept governing macroscopic collections of particles and how their energies are spread out in a system in equilibrium. Measuring that parameter is something different. A thermometer is a seemingly straight-up measurement device for temperature, but for a hot cathode in a vacuum chamber different methods are used. For the latter, a pyrometer is typically used, by which the energy distribution of the emitted light from a cavity is measured as a function of wavelength $\lambda = 2\pi/k$ where for light, $k = c/\omega$. The energy density in each volume element of phase space is $2\epsilon_\omega d^3k/(2\pi)^3$, where the 2 out front accounts for the different polarizations. The energy depends on the magnitude of $k$, not the direction, and so it is convenient to integrate over the angular components and let $d^3k = 4\pi k^2 dk$. Letting $\hbar\omega = \epsilon_\omega = k_B T$ for the harmonic modes, modeling the light frequencies and remembering that $\hbar\omega = \hbar kc$ for light, then it follows that the energy density $u(\omega)$ is the Raleigh–Jeans formula for low frequencies

$$u(\omega)d\omega = \frac{\hbar\omega^3}{\pi^2 c^3} e^{-\hbar\omega/k_B T} d\omega \tag{4.38}$$

Problematically, the integral of this formula over all frequencies is infinite (the ultraviolet catastrophe). For high frequencies, Wein found instead

$$u(\omega)d\omega = \epsilon_\omega \frac{2}{(2\pi)^3} 4\pi k^2 dk = \frac{k_B T\omega^2}{\pi^2 c^3} d\omega \tag{4.39}$$

Planck found that these formulae could be reconciled if he used the trick that the energy of the oscillators modeling light was assumed to be quantized rather than continuous, for which he could then show

$$u(\omega)d\omega = \frac{\omega^2}{\pi^2 c^3} \frac{\hbar\omega}{e^{\hbar\omega/k_B T} - 1} d\omega \tag{4.40}$$

Einstein realized it was no trick: light was quantized [76], and he was able to account for photoemission phenomena from that bold leap and a particle conception of photons. The total energy $U$ in the cavity is found by integrating over all frequencies, or

$$U = \frac{(k_B T)^4}{\pi^2 \hbar^3 c^3} \int_0^\infty \frac{x^3}{e^x - 1} dx = \frac{(k_B T)^4}{\pi^2 \hbar^3 c^3} \Gamma(4)\zeta(4) \tag{4.41}$$

where the $\Gamma(p)$ and $\zeta(p)$ functions of Appendices A.1 and A.2. The power $P$ radiated in all directions is then $P = (c/4)U$: if $dU$ is the amount radiated into a solid angle $2\pi \sin\theta \, d\theta$ in a unit time, then from Lambert's cosine law, a $\cos\theta$ accounts for diffuse radiation, so that $2\pi \int_0^{\pi/2} \sin\theta \; \cos\theta \, d\theta = \pi$ is (1/4) of $2\pi \int_0^{\pi} \sin\theta \, d\theta = 4\pi$ of what would be if there was no cosine dependence (notice the upper limit of integration, accounting for the fact that the flow is only forward, not backward as well). The flow of energy through a spherical surface, often written as $\sigma_{sb} T^4$, identifies the Stefan–Boltzmann constant as

$$\sigma_{sb} = \frac{c}{4} \frac{k_B^4}{\pi^2 \hbar^3 c^3} \Gamma(4)\zeta(4) = \frac{k_B^4 \pi^2}{60\hbar^3 c^2} = 3.53917 \times 10^{-22} \frac{\text{eV}}{\text{K}^4 \text{ fs nm}^2} \tag{4.42}$$

or, as it is more conventionally expressed, $\sigma_{sb} = 5.67037 \times 10^{-8}$ W / m$^2$ K$^4$.

The black body radiation measurement, however, must contend with the properties of actual materials making up the emitter and measuring device. A pyrometer measures gray body emission and compares it to black body emission to find $T$. In practice, pyrometers use a few wavelengths to

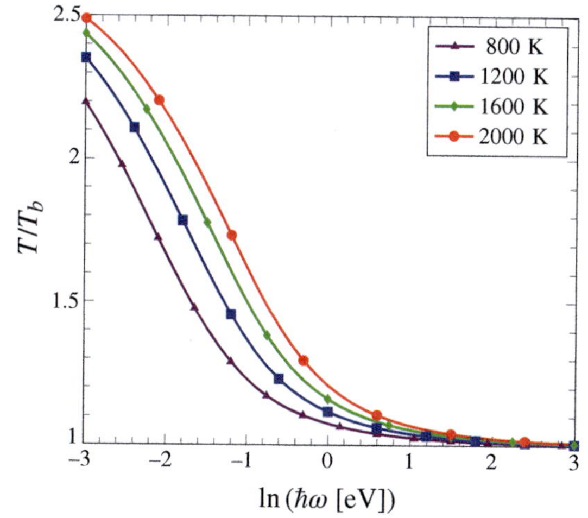


**Figure 4.4** Relation of $T$ to $T_b$ as a function of photon energy as per Eq. (4.43) using parameters typical of a dispenser cathode of $(t, r, \epsilon) = (0.93, 0.76, 0.52)$ [43], the product of which is $tr\epsilon = 0.36754$.

make a measurement (dual wavelength pyrometers could choose 700 nm and 800 nm, for example). If $T_b$ is the "brightness" temperature measured by the pyrometer, then equating the gray body radiation spectrum to the black body (Planck's law) spectrum [70] and eliminating common factors gives

$$\frac{tr\epsilon}{\exp(\hbar\omega/k_B T) - 1} = \frac{1}{\exp(\hbar\omega/k_B T_b) - 1} \tag{4.43}$$

where $(t, r, \epsilon)$ are the transmissivity of the viewport, reflectivity of the mirror, and the emissivity of the cathode, respectively. Solving Eq. (4.43) is a straightforward exercise: the behavior of $T_b$ is shown in Figure 4.4 (see Algorithm B.2.7).

## 4.7 Exercises

**Exercise 4.1** Calculate the thermal wavelength $\lambda_T$ of a neon atom Ne in units of the Bohr radius. Assume the mass of a neon atom is $M = 20.1797u$, where the atomic mass unit $u = 1822.89\ m$. Calculate for both room temperature (300 K) and similar to that of a plasma in a neon lamp (2000 K).

**Solution** $0.424a_o$ for $T = 300$ K and $0.164a_o$ for $T = 2000$ K.

**Exercise 4.2** For a one-dimensional model, a Maxwell–Boltzmann distribution in velocity $v$ is of the form $f(v) = Ne^{-mv^2/2k_BT}$. Find $N$ such that $\int_{-\infty}^{\infty} f(v)dv = 1$. Find the root mean squared $v_{rms} = \sqrt{\langle v^2 \rangle}$ and evaluate $v_{rms}/c$ for an electron at room temperature $T = 300$ K.

**Solution** $N = \sqrt{m/2\pi k_B T}$. $v_{rms}/c = \sqrt{k_B T/mc^2} = 2.2492 \times 10^{-4}$

**Exercise 4.3** How many distinct ways are there to arrange the letters of the word "Avogadro" assuming order is not important?

**Solution** 10080.

**Exercise 4.4** In the evaluation of Eq. (4.27), by letting $e^{\beta(\mu-E)} = x < 1$, then a series of the form $S = \sum_{n=0}^{\infty} x^n$ is encountered. Show that $S = 1 + xS$. Show that such a relation can be solved for both $S$ and $\partial_x S$.

**Solution** $S = 1/(1-x)$. $\partial_x S(x) = -S(x)[1 - S(x)]$

**Exercise 4.5** What is the number density $\rho$ for a zero-temperature electron gas when the Fermi level $\mu = 1$ eV? When $\mu = 5$ eV?

**Solution** $\rho(\mu = 1\,\text{eV}) = 4.54\ \text{nm}^{-3}$. $\rho(\mu = 5\,\text{eV}) = 50.8\ \text{nm}^{-3}$

**Exercise 4.6** (hard) Using the Fermi–Dirac distribution of Eq. (4.35) but assuming $g = 1$, approximate $-f_{FD}(E)\partial_E f_{FD}(E)$ by a Gaussian of the form $a \exp[-b(E-\mu)^2]$. What are $a$ and $b$?

**Solution** $a = \beta/4$. $b = \beta^2/4$.

**Exercise 4.7** Assuming $g = 1$ again, show that $-f_{FD}(E)\partial_E f_{FD}(E)$ is the same as $h(E) \equiv \beta f_{FD}(E)[1 - f_{FD}(E)]$. Evaluate

$$\beta^{-1}\int_{-\infty}^{\infty} h([\mu + (x/\beta)])dx$$

How does this compare to a Dirac $\delta$-function?

**Solution** The integral is 1. $h(E) = \delta(E-\mu)$ in the $\beta \to \infty$ limit.

**Exercise 4.8** Using the three-dimensional Boltzmann distribution of Eq. (4.8), calculate the rms velocity $\sqrt{\langle v^2\rangle}$ in terms of the one-dimensional $\sqrt{\langle v_x^2\rangle}$ of Exercise 4.2. Why is this answer expected?

**Solution** $\langle v^2\rangle = 3\langle v_x^2\rangle$.

# 5

# Light

*Tomorrow, and tomorrow, and tomorrow*
*Creeps in this petty pace from day to day*
*To the last syllable of recorded time;*
*And all our yesterdays have lighted fools*
*The way to dusty death. Out, out, brief candle!*

W. Shakespeare, *Macbeth* V.5

## 5.1 Index of Refraction

Light traveling between two points takes that path which is stationary with respect to small variations of that path [127], or, as more popularly understood and known as Fermat's principle, light travels in a manner to minimize the time it travels. Let light travel between two points, $A$ and $B$, equidistant from a mirror along the $\hat{x}$-axis, such that the $(x, y)$ coordinates of $A$ are $(0, h)$ and $B$ are $(D, H)$, and the point $C$ is at $(x, 0)$. Since light travels with the same speed on both legs of its journey, the shortest time requirement is equivalent to the shortest path. The path length is the sum of the distances from $A$ to $C$ (call it $l(x)$) plus $C$ to $B$ (call it $L(x)$), or $P(x) = l(x) + L(x)$, and so

$$P(x) = \sqrt{h^2 + x^2} + \sqrt{H^2 + (D - x)^2} \tag{5.1}$$

Small variations $\Delta$ in $x$ make equal and opposite variations in $l(x)$ and $L(x)$, and so

$$\frac{l(x + \Delta) - l(x - \Delta)}{2\Delta} = \frac{L(X + \Delta) - L(X - \Delta)}{2\Delta} \tag{5.2}$$

an equation which can be manipulated, in the limit $\Delta \to 0$, to give $dl/dx + dL/dx = 0$ or

$$\frac{x}{\sqrt{h^2 + x^2}} = \frac{D - x}{\sqrt{H^2 + (D - x)^2}} \tag{5.3}$$

Introducing $\tan\varphi = h/x$ and $\tan\theta = H/(D - x)$ gives

$$\frac{1}{2\sqrt{1 + \tan^2\varphi}} = \frac{1}{2\sqrt{1 + \tan^2\theta}} \tag{5.4}$$

from which it is concluded that $\varphi = \theta$, that is, the angle of incidence equals the angle of reflection.

*Fundamentals of Electron Emission Physics*, First Edition. Kevin L. Jensen.

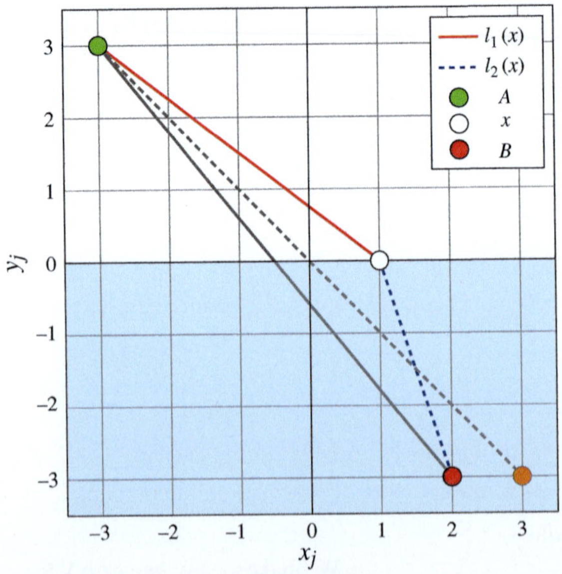


**Figure 5.1** Two representative paths for light to travel from $A$ (green dot in clear medium) to $B$ (red dot in blue medium): gray line = shortest path and red/blue line = shortest time. Dashed line shows path if the velocity of light does not change. Position $x$ denoted by ∘.

If the mirror is replaced by a transparent material medium, then the velocity of light in that medium can be different from that in air (or vacuum). Now let $B$ be below the $\hat{x}$-axis. Take the origin of the coordinate system $(x, y) = (0, 0)$ to be at the midpoint along the $\hat{x}$-axis between $A$ and $B$ and let $A$ be at $(x_1, y_1)$ and $B$ be at $(x_2, y_2)$, where the mediums of $A$ and $B$ are designated by "1" and "2" respectively, as schematically shown in Figure 5.1. Now the total time $t$ is the sum of the amount of time in medium 1 $(l_1(x)/v_1)$ and medium 2 $(l_2(x)/v_2)$. Taking the speed of light in vacuum ($c$) as a reference point, define the index of refraction $n_j$ such that $v_j = c/n_j$, and so if the light intersects the $\hat{x}$-axis at $X$

$$t = t_1 + t_2 = \frac{n_2}{c} l_2(x) + \frac{n_1}{c} l_1(x) \tag{5.5}$$

From the Pythagorean theorem, the path lengths are $l_j(x)^2 = (x_j - x)^2 + y_j^2$ for $j \in (1, 2)$. The smallest value of $t$ occurs when $x_o$ is such that

$$\left(\frac{n_1}{c}\right) \frac{d}{dx} l_1(x)\Big|_{x=x_0} = -\left(\frac{n_2}{c}\right) \frac{d}{dx} l_2(x)\Big|_{x=x_0} \tag{5.6}$$

Evaluation and simplification yield

$$\frac{n_1(x_o - x_1)}{c\, l_1(x_o)} = \frac{n_2(x_2 - x_o)}{c\, l_2(x_o)} \tag{5.7}$$

Defining $\theta_j$ as the angle between $l_j(x)$ and a vertical line through $x$ and canceling common factors of $c$ then gives what is commonly known as the law of refraction, *aka* Snell's law, although Thomas Harriot had deduced it some two decades prior [7]

$$n_1 \sin\theta_1 = n_2 \sin\theta_2 \tag{5.8}$$

Notably, this equation does not depend on the actual speed of light $c$, but rather the relative speed of light in the two media, or $n_2/n_1$.

The impact carries through to the propagation of light in a medium. The wave equation governing the propagation of light is (along the direction of propagation, or 1D, and focusing on the electric field $\mathcal{E}$ for simplicity)

$$\left[\frac{d^2}{dx^2} - \frac{1}{c^2}\frac{d^2}{dt^2}\right]\mathcal{E}(x,t) = 0 \tag{5.9}$$

for which the solution is $\mathcal{E}(x,t) = \mathcal{E}_0 \exp[i(kx - \omega t)]$ and the velocity relation is $c = \omega/k$. As the energy $E = \hbar\omega = pc$ of the photon is constant, in a medium, the momentum changes to $p = \hbar k \rightarrow \hbar k/n$.

## 5.2 Relativistic Effects

The speed of light in a medium with an index of refraction $n$ is then $c/n$, where $c$ is its velocity in vacuum. The Michelson–Morley experiment of 1887 attempted to find the change of light velocity with respect to the presumed background ether but returned a null result. Albert Einstein's explanation of that null result changed the concept of time.

A particle bouncing off of a wall (or reflecting off a mirror) travels there and back with a velocity that, with malice of forethought, shall be called $c$. Its path is traced by the dashed red/blue dashed lines in the $\hat{y}$-direction, and it travels a total distance of $2l_o$. If the platform from which the particle is launched is moving in the $\hat{x}$-direction with a velocity $v = \beta c$, then the path length is instead $2l$. The particle takes a time $2l_o/c$ to go round trip. If the platform is moving at a speed $v = \beta c$, but the launch of the particle is such that from the perspective of someone moving with the platform, the particle still travels with a velocity $c$ with respect to the rest frame, then from the perspective of the moving platform, the particle is traveling slower with a velocity $v/c = \sqrt{c^2 - \beta^2 c^2}/c = \sqrt{1-\beta^2}$ because velocities combine vectorially. The transit time in the moving frame is then

$$\Delta t' = \frac{2l_o}{c\sqrt{1-\beta^2}} \tag{5.10}$$

To an observer not moving with the platform, the path length is longer and so the transit time given by $2l$ in Figure 5.2, and is from the Pythagorean theorem

$$\Delta t = \frac{2l}{c} = \frac{2\sqrt{l_o^2 + (\beta c \Delta t/2)^2}}{c} \tag{5.11}$$

which, when solved for $\Delta t$ shows that $\Delta t'$ and $\Delta t$ satisfy the same equation. That is, the transit time for observers in the unmoving and moving frames should be the same, and both observers seem to agree on how long the particle takes to go there and back again.

But "seems" is not the same as "is": the finding of Michelson and Morley indicated that the observers instead only agreed on the velocity of the particle, which implies they must disagree on what the transit time was. If both observers measure the same velocity $c$ for the particle, that is the same as demanding that the ratio of the path length with the transit time is the same, or

$$c = \frac{2l_o}{\Delta t'} = \frac{2\sqrt{l_o^2 + (\beta c \Delta t/2)^2}}{\Delta t} \tag{5.12}$$

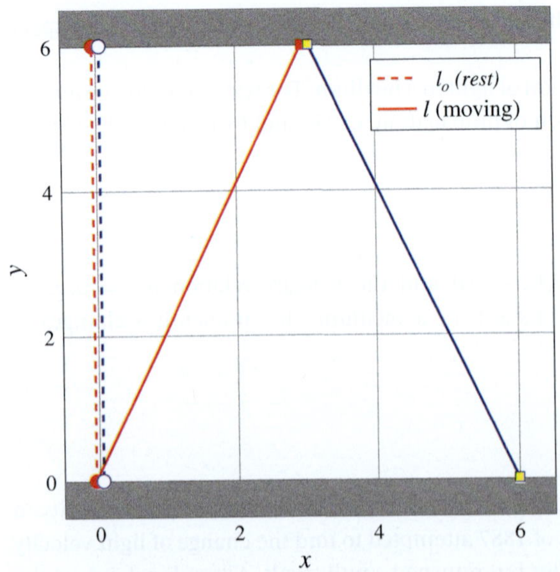


**Figure 5.2** Path of light bouncing off a mirror (dashed) in a stationary frame and (solid) in a moving frame traveling to the right at $v = \beta c$.

The distance to the mirror as measured in the prime frame is $l_o = c\Delta t'/2$. Putting those into Eq. (5.12) results in

$$c = \frac{c}{\Delta t}\sqrt{\beta^2\Delta t^2 + \Delta t'^2} \tag{5.13}$$

Solving for $\Delta t$ gives

$$\Delta t = \frac{\Delta t'}{\sqrt{1-\beta^2}} \equiv \gamma\Delta t' \tag{5.14}$$

where the factors $\gamma$ and $\beta$ become the usual relativistic factors of the Lorentz transformation. The stationary observer is forced to conclude that the moving observer experiences time passing by more slowly, an effect known as time dilation. The effect can be measured: muons (or mu mesons) have a half-life of 2 $\mu$s, but when accelerated to $\beta = 1 - \delta$ with $\delta = 10^{-4}$, their lifetimes are $1/\sqrt{1-\beta^2} \approx 1/\sqrt{2\delta} \approx 70.7$ times longer [374]. Similar dilations occur for length.

If the measured velocity of light is the same in all reference frames (for the sake of argument, let two of them be called the "primed" and "unprimed" frames), then it follows that

$$c = \frac{d}{dt}|\vec{r}| = \frac{d}{dt'}|\vec{r}'| \tag{5.15}$$

On a Cartesian grid, three-dimensional vector lengths are obtained by summing the square of their components, and so that the equation can be rewritten using the infinitesimal term $dr^2 = dx^2 + dy^2 + dz^2$ as

$$ds^2 = c^2dt^2 - dr^2 = c^2dt'^2 - dr'^2 = ds'^2 \tag{5.16}$$

where $ds$ is an invariant quantity (the same in different frames) that separates two events, and is called a world line element that relates time and space increments [9]. Other invariants exist, such as those joining energy and momentum. Their similar relation is suggestively indicated as follows. Force is the differential change of momentum in an infinitesimal time, or $F = dp/dt$, but for particles moving near the speed of light, $v = \beta c$ is not departing much from $c$ and so $\beta \approx 1$. Something else is changing, and it is mass, so rewrite the force as $F = d[(\gamma m_o)(\beta c)]/dt$. The invariant quantity is $m_o$, and so the analogous energy–momentum equation to the time-space one of Eq. (5.16) is

$$m_o^2 = m^2 - (p/c)^2 \tag{5.17}$$

which can be put in a more familiar form by letting $E = mc^2$ to find

$$E^2 = (m_o c^2)^2 + (cp)^2 = (m_o c^2)^2 + (\gamma\beta m_o c^2)^2 = (\gamma m_o c^2)^2 \tag{5.18}$$

So as with Eq. (5.14), mass too is affected by velocity. The momentum of a particle, nonrelativisitically given as $p = mv$, is therefore altered by $m \to \gamma m$ and $v \to \beta c$ so that the familiar relativistic relations

$$E = \gamma m_o c^2; \quad p = \gamma\beta m_o c \tag{5.19}$$

result. As force is applied to a particle, the increase in energy increases the momentum term, and so kinetic energy is $KE = (\gamma - 1)m_o c^2$, which becomes $KE \approx (\beta^2/2)m_o c^2 = mv^2/2$ and recovers the nonrelativistic limit. Because the photon is massless, the energy of motion is $E = \hbar\omega$, and so the propagation of a light wave is (using $A$ for amplitude for pedagogical reasons)

$$A(x, t) = A_o \exp[i(kx - Et/\hbar)] \tag{5.20}$$

as the solution to Eq. (5.9).

# 6

# Quantum Mechanics

*Until I know this sure uncertainty,*
*I'll entertain the offer'd fallacy.*

W. Shakespeare, *Comedy of Errors* II.2

## 6.1 Wave Mechanics

As de Broglie postulated in his Ph.D. thesis of 1924 (and was proven by 1927), particles also act as waves. It follows that their wave motion should in some way resemble Eq. (5.20), but now $\hbar\omega = E = \hbar^2 k^2/2m$ as the kinetic energy of a massive particle. With that substitution, the differential equation for massive particles that replaces the wave equation of Eq. (5.9) is

$$\left[\frac{\hbar^2}{2m}\frac{d^2}{dx^2} + i\hbar\frac{d}{dt}\right] A(x,t) = 0 \tag{6.1}$$

which differs from the electromagnetic wave equation for photons because the time derivative is only to the first power rather than the second, but it also differs from the diffusion equation of Eq. (4.21) in that the coefficient of the time derivative is now imaginary, not real. It is sometimes alternately said that Schrödinger's equation is the diffusion equation where $t \to it$ (time becomes imaginary). As a result, whereas concentrations of particles "disperse to naught,"[1] the waves still oscillate for particles above barriers (that is, not tunneling). Schrödinger would reconfigure Eq. (5.9), apply it to the wave function of the electron, and reintroduce the effects of spatially varying potential energy $V(x)$ to create his celebrated equation

$$i\hbar\frac{d}{dt}\psi(x,t) = \left[-\frac{\hbar^2}{2m}\frac{d^2}{dx^2} + V(x)\right]\psi(x,t) \equiv \hat{H}\psi(x,t) \tag{6.2}$$

where the hat notation on $\hat{H}$ (the "Hamiltonian") is what distinguishes it as an operator which, when it acts on $\psi(x,t)$, returns an eigenvalue that is energy, or $\hat{H}\psi = E\psi$, as it is conventionally expressed. Solutions to quantum mechanical problems then revolve around finding constants of the motion, like $E$ and others, quantities that are reflected in the symmetries and independences of the Hamiltonian. The three-dimensional Coulomb potential $V(\vec{r}) \propto 1/r$ results in energy, squared

1 The phrasing is Shakespeare's, and refers to how glory fades (*Henry 6.1* I.2) like the ripples in a circle of water, but in a curious irony, water waves oscillate and particles disperse, another beautiful metaphor ruined by physics exactitude.

*Fundamentals of Electron Emission Physics*, First Edition. Kevin L. Jensen.

angular momentum $\mathcal{L}^2$, and the $z$-component of angular momentum $\mathcal{L}_z$ being the constants of the motion. For one-dimensional problems to which Eq. (6.2) applies, however, none is more illuminating than the harmonic oscillator for which $V(x) = \gamma x^2/2$, where Schrödinger's equation becomes

$$\left(-\frac{\hbar^2}{2m}\frac{d^2}{dx^2} + \frac{1}{2}\gamma x^2\right)\psi(z) = E\psi_E(x) \tag{6.3}$$

The equation is still a bit busy, so make some powerfully simplifying modifications. The equation is converted to dimensionless form by introducing $\gamma/2 = (\hbar^2 k_o^2/2m)k_o^2$ and $E = \hbar^2 k^2/2m$ to get

$$\left(-\frac{1}{k_o^2}\frac{d^2}{dx^2} + (k_o x)^2\right)\psi(x) = \left(\frac{k}{k_o}\right)^2 \psi(x) \tag{6.4}$$

The equation can be made even simpler with two additional changes: first, let $(k/k_o)^2 \equiv n + (1/2)$, where the choice of notation reflects the clairvoyant suspicion that $n$ will prove itself to be an integer; and second, notice that wherever $x$ appears, a $k_o$ is stuck to it, so that it is extremely convenient to work in units where $k_o = 1$, which is the same as $k_o x \to x$. As a result, the differential equation to be solved for the harmonic oscillator is

$$\left(-\frac{d^2}{dx^2} + x^2\right)\psi_n(x) = (2n+1)\,\psi_n(x) \tag{6.5}$$

where now the wave function is explicitly labeled by its eigenvalue, as is commonly done. If $n = 0$, then the equation looks very familiar to ones that arise when $\psi(x)$ is a Gaussian: in fact, it is well known that

$$\left(-\frac{d^2}{dx^2} + x^2\right)e^{-x^2/2} = e^{-x^2/2} \tag{6.6}$$

If the differential term $d/dx$ were called something like $D$, then the left-hand side "multiplier" of $\psi$ looks like $(x^2 - D^2)$, and the urge to write $(x - D)(x + D)$ as algebra would suggest becomes powerful, but that is not possible: $D$ is an *operator* and acts on everything to the right of it, and so order matters: $(x - D)(x + D)$ gives something different than $(x + D)(x - D)$, even though they would otherwise be the same if $D$ were just some variable. Order matters for another kind of mathematics, the mathematics of matrices, leading to the formulation of quantum mechanics created by Heisenberg, and that formulation will use the notation on which all that follows is dependent.

## 6.2 Matrix Mechanics

If $\mathbb{M}$ is a matrix and $\vec{v}$ a column vector, then for $\lambda$ a (possibly complex) number, the equation $\mathbb{M} \cdot \vec{v} = \lambda\vec{v}$ is an eigenvalue equation familiar from linear algebra. Problems in classical dynamics, such as rotational motion, can be solved by finding vectors for which $\mathbb{M}$ is diagonal [111]. For non-diagonal matrices, it is well known that the order in which the matrices are multiplied before multiplying a vector matters, and in that way, matrices resemble the operators of the previous section. Heisenberg developed the matrix formalism that treated quantum mechanics, but it is the notation of P. A. M. Dirac that will become very useful here.

The first step is to generalize the vectors by letting the column vectors be represented as $|v\rangle$ and row vectors as $\langle v|$. For example, if the unit vectors for the $(x, y, z)$ axes, normally called $(\hat{\imath}, \hat{\jmath}, \hat{k})$, are instead represented as $|i\rangle$, $|j\rangle$, $|k\rangle$ (e.g. $\langle i| = (1, 0, 0)$, $\langle j| = (0, 1, 0)$, and $\langle k| = (0, 0, 1)$), then $\vec{v}$ and its dot product with itself, $\vec{v} \cdot \vec{v}$ are represented as

$$\begin{aligned} |v\rangle &= v_x|i\rangle + v_y|j\rangle + v_z|k\rangle \\ \langle v|v\rangle &= v_x^2 + v_y^2 + v_z^2 \end{aligned} \tag{6.7}$$

in Cartesian coordinates. If $\mathbb{M}$ is an $N \times N$ matrix, then there can be $N$ different eigenvectors $|v_j\rangle, j \in (1, \cdots, N)$, such that $\mathbb{M}|v_j\rangle = \lambda_j|v_j\rangle$ and $\langle v_j|v_k\rangle = \delta_{jk}$ (the Kronecker delta function) when the $|v_j\rangle$ are normalized to unity (that is, $\langle v_j|v_j\rangle = 1$): the $|v_j\rangle$ are then said to be an orthogonal basis set that can be used to express any vector in that space, or if $|w\rangle$ is some $N$-component vector, then

$$|w\rangle = \sum_{j=1}^{N} |v_j\rangle\langle v_j|w\rangle \equiv \sum_{j=1}^{N} C_j|v_j\rangle \tag{6.8}$$

where $C_j$ is the dot product between $|w\rangle$ and $\langle v_j|$, or $C_j = \langle v_j|w\rangle$. Observe that the dot product of $|w\rangle$ with itself says something important about $|v_j\rangle\langle v_j|$: first, that the sum over all $|C_j|^2 = \langle w|v_j\rangle\langle v_j|w\rangle = |\langle v_j|w\rangle|^2$ is unity, where by convention $\langle v_j|w\rangle$ is the complex conjugate of $\langle w|v_j\rangle$ (the product of a bra with a ket is a number); and second, that $\mathbb{I}_v \equiv \sum_{j=1}^{N}|v_j\rangle\langle v_j|$ acts as an identity matrix. This is trivially demonstrated using the basis states ($|i\rangle$, $|j\rangle$, $|k\rangle$). The product of a ket with a bra is a matrix: for the general case, if $|a\rangle$ is a 3 row - 1 column vector, and $\langle b|$ is a 1 row - 3 column vector, then

$$|a\rangle\langle b| = \begin{pmatrix} a_1 \\ a_2 \\ a_3 \end{pmatrix} \begin{pmatrix} b_1 & b_2 & b_3 \end{pmatrix} = \begin{bmatrix} a_1b_1 & a_1b_2 & a_1b_3 \\ a_2b_1 & a_2b_2 & a_2b_3 \\ a_3b_1 & a_3b_2 & a_3b_3 \end{bmatrix} \tag{6.9}$$

and so

$$\begin{aligned} \mathbb{I}_c &= |i\rangle\langle i| + |j\rangle\langle j| + |k\rangle\langle k| \\ &= \begin{bmatrix} 1 & 0 & 0 \\ 0 & 0 & 0 \\ 0 & 0 & 0 \end{bmatrix} + \begin{bmatrix} 0 & 0 & 0 \\ 0 & 1 & 0 \\ 0 & 0 & 0 \end{bmatrix} + \begin{bmatrix} 0 & 0 & 0 \\ 0 & 0 & 0 \\ 0 & 0 & 1 \end{bmatrix} = \begin{bmatrix} 1 & 0 & 0 \\ 0 & 1 & 0 \\ 0 & 0 & 1 \end{bmatrix} \end{aligned} \tag{6.10}$$

The bra-ket notation, though, is much more powerful than a different way to represent vectors preferred by physicists. In fact, kets are not restricted to being just vectors; they can be constructed to represent *functions* as well. As an example, Legendre polynomials [6] of Appendix A.4 are an orthogonal set of functions, and any polynomial of degree $n$ can be recast as a sum over these polynomials. An easy example is afforded by considering the simple polynomial

$$f(x) = x^2 + 2x + 1 \tag{6.11}$$

If the Legendre functions take on the role of the "basis" and the range of values of the argument $\langle x|$ takes over the role of the vector components, then $P_n(x) = \langle x|P_n\rangle$ serves as a notation for the Legendre polynomial [6] of degree $n, f(x) = \langle x|f\rangle = x^2 + 2x + 1$ takes over the role of the vector $|w\rangle$. The coefficients of a summation expressing $f(x)$ as a sum over Legendre polynomials $|P_n\rangle$ are then calculated by

$$C_n = \frac{2n+1}{2}\langle P_n|f\rangle = \frac{2n+1}{2}\int_{-1}^{1}\langle P_n|x\rangle\langle x|f\rangle\, dx \tag{6.12}$$

where the $(2n+1)/2$ is a normalization factor that arises because of Eq. (A.22). Calculation shows $(C_0, C_1, C_2) = (4/3, 2, 2/3)$ so that

$$f(x) = \sum_{j=0}^{2} C_n P_n(x) = \frac{2}{3}P_2(x) + 2P_1(x) + \frac{4}{3}P_0(x) \tag{6.13}$$

The sum should extend to $\infty$, but $C_n$ for $n > 2$ is zero. The exercise reveals that something profound is afoot, and that something is

$$\mathbb{I} = \int_{-1}^{1} |x\rangle\langle x|\, dx \tag{6.14}$$

acts like the unity matrix. It is what allows switching from one representation to another in the manner of Eq. (6.8). One can intuit that Fourier transforms are yet another example of such a way of doing things analogous to Legendre polynomials. Extending the integration limits to $\pm\infty$ and taking $x$ to be the position variable is the path to quantum mechanics, with the only caveat that the basis states must be normalized: for the Legendre example, this would entail creating $|\phi_n\rangle = \sqrt{(2n+1)/2}\,|P_n\rangle$ to be the normalized ket such that $\langle\phi_n|\phi_m\rangle = \delta_{nm}$. The kets are representations and $|x\rangle$ in particular is the position-space representation, the wave number $k$ is the conjugate variable to $x$ and $|k\rangle$ is the $k$-space representation. Time $t$ and energy $E$ are another pair of conjugate variables. An example of their usage is well illustrated by the harmonic oscillator problem.

The first step is to recast Eq. (6.5) in terms of the bra-ket matrix notation, and letting $\psi_n(x) = \langle x|\psi_n\rangle$ seems natural. Notice, however, that carrying around $\psi$ in the ket is superfluous: it is notationally easier to call the wave function ket $|n\rangle$ instead, so that $\psi_n(x) \equiv \langle x|n\rangle$, where $|n\rangle$ is the basis states of the harmonic oscillator, and $|x\rangle$ is the position representation. How should $x\psi_n(x)$ then be understood? The factor $x$ looks like just a variable representing position, but instead, it should now be interpreted as the value that results when the operator $\hat{x}$ acts on position representation ket, or $\hat{x}|x\rangle = x|x\rangle$, where here and below, anything with a carat above it, like $\hat{x}$, must be understood as an operator that returns a quantity by acting on a wave function. And lastly, how should $(d/dx)\psi_n(x)$ be understood? For that, introduce a new operator $\hat{k}$ such that its action on the ket $|x\rangle$ gives the product of $i$ with the derivative operator, or

$$\hat{k}|x\rangle = \left(i\frac{d}{dx}\right)|x\rangle = \left[\langle x|\left(-i\frac{d}{dx}\right)\right]^{\dagger} \tag{6.15}$$

where something new is encountered: operators can act to the left, on bras, not just to the right on kets, but when they do, the complex conjugate is required. That requirement means that wave functions can be defined up to an overall phase $\varphi$ without changing the density: if $|\psi'\rangle = e^{i\varphi}|\psi\rangle$, then $\rho = \langle\psi'|\psi'\rangle = (\langle\psi|e^{-i\varphi})(e^{i\varphi}|\psi\rangle) = \langle\psi|\psi\rangle$ as it should. The inclusion of $i = \sqrt{-1}$ into $\hat{k}$ was sneaky, but the need to do so is obvious when comparing

$$\left(\hat{k}^2 + \hat{x}^2\right)|n\rangle = (2n+1)|n\rangle \tag{6.16}$$

to the harmonic oscillator differential equation: the inclusion of $i$ in $\hat{k}|x\rangle = i(d/dx)|x\rangle$ makes $\hat{k}^2|x\rangle = -(d^2/dx^2)|x\rangle$.

Seemingly $(\hat{k}^2 + \hat{x}^2)$ should be the same as $(i\hat{k} + \hat{x})(-i\hat{k} + \hat{x})$ but that is not the case: working out the product of the later,

$$\left(i\hat{k} + \hat{x}\right)\left(-i\hat{k} + \hat{x}\right)|n\rangle = \left(\hat{k}^2 + \hat{x}^2 + i\hat{k}\hat{x} - i\hat{x}\hat{k}\right)|n\rangle \tag{6.17}$$

The term $i\hat{k}\hat{x} - i\hat{x}\hat{k}$ does not vanish, but gives rise to

$$i\langle x|\hat{k}\hat{x} - \hat{x}\hat{k}|n\rangle = \frac{d}{dx}[x\psi_n(x)] - x\frac{d}{dx}\psi_n(x) = \psi_n(x) \tag{6.18}$$

It is seen that $i(\hat{k}\hat{x} - \hat{x}\hat{k}) = 1$. The commutator notation $[A, B] = AB - BA$, renders it as

$$[\hat{x}, \hat{k}] = \hat{x}\hat{k} - \hat{k}\hat{x} = i \tag{6.19}$$

and is a key ingredient of Heisenberg's uncertainty relation in that eigenstates of momenta are not at the same time eigenstates of position. The commutator of powers of $\hat{k}^n$ can be found to be

$$\begin{aligned} \left[\hat{x}, \hat{k}^n\right] &= \hat{x}\hat{k}^n - \hat{k}\hat{x}\hat{k}^{n-1} = [\hat{x}, \hat{k}]\hat{k}^{n-1} + \hat{k}\hat{x}\hat{k}^{n-1} - \hat{k}^n\hat{x} \\ &= i\hat{k}^{n-1} + \hat{k}[\hat{x}, \hat{k}^{n-1}] = in\hat{k}^{n-1} \end{aligned} \tag{6.20}$$

where the last step is by iteration. And so, Eq. (6.17) becomes

$$\left(\hat{k}^2 + \hat{x}^2\right)|n\rangle = \left[(i\hat{k} + \hat{x})(-i\hat{k} + \hat{x}) - 1\right]|n\rangle \tag{6.21}$$

Define a new operator by $\hat{a} \equiv (i\hat{k} + \hat{x})$, so that the operator in square brackets becomes $(\hat{a}\hat{a}^\dagger - 1)$. Using $[\hat{a}, \hat{a}^\dagger] = 2$ as can be shown by Eq. (6.19), it follows

$$\left(\hat{a}\hat{a}^\dagger - 1\right)|n\rangle = \left(\hat{a}^\dagger\hat{a} + 1\right)|n\rangle = (2n + 1)|n\rangle \tag{6.22}$$

The question of what $\hat{a}$ by itself does to $|n\rangle$ is answered by considering

$$\begin{aligned} \left(\hat{a}^\dagger\hat{a} + 1\right)\hat{a}|n\rangle &= \left(\hat{a}\hat{a}^\dagger - [\hat{a}, \hat{a}^\dagger] + 1\right)\hat{a}|n\rangle \\ &= \hat{a}\left(\hat{a}^\dagger\hat{a} - 1\right)|n\rangle = \hat{a}(2n - 1)|n\rangle \end{aligned} \tag{6.23}$$

It follows that $\left(\hat{a}^\dagger\hat{a} + 1\right)\hat{a}|n\rangle = (2n - 1)\hat{a}|n\rangle$ or the action of $\hat{a}$ on $|n\rangle$ results in a state with an eigenvalue of $(2n - 1)$. The same arguments show that $\hat{a}^\dagger$ acting on $|n\rangle$ results in a state with an eigenvalue of $(2n + 1)$. This means that the eigen-energies of Eq. (6.4) are $E_n = [n + (1/2)]\hbar^2k_0^2/2m \equiv [n + (1/2)]\hbar\omega$. As a result, $\hat{a}$ is often called the "annihilation operator" and $\hat{a}^\dagger$ the "creation operator" (or, less dramatically, the lowering and raising operators [306]), but their actions are analogous to the angular momentum operators that lead to the quantization of the angular momentum terms $\mathcal{L}^2$ and $\mathcal{L}_z$ in the hydrogen atom model which give rise to the Legendre polynomials [154]. The harmonic case gives rise to the Hermite orthogonal polynomials [6] instead. Knowing that $\hat{a}^\dagger|x\rangle = (i\hat{k} + \hat{x})|x\rangle\langle x|0\rangle$ and knowing $\hat{k} = i(d/dx)$ gives

$$\langle x|n\rangle = \langle x|(\hat{a}^\dagger)^n|0\rangle \equiv H_n(x)\exp(-x^2/2) \tag{6.24}$$

as reflected in the first form of Eq. (A.26). The first few $H_n(x)$ are given by Eq. (A.27), in which the second form shows more directly even and odd parity, that is, how the sign of a function changes if the sign of its argument changes: when $n$ is even then $H_n(-x) = H_n(x)$, and when it is odd then $H_n(-x) = -H_n(x)$.

## 6.3 Representations

The orthogonality of the Legendre or Hermite or any of the other orthogonal polynomials that populate quantum mechanics is expressed by $\langle\psi_n'|\psi_n\rangle = \delta_{nn'}$. But what happens when two operators like $\mathbb{I}^2$ of Eq. (6.14) are encountered and the domain of $x$ is $\pm\infty$? It implies

$$\mathbb{I}^2 = \int_{-\infty}^{\infty}\int_{-\infty}^{\infty} |x\rangle \langle x|x'\rangle \langle x'|\, dxdx' = \int_{-\infty}^{\infty} |x\rangle\langle x|\, dx = \mathbb{I} \tag{6.25}$$

which demands $\langle x|x'\rangle = \delta(x - x')$, or the Dirac $\delta$-function. Similarly, $\mathbb{I}^3$ can be perversely represented as

$$\mathbb{I}^3 = N \iiint |x\rangle\langle x|k\rangle\langle k|x'\rangle\langle x'|\, dx\, dk\, dx' = \mathbb{I} \tag{6.26}$$

where the limits of integration are understood to be $\pm\infty$ but not shown for visual simplicity, and the factor $N$ will have to be determined once $\langle x|k\rangle$ is figured out. This is seen to demand

$$N \int_{-\infty}^{\infty} \langle x|k\rangle\langle k|x'\rangle dk = \delta(x' - x) \tag{6.27}$$

for a continuous basis. A discrete basis proceeds analogously (see Chapter 8 of Ref. [154]), and, in fact, is a more rigorous introduction to the relationship between $|k_n\rangle$ and $|x_n\rangle$, but for expediency here, familiarity with Fourier transforms and its relation to harmonic functions suggests right away that $\langle k|x\rangle = e^{ikx}$ and $N = 1/2\pi$ satisfy not only the Dirac $\delta$-function demand, but also recovers the action of $\hat{k}$ on $|x\rangle$ by

$$\begin{aligned}\hat{k}|x\rangle &= \frac{1}{2\pi}\int_{-\infty}^{\infty} \hat{k}|k\rangle\langle k|x\rangle dk \\ &= \frac{1}{2\pi}\int_{-\infty}^{\infty} ke^{ikx}|k\rangle dk = -i\frac{d}{dx}|x\rangle\end{aligned} \tag{6.28}$$

As a result, $\langle x|k\rangle\langle k|x'\rangle \propto e^{ik(x'-x)}$ and Eq. (6.27) are recovered. In fact, $e^{\pm ikx}$ are the energy eigenstates for Eq. (6.2) when $V(x) = 0$. When $V(x)$ is not zero, other eigenstates arise that can be considerably more complex. As a result, most pedagogical treatments consider force-free motion so as to benefit from the simplicity of the plane wave eigenstates. When more complexity is needed, space can be separated into regions supporting different eigenstates, for which the wave function for a given energy is found by matching wave function and its first derivative at the boundaries between those regions. That methodology forms the basis behind the Transfer Matrix Approach [154].

## 6.4 Wave Functions

Once a wave function $|\Psi\rangle$ that satisfies Eq. (6.2) is found, the probability density $\rho(x)$ of finding a particle at a position $x$ is obtained from its absolute square. From the demand that the particle must be found somewhere (i.e. the wave function is normalized) then

$$1 = \langle\Psi|\Psi\rangle = \int_{-\infty}^{\infty} \langle\Psi|x\rangle\langle x|\Psi\rangle dx \equiv \int_{-\infty}^{\infty} \rho(x)dx \tag{6.29}$$

If the wave function is the weighted sum over energy eigenstates, then

$$\langle x|\Psi\rangle = \Psi(x, t) = \langle x| \sum_n c_n|\psi_n\rangle e^{i\omega_n t} = \sum_n c_n\psi_n(x, t) \tag{6.30}$$

and as a result, Eq. (6.29) becomes at ($t = 0$)

$$1 = \int_{-\infty}^{\infty} \sum_{n,l} c_n^{\dagger} c_l \langle \psi_n | x \rangle \langle x | \psi_l \rangle dx = \sum_{n,l} c_n^{\dagger} c_l \, \delta_{nl} \tag{6.31}$$

or the sum over all the weights $|c_n|^2 \equiv C_n$ is unity because the wave functions are normalized. The eigenstate $\psi_n(x, t)$ can be recast as the action of a propagation operator $\hat{U}(\hat{t})$ on $|\psi_n\rangle$ such that (see Ref. [154] for an extended treatment)

$$|\psi_n(t)\rangle = \hat{U}|\psi_n(0)\rangle = e^{i\omega_n t}|\psi_n(0)\rangle \tag{6.32}$$

For an eigenstate

$$\langle \psi_n(t)|\psi_n(t)\rangle = \langle \psi_n(0)|\hat{U}^{\dagger}\hat{U}|\psi_n(0)\rangle = \langle \psi_n(0)|\psi_n(0)\rangle \tag{6.33}$$

which demands that $\hat{U}$ is unitary, or $\hat{U}^{\dagger}\hat{U} = \mathbb{I}$ where $\mathbb{I}$ is the identity operator. For that state, the energy $E$ does not change, and so that entails

$$\begin{aligned} E = \langle \psi_n(t)|\hat{H}|\psi_n(t)\rangle &= \langle \psi_n(0)|\hat{U}^{\dagger}\hat{H}\hat{U}|\psi_n(0)\rangle \\ &= \langle \psi_n(0)|\hat{U}^{\dagger}\left([\hat{H}, \hat{U}] + \hat{U}\hat{H}\right)|\psi_n(0)\rangle \\ &= \langle \psi_n(0)|\hat{U}^{\dagger}[\hat{H}, \hat{U}]|\psi_n(0)\rangle + \langle \psi_n(0)|\hat{H}|\psi_n(0)\rangle \\ &= \langle \psi_n(0)|\hat{H}|\psi_n(0)\rangle = E \end{aligned} \tag{6.34}$$

and that means the energy does not change only if the commutator of $\hat{U}$ with the Hamiltonian $\hat{H}$ vanishes, or $[\hat{H}, \hat{U}] = 0$. As a result, an operator that commutes with $\hat{H}$ is time-independent, for otherwise, a general operator $\hat{O}$ would satisfy

$$i\hbar \frac{\partial}{\partial t}\hat{O} = [\hat{H}, \hat{O}] \tag{6.35}$$

Consider two applications in turn: first, a superposition of states to create the wave function $|\Psi\rangle$; and second, the summation over many filled states. For the superposition case, find the density operator $\hat{\rho}$ by

$$\begin{aligned} \langle \Psi|\Psi\rangle &= \int_{-\infty}^{\infty} \langle \Psi|x\rangle\langle x|\Psi\rangle dx = \int_{-\infty}^{\infty} \langle x|\Psi\rangle\langle \Psi|x\rangle dx \\ &= \int_{-\infty}^{\infty} dx \langle x| \left\{ \sum_n f_n \, \hat{U}|\psi_n\rangle\langle\psi_n|\hat{U}^{\dagger} \right\} |x\rangle \equiv \int_{-\infty}^{\infty} dx \langle x|\hat{\rho}(t)|x\rangle \end{aligned} \tag{6.36}$$

where (for reasons that become obvious in the continuum limit), renaming the weighting factor $f_n$ instead of $C_n$ is advantageous. And so, a density operator $\hat{\rho}(t)$ (or density matrix) is defined as

$$\hat{\rho}(t) \equiv \sum_n f_n |\psi_n(t)\rangle\langle\psi_n(t)| = \sum_n f_n \hat{\rho}_n(t) \tag{6.37}$$

Motivated by knowing that Eq. (4.17) holds, then the time evolution of number density $\hat{\rho}_n(t)$ allows for a current density operator $\hat{j}_n$ to be defined using Eq. (6.35) to get

$$i\hbar \frac{\partial}{\partial t}\hat{\rho}_n(t) = [\hat{H}, \hat{\rho}_n] = \frac{\hbar^2}{2m}\left[\hat{k}^2, \hat{\rho}_n\right] \tag{6.38}$$

where $V(\hat{x})$, being the potential energy operator, commutes with $\hat{\rho}_n$ because $\langle x|V(\hat{x})\hat{\rho}_n(t)|x\rangle = \langle x|\hat{\rho}_n(t)V(\hat{x})|x\rangle$ because $V(\hat{x})$ operates to the left or right, respectively, and gives the same result as $V(x)$. Eq. (6.38) becomes

$$
\begin{aligned}
i\hbar\frac{\partial}{\partial t}\hat{\rho}_n(t) &= \left[\hat{H}, \hat{\rho}_n\right] = -\frac{\hbar^2}{2m}\left[\hat{k}^2, \hat{\rho}_n\right] \\
&= -\frac{\hbar^2}{2m}\left\{\hat{k}\left(-i\frac{\partial}{\partial \hat{x}}|\psi_n\rangle\right)\langle\psi_n| + |\psi_n\rangle\left(-i\frac{\partial}{\partial \hat{x}}|\psi_n\rangle\right)^{\dagger}\hat{k}\right\} \\
&= -i\frac{\hbar^2}{2m}\frac{\partial}{\partial \hat{x}}\left[\hat{k}\hat{\rho}_n + \hat{\rho}_n\hat{k}\right]
\end{aligned}
\tag{6.39}
$$

which means that $\hat{j}_n$ is to be identified with

$$
\hat{j}_n \equiv \frac{\hbar}{2m}\left[\hat{k}\hat{\rho}_n + \hat{\rho}_n\hat{k}\right] \equiv \frac{\hbar}{2m}\{\hat{k}, \hat{\rho}_n\}
\tag{6.40}
$$

The curly brackets term defines the anti-commutator. Sandwiching it between $|x\rangle$ kets, the current density $j_n$ is in its more conventional representation is

$$
j_n(x) = \frac{\hbar}{2mi}\left[\psi_n^{\dagger}(x)\partial_x\psi_n(x) - \psi_n(x)\partial_x\psi_n^{\dagger}(x)\right]
\tag{6.41}
$$

If the wave function eigenstates are plane waves so that the incident wave onto a barrier is $Ne^{ikx}$, and far to the left or right the potential is flat and at the same level, the reflected wave is $r(k)Ne^{-ikx}$, and the transmitted wave is $t(k)Ne^{ikx}$ (where $N$ is a normalization factor) then it follows that if the incident current density $j_{inc}$ is the sum of the reflected and transmitted current densities, then dividing throughout by $j_{inc}$ gives

$$
1 = |r(k)|^2 + |t(k)|^2
\tag{6.42}
$$

If a transmission coefficient $D(k)$ is defined such that $j_{trans} = D(k)j_{inc}$ for an eigenstate, then $D(k) = j_{trans}/j_{inc} = |t(k)|^2$. When the density $\rho(x)$ is continuous and the current density is constant, this imposes the following two constraints on a wave function that spans two (or more) separate regions: where the regions join (the interface - let it be at $x_o$), these requirements demand

1. (density) the wave function $\psi_n(x)$ is continuous across $x_o$, or $\psi_n(x_o - \epsilon) = \psi_n(x_o + \epsilon)$, and
2. (current density) the derivative of the wave function is continuous across $x_o$, or $\partial_x\psi_n(x_o - \epsilon) = \partial_x\psi_n(x_o + \epsilon)$

where $\epsilon$ is an infinitesimal distance ($\epsilon \to 0$).

For the summation of many states, consider the states to be in a well with infinite walls. At zero temperature in the case of fermions, all the lowest eigenstates are filled (or filled with two electrons if the spin of the electron is included), and the energy of the highest filled or $N^{th}$ level defines the Fermi energy. Consider first a one-dimensional model of a well of width $L$ with infinite walls so that $V(x) = V_o\Theta[L^2 - (2x)^2]$, where $\Theta(s)$ is the Heaviside step function (zero for negative and one for positive argument). That is, the potential is zero for $-L/2 < x < L/2$ and $V_o$ otherwise. Taking the limit $V_o \to \infty$ means the wave function does not penetrate the walls, and so the boundary conditions become $\psi_n(\pm L/2) = 0$. Elementary quantum mechanics then shows that the wave functions are sines and cosines of the form

$$
\psi_n(x) = \sqrt{\frac{2}{L}} \times \begin{cases} \cos(k_n x) & \text{(even)} \\ \sin(k_n x) & \text{(odd)} \end{cases}
\tag{6.43}
$$

where "even/odd" refers to the parity of $\psi_n(\pm x) = \pm\psi_n(x)$, with the "+" case being even and the "−" case being odd. To distinguish them, designate them as $\psi_n^{\pm}(x)$. To enforce the boundary conditions, the eigenvalues $k_n$ change for each case, with $k_n = (2n-1)\pi/L$ for the even case, and $k_n = 2n\pi/L$ for the odd case, with $n \in (1, 2, 3, \cdots)$. The first few are shown in Figure 6.1. Their sum for $N = 4$ and $N = 8$, along with the sum of the two parity cases, are shown in Figure 6.2 (see Algorithm B.2.1). Classically, one might expect for a large number of particles, the probability density would be uniform across the well, and the sum of the parity solutions shows that to be occurring, but when only the summation of one parity is used, then there is an unphysical gap (odd) or surplus (even) of probability density near the origin. Parity is a property of quantum systems, not classical, and any non-degenerate eigenstate has either an even or an odd parity. Because all of the potential energies upon which the emission equations will be based do not show a handedness, parity $\hat{\Pi}$ and the Hamiltonian $\hat{H}$ "commute" or $[\hat{\Pi}, \hat{H}] = 0$, and this means parity is conserved for processes involving interactions that do not depend on it.

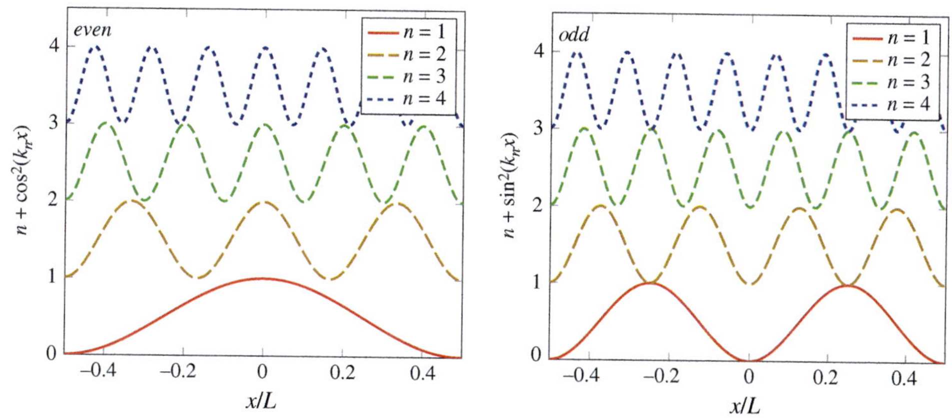


**Figure 6.1** $|\psi_n(x)|^2$ of Eq. (6.43) for the lowest eigenstates in a well of width $L$ with infinite walls at $\pm L/2$ for (*left*) even parity / cosine functions for $k_n = (2n-1)\pi/L$ and (*right*) odd parity / sine functions for $k_n = 2n\pi/L$. The curves are offset from each other vertically by the addition of $n$.

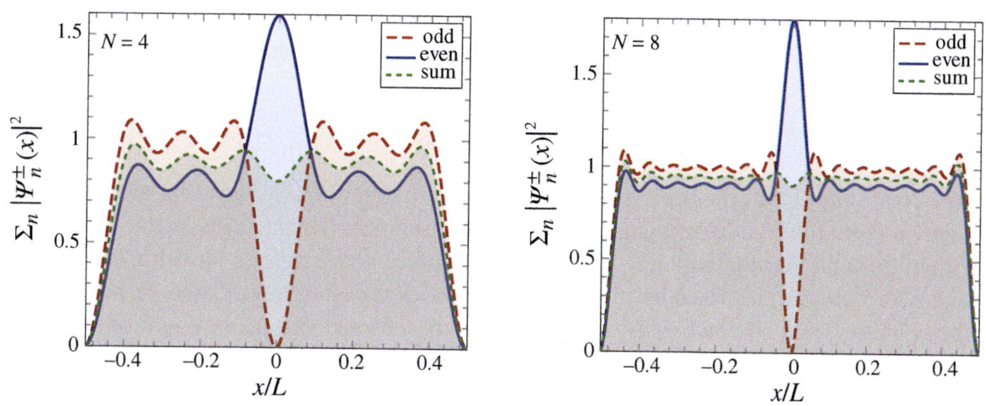


**Figure 6.2** $\sum_n |\psi_n^{\pm}(x)|^2$ of Eq. (6.43) for the lowest eigenstates in a well of width $L$ with infinite walls at $\pm L/2$ for (*left*) the sum of $N = 4$ lowest even and odd parity eigenstates, and (*right*) the sum of $N = 8$ lowest. For both, the weights are $C_n = 1$ for all $n$.

The last example used the appealing choice $C_n = 1$ for all $n$ for the one-dimensional problem. The physics of materials is three-dimensional, and energy $E$ is the sum of the square of three momentum components, or

$$E = \frac{\hbar^2}{2m}\left(k_x^2 + k_y^2 + k_z^2\right) = \frac{\hbar^2 k^2}{2m} \tag{6.44}$$

therefore, eigenstates with different momentum components can nevertheless have the same energy, and so the last particle in the three-dimensional well can still be at the Fermi level but different momentum components than another particle at the Fermi level. A cubic volume has quantized momenta in each of its Cartesian directions: Schrödinger's equation is separable in Cartesian coordinates when the potential energy is constant or zero. A flat surface is such that if $\hat{x}$ is normal to it, then waves normal the surface depend on $k_x$ and are constrained by the surface barrier, but the waves in the plane of the surface depend on $k_y$ or $k_z$, and the surface barrier does not constrain them. The $C_n$ are then inferred by integrating the Fermi–Dirac distribution over all the transverse momenta $k_\perp$, or

$$\rho = \frac{g}{2\pi}\int f(k_x)dk_x = \frac{g}{2\pi}\int dk_x \left[\frac{1}{(2\pi)^2}\int f_{FD}[E(k)]2\pi k_\perp dk_\perp\right] \tag{6.45}$$

where the term in the square brackets is now identified as the supply function and the form used to approximate an electron gas in a metal by Sommerfeld. Realizing that $E(k) = E_x + E_\perp$ and that $k_\perp dk_\perp = (m/\hbar^2)dE_\perp$, then the integration is of the form $\int du/(1 + e^u) = \ln(1 + e^{-u})$ and is straightforward

$$f(k_x) \equiv \frac{m}{\pi\beta\hbar^2}\ln\left[1 + e^{\beta(\mu - E_x)}\right] \tag{6.46}$$

where $E_x = \hbar^2 k_x^2/2m$. For positive $k_x$, then $f(k_x)$ describes the distribution of particles moving towards the surface, and for negative $k_x$, it describes the distribution moving away. In the zero-temperature limit, $f(k_x)$ is

$$\lim_{T\to 0} f(k_x) = \frac{1}{2\pi}\left(k_F^2 - k_x^2\right)\Theta(k_F^2 - k_x^2) \tag{6.47}$$

This form allows $C_n$ for the quantum problem to be determined, but for narrative simplicity and to make matters easy, consider only the odd-parity eigenstates (it allows the origin to represent the location of the surface), and so replace $k_F = 2\pi N/L$ and $k_x = 2\pi n/L$. To make comparisons for different Fermi levels, it is further convenient to normalize the coefficients by $\rho_o = k_F^3/3\pi^2$, and so, let $C_n$ be given by

$$C_n \equiv \frac{3}{2N^3}\left(N^2 - n^2\right) \tag{6.48}$$

In the continuum limit, the representation continues to describe electrons that are prevented from moving into the positive $x$ region; that is, an impenetrable barrier exists at the surface. To show that, imagine that the region to the left of the surface is a large box of width $L/2$, so that the eigenstates are characterized by incoming plane waves traveling towards the surface (to the right) having the form $e^{ik_n x}$, where $k_n = 2\pi n/L$; they are reflected and plane waves travel to the left with $k_{-n} = 2\pi(-n)/L = -2\pi n/L$ traveling in the $-\hat{x}$ direction: a wave that hits an impenetrable wall is reflected, causing the wave $e^{ik_n x}$ to be reflected into $e^{ik_{-n}x} = e^{-ik_n x}$. The energy of the wave is $E(k_{\pm n}) = \hbar^2 k_n^2/2m$, so it is convenient to treat $n$ as positive and put the sign on the direction the wave is traveling. Because the wave function vanishes at the infinite barrier (there is no probability of finding any part of the wave function inside an infinite barrier), then the sum of the wave function and its reflection vanish at $x = 0$, and so

$$\langle x|\psi_n\rangle = e^{ik_n x} - e^{-ik_n x} = 2i\sin(k_n x) \tag{6.49}$$

as expected. The factor of $2i$ can be absorbed into the definition of the weighting attached to $\psi_n(x)$. Now the continuum limit can be approached by taking $L \to \infty$ and letting $k_n \to k$, where because the problem is now one-dimensional again, carrying around the $x$ subscript on $k_x$ is superfluous and so suppressed henceforth. Because the wave functions are normalized, as $L \to \infty$,

$$\rho_o = \frac{g}{2\pi} \int_{-k_F}^{k_F} f(k) \left\{ \frac{4}{L} \int_{-L/2}^{0} [\sin(kx)]^2 dx \right\} dk \tag{6.50}$$

It is seen that the density to the left of the wall at $x = 0$ is no longer constant and equal to $\rho_o = k_F^3/3\pi^2$ but rather has a variation with $x$ that behaves as

$$\begin{aligned} \frac{\rho(x)}{\rho_o} &= \frac{g}{2\pi} \int_{-k_F}^{k_F} \left( \frac{k_F^2 - k^2}{2\pi} \right) [\sin(kx)]^2 dk = \xi(2k_F x) \\ \xi(s) &\equiv 1 + 3\frac{\cos(s)}{s^2} - 3\frac{\sin(s)}{s^3} \end{aligned} \tag{6.51}$$

The behavior of $\xi(2k_F x)$ is called "Friedel oscillations" at the surface of a metal [217], and is a consequence of the wave nature of the electron coming up against an infinite barrier at the surface. Because of the manner in which $\xi(s)$ was found, it is clear that this continuum problem has a connection to the discrete version, as shown in Figure 6.3 (see Algorithm B.2.2), where even for the addition of a small number of states as shown, the emerging behavior of $\xi(s)$ is clearly seen: as $N$ gets larger, $\xi(2k_F x)$ pushes closer to the origin, and the agreement with the discrete model becomes better further away from the barrier at the origin. For finite barriers, the wave nature of the electrons penetrates into the barrier region, and the variation in density and its penetration into the barrier is implicated in the generation of the image charge modifications to the barrier [217] and causes the origin of the background positive charge due to the charged metal atoms to be different than the origin of the electron gas [136, 146, 149], a topic that shall reappear in Section 11.5.

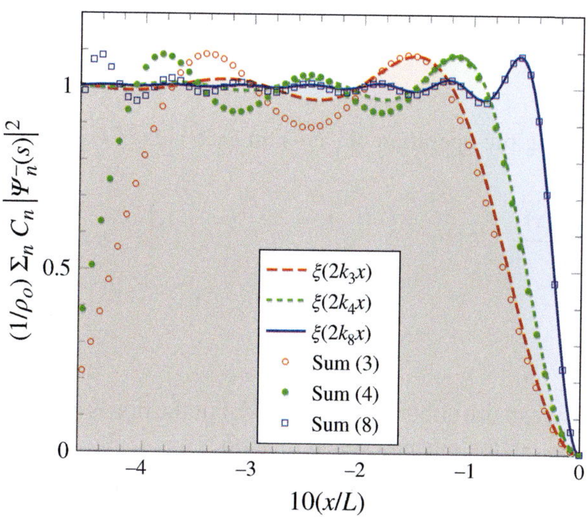


**Figure 6.3** Comparison of the discrete summation of energy states (Sum($N$), symbols) in a well weighted by $C_n$ given by Eq. (6.48) with the continuum Eq. (6.51) ($\xi(2k_N x)$, lines). An infinite barrier is taken to be at the origin ($x = 0$).

## 6.5 Tunneling Through a Delta-Barrier

Return now to a time-dependent $|\Psi(t)\rangle$ constructed from a superposition of eigenstates $|\psi_n(t)\rangle$ as in Eq. (6.30). The simplest single-parameter barrier that entails tunneling for all eigenstates is the $\delta$-function barrier at the origin [39, 299] for which

$$V(x) = \frac{\hbar^2}{2m}\gamma\delta(x) \tag{6.52}$$

where $\gamma$ is a strength parameter with units of [1/nm] (recall that $\delta(x)dx$ is dimensionless, so $\delta(x)$ has units of [1/nm]). There is some irony that this simplest of models is also an exception to one of the rules of continuity: although the wave function is continuous across the barrier at $x = 0$, its gradient does not seem to be: in fact, there are two interfaces at the barrier, as can be seen using a delta-function sequence $\delta_l(x)$ to model it [181, 182] (e.g. a rectangular barrier of decreasing width $2l$ and increasing height $V_o/l$ in the limit $l \to 0$). A rectangular barrier clearly has two interfaces where the gradient of the wave function must be held continuous, but the choice of a delta-function barrier obfuscates this. Proceed, therefore, using a slightly modified approach. Define $\mathbb{L}[g(x)]$ such that

$$\mathbb{L}[g(x)] \equiv \lim_{\epsilon\to 0}\int_{-\epsilon}^{\epsilon} g(x)dx \tag{6.53}$$

and use it on the various parts of Schrödinger's equation, rewritten by dividing by $\hbar^2/2m$ throughout

$$\left[-\frac{d^2}{dx^2} + \gamma\delta(x)\right]\psi_k(x) = k^2\psi_k(x) \tag{6.54}$$

The wave function on the left side of the barrier (normalized to the incident wave) is $\psi_k^-(x) = e^{ikx} + r(k)e^{-ikx}$ for $k > 0$, and on the right, is $\psi_k^+(x) = t(k)e^{ikx}$. The reflection $r$ and transmission $t$ coefficients are found by first demanding continuity of wave function at $x = 0$, which sets

$$1 + r = t \tag{6.55}$$

A second relation is then had by considering the operation of $\mathbb{L}(\cdots)$ on each side of Eq. (6.54) to find

$$\begin{aligned}
\mathbb{L}\left(-\frac{d^2\psi_k}{dx^2}\right) &= -\left.\frac{d\psi_k^+(0)}{dx}\right|_{x=0} + \left.\frac{d\psi_k^-(x)}{dx}\right|_{x=0} \\
\mathbb{L}[\gamma\delta(x)\psi_k(x)] &= \gamma\psi_k(0) \\
\mathbb{L}[k^2\psi_k(x)] &= 2\epsilon k^2\psi_k(0) \to 0
\end{aligned} \tag{6.56}$$

where the second and third of these equations can use either $\psi_k^+(x)$ or $\psi_k^-(x)$, but the first is chosen. From $d\psi_k^-(0)/dx = ik(1-r)$ and $d\psi_k^+(0)/dx = ikt$, the second needed relation is

$$-ikt + ik(1-r) + \gamma t = 0 \tag{6.57}$$

Solving for $r$ and $t$ give

$$t(k) = \frac{2k}{2k + i\gamma}, \quad r(k) = -\frac{i\gamma}{2k + i\gamma} \tag{6.58}$$

Next, let the weighting for each eigenstate characterized by $k$ be $C(k)$ and choose it for an initial Gaussian wave packet centered in energy $E_o = \hbar^2 k_o^2/2m = 1$ eV so that

$$C(k) = \exp\left\{-[L(k-k_o)]^2\right\} \tag{6.59}$$

where $L = 1$ nm is a convenient choice. At $t = 0$, let the center of the wave packet in the absence of a barrier be at the origin: a Gaussian wave packet will have a larger width both earlier ($\pm t_{max} = \pm Wm/\hbar k_o$) and later in time by virtue of how $\Psi(x, t)$ is constructed.

With these choices, the motion of a wave packet tunneling through a delta-function barrier can be simulated exactly, and its interactions with that barrier visualized. To make the simulation computationally fast and simple, a discrete number of eigenstates $N_k = 48$ is used, spaced such that $C(k_{max}) \ll 1$, where $k_j = k_{max}(2j - N_k - 1)/(N_k - 1)$. However, a finite number of discrete basis states means that $\rho(x, t)$ will spatially repeat: the more closely spaced the eigenstates used, the further apart those repeating packets will appear. Therefore, chose a large value of $N_k$, a small value of $\Delta k = 2k_{max}/(N_k - 1)$, and pick values of $(L, W, N_k)$ that insure that the repeating quasi-Gaussian packets will be far from each other, far enough that the tunneling of the center packet will be concluded before the next packet intrudes on the viewing region $-W < x < W$. The wave function is then

$$\Psi(x,t) = \langle x|\Psi\rangle = \sum_{n=1}^{N_k} C(k)e^{i\omega_n t}\begin{cases} e^{ik_n x} + r_n e^{-ik_n x} & (x<0) \\ t_n e^{ik_n x} & (x>0) \end{cases} \tag{6.60}$$

where $\hbar\omega_n = \hbar^2 k_n^2/2m$, $r_n = r(k_n)$ and $t_n = t(k_n)$. Note that time $t$ is different than transmission coefficient $t_n$.

The behavior of $\rho(x, t) = |\langle x|\Psi\rangle|^2$ is shown three ways in Figures 6.4, 6.5, and 6.6 (see Algorithm B.2.8). In the first, three different snapshots at different times are overlayed: the incident (*inc*) wave packet approaching the barrier (black vertical line in center) from the left is red and shown

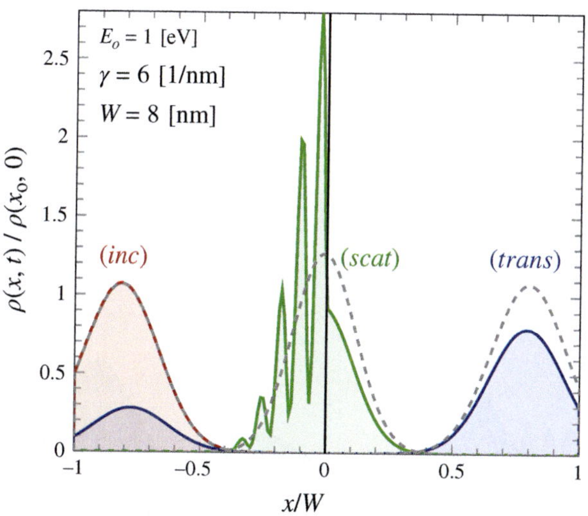


**Figure 6.4** Gaussian wave packet tunneling through a $\delta$-function barrier. The red (*inc*) packet is just after the start time; the green (*scat*) packet is both tunneling and being reflected by the barrier (black line); and the blue (*trans*) on the right-hand side is the portion that has tunneled through, where a smaller blue portion on the left is what has been reflected. Gray dashed lines are the packet in the absence of a barrier.

at an earlier time ($-t_{max}$); the scattered (*scat*) wave packet is shown at ($t \approx 0$) hitting the barrier, with a portion tunneling through to the right of the barrier, and the still incoming part of the wave packet interfering with the reflected part of the wave packet causing the ripples in $\rho$ just to the left of the barrier is blue and shown at a later time ($t_{max}$), with part of the packet moving to the right, and the reflected part of the packet moving to the left. Superimposed in the background is a dashed gray line showing the same wave packet moving to the right when there is no barrier: the gray dash is therefore not visible behind the incident red packet because they overlap, but it can be seen behind the scattered packet, and is also larger than and slightly ahead of the transmitted packet. A zero-thickness barrier has introduced a delay in transmission, and makes apparent a tunneling delay time [181, 182]. The second figure, Figure 6.5, shows more of the time slices between $-t_{max} < t < t_{max}$, again following the color scheme of red being incident, green being near $t \approx 0$, and blue being transmitted, but with more gradations in between.

When even more frames of each time step are stitched together and viewed sequentially (a movie), then the passage of the wave packet through the barrier shows that the oscillations prior to the barrier are in the form of standing waves: they represent the reflected wave constructively and destructively interfering with the incident wave. Water waves hitting a concrete breaker or standing

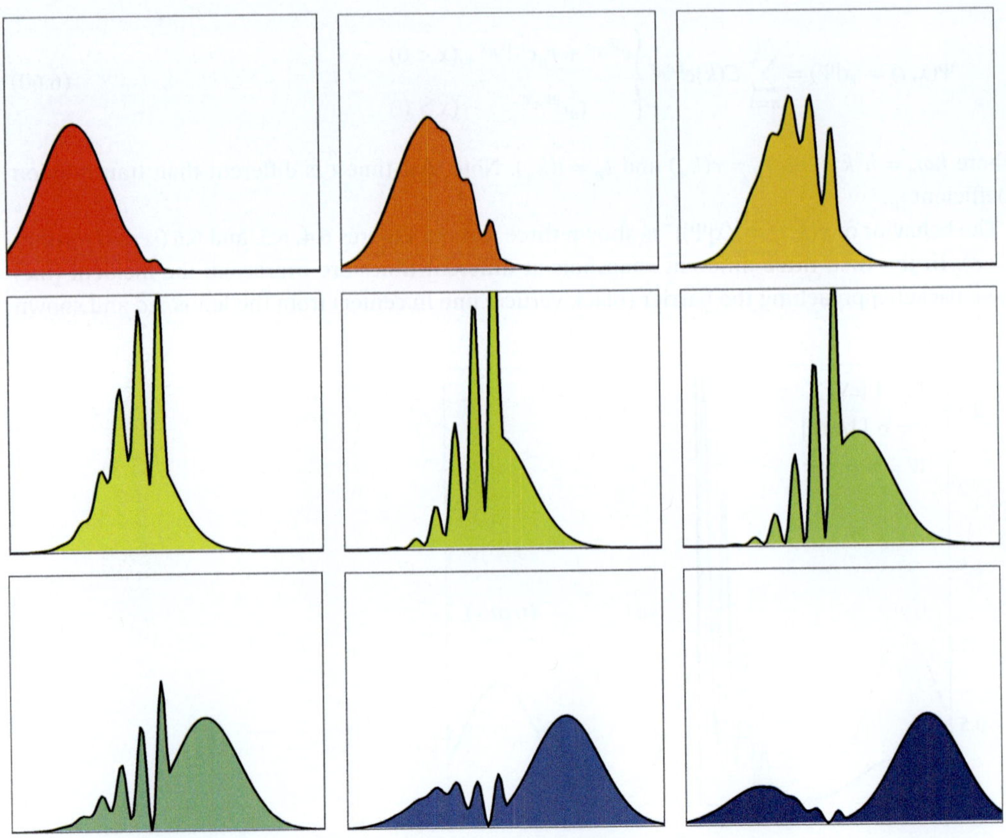

**Figure 6.5** Snapshots of the evolution of the Gaussian wave packet incident on a $\delta$-function barrier. When more of the slices are put together, the waterfall plot of Figure 6.6 results.

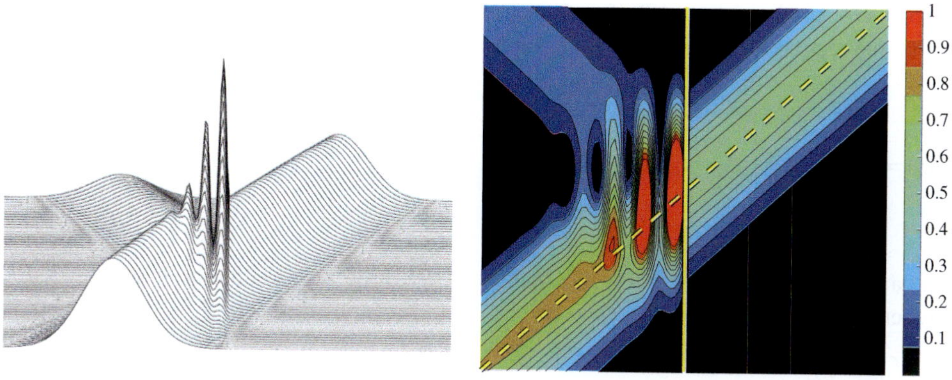


**Figure 6.6** (*left*) Waterfall plot and (*right*) contour plot of the time evolution of the wave packet shown in Figure 6.4. On the right, the yellow vertical line is the location of the $\delta$-barrier, and the dashed yellow line is the path of the maximum of the equivalent free (no barrier) wave packet.

waves created in a wave tank behave much the same way. Those many frames can be shown in a single image in one of two ways: as a waterfall plot, Figure 6.6 shows the standing waves more clearly in Figure 6.6 (left), and as a contour plot in Figure 6.6 (right): the standing waves are spikes in the former, and red regions in the latter. In the contour plot (horizontal axis is position, vertical axis is increasing time), a solid yellow line shows the location of the barrier, and a dashed yellow line shows the path of the maximum of the incident packet. The dashed yellow line is perceptibly below the center of the transmitted packet to the right of the barrier, and shows that the transmitted packet is slightly delayed in its passage through.

Observe that these results show behavior that is revealed by numerical time evolution simulations, but there is an important difference. Time-evolving Schrödinger's equation using finite difference methods, as is often done, cannot resolve very high-frequency components and accrues numerical errors over time that undermine the visualization of the transmitted/reflected packets, but the present approach is an exact method in that the behavior of $\rho(x, t)$ can be calculated exactly at any time, and it does not require the numerical solution of earlier times to be performed first. The delta function barrier facilitated the ease of doing so by providing analytic $r(k)$ and $t(k)$: the ease of a one-parameter barrier model and the straightforwardness of Eq. (6.60) to numerically model makes the $\delta$-function barrier a good model for pedagogical needs and numerical experimentation.

## 6.6 Tunneling Through a Rectangular Barrier

The simple rectangular barrier is the most often-considered barrier for which to introduce the methodology of treating tunneling [39, 84, 306] in a manner that can be generalized to the Transfer Matrix Approach (TMA) [33, 154, 164, 165, 251, 242], especially when the TMA resorts to the usage of Airy functions rather than plane wave eigenstates. Instead of a single interface where the wave function and its first derivative must be matched, now there are two: the left side of the barrier and the right. These boundaries serve to define three regions: Region 1, or to the left of the barrier;

Region 2, or inside the barrier; and Region 3, or to the right of the barrier. The wave function in each of the three regions is then for the case $k > k_o$ (fly-over)

$$\psi_k(x) = \begin{cases} e^{ikx} + r\,e^{-ikx} & \text{Region1} \\ a\,e^{i\kappa x} + b\,e^{-i\kappa x} & \text{Region2} \\ t\,e^{ikx} & \text{Region3} \end{cases} \tag{6.61}$$

where the barrier is $V_o \equiv \hbar^2 k_o^2/2m$. For $k > k_o$, solutions to Schrödinger's equation in Region 2 are compactly rendered as $\psi_2(x) \propto \exp(\pm i\kappa x)$ where

$$\kappa \equiv \sqrt{|k_o^2 - k^2|} \tag{6.62}$$

and the subscript on $\psi_n$ denotes the region, as in Figure 6.7. Although interest is in the case of $k < k_o$ (tunneling), there the replacement $\kappa \to i\kappa$ can be simply made, an advantage of the way $\kappa(k)$ is defined through its definition of always being real and positive.

The two equations at each $(x_1, x_2)$ for the first and second boundaries relate the coefficients of the wave function and its first derivative. The boundary conditions at $x = x_1$ are then

$$\begin{aligned} e^{ikx_1} + r\,e^{-ikx_1} &= a\,e^{\kappa x_1} + b\,e^{-\kappa x_1} \\ ik\,e^{ikx_1} - ikr\,e^{-ikx_1} &= \kappa a\,e^{\kappa x_1} - \kappa b\,e^{-\kappa x_1} \end{aligned} \tag{6.63}$$

and at $x = x_2$

$$\begin{aligned} a\,e^{\kappa x_2} + b\,e^{\kappa x_2} &= t\,e^{ikx_2} \\ \kappa a\,e^{\kappa x_2} - \kappa b\,e^{-\kappa x_2} &= ikt\,e^{ikx_2} \end{aligned} \tag{6.64}$$

where $t_1 = 1$ (incoming wave from the left) and $r_3 = 0$ (no incoming wave from the right), and for convenience, the left and right regions do not use subscripts for $(k, t, r)$ (remember to distinguish between $t$ as time and $t$ as transmission coefficient, which one is in use is made apparent by context). Such is the manner by which the problem is often presented, but (to the distaste of physicists) it is difficult to generalize and (even worse) inelegant. Elegance is made possible by rendering the equations as matrix equations and subscripting terms by the regions in which they are found. These

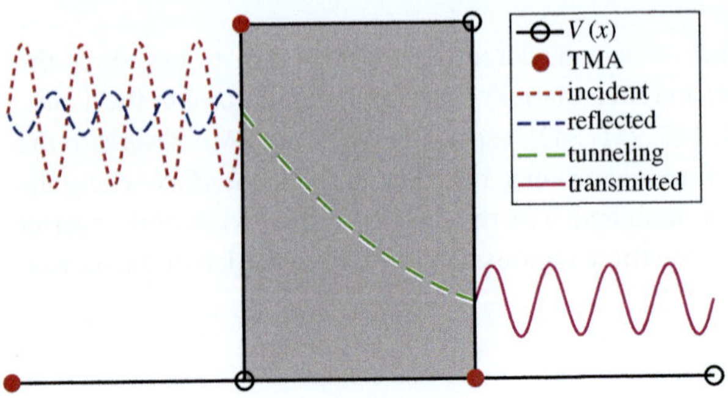


**Figure 6.7** Rectangular barrier, showing incident, reflected, tunneling (barrier), and transmitted waves. $x_1$ corresponds to the left side of the barrier, and $x_2$ to the right. The dots are where $V(x)$ changes in height or slope or both; the Transfer Matrix Approach uses the red dots and the linear line segments to their right.

two equations can therefore be more compactly expressed by defining a matrix of propagation terms $\mathbb{M}$ and a vector of coefficients $|\xi\rangle$, defined by

$$\mathbb{M}_n(x) \equiv \begin{bmatrix} e^{ik_n x} & e^{-ik_n x} \\ ik_n e^{ik_n x} & -ik_n e^{-ik_n x} \end{bmatrix}, \quad |\xi_n\rangle = \begin{pmatrix} t_n \\ r_n \end{pmatrix} \tag{6.65}$$

with the understanding that $(t_1, r_1) = (1, r)$, $(t_3, r_3) = (t, 0)$, $k_1 = k_3 = k$, and $k_2 = \kappa$. The reasons for doing so are that four equations now become two matrix equations

$$\begin{aligned} \mathbb{M}_1(x_1)|\xi_1\rangle &= \mathbb{M}_2(x_1)|\xi_2\rangle \\ \mathbb{M}_2(x_2)|\xi_2\rangle &= \mathbb{M}_3(x_2)|\xi_3\rangle \end{aligned} \tag{6.66}$$

The solution is then obtained by expressing $|\xi_1\rangle$ in terms of $|\xi_3\rangle$ using matrix operations. To that end, introduce the product of matrices

$$\mathbb{S}_n(x) \equiv \mathbb{M}_n(x)^{-1} \cdot \mathbb{M}_{n+1}(x) \tag{6.67}$$

so that

$$|\xi_1\rangle = \mathbb{S}_1(x_1) \cdot \mathbb{S}_2(x_2)|\xi_3\rangle \tag{6.68}$$

It is now appreciated that the method generalizes to $n$ rectangular barriers (or wells or gaps) where a matrix $\mathbb{S}_j(x_j)$ is associated with each change in the height of a region: if only the height of a region changes, then the method is plane wave TMA (regions are rectangles), whereas if the slope changes (regions are trapezoids), then the method is Airy TMA [33, 154, 166]. For simplicity here, only the plane wave approach is treated. Specifying the $x_n$ allows the $\mathbb{S}_n(x)$ to be specified. For $k > k_o$ (fly-over) for which $k_2 = i\kappa$ in Region 2 and the choices $x_1 = 0$ and $x_2 = L$, then

$$\mathbb{S}_1(0) = \frac{1}{2k} \begin{bmatrix} k+\kappa & k-\kappa \\ k-\kappa & k+\kappa \end{bmatrix} \tag{6.69}$$

$$\mathbb{S}_2(L) = \frac{1}{2\kappa} \begin{bmatrix} (k+\kappa)e^{i(k-\kappa)L} & (k-\kappa)e^{-i(k+\kappa)L} \\ (k-\kappa)e^{i(k+\kappa)L} & (k+\kappa)e^{-i(k-\kappa)L} \end{bmatrix} \tag{6.70}$$

Using $\langle\xi_1| = [1, r(k)]$ and $\langle\xi_3| = [t(k), 0]$ then results in

$$\begin{pmatrix} 1 \\ r(k) \end{pmatrix} = \frac{t(k)}{4k\kappa} e^{ikL} \begin{pmatrix} 4k\kappa \cos(\kappa L) - 2i(k^2 + \kappa^2)\sin(\kappa L) \\ -2i(k^2 - \kappa^2)\sin(\kappa L) \end{pmatrix} \tag{6.71}$$

where $e^{\pm i\kappa L} = \cos(\kappa L) \pm i \sin(\kappa L)$ has been used. The two components of the vector on each side must be equal, and so

$$r(k) = \frac{-i(k^2 - \kappa^2)\sin(\kappa L)}{2k\kappa\cos(\kappa L) - i(k^2 + \kappa^2)\sin(\kappa L)} \tag{6.72}$$

$$t(k) = \frac{2k\kappa}{2k\kappa\cos(\kappa L) - i(k^2 + \kappa^2)\sin(\kappa L)} e^{-ikL} \tag{6.73}$$

where the extension to $k < k_o$ (tunneling) is accomplished by the replacement $\kappa \to i\kappa$ (see Eq. (11.49)). Recalling the transmission probability $D(k) = |t(k)|^2$ when Regions 1 and 3 are flat and at the same level, then for the rectangular barrier

$$D_{rec}(k > k_o) = \left\{1 + \frac{1}{4}\left(\frac{k}{\kappa} - \frac{\kappa}{k}\right)^2 \sin^2(\kappa L)\right\}^{-1} \tag{6.74}$$

$$D_{rec}(k < k_o) = \left\{1 + \frac{1}{4}\left(\frac{k}{\kappa} + \frac{\kappa}{k}\right)^2 \sinh^2(\kappa L)\right\}^{-1} \tag{6.75}$$

These cases can be expressed in the single equation

$$\begin{aligned} D_{rec}(k) &\equiv \left\{1 + \frac{k_o^4}{k^2(k_o^2 - k^2)}\sinh^2[L\sqrt{k_o^2 - k^2}\right\}^{-1} \\ &= \left\{1 + \left[\frac{k_o^2}{k\kappa}\sinh(\kappa L)\right]^2\right\}^{-1} \end{aligned} \tag{6.76}$$

if one remembers Eq. (6.62) and $\kappa \to i\kappa$ when $k > k_o$. The pedagogical utility of $D_{rec}(k)$ is quite high. Consider four examples.

First, in the tunneling regime, and in anticipation of methods that will be used for more general barriers, observe that for $\kappa L \gg 1$, then $4\sinh^2(\kappa L) \approx e^{2\kappa L}$. As a result, $D(k)$ can be compactly represented as

$$\begin{aligned} D_{rec}(k) &\approx \frac{1}{1 + P(k)e^{\theta_o(k)}} \\ P(k) &= \left(\frac{k_o^2}{2k\kappa}\right)^2, \quad \theta_o(k) \equiv 2\kappa L \end{aligned} \tag{6.77}$$

Clearly, $P(k)$ is a factor of order unity: when it is replaced by unity, which is very convenient for general barriers as shall be seen, then the approximation to $D(k)$ is called the Kemble approximation, and the canonical equations will be seen to make use of it. Further, observe the subscript on $\theta_{rec}(k) \to \theta_o(k)$: this term will have a privileged position in the treatment of more general potentials, as the effect of changing the shape of the potential will simply append a dimensionless constant, or shape factor in Section 9.1, that is less than unity.

Second, and again in the tunneling regime, observe that when $\kappa L \gg 1$, then if $n$ rectangular barriers of different height but (for convenience) the same width are back-to-back, then

$$D_{total}(k) \approx e^{-2\kappa_1 \Delta L} e^{-2\kappa_2 \Delta L} \dots e^{-2\kappa_n \Delta L} = \exp\left[-2\sum_{j=1}^{n} \kappa_j \Delta L\right] \tag{6.78}$$

but the summation looks like an integration in the limit $\Delta L \to 0$, and so

$$\theta(k) = 2\int_0^L \kappa(x)dx \tag{6.79}$$

a formula that shall be revisited by other means in Eq. (6.93), where this $\theta(k)$ is identified with the Gamow factor. It relies on the fact that for a general potential composed of piece-wise flat segments, the value of $k_o \to k_o(x)$ becomes a function of position, which is the same as saying that $V(x) = \hbar^2 k_o(x)^2/2m$ is likewise a function of position. In the special case where $n$ identical barriers occur such that $\kappa$ is constant for all the segments and $L = n\Delta L$, then $\theta_o = 2\kappa L$ is simply recovered by the integral defining $\theta(k)$.

Third, a step barrier is likely an even simpler quantum mechanical problem than the rectangular barrier, as it only involves $S_1(0)$. A simple evaluation shows

$$D_{step}(k > k_o) = \frac{4k\varkappa}{(k+\varkappa)^2} \tag{6.80}$$

where now $\varkappa = \sqrt{k^2 - k_o^2}$ (remember Eq. (6.62)) and yet this limiting case does not appear to follow easily from $\lim_{L\to\infty} D_{rec}(k)$. It is a consequence of the fact that the second interface matters: a step function does not have one, and a rectangular barrier does, and at that second barrier, a reflection occurs. Therefore, to get from the later to the former, some handwaving about "averages" over the sine-function is sometimes made, arguments that are on some level intuitively unappealing because pure waves extend to infinity, but in support of them, the behavior of $D_{step}(k)$ can be seen to lurk in the behavior of $D_{rec}(k)$ as $L$ increases in a manner shown in Figure 6.8 (see Algorithm B.2.9).

Fourth and lastly, the quality of the Kemble approximation for $D_{rec}(k)$ is indicated by how well the relation $D_{rec}(k) \approx 1/(1+e^{\theta})$ holds, which is tantamount to the relation

$$2\sqrt{k_o^2 - k^2}L \Leftrightarrow 2\ln\left[\frac{k_o^2}{2k\kappa}\sinh(L\kappa)\right] \tag{6.81}$$

The ratio of the two sides is shown in Figure 6.9 for various $L$. The behavior is a consequence of having approximated $P(k) \approx 1$ in Eq. (6.77). Clearly, that approximation becomes lacking at $k \approx k_o$, where the Kemble form of $D_{rec}(k)$ suggests $D_{rec}(k_o) = 1/2$ but Eq. (6.77) shows it to be instead

$$D_{rec}(k_o) = \left\{1 + \left(\frac{k_o L}{2}\right)^2\right\}^{-1} \tag{6.82}$$

which is 1/2 only for $k_o L = 2$ when $P(k)$ is approximated by unity.

In the development of the canonical emission equations, with the exception of the triangular barrier Fowler Nordheim equation, the approximation that $P(k) \to 1$ is generally made. This allows for a convenient and analytical equation that is generally independent of the Fermi level $\mu$ for tunneling, hence its appeal, but it does have consequences that will need examination, and motivates efforts to retain it in generalized FN equations [86].

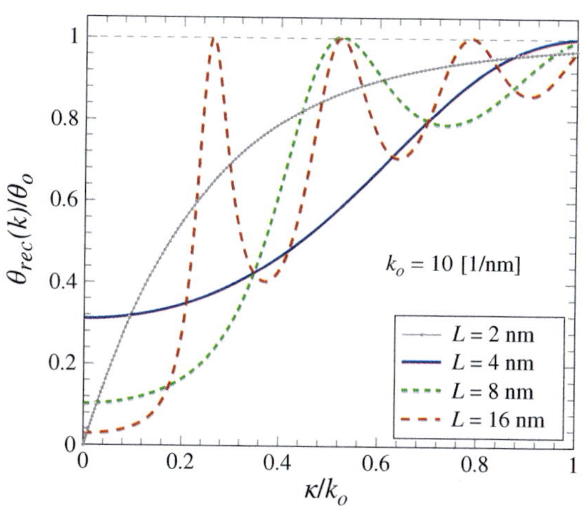


**Figure 6.8** $\theta(k)$ for $k > k_o$ and various $L$ as a function of $\varkappa/k_o$. $V_o$ is such that $k_o = 10$ [1/nm]. Solid gray line is $D_{step}(k)$ for a step barrier (Eq. (6.80)). Dashed gray line is unity.

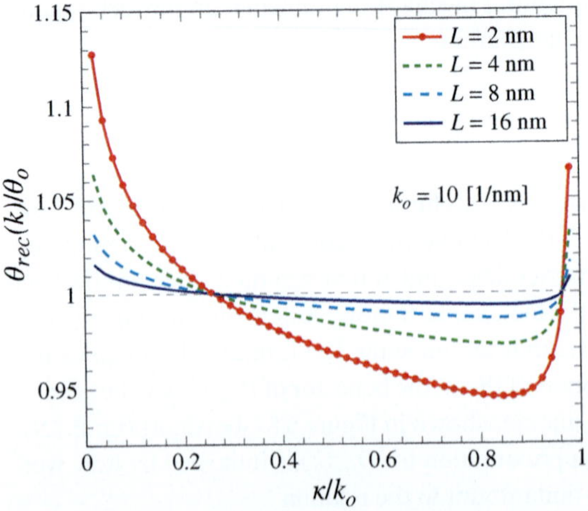


**Figure 6.9** Ratio of $\theta(k)$ (right side of Eq. (6.81) to $\theta_o(k) = 2\kappa(k)L$ (left side of Eq. (6.81)) for $k < k_o$ and various $L$. $V_o$ is such that $k_o = 10$ [1/nm]. Dashed gray line is unity.

## 6.7 Other Quantum Representations

The statistical mechanics approach to current density via a distribution function and the quantum mechanical approach via a wave function seem unconnected, but in fact a correspondence can be shown in two ways using alternate formulations that are entirely equivalent to the Schrödinger/Heisenberg formulations in their predictions, but rely on a different viewpoint. The first, called the Bohm formulation, is sometimes called the "pilot-wave" model because it interprets the dynamics of a particle as affected by a quantum potential energy $\Omega(x)$ plus the classical potential energy $V(x)$. The two potential energies then specify the force which governs the dynamics of the particle's trajectory [57, 349]. The second uses the Wigner distribution function, recasts the classical phase space $f(x, k, t)$ into a function that acts like a probability function and incorporates quantum mechanical effects but at the expense of introducing the notion of "negative probabilities" [36, 81, 96, 130]. Both result in a quantum "continuity equation."

The Bohm formulation begins with Schrödinger's equation (Eq. (6.2))

$$\left[-\frac{\hbar^2}{2m}\frac{\partial^2}{\partial x^2} + V(x)\right]\psi(x,t) = i\hbar\frac{\partial}{\partial t}\psi(x,t) \tag{6.83}$$

where $\psi(x, t)$ is superposition of eigenstates rather than an energy eigenstate. The wave function is then recast as a combination of two real functions $R$ and $S$ such that

$$\psi(x,t) = R(x,t)\exp[iS(x,t)] \tag{6.84}$$

After insertion into Eq. (6.83) and expansion, all the real terms are brought to one side of the equation, and all the imaginary terms to the other. Because $R$ and $S$ are real, one side of the resulting equation is purely real and the other purely imaginary, and so both sides must separately equal zero. Therefore

$$\begin{aligned} 0 &= \frac{\hbar^2}{2m}\left[(\partial_x S)^2 - \frac{1}{R}\partial_x^2 R\right] + V(x) + \hbar\partial_t S \\ 0 &= i\hbar\frac{1}{R}\partial_t R + i\frac{\hbar^2}{2mR}\left[2\partial_x R\,\partial_x S + R\partial_x^2 S\right] \end{aligned} \tag{6.85}$$

A few replacements for conditions where $k > k_o$ render these equations in a more familiar form:

1. From the identification of density as $\rho = |\psi(x,t)|^2$, it follows $\rho = R^2$;
2. Generalizing the form of Eq. (6.41) for current density,

$$J(x,t) = \frac{q\hbar}{2mi}\left[\psi^\dagger \partial_x \psi - \psi \partial_x \psi^\dagger\right] \\ = q\rho(x,t)\frac{\hbar}{m}\frac{\partial}{\partial x}S(x,t) \tag{6.86}$$

   Comparing this to the oft-used approximation $J = q\rho\langle v\rangle$, then $(\hbar/m)\partial_x S$ is to be identified as a velocity;
3. From that definition of velocity, it is clear that the quantity $(\hbar^2/2m)(\partial_x S)^2$ has the form of a kinetic energy, suggesting that $(\hbar^2/2mR)\partial_x^2 R$ resembles a potential energy, and so, the Quantum Potential is defined as

$$\Omega(x,t) = -\frac{\hbar^2}{2mR}\frac{\partial^2}{\partial x^2}R(x,t) \tag{6.87}$$

These substitutions along with replacing the $\partial_x^2 S$ term with a $\partial_x J$ term using Eq. (6.86), simplifies Eq. (6.85) into

$$0 = q\partial_t \rho + \partial_x J \\ 0 = \hbar\partial_t S + \frac{\hbar^2}{2m}(\partial_x S)^2 + \Omega(x,t) + V(x) \tag{6.88}$$

where common factors have been divided out. The first of these two new equations is the continuity equation first experienced in Eq. (4.17). The second one looks formidable, but observe that for steady state conditions, $\partial_t S$ does not appear, and so the velocity term can be rewritten as

$$(\partial_x S)^2 = \frac{2m}{\hbar^2}\left[E - V(x) - \Omega(x)\right] \tag{6.89}$$

from which $S(x)$ is then given by

$$S(x) = \frac{\sqrt{2m}}{\hbar}\int \sqrt{E - V(x) - \Omega(x)}\, dx \tag{6.90}$$

In dealing with the rectangular barrier, as $k$ went from above the barrier to below it, $\kappa \to i\kappa$ in the matrix equation. Therefore, even though the wave function $\psi(x)$ is real under the barrier and so $S(x)$ should be zero there, it is useful instead to let $S(x)$ become imaginary there, and so for $k < k_o$, let

$$iS(x) = -\frac{\sqrt{2m}}{\hbar}\int^{x} \sqrt{V(x') + \Omega(x') - E}\, dx' \tag{6.91}$$

The quantum potential term $\Omega(x)$ is small if the density varies slowly, and so for the tunneling problems here, where the tunneling barrier occurs for $x_1 \le x \le x_2$, it is convenient and common to neglect $\Omega(x)$ in the integrand and allow $-iS(x)$ to be given by half the Gamow tunneling factor

$$\theta(E) \equiv 2\frac{\sqrt{2m}}{\hbar}\int_{x_1}^{x_2} \sqrt{V(x) - E}\, dx = 2\int_{x_1}^{x_2} \kappa(x) dx \tag{6.92}$$

where $V(x_j) = E$ identify the roots $x_j$ where the integrand vanishes. Such a viewpoint is for visualization: because the wave function is real and a sum of exponentially growing and decaying solutions, $S(x) = 0$ for $E < V_o$ in the tunneling barrier, but pedagogically, asserting $S \to iS$ is in keeping with the usual simple rectangular barrier model, often encountered, where $\kappa \to i\kappa$. Eq. (6.92) can also be found by more conventional means as done by Schiff [306] by a perturbation approach to find

$$\psi(x) \approx \frac{1}{\sqrt{\kappa(x)}} \exp\left[ -\int_{x_1}^{x_2} \kappa(x)dx \right] \tag{6.93}$$

a form that, while pedagogically illuminating, is superseded by the Transfer Matrix Approach to be considered later, but which also acts as a motivation for the Gamow factor approach to so-called Wentzel–Kramers–Brillouin or WKB-like models. It is seen that $\theta_o(E)$ of Eq. (6.77) is the constant-$\kappa$ form of Eq. (6.92) as anticipated by Eq. (6.79).

The Wigner distribution function (WDF) approach, in contrast, starts from the density operator of Eq. (6.37) to define

$$\rho(x, t) \equiv \frac{1}{2\pi} \int_{-\infty}^{\infty} f(x, k, t)dk \tag{6.94}$$

but now the non-locality of quantum mechanics is contained in the redefinition of how $f(x, k, t)$ is formed to be [154]

$$f(x, k, t) = \frac{1}{\pi} \int_{-\infty}^{\infty} e^{2iky} \langle x - y|\hat{\rho}(t)|x + y\rangle dy \tag{6.95}$$

Recalling the commutator relation of Eq. (6.38) and the anti-commutator relation of Eq. (6.40) that comes from taking the commutator of the density operator with the Hamiltonian, the time derivative of $f(x, k, t)$ is

$$\begin{aligned} \frac{\partial}{\partial t} f(x, k, t) = -\frac{1}{2} \int_{-\infty}^{\infty} e^{2iky} dy \Big[ &\langle x - y| \frac{\hbar}{2m} \partial_{\hat{x}} \{\hat{k}, \hat{\rho}(t)\} |x + y\rangle \\ &+ \langle x - y| \frac{i}{\hbar} [V(\hat{x}), \hat{\rho}(t)] |x + y\rangle \Big] \end{aligned} \tag{6.96}$$

The right-hand side has terms that take getting used to. In detail,

$$\begin{aligned} \langle x - y|\partial_{\hat{x}} \{\hat{\rho}(t), \hat{k}\} |x + y\rangle &= k \left(\partial_{x-y} + \partial_{x+y}\right) \langle x - y|\hat{\rho}(t)|x + y\rangle \\ &= 2k\partial_x \langle x + y|\hat{\rho}(t)|x - y\rangle \end{aligned} \tag{6.97}$$

where one must be aware that $\hat{x}|x_\pm\rangle = x_\pm|x_\pm\rangle$ for $x_\pm = x \pm y$. Using that $x_\pm$ notation simplifies the next term, allowing it to be compactly rendered as

$$\langle x_-| [V(\hat{x}), \hat{\rho}(t)] |x_+\rangle = (V(x_-) - V(x_+)) \langle x_-|\hat{\rho}(t)|x_+\rangle \tag{6.98}$$

where one must remember that operators can act to the left (on bras) as well as the right (on kets). The integrations over $k$ can now be performed: observing that an integration of $e^{ikz}$ over all $z$ is proportional to a Dirac delta function $\delta(z)$, introduce the function

$$\mathcal{V}(x, k-k') = -\frac{i}{\pi\hbar}\int_{-\infty}^{\infty} e^{2i(k-k')y}\{V(x+y) - V(x-y)\}\, dy \tag{6.99}$$

The notation slightly departs from previous work [36, 158, 159] so that it is apparent that $\mathcal{V}(x, k-k')$ is a different creature than $V(x)$, on which it depends: it describes how far out in $k$-space quantum effects extend as a consequence of the abruptness of $V(x)$. Lastly, the time evolution equation for the Wigner function is

$$\frac{\partial}{\partial t} f(x,k,t) = -\frac{\hbar k}{m}\frac{\partial}{\partial x} f(x,k,t) + \int_{-\infty}^{\infty} \mathcal{V}(x,k-k') f(x,k',t)\, dk' \tag{6.100}$$

By virtue of Eq. (6.99), the impact of quantum effects on $f(x, k, t)$ is non-local, because the potential energy a distance of $\pm y$ away from $x$ intrudes. The exception is for polynomial $V(x)$ which is at most quadratic in $x$, in which case the phase space points follow ballistic trajectories (i.e. $x(t) = x_o + v_o t + (g/2)t^2$).

Some $V(x)$ are such that the integral in $\mathcal{V}(x, k)$ can be evaluated analytically [171]. Consider a few simple types: $\delta$-function ($V_d$), rectangular ($V_r$), parabolic ($V_p$), Gaussian ($V_g$), and triangular ($V_t$), for which

$$V_d(x) = 2V_o\lambda\delta(x) \tag{6.101}$$
$$V_r(x) = V_o\,\Theta(\lambda^2 - x^2) \tag{6.102}$$
$$V_p(x) = V_o\left[1-(x/\lambda)^2\right]\Theta(\lambda^2 - x^2) \tag{6.103}$$
$$V_g(x) = V_o \exp\left[-(x/\lambda)^2\right] \tag{6.104}$$
$$V_t(x) = V_o\,[1-(x/\lambda)]\,\Theta(x)\Theta(\lambda - x) \tag{6.105}$$

where $\Theta(x)$ is the Heaviside step function, $V_o$ gives a measure of the strength of the barrier, $\lambda$ is a parameter governing its extent, and $\delta(x)$ is the Dirac $\delta$-function. The first is the simplest, giving

$$\mathcal{V}_d(x,k) = -\frac{4\lambda V_o}{\pi\hbar}\sin(2kx) \tag{6.106}$$

For the rest of the special cases, they are chosen to allow for the effects of different kinds of behavior where $V(x)$ changes, or not, in some way:

1. continuity in $V(x)$ wherein $V(x+\epsilon) = V(x-\epsilon)$ or discontinuity wherein $V(x+\epsilon) \neq V(x-\epsilon)$ as $\epsilon \to 0$
2. discontinuity in slope such that $\partial_x V(x+\epsilon) \neq \partial_x V(x-\epsilon)$, and
3. symmetric behavior such that $V(+|x|) = V(-|x|)$.

In terms of those distinctions, and labeling the cases by the subscript on $V_c(x)$, then for $c \in (d, r, p, g, t)$

1. the cases $c = (r, t)$ are discontinuous whereas cases $(p, g)$ are continuous,
2. the cases $c = (r, p, t)$ are abrupt whereas case $(g)$ is smooth, and
3. the cases $c = (r, p, g)$ are symmetric whereas case $(t)$ is asymmetric.

It can be shown that the symmetric cases result in $V_c(x,k)$ given by (for $c \in (d, r, p, g)$)

$$\frac{V_s(x,k)}{V_o N_c} = -\eta_c(2k\lambda)\sin(2kx) \tag{6.107}$$

where $k - k' \to k$ for notational simplicity, whereas the antisymmetric case contains an additional term such that

$$\frac{V_t(x,k)}{V_o N_t} = \eta_t(2k\lambda)\sin(2kx) - \eta_t'(2k\lambda)\cos(2kx) \tag{6.108}$$

where the $(t)$ subscript is used instead of $c$ because it is the only asymmetric case under consideration: the right triangle barrier is the one Fowler and Nordheim used in their study of field emission [95, 154]. The $N_c$ term in Eq. (6.108) is

$$N_r = \frac{4\lambda}{\pi\hbar};\ N_p = \frac{8\lambda}{3\pi\hbar};\ N_g = \frac{2\lambda}{\sqrt{\pi}\hbar};\ N_t = \frac{2\lambda}{\pi\hbar} \tag{6.109}$$

$$\begin{aligned}
\eta_r(\varphi) &= \frac{\sin\varphi}{\varphi};\ \eta_g = e^{-\varphi^2/4} \\
\eta_p(\varphi) &= \frac{3}{\varphi^2}(\sin\varphi - \varphi\cos\varphi) \\
\eta_t(\varphi) &= \frac{1-\cos\varphi}{\varphi^2};\ \eta_t'(\varphi) = \frac{\varphi - \sin\varphi}{\varphi^2}
\end{aligned} \tag{6.110}$$

The behavior is shown in Figure 6.10. For the symmetric cases,

$$\frac{N_c\hbar}{4\lambda}\int_{-\infty}^{\infty}\eta_c(\varphi)d\varphi = 1 \tag{6.111}$$

holds, so that the $N_c\hbar\eta_c(2k\lambda)/(4\lambda)$ are seen to be $\delta$-function sequences [37] such that in the limit that $\lambda \to 0$, they act as a delta function.

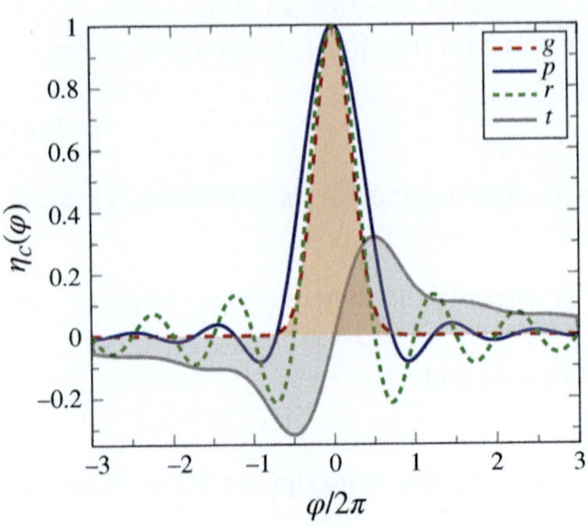


**Figure 6.10** The $\eta_c(\varphi)$ functions given by Eqs. (6.107) and (6.108). The Gaussian curve $g$ (orange fill) is smooth and continuous, and therefore does not spread out in $k$-space; the triangular $t$ (or FN) barrier (gray fill) is both discontinuous in height and slope, and additionally is asymmetrical, and so is seen to spread out significantly. After Ref. [171] / American Physical Society.

The rapidity with which $V_c(x, k)$ decreases with increasing $k$ greatly affects oscillations in $f(x, k, t)$ for large $\pm|k|$, which also depends on $N_c$, e.g. $V_r(x, k)/V_oN_r$ diminishes with increasing $\varphi = 2k\lambda$ but $V_d(x, k)/V_oN_r$ does not, and gives an idea of what the effects of "abrupt" barriers are, effects that are intimately related to wave propagation. When barriers are abrupt or discontinuous in height or slope, $\mathcal{V}(x, k)$ expands in phase space and causes more momentum states to come into play in the evolution of the quantum distribution function, a subtle but important property of quantum systems.

A Gaussian wave packet holds a special place in quantum mechanical discussions, and often is used to show Heisenberg's uncertainty relation. The WDF provides a convenient way to do so. The minimum uncertainty wave function gives rise to a gaussian density $\rho(x)$ and is

$$\Psi(x) = \left(\frac{2}{\pi a^2}\right)^{1/2} \exp\left[ik_o x - \frac{(x - x_o)^2}{2a^2}\right] \tag{6.112}$$

where $a$ is governs the spatial extent of the packet. When $\Psi(x)$ is put into Eq. (6.95), then $f(x, k)$ becomes [171]

$$f(x, k) = \frac{2}{\sqrt{\pi}\hbar a} \exp\left[-\frac{(x - x_o)^2}{a^2} - a^2(k - k_o)^2\right] \tag{6.113}$$

where $(x_o, k_o)$ is the location of the phase space maximum of $f(x, k)$. For $V(x)$ linear in $x$, like gravity, then the wave packet evolves as

$$f(x, k, t) = f\left(x - vt - (g/2)t^2, v - gt, 0\right) \tag{6.114}$$

where $v = \hbar k/m$. Although a free wave packet has $g = 0$, consider the more general nonzero case in the example here. The evolution of Eq. (6.114) for $g = 4$ (in arbitrary units where $a = 1$, $x_o = 2$, and $k_o = 2$) is then shown in Figure 6.11. The boundaries of the rectangle enclosing the $(t = 0)$ starting case follow the same ballistic equations, and so the rectangle expands in size, even though the parallelogram set by the trajectories does not change in area and contains the same number of phase space points within it that the initial rectangle did (particles are not created or destroyed). The size of the rectangular area enclosing the parallelogram grows as $t \to \pm\infty$, with its minimum area occurring at $t = 0$.

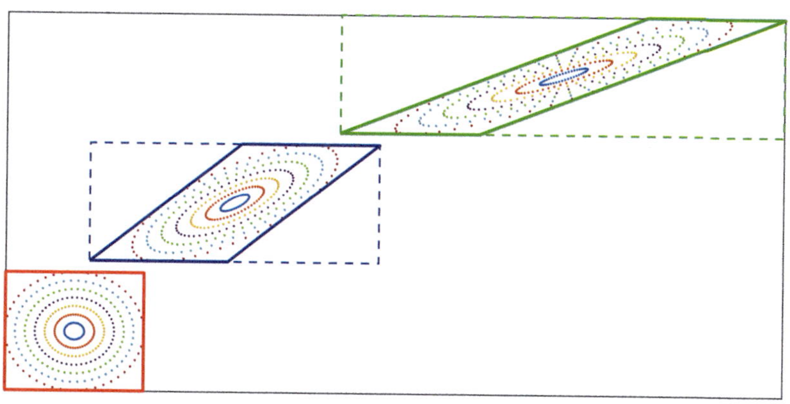

**Figure 6.11** The expansion of the contour lines of a Gaussian Wigner wave packet $f(x, k, t)$ in scaled units for $t = [0, 1, 2]\Delta t$, $g = 4$ and $\Delta t = 1.1$. $x$ is in units of $a$, and $k$ in units of $1/a$. Notice the similarity to Figure 4.2. The acceleration $g > 0$ for visual clarity in the example.

Returning to the free case $g = 0$, then the parallelogram area is the same over time as the initial rectangle. However, the area of phase space $\Delta x \Delta k$ is the area bounded by the rectangles with the dashed lines in Figure 6.11, and even though $\Delta k = k_{max}(t) - k_{min}(t) = k_{max}(0) - k_{min}(0)$ would not change, $\Delta x = x_{max}(t) - x_{min}(t)$ does as the parallelogram shears. The width of the box $\Delta x(t) \geq a$ and $\Delta k = 1/a$ is shown as the rectangles with the dashed perimeters. And that means $\Delta x(t)\, \Delta k \geq 1$, with the equality occurring at $t = 0$. A more formal argument, based on the spread of Gaussian functions and their minimization[2] changes the 1 to 1/2 and is more specific about what $\Delta x$ and $\Delta k$ are. And so, Heisenberg's Uncertainty Principle is seen to be embedded in the Wigner wave packet model

$$\Delta x \Delta p \geq \hbar/2 \tag{6.115}$$

where $p = \hbar k$ is momentum. A similar equation holds for energy and time of the form

$$\Delta E \Delta t \geq \hbar/2 \tag{6.116}$$

## 6.8 Exercises

**Exercise 6.1** Using Eq. (6.24), find $H_3(x)$.

**Solution** Eq. (A.27).

**Exercise 6.2** Using Eq. (6.12) and letting $f(x) = \sin(\pi x)$, find $C_n$ for the Legendre polynomials for $n \in (0, 1, 2, 3)$.

**Solution** $C_n = 0$ for all even $n$. $C_1 = 3/\pi$. $C_3 = 7(\pi^2 - 15)/\pi^3$.

**Exercise 6.3** Show that $[\hat{a}, \hat{a}^\dagger] = 2$.

**Exercise 6.4** The de Broglie wavelength $\lambda_b$ is the ratio of Planck's constant $2\pi\hbar$ with a particle's momentum. Assuming an electron is at the Fermi level $\mu = 7$ eV of copper, find $\lambda_b$ in terms of $k_F$.

**Solution** $\lambda_b = 2\pi/k_F = 0.464$ nm. $k_F = 13.55\ \text{nm}^{-1}$.

**Exercise 6.5** Show Eq. (6.86).

**Exercise 6.6** Derive the continuity Eq. (4.17) from Eq. (6.100), with density $\rho$ defined by Eq. (6.94), $V(x)$ non-zero only over a finite range, and $J = q\rho\langle v \rangle$. Remember that momentum is $\hbar k = mv$.

**Exercise 6.7** Show Eq. (6.111) for the symmetric potentials $c \in (r, p, g) =$ rectangular, parabolic, and Gaussian.

2 See, for example, Prob. 4.1 of Ref. [39].

**Exercise 6.8** Consider a field emitted electron tunneling through a triangular barrier of the form $V(x) = \mu + \Phi - Fx$. Let the uncertainty in time $\Delta t = \Delta x / v_F$ where $v_F$ is the velocity of an electron with a kinetic energy of $\mu$ and $\Delta x$ is the width of the barrier to it. For copper-like parameters ($\mu = 7$ eV, $\Phi = 4.5$ eV), calculate the value of $F$ required for Heisenberg's uncertainty relation Eq. (6.116) to hold if the uncertainty in $\Delta E \approx (1/2)$ eV.

**Solution** $F = 4.648$ eV/nm.

# 7

# Solid State Physics

*Bring me to the test,*
*And I the matter will reword; which madness*
*Would gambol from.*

W. Shakespeare, *Hamlet* III.4

## 7.1 Conductivity

If a battery is connected to opposite sides of a block of material of length $L$ and cross-sectional area $A$, then generally current $I$ will flow through the block in proportion to the voltage $\mathcal{V}$.[1] The constant of proportionality is the resistance $\mathcal{R}$, such that $\mathcal{V} = I\mathcal{R}$. The ratio of the voltage to the length is a measure of the electric field $\mathcal{E}$ across the block, and the ratio of the current to the area is a measure of the current density $J$, and so

$$\frac{\mathcal{V}}{L} = \left(\frac{I}{A}\right)\left(\frac{\mathcal{R}A}{L}\right) \Rightarrow J \equiv -q\sigma\mathcal{E} = q\sigma\frac{d\varphi}{dx} \tag{7.1}$$

where the "electrical" conductivity $\sigma = L/A\mathcal{R}$ and $\mathcal{E} = F/q$. It is the second form of interest here, concerning as it does current density $J$. Electrical conductivity is seen to be the coefficient of the gradient of the potential $\varphi(x)$. The inverse conductivity $1/\sigma$ is known as the resistivity, and its behavior as a function of temperature for various metals is shown in Figure 7.1.

Intuitively, there seems to be a commonality among the curves, and in fact there is. What causes resistance in metals is electrons colliding the vibrating lattice of metal ions. Lattice vibrations can be represented as particles called phonons, the lower-frequency ones being called acoustic phonons, wherein adjacent atoms are more or less moving in phase (in the same direction), compared to the higher-frequency ones called optical phonons, wherein adjacent atoms are moving out of phase (in opposite directions). If the atoms are thought of as being in a box, then there are only a finite number of vibrations $N$ that can occur [154, 203]: putting in a factor of 3 to account for three dimensions, then

$$3N = \int_0^{\omega_D} D(\omega)d\omega \tag{7.2}$$

1 Do not confuse $\mathcal{V}$ here with a similar symbol in the WDF discussion: in both cases, the intent is to retain the meaning of $V(x)$ as a potential energy; here $\mathcal{V}$ is a voltage.

*Fundamentals of Electron Emission Physics*, First Edition. Kevin L. Jensen.

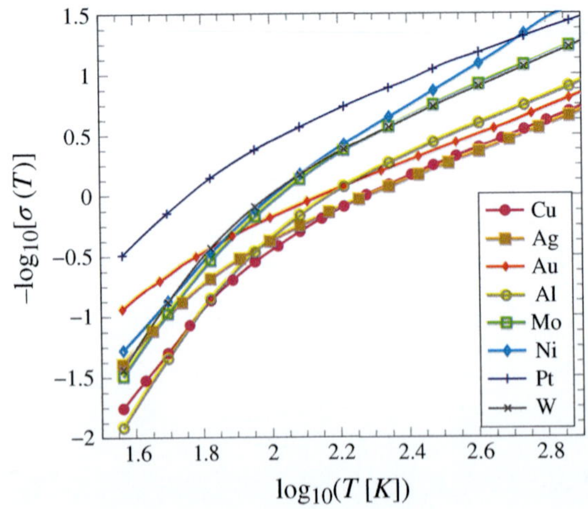


**Figure 7.1** Resistivity $1/\sigma$ in units of [$\mu\Omega$-cm] as a function of temperature $T$ [K] for various common metals using standard values [113, 249].

where $\omega_D$ is the Debye frequency and is the maximum frequency of oscillations supported by the vibrating lattice of atoms. Phase space arguments indicate that the distribution $D(\omega)$ is given by

$$D(\omega) = \frac{4\pi V}{(2\pi)^3}\left(\frac{1}{c_l^3} + \frac{2}{c_t^3}\right)\omega^2 d\omega \tag{7.3}$$

where $c_l$ and $c_t$ are the longitudinal and transverse sound velocities, respectively, $V$ is the volume of the block of material, and the factor of $4\pi$ comes from integration over all angles. If $c_l = c_t$, then it simply follows

$$3N = \frac{V}{2\pi^2}\left(\frac{\omega_D}{c_l}\right)^3 \tag{7.4}$$

Thus, knowing the sound velocity and the density $N/V$ of a metal, the Debye frequency can be calculated. A more useful parameter, however, is the Debye temperature, whereby $T_D \equiv \hbar\omega_D/k_B$. The interest in $T_D$ here is precisely because its use in plotting $\sigma(T_D)/\sigma(T)$ versus $T/T_D$ results in the kind of universal curve shown in Figure 7.2 using values of $T_D$ from Table 32.2 of Ref. [154]. The red line on which all these datasets fall is the Bloch–Grüneisen (BG) relation [135, 358–360] that will be encountered when acoustic scattering is discussed in detail: here, $T_D$ is evaluated by the best value to achieve that correspondence, due to the ambiguity associated with differing sound velocities and the peculiarities of metal crystal lattices. Still, the agreement is notable, and portends deep underlying physics governing the phenomena of electron transport in a Sommerfeld metal, representing one of the early success stories of solid-state physics.

Understanding the connection between charge and heat flow in solids was an intriguing problem investigated by Drude, [66] who explored the connection between the electrical $\sigma$ and thermal $\kappa$ conductivities of metals: as temperature increased, the ratio $L = \kappa/\sigma T$ appeared to converge to the same value regardless of metal. Drude argued that electrical current as the product of the charge of the carrier with the number density of carriers $\rho$ and their average velocity $\langle v \rangle$. Because the charged carriers accelerated due to an electric field, their velocity increased until they suffered a scattering event. The probability of scattering was proportional to the amount of time passing since their last

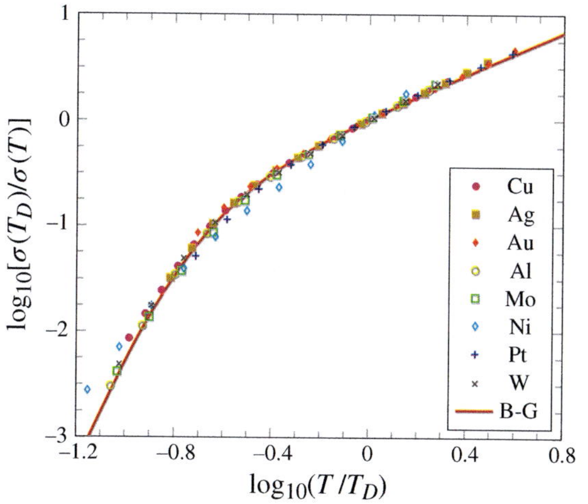


**Figure 7.2** Ratio of $\sigma(T_D)/\sigma(T)$ versus the ratio $T/T_D$ where $T_D$ is the Debye temperature evaluated by a best fit of the data to the Bloch–Grüneisen relation ([154] / John Wiley & Sons).

scattering event. Drude assumed the probability $P(t)$ that an electron suffered a collision in time $t$ went as $P(t) \propto \exp(-t/\tau)$, where $1/\tau$ is the scattering rate, and $\tau$ itself was the "relaxation time" (scattering caused electrons to lose energy, and so relax back to an equilibrium state). Thus, the average velocity is the integral of $v(t) = Ft/m$ with the probability of collision, or

$$\langle v \rangle = \frac{\int_0^\infty (Ft/m) e^{-t/\tau} dt}{\int_0^\infty e^{-t/\tau} dt} = \frac{F\tau}{m} \tag{7.5}$$

and so Drude concluded

$$\sigma = \frac{q^2 \tau}{m} \rho \tag{7.6}$$

A copper value of $1/\sigma = 1.723 \times 10^{-6}$ Ω-cm = 0.002761 eV fs nm / q$^2$ suggests $\tau = 24.49$ fs, useful below in considering the components of $\tau$ in Section 7.4.

In contrast to electrical current, the flow of heat is a flow of energy carried by the electrons, and heat will flow from hotter to colder materials because the average velocity on the hot side is greater. As a result, the $q$ in $J$ is replaced by the energy the electron carries, or $E$. The effect complicates the coefficient of "thermal" conductivity when a temperature gradient leads to a net flow of energy. If a difference in temperature on the two sides of a block of material leads to a flow, or "current," of energy $J_T$, then that flow will be [154]

$$J_T = \rho \left[ \langle E_x^+ v_x^+ \rangle - \langle E_x^- v_x^- \rangle \right] \approx 2\rho l \frac{d}{dx} \langle E_x v_x \rangle \tag{7.7}$$

where $l = \langle v_x \rangle \tau$ is the mean free path that electrons travel between scattering events, and $v_x^\pm$ is the average velocity of electrons traveling in the $\pm$-direction from their respective boundaries, for which electrons coming from the left are characterized by a distribution using the left boundary temperature, and similarly for the right. Given the gradient in temperature, then this becomes

$$J_T = 2l\rho \left( \frac{d\langle E_x v_x \rangle}{dT} \right) \frac{dT}{dx} \tag{7.8}$$

Calculation of the averages is required to find $\kappa$, or the thermal conductivity. Drude, reasonably for his time, assumed that the distribution in energy of the charge carriers was Maxwell–Boltzmann, or $D(E_x) = \exp(-E_x/k_B T) = \exp(-\beta E_x)$ in his efforts to compare the electrical $\sigma$ and thermal $\kappa$ conductivities of metals. Using the MB distribution, then averages like

$$\langle v_x^+ \rangle = \frac{\int_0^\infty v_x e^{-\beta E_x} dv_x}{\int_0^\infty e^{-\beta E_x} dv_x} = \sqrt{\frac{2k_B T}{\pi m}} \tag{7.9}$$

for $E_x = mv_x^2/2$. Note how the limits of integration are specific to the forward ($+\hat{x}$) direction. Consequently, $\langle v_x^\pm \rangle = [2k_B T_\pm/\pi m]^{1/2}$ and $\langle E_x^\pm v_x^\pm \rangle = [2k_B^3 T_\pm^3/\pi m]^{1/2}$ result. Finally, letting $l \approx \langle v_x^\pm \rangle \tau$ results in

$$\kappa \approx \frac{6\tau k_B^2 T}{\pi m} \rho \tag{7.10}$$

The ratio of $\kappa$ to the product $\sigma T = q^2 T\tau\rho/m$, or the Lorentz number, that Drude proposed requires that the scattering time $\tau$ of electrical conduction is the same as that of thermal conduction, a reasonable expectation given that electrons carry both charge and energy. The distribution used to evaluate the averages, however, is Fermi–Dirac, not Maxwell–Boltzmann. When the correct distribution is used [154], the ratio of thermal to electrical conductivity becomes

$$L \equiv \frac{\kappa}{\sigma T} = \frac{\pi^2}{3} \left( \frac{k_B}{q} \right)^2 \equiv L_{wf} \tag{7.11}$$

where the subscript on $L_{wf}$ denotes the Wiedemann–Franz law. $L_{wf}$ is seen to be combinations of physical constants and numbers, and therefore material-independent. As temperature increases, the $L$ ratio for many metals converges to $L_{wf}$, another universal behavior traceable to electrons satisfying the Fermi–Dirac distribution relation of Eq. (4.35). Although Drude was undoubtedly nonplussed by his estimate being about 16% too large (the percentage difference between $\pi/12$ and $\pi^2/3$, where the latter is due to integrals like Eq. (13.57)), the resolution was a boon for quantum and statistical mechanics applied to the properties of metals.

## 7.2 Metals and Semiconductors

There is more than just metals: there are materials that pass current, just not as well, and some barely at all. Conductivity distinguishes between metals and insulators (and semiconductors seemingly share attributes of both). Because insulators can pass current at increased temperatures, it is preferable here to speak of semiconductors to emphasize that current is flowing. The difference between metals and semiconductors is how electrons get into the conduction band where they can

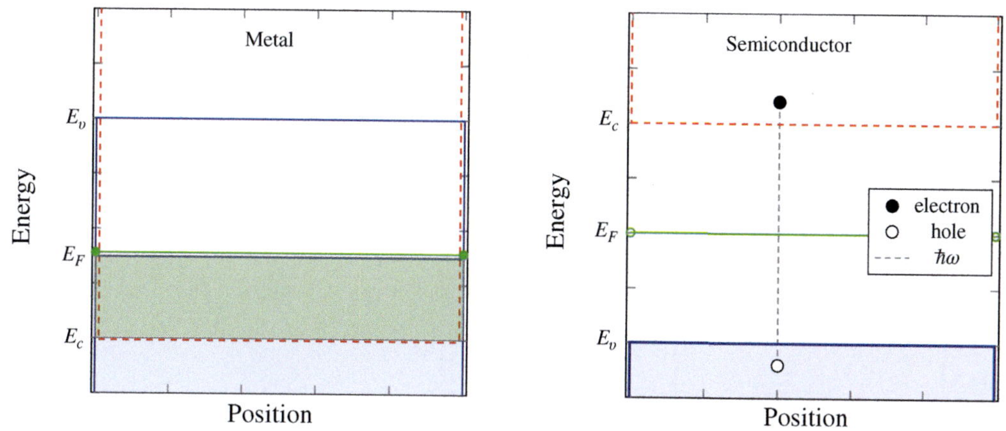


**Figure 7.3** (*left: metal*) Top of valence band $E_v$ is above bottom of conduction band $E_c$ and mobile electrons fill up to Fermi level $E_F$ (dark region). (*right: semiconductor*) Top of valence band is below bottom of conduction band. For an intrinsic (undoped) semiconductor, $E_F$ lies half way between them. The bandgap $E_g$ is the energy difference between $E_c$ and $E_v$. An electron • that absorbs a photon $\hbar\omega > E_g$ and is elevated from the valence band to the conduction band and leaves behind a hole ∘.

readily flow in response to fields. As in Figure 7.3, for metals, the bottom of the conduction band $E_c$ is below the top of the valence band $E_v$, and so electrons can fill up the conduction band to the Fermi level naturally. In semiconductors, $E_c$ is above $E_v$ and so electrons cannot get into the conduction band unless promoted there by thermal processes, photoexcitation, or collisions with high-energy primary electrons. The difference $E_g = E_c - E_v$ is the bandgap.

Electrons in the conduction band flow in response to applied fields, the degree to which they do being temperature-dependent. The distinction between a metal and an insulator is therefore that as $T \to 0$ K, the conductivity of a metal is finite, whereas that of an insulator goes to 0. At finite temperatures, insulators can pass current and are therefore semi-conducting, or semiconductors, albeit that pure semiconductors may not conduct very well, whereas doped semiconductors do. For $T > 0$, the inverse conductivity, or resistivity $\tilde{\rho} = 1/\sigma$ (where a tilde is used to distinguish resistivity $\tilde{\rho}$ from number density $\rho$ herein), further distinguishes them [10, 35, 131, 203]

$$\tilde{\rho}(T)|_{metal} \approx \tilde{\rho}(T_o) + \alpha(T - T_o) \tag{7.12}$$

$$\tilde{\rho}(T)|_{semi} \approx R_o \exp\{\alpha/T\} \tag{7.13}$$

where the behavior of $\tilde{\rho} = 1/\sigma$ is starkly different between them, and $\alpha$ is just a parameter (not the fine structure constant here). That Eq. (7.12) is good for metals is indicated in Figure 7.4. The form of the resistivity for metals says something about how scattering behaves, and the bandgap of semiconductors gives a clue about why Eq. (7.12) is so different from Eq. (7.13).

For a metal, even at $T = 0$ K, there are electrons in the conduction band, in contrast to pure un-doped semiconductors for which electrons only get there by thermal excitations from the valence band (in the absence of photo-excitation and secondaries). For a simple Sommerfeld metal, recall that $\mu$ governed the electron density $\rho$ and allows Eq. (4.36) to be a good approximation to the number density of electrons defined by Eq. (4.36) in a cold metal because $\mu/k_B T \equiv \beta\mu \gg 1$. As the temperature increases, though, the number of electrons does not change, and so (if the volume does not change either) $\rho = N/V$ does not change. That means $\mu$ must be temperature-dependent.

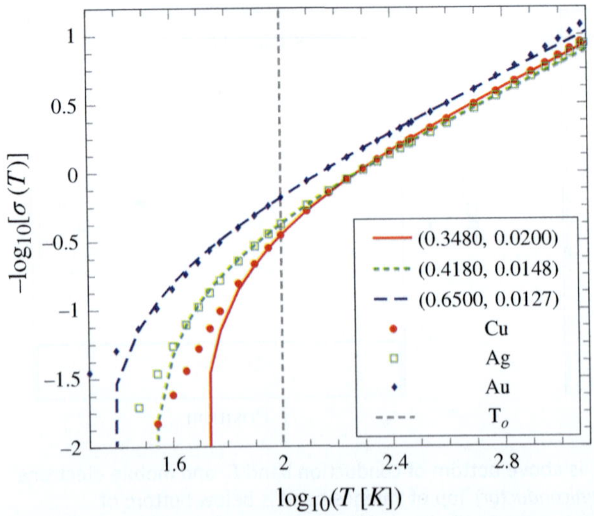


**Figure 7.4** Symbols: Resistivity ($\tilde{\rho} = 1/\sigma$) of copper (Cu), silver (Ag), and gold (Au); Adapted from [249]. Lines: linear fit $\tilde{\rho}_o[1 + \alpha(T - T_o)]$ labeled by $(\tilde{\rho}_o, \alpha)$. Units of $\tilde{\rho}$ are [$\mu\Omega$-cm] as per Matula.

The number density $\rho$ integral can be recast in terms of energy by introducing a density of states $D(E)$ and assuming $E = \hbar^2 k^2/2m$ to find

$$\rho = \int_0^\infty D(E) f_{FD}(E) dE \tag{7.14}$$

$$D(E) = \frac{m}{\pi^2 \hbar^3}\sqrt{2mE} \equiv N_c \sqrt{E} \tag{7.15}$$

$$f_{FD}(E) = \{1 + \exp[\beta(E - \mu)]\}^{-1} \tag{7.16}$$

where the density of states (DOS) shown is for three dimensions, $\beta = k_B T$, the spin term g of the electron is set to 2, and the new term $N_c$ is defined by

$$N_c = \frac{\sqrt{2m^3}}{\pi^2 \hbar^3} r_e^{3/2} = 6.81217 r_e^{3/2} \tag{7.17}$$

where the right-hand side is in units of [$\mathrm{eV}^{-3/2}\ \mathrm{nm}^{-3}$] and the dimensionless term $r_e$ is the effective mass (ratio of mass with electron rest mass) that for metals is often approximated by unity, but which for semiconductors can be quite interesting (Section 7.3). The term $\mu$ is, as in Section 4.5, the chemical potential [203, 289], but often in the discussion of metals, it can be called the Fermi energy $E_F$, even though that is not quite correct: the actual correspondence is $\lim_{T\to 0} \mu(T) = E_F$, but since the chemical potential for metals is large, its temperature dependence is very weak, and so being insistent about the distinction borders on fastidious. For semiconductors, where $\mu$ tends to be smaller (and even negative), knowing the temperature dependence of $\mu(T)$ is far more important. The trick to calculating it is to realize that the number of electrons in a volume does not change when the density does, and so $\rho(0) = \rho(T)$ if the volume is temperature-independent (which it is not, but assume it is negligible). Here, the convention is to call the Fermi energy $E_F = \mu(T = 0) \equiv \mu_o$. Density is [26]

$$
\rho = \frac{N_c}{\beta^{3/2}} F_{1/2}(\beta\mu)
$$

$$
F_p(u) \equiv \int_0^\infty \frac{x^p}{e^{x-u}+1} dx \tag{7.18}
$$

where $F_p(u)$ is the Fermi function and so $\rho(0) = \rho(T)$ gives a means of determining what the temperature dependence of $\mu(T)$ is. The first step is to observe

$$
\lim_{\beta\to\infty} \int_0^\infty \frac{E^p}{e^{\beta(E-\mu)}+1} dE = \frac{\mu_o^{p+1}}{p+1} \tag{7.19}
$$

which is a bit more subtle than it looks: except in the limit, the integral is not analytic and so $\mu(T)$ cannot be obtained analytically from it: to do so, some numerical subterfuge is required. Observe that the density integral can be recast for $p = 1/2$ and after a change in integration variable as

$$
\frac{2}{3}\mu_o^{3/2} = \beta^{-3/2} \int_{-u}^\infty \frac{\sqrt{x+u}}{e^x+1} dx \tag{7.20}
$$

where $u \equiv \beta\mu$. Now break up the integral into regions where $x < 0$ and $x > 0$, and for the former case, change $x \to -x$. Now all the integrals are for $x > 0$, with some of the integrals ranging from $x = 0$ to $x = u$, and the remaining from $x = u$ to $x \to \infty$. Lastly, use $(e^{-x}+1)^{-1} = 1 - (e^x+1)^{-1}$, which allows the leading order integral $\int_0^u \sqrt{x}dx$ (from the first term) to be found exactly. Putting everything back together gives

$$
\int_{-u}^\infty \frac{\sqrt{u+x}}{e^x+1} dx = \frac{2}{3}u^{3/2} + \int_0^u \frac{\sqrt{u+x} - \sqrt{u-x}}{e^x+1} dy + \int_u^\infty \frac{\sqrt{u+x}}{e^x+1} dy \tag{7.21}
$$

For purposes of discussion below, call the full integral on the left-hand side $I_u$, and the three integrals on the right-hand side $I_a$, $I_b$, and $I_c$, respectively, so that $I_u = I_a + I_b + I_c$. Although Eq. (7.21) is exact for $I_u$, the last integral $I_c$ for any metal-like value of $\mu$ is exponentially small and therefore ignorable over any reasonable temperature range for a metal, but for semiconductors with a small $\mu_o$, then $\mu$ can become very small. When it vanishes, then only $I_c$ survives: for example, for $T = 1000$ K and $\mu_o = 0.087154$ eV (corresponding to $\rho = 1.1685 \times 10^{20}$ cm$^{-3}$, a hefty doping density), then $\mu(T) = 0$ eV. Small doping densities then can require that $\mu$ even become negative at laboratory temperatures. The temperature $T_m$ at which $u = \mu(T_m)/k_B T_m = 0$ occurs when the first two terms of Eq. (7.21) vanish, leaving

$$
\frac{2}{3}\mu_o^{3/2} = N_o (k_B T_m)^{3/2} \Rightarrow T_m = (2/3N_o)^{2/3}(\mu_o/k_B)
$$

$$
N_o = \int_0^\infty \frac{\sqrt{x}}{e^x+1} dy = \frac{\sqrt{\pi}}{2}\left(1 - \frac{1}{\sqrt{2}}\right)\zeta\left(\frac{3}{2}\right) = 0.678094 \tag{7.22}
$$

The last integral in Eq. (7.21) is therefore generally small, except when $u$ approaches zero, and if $\mu_o > (1/4)$ eV, then $T_m > 2868$ K, an unusually high temperature.

Even though $I_c$ of Eq. (7.21) is generally small, it is clearly not always negligible. Its approximate size can be estimated from

$$I_c > \sqrt{2u}\int_u^{\infty}\frac{dy}{e^x+1}dy = \sqrt{2u}\ln\left(e^{-u}+1\right) \tag{7.23}$$

and using that approximation is sufficient to stay abreast of when the last integral matters: clearly, it matters when $I_a \sim I_c$. For $u = 1$, then $I_a = 2/3$, whereas the approximation to $I_c = \sqrt{2}\ln(1+e^{-1}) \approx 0.44301$ (its exact value is 0.54331).

Estimating $\mu(T)$ therefore requires examining the middle integral. Profitable use can be made of the Taylor series expansion to get

$$\frac{\sqrt{u+x}-\sqrt{u-x}}{\sqrt{u}} = \frac{x}{u}+\frac{x^3}{8u^3}+\frac{7x^5}{128u^5}+O(x^6) \tag{7.24}$$

Again, use the fact that $u = \beta\mu$ is typically rather large for common temperatures and metallic chemical potentials (e.g. even for $T = 2000$ K and $\mu = 1$ eV, $u \approx 5.8$ and so $e^{-u} \approx 0.003$). Under such conditions, the middle integral is well approximated by extending the upper limit to $\infty$ and using the definition of the $\zeta(n)$ function for $n \in (2, 4, 6)$ to get

$$I_b \approx \frac{2}{3}u^{3/2}\left[\frac{3\zeta(2)}{4u^2}+\frac{63\zeta(4)}{64u^4}+\frac{9765\zeta(6)}{1024u^6}\right] \tag{7.25}$$

Putting in the values of $\zeta(n)$ then gives the approximation

$$F_{1/2}(u) \approx \frac{2}{3}u^{3/2}\left[1+\frac{\pi^2}{8u^2}+\frac{7\pi^4}{640u^4}+\frac{31\pi^6}{3072u^6}+\delta f_r(u)\right] \tag{7.26}$$

where $\delta f_r(u) \approx 0$ for metal-like parameters but if the third integral of Eq. (7.21) matters, that can be approximated using Eq. (7.23) by

$$\delta f_r(u) = \frac{3\sqrt{2}}{2u}\ln\left(e^{-u}+1\right) \tag{7.27}$$

An equation like Eq. (7.26) is ripe for trickery. Divide both sides by $u^{3/2} = (\beta\mu)^{3/2}$, let $\mu = \mu_o/y$, and define the dimensionless term $r \equiv (\pi/\beta\mu_o)^2$ to finally obtain

$$y = \left[1+\frac{1}{8}ry^2+\frac{7}{640}\left(ry^2\right)^2+\frac{31}{3072}\left(ry^2\right)^3+\delta f_r\left(\frac{\pi}{y\sqrt{r}}\right)\right]^{2/3} \equiv f_r(y) \tag{7.28}$$

In that form, the iterative methods of Appendix A.10 can be employed: start with an approximation to $y$, evaluate the right-hand side using it, and repeat until the value of $y$ converges to an acceptable accuracy. If the first estimate is good, only a few iterations are required. If $r$ becomes too large however, iteration no longer works for Eq. (7.28): $y$ races to infinity so that $u$ heads to zero, and a more rugged method like bisection to find when $y - f_r(y) = 0$ and $u$ is small (or negative) is necessary. For present purposes, though, iteration works fairly well to examine metals.

A method to find the temperature variation of $\mu(T)$ is then to start at $T = 0$ so that $\mu(0) = \mu_o$ and $y_0 = 1$ by definition. The value of $y_j$ can be found for $T_j = j\Delta T$ by using $T_{j-1}$ as the first guess of the $j^{th}$ iteration process to find $y_j$. After the $y_j$ are all found, then $\mu(T_j) = \mu_o/y_j$. Doing so gives Figure 7.5 (see Algorithm B.2.10), where the symbols (•) include $\delta f_r(y)$, the circles (∘) neglect

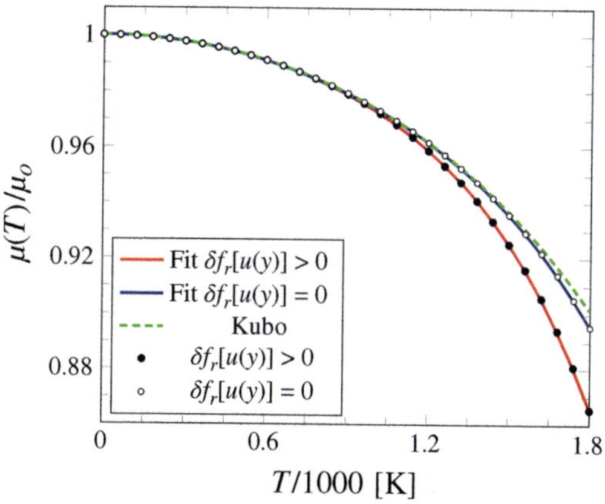


**Figure 7.5** Temperature dependence of $\mu(T)/\mu_o$ when $\mu_o = 0.5$ eV and $0 < T < 1800$ K. "Kubo" is the approximation of Eq. (7.29). Fits (lines) are based on $\sum_{j=0}^{3} P_j r^j$ with $P_j$'s given in Table 7.1. Symbols are solutions to Eq. (7.28) for $\delta f_r(u) > 0$ (•) and $\delta f_r(u) = 0$ (∘) with $u = \pi/y\sqrt{r}$.

**Table 7.1** Coefficients $P_j$ of $\mu(T)/\mu_o = \sum_{j=0}^{3} P_j r^j$. $P_j^k$ are the coefficients of Eq. (7.29); $P_j/P_j^k$ are the ratio of the coefficients for the $\delta f > 0$ line to the Kubo coefficients, and $P_j^o/P_j^k$ are the ratio of the coefficients for the $\delta f = 0$ line to the Kubo coefficients.

| $j$ | $P_j^k$ | $P_j/P_j^k$ | $P_j^o/P_j^k$ |
|---|---|---|---|
| 0 | 1 | 1.0000 | 1.0000 |
| 1 | −1/12 | 1.0556 | 1.0295 |
| 2 | −1/80 | 0.7644 | 0.0166 |
| 3 | −247/25920 | 5.6309 | 2.7872 |

$\delta f_r(u)$, and the dashed line is the well-known approximation from statistical mechanics [154, 209] which Kubo represented as

$$\frac{\mu}{\mu_o} \sim 1 - \frac{1}{12}r - \frac{1}{80}r^2 - \frac{247}{25920}r^3 \tag{7.29}$$

It is clear that the Kubo results are good for low temperature, but for more practical temperatures, the coefficients require modification. The polynomial fits use the coefficients of Eq. (7.29) in $\mu(T)/\mu_o = \sum_{j=0}^{3} P_j r^j$ as determined by least squares fits to ∘ data ($\delta f_r(y) = 0$) and • data ($\delta f_r(y) > 0$) determined by solutions of Eq. (7.28). Importantly, these fit factors change as the $r$-range over which they are calculated changes.

Semiconductors, by comparison, add further difficulties to consider. When an electron jumps across the bandgap $E_g$ through collisions and thermal agitation, its absence in the valence band is a vacancy. Moving vacancies can be thought of as particles, "holes," that move with velocities and masses similar to (but not necessarily equal to) those of the electrons, but which are positively charged. The flow of holes is a counter-current to the flow of electrons. Two charge carriers are possible, and which one dominates depends on circumstances. The hole distribution has parallels to the electron distribution. If the valence band is filled up to $E_v$, then

$$\rho = \int_0^{E_v} \mathcal{D}(E)dE \tag{7.30}$$

An increase in temperature thermally excites some electrons into the conduction band, and so

$$\rho = \int_0^{E_v} \mathcal{D}(E)f_{FD}(E)dE + \int_{E_c}^{\infty} \mathcal{D}(E)f_{FD}(E)dE \tag{7.31}$$

The total number of electrons does not change as the temperature increases. Assume the volume does not change either (it does, and so affects the bandgap [154] – but again, neglect it) then

$$\int_0^{E_v} \mathcal{D}(E)\left(1 - f_{FD}(E)\right) dE = \int_{E_c}^{\infty} \mathcal{D}(E)f_{FD}(E)dE \tag{7.32}$$

At this point, acknowledging that holes and electrons are not mirror images is required: their masses differ, and therefore, subscripts of "$h$" and "$e$" are appended to terms to distinguish which corresponds to holes, and which to electrons, and so density of states $\mathcal{D}_e(E)$ and $\mathcal{D}_h(E)$ and masses $m_e$ and $m_h$ acquire subscripts.

The Fermi–Dirac function is of the form $f(s) = 1/(e^s + 1)$, from which it is simple to show that $f(-s) = 1 - f(s)$. Using that,

$$\begin{aligned}\rho_e(T) &= \frac{\sqrt{2}}{\pi^2\hbar^3}(m_e k_B T)^{3/2} F_{1/2}\left[\beta(\mu - E_c)\right]\\ &\equiv \frac{N_c}{\beta^{3/2}} F_{1/2}\left[\beta(\mu - E_c)\right]\end{aligned} \tag{7.33}$$

$$\begin{aligned}\rho_h(T) &= \frac{\sqrt{2}}{\pi^2\hbar^3}(m_h k_B T)^{3/2} F_{1/2}\left[\beta(E_v - \mu)\right]\\ &\equiv \frac{N_v}{\beta^{3/2}} F_{1/2}\left[\beta(E_v - \mu)\right]\end{aligned} \tag{7.34}$$

where $N_v$ is like $N_c$ but for holes. For intrinsic (undoped) semiconductors, few electrons are in the conduction band: whereas $\rho \sim 8.41 \times 10^{22}$ cm$^{-3}$ = 84.1 nm$^{-3}$ for metals like copper, intrinsic semiconductors like intrinsic silicon at room temperature have densities closer to $10^{10}$ cm$^{-3}$ = $10^{-11}$ nm$^{-3}$. To make that happen, then $\mu$ has to be negative, and it is thereby between the conduction band minimum $E_c$ and above the valence band $E_v$, or right in the bandgap. Let $m_e = r_e m$ and $m_h = r_h m$, where $r_e$ and $r_h$ are the effective mass ratios for electrons and holes respectively. Demanding $\rho_e(T) = \rho_h(T)$ and using the approximation that the Fermi function for large and negative $\beta\mu$ is

$$F_p(-u) = \int_0^{\infty} \frac{x^p}{e^{x+u} + 1} dx \approx e^{-u} \int_0^{\infty} x^p e^{-x} dx = \Gamma(p+1)e^{-u} \tag{7.35}$$

then it follows

$$r_e^{3/2} \exp\left[\beta(E_v - \mu)\right] = r_h^{3/2} \exp\left[\beta(\mu - E_c)\right] \tag{7.36}$$

Isolating $\mu$ then shows

$$\mu(T) = \frac{1}{2}(E_c + E_v) - \frac{3}{4}k_B T \ln\left(\frac{r_h}{r_e}\right) \tag{7.37}$$

At room temperature, $k_B T \approx 0.025852$ eV. If the electron and hole masses are comparable, then the log term is small. As a consequence, the Fermi level does not depend on $T$ for intrinsic semiconductors to a good approximation, rather, it is located midway between the conduction and valence band edges. The product of the electron density $\rho_e$ with the hole density $\rho_h$ is then very simple to leading order. Keeping only the leading order terms of $F_{1/2}(x)$ and writing the bandgap as $E_g \equiv E_c - E_v$ results in

$$\rho_e \rho_h \approx \left(\frac{m\sqrt{r_e r_h}}{4\pi\beta\hbar^2}\right)^3 e^{-\beta E_g} = (r_e r_i)^{3/2}\left(\frac{2e^{-\beta E_g/2}}{\lambda_T^3}\right)^2 \equiv \rho_i^2 \tag{7.38}$$

where the thermal wavelength $\lambda_T^2 = 2\pi\beta\hbar^2/m$. When $\rho_e = \rho_h$ and the electron and hole masses are the same as well, then $\rho_i$ is the intrinsic carrier density. Given that conductivity is proportional to the carrier density even in the Drude model of Eq. (7.6), then Eq. (7.13) then makes sense, indicating that $\alpha \to E_g/2k_B T$. The measured data of Hoffman et al. (lines 1 and 6 of Figure (2) of Ref. [131]) can be used to estimate that $E_g = 1.2$ eV as the bandgap of intrinsic silicon at room temperature, as shown in Figure 7.6.

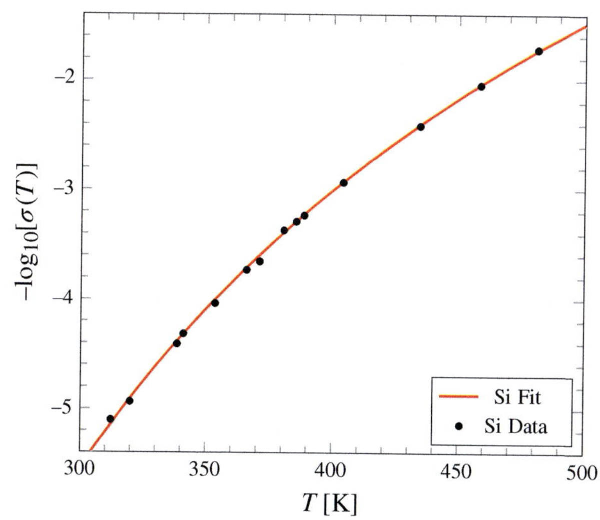


**Figure 7.6** Symbols: Resistivity ($\tilde{\rho} = 1/\sigma$) of intrinsic silicon (Si data) from [131] / Elsevier. Lines: ($R_o$, $\alpha$) determined from linear fit of $1/T$ vs. $-\ln(\sigma)$ (Si fit). Units of $\sigma$ are [1/Ω-cm]. Fit gives $2k_B\alpha = 1.2$ eV, equal to the bandgap $E_g$ of intrinsic silicon at room temperature, and $R_o = 3.91 \times 10^4$ Ω-cm.

## 7.3 Alpha Semiconductor Model

If energy is measured, not from the bottom of the conduction band as needed in the emission equations, but rather from the center of the bandgap itself, and if the symbol for that energy is called $W$, then for semiconductors for which the electron has an effective mass $m_n$

$$W^2 = \left(\frac{E_g}{2}\right)^2 + E_g \frac{\hbar^2 k^2}{2m_n} \tag{7.39}$$

This is a hyperbolic dispersion relation [154], and it hearkens back to similar formulae encountered in the relativistic relation of Eq. (5.18). Therefore, see what happens if it is actually shoehorned into such a relativistic form. To that end, replace $W \equiv \gamma m_n (\alpha_o c)^2$ where $\gamma$ is a dimensionless coefficient, and $\alpha_o$ a dimensionless ratio of a velocity term with the speed of light, both of which are specified through further arguments. Now Eq. (7.3) can be rewritten as

$$\gamma^2 = \left[\frac{E_g}{2m_n(\alpha_o c)^2}\right]^2 + E_g \frac{\hbar^2 k^2}{2m_n^3(\alpha_o c)^4} \tag{7.40}$$

This will look just like the relativistic relation $\gamma^2 = 1 + \beta^2\gamma^2$ if $\alpha_o$ is defined so that the first term on the right-hand side is unity and $\beta$ is defined so that the second term is equivalent to $\beta^2\gamma^2$. Those requirements then set

$$\frac{E_g}{2} \equiv m_n(\alpha_o c)^2; \quad \hbar k \equiv \gamma m_n(\beta\alpha_o c) \tag{7.41}$$

and now the interpretations are clear: $\gamma m$ acts like a relativistic mass for a particle that has a velocity $\beta\alpha_o c$, $E_g/2$ acts like a rest energy, and $\alpha_o c$ acts like a limiting upper velocity. For the III-V compounds and some others, as in Table 7.2, $\alpha_o c$ is approximately constant, and so the electron dispersion relation has the form of a relativistic particle [141, 168, 169]. There is a buy-one-get-one-free flavor to the relations if one considers the ratio $E_g/R_y$, where $R_y$ is the familiar Rydberg energy of Eq. (2.6) expressed as $R_y = (1/2)mc^2$ with $\alpha$ being the fine structure constant: if $\alpha_o \equiv \alpha/2$, then [154]

$$\frac{E_g}{R_y} = \frac{m_n}{m} \tag{7.42}$$

**Table 7.2** Room temperature bandgap $E_g$(300 [K]) and effective mass of various semiconductors that follow the alpha-semiconductor model (see Figure 7.7) [144, 168, 337].

| Index | Semiconductor | $E_g$ | $E_g/R_y$ | $m_n/m$ | $m/m_n$ |
|---|---|---|---|---|---|
| 1 | InSb | 0.17 | 0.0125 | 0.0150 | 66.667 |
| 2 | InAs | 0.36 | 0.0265 | 0.0230 | 43.478 |
| 3 | GaSb | 0.72 | 0.0529 | 0.0420 | 23.810 |
| 4 | InP | 1.35 | 0.0992 | 0.0700 | 14.286 |
| 5 | GaAs | 1.42 | 0.1044 | 0.0710 | 14.085 |
| 6 | CdTe | 1.50 | 0.1102 | 0.1100 | 9.091 |
| 7 | AlSb | 1.58 | 0.1161 | 0.1200 | 8.333 |
| 8 | CdSe | 1.70 | 0.1249 | 0.1300 | 7.692 |

| Index | Semiconductor | $E_g$ | $E_g/R_y$ | $m_n/m$ | $m/m_n$ |
|---|---|---|---|---|---|
| 9 | GaP | 2.25 | 0.1654 | 0.1300 | 7.692 |
| 10 | ZnTe | 2.28 | 0.1676 | 0.1500 | 6.667 |
| 11 | AlP | 2.42 | 0.1779 | 0.2120 | 4.717 |
| 12 | CdS(W) | 2.49 | 0.1830 | 0.2000 | 5.000 |
| 13 | ZnSe | 2.58 | 0.1896 | 0.1700 | 5.882 |
| 14 | ZnO | 3.35 | 0.2462 | 0.2700 | 3.704 |
| 15 | ZnS(W) | 3.78 | 0.2778 | 0.2870 | 3.484 |

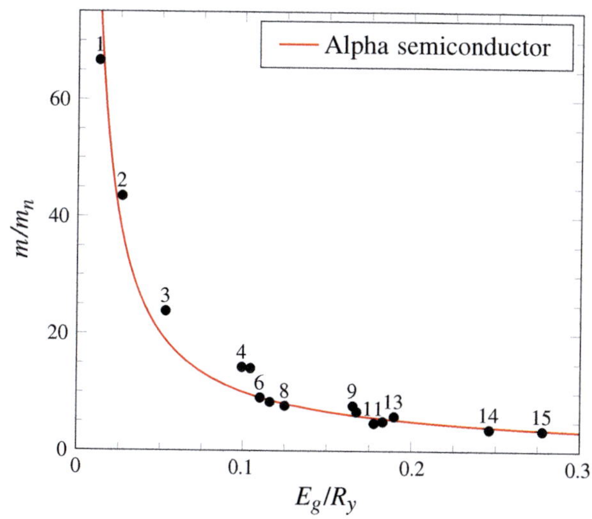


**Figure 7.7** The ratio of $E_g/R_y$ compared to the electron effective mass ratio $m_n/m_o$ and presented in a form that suggests the Gudermannian function associated with hyperbolic dispersion relations. The numbers near the data points are from Table 7.2 (some numbers suppressed for visual clarity). The red line is Eq. (7.42).

a form which is compared to room temperature bandgaps for a variety of semiconductors in Figure 7.7. The form shown is intended to bring to mind the hyperbolic dispersion relations behind the model, relations which involve the Gudermannian function that appears in a variety of applications, from relativity to Mercator projection maps of the Earth's surface (such maps render loxodromes as straight lines and so were essential to early navigation [7] in the time of Shakespeare and Thomas Harriot).

## 7.4 Scattering

How electrons transport through a material, and therefore how electrons scatter, is exceedingly important. A measure of that importance is the transition of a material from metal-like behavior to insulator-like behavior as $T \to 0$, that is, the metal-insulator transition (MIT), particularly in disordered systems such as Si:Au or Ge:Au. The latter was shown to go through the MIT as it was annealed, which was thought to be due to a clustering of the gold atoms and a reduction in the

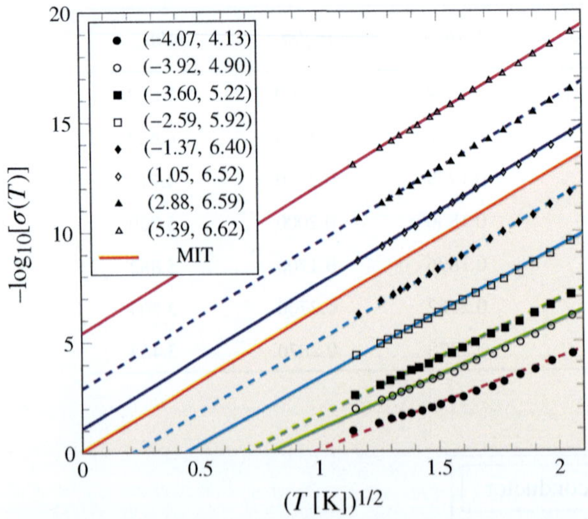


**Figure 7.8** The metal-insulator transition (MIT) for $Ge_{0.84}Au_{0.16}$ based on data digitally extracted from Figure 1 of Ref. [271] / American Physical Society, where the legends refer to the $(\sigma_0, \sigma_1)$, or slope-intercept, values of $\sigma(T) = \sigma_0 + \sigma_1 T^{1/2}$. All of the best fit slopes are found to be near the 6 [$1/\Omega$ cm $K^{1/2}$] value of the reference. The shaded area corresponds to the insulator region.

number of conduction electrons [271]. Unlike metals at higher temperatures, the conductivity of these materials near absolute zero behaves as $\sigma(T) \approx \sigma_0 + \sigma_1 T^p$, where $p$ is not necessarily an integer, due to more complex scattering interactions than those on display below, but the behavior nevertheless illustrates a point: the boundary between metal and insulator, or when MIT occurs, is when $\sigma_0 \to 0$, as shown in Figure 7.8.

More pedestrian materials like copper and silicon, however, serve the needs of the present discussion to examine scattering at higher temperatures. The Drude model, although successful in relating conductivity / to scattering, was marred a bit because of the use of the Maxwell–Boltzmann distribution rather than the Fermi–Dirac distribution to model the statistics of the electron gas. Conductivity is now reconsidered using the proper account. Electrons in bulk material are characterized by a distribution $f(\vec{x}, \vec{k})$. Using it, current density (taken to be in the $\hat{x}$–direction) is

$$J_x = \frac{2q}{(2\pi)^3} \int d\vec{k} \left(\frac{\hbar k_x}{m}\right) f(\vec{x}, \vec{k}) \tag{7.43}$$

Now assume that

**i.** the number density is stationary and current density is constant ($\partial_t f = \partial_x f = 0$), and
**ii.** the difference $f(\vec{x}, \vec{k}) - f_{FD}(E)$ (with $E = \hbar^2(k_x^2 + k_\perp^2)/2m$) is small.

Because the FD distribution is symmetric in each momentum component, the integral over $dk_x$ of $(\hbar k_x/m) f_{FD}(E_k)$ vanishes because the integral over all $k_x$ of odd and even functions vanishes. Thus, $f \to f - f_{FD}$ is permissible inside the integral. With these assumptions, the relaxation time approximation to Boltzmann's transport equation encountered in Eq. (4.16) is

$$\partial_t f + \frac{\hbar k_x}{m} \frac{\partial}{\partial x} f - \frac{F}{\hbar} \frac{\partial}{\partial k_x} f = -\frac{1}{\tau}(f - f_{FD}) \tag{7.44}$$

where the right-hand side indicates that the perturbed distribution relaxes back to Fermi–Dirac distribution over a timescale governed by $\tau$. Using it results in

$$J_x = \frac{2q}{(2\pi)^3} \int d\vec{k} \left(\frac{\hbar k_x}{m}\right) \tau(E) \left(\frac{F}{\hbar}\right) \frac{\partial}{\partial k_x} f_{FD}(E) \tag{7.45}$$

where $\tau(E)$ is the energy-dependent relaxation time, $F = q\mathcal{E}$ and $\mathcal{E}$ is the electric field in the $\hat{x}$–direction, and the swapping out of $f$ with $f_{FD}$ is permissible to leading order because their difference is presumed small. Observe that Eq. (7.1) entails $J = \sigma F/q$, and so eliminate $F$ from both sides, multiply through by $q$, and treat $\sigma$ instead. Now comes much trickery. Evaluate

$$\partial_x f_{FD}(E) = f_{FD}(1 - f_{FD})\beta \frac{\hbar^2 k_x}{m} \tag{7.46}$$

All the factors of $\hbar$, $\pi$, and $m$ can get conveniently swept up in going from a $k$-integration to an $E$-integration, where $E = \hbar^2 k^2/2m$ and noting from Eq. (7.14) that

$$\rho = \frac{2}{(2\pi)^2} \int d\vec{k} f_{FD}[E(k)] = \int D(E) f_{FD}(E) dE \tag{7.47}$$

and also, that

$$\int d\vec{k}\, k_x^2 f_{FD}[E(k)] = \frac{1}{3} \int d\vec{k}\, k^2 f_{FD}[E(k)] \tag{7.48}$$

because $k^2 = k_x^2 + k_y^2 + k_z^2$ and the same equation would be true were $k_x^2$ replaced with one of the other components. The final trick is to introduce

$$\delta_l\left[\beta(E-\mu)\right] \equiv \delta_l[s(E)] = f_{FD}(E)\left(1 - f_{FD}(E)\right) \tag{7.49}$$

where $s(E) = \beta(E - \mu)$. The notation intentionally suggests the Dirac delta function relation because $\int_{-\infty}^{\infty} \delta_l(s) ds = 1$ (i.e. it is a delta function sequence). Combining all of these subtle tricks together finally results in

$$\sigma = \frac{2q^2}{3m} \beta \int_0^{\infty} E D(E) \tau(E) \delta_l[s(E)] dE \tag{7.50}$$

The importance of this equation is because generally $\beta$ is large: even at $T = 2000$ K, $\beta = 5.8\ \text{eV}^{-1}$. As a result, over the energy ranges of the rest of the integral for typical metals, $\delta_l[s(E)]$ really does mimic a Dirac delta function by being very sharply peaked. Thus

$$\begin{aligned} \beta \int_0^{\infty} E D(E) \tau(E) \delta_l[s(E)] dE &\approx \mu D(\mu) \tau(\mu) \int_0^{\infty} \delta_l(s)\, ds \\ &\approx \mu D(\mu) \tau(\mu) \end{aligned} \tag{7.51}$$

where in the last step, the integration can be extended to $-\infty$ without appreciable error for general conditions. Inserting into Eq. (7.50), and letting $\rho_o \equiv k_F^3/3\pi^2$ from the zero-temperature relation of Eq. (2.15), finally gives (compare Eq. (7.6), but let $\rho_o$ replace density so as not to be confused with resistivity)

$$\sigma = \frac{q^2}{m} \rho_o\, \tau(\mu) \tag{7.52}$$

confirming Drude's intuition. The temperature-dependence of $\sigma(T)$ therefore resides in the relaxation time $\tau$: for metals, all the interesting effects happen near $\mu$, so carrying around the $\mu$ argument is tedious and so simply assumed: $\tau$ is the relaxation time at the Fermi level. Now the variation of that relaxation time as a function of temperature takes center stage. The relaxation time $\tau$ generally has electron–phonon ($\tau_{ep}$), electron–electron ($\tau_{ee}$), and ionized impurity ($\tau_{ii}$) components that are combined via Matthiessen's Rule

$$\frac{1}{\tau} = \frac{1}{\tau_{ep}} + \frac{1}{\tau_{ee}} + \frac{1}{\tau_{ii}} \tag{7.53}$$

which makes intuitive sense: resistivity (the inverse of conductivity) tend to add, and so reciprocal relaxation times will add as well. At sufficiently high temperatures, the first two dominate and are represented by [149, 161]

$$\frac{1}{\tau} \approx A_{ee}T^2 + B_{ep}T \tag{7.54}$$

At room temperature, *ep* scattering dominates.

Evaluating the relaxation time for electron scattering with acoustic phonons (*ep*) is cumbersome [149, 359, 360, 382]. Acoustic phonons are vibrations due to sound waves, and so the phonons follow a Bose–Einstein distribution which introduces factors of $1/(e^{\beta\hbar\omega} - 1)$ into an evaluation of their scattering rates. Quantum mechanical accounts of electrons scattering off of lattice vibrations generally come to a form of $\tau_{ep}$ resembling [298]

$$\tau_{ep}(E) = \frac{\pi\hbar^4 c_L}{\sqrt{2}\,\Xi^2 m^{3/2} k_B T \sqrt{E}} \tag{7.55}$$

where the various factors are defined in Table 7.3. Note that Ridley [298] sets $\rho_i$ as the mass density, hence his formula hides the atomic mass term $M$: to keep notation consistent, $\rho_i$ is treated as a number density here. A few simplifying approximations help. First, for a simple metal such that each atom contributes one electron to the sea of electrons making up the conduction current, then $\rho_i \approx \rho$, but generally that is not true, although it is convenient for copper. Second, the combination of terms

$$\lambda_o \equiv \frac{3\Xi^2}{4\mu M v_s^2} \tag{7.56}$$

is dimensionless. The simple relationship of Bohm and Staver [27, 149] suggested the approximation $\Xi = 2\mu/3$. When evaluated at the Fermi level, this puts the equation for $\tau_{ep}(\mu)$ into the more pleasing form [149, 272]

$$\tau_{ep}(\mu) = \frac{\hbar}{2\pi\lambda_o k_B T} \tag{7.57}$$

It is seen that in this simple form, the conductivity of a metal $\sigma \approx q^2\tau_{ep}\rho/m$ can suggest the value of $\lambda_o$ if $\tau_{ep}$ dominates the other scattering mechanisms for $E = \mu$ at room temperature.

Lest one erroneously conclude that the acoustic relaxation time for copper of Table 7.3 is indicative of materials in general, consider the contrary case of a good semiconductor like GaAs [47], for which $\Xi = 7.01$ eV, $M\rho_i = 5360$ kg/m$^3$, and $v_s = 5240$ m/s in [MKSA]. Using these values, then for GaAs, $\tau_{ep} = 6290$ fs, or about 6.3 ps, for an electron with an energy of $E = 50$ meV. The longer scattering times in semiconductors compared to metals are reflected in their higher quantum efficiency.

**Table 7.3** Parameters for the modeling of copper (Cu) scattering terms at the Fermi level $\mu$ at Room Temperature. Terms are expressed in emission [efnq] units: in conventional units, $v_s = 4760$ [m/s], $M = 63.546$ [g/mole], and $\sigma(T) = 0.5804$ [1/$\mu\Omega$ cm] at room temperature. Fundamental constants are as in Table 2.2. Ion number density $\rho_i \approx \rho$ for simplicity in evaluating $c_L$ (compare to exact value for copper of $\rho_i|_{Cu} = 84.7228$ nm$^{-3}$ based on a copper density of 8.94 g/cm$^3$).

| Term | Symbol | Equation | efnq |
|---|---|---|---|
| Temperature | $T$ | - | 300 K |
| Fermi Level | $\mu$ | - | 7 eV |
| Fermi wave number | $k_F$ | $\sqrt{2m\mu}/\hbar$ | 13.5546 nm$^{-1}$ |
| Sound Velocity | $v_s$ | - | 0.00476 nm/fs |
| Effective mass | $r_e$ | $m/m_o$ | 1 |
| Atomic Mass | $M$ | - | 658608 eV (fs/nm)$^2$ |
| Number density | $\rho$ | Eq. (4.36) | 84.1088 nm$^{-3}$ |
| Elastic Constant | $c_L$ | $M\rho_i v_s^2$ | 1255.11 eV/nm$^3$ |
| Deformation Potential | $\Xi$ | - | 4.8 eV |
| Conductivity | $\sigma$ | Eq. (7.6) | 362.247 q$^2$ / eV fs nm |
| Screening wave number | $q_o$ | Eq. (7.62) | 11.9079 nm$^{-1}$ |
| *ee* term | $A_o$ | Eq. (7.64) | 0.850921 |
| *ep* term | $\lambda_o$ | Eq. (7.56) | 0.165426 |
| Electron–Phonon | $B_{ep}$ | $2\pi\lambda_o k_B/\hbar$ | $1.36079 \times 10^{-4}$ 1/K fs |
| Electron–Electron | $A_{ee}$ | $k_B^2 A_o/\hbar\mu$ | $1.37142 \times 10^{-9}$ 1/K$^2$ fs |
| Relaxation time | $\tau$ | Eq. (7.54) | 24.4217 fs |

At low temperatures, however, departures occur from $\tau \propto 1/T$ as can be expected from Figure 7.2. In particular, for electrons at the Fermi level,

$$\frac{1}{\tau_{ep}(\mu)} = \frac{8\pi\lambda_o k_B T_D}{\hbar}\left(\frac{T}{T_D}\right)^5 W_-\left(5, \frac{T_D}{T}\right) \tag{7.58}$$

where $T_D$ is the Debye temperature as before. The Bloch–Grüneisen function $W_-(5, x)$ is a specific case of the more general function [149, 161]

$$W_\pm(n, x) = \int_0^x \frac{s^n}{(e^s \pm 1)(1 \pm e^{-s})} ds \tag{7.59}$$

that appears when evaluating parameters for particles following Fermi–Dirac (+) or Bose–Einstein (−) statistics, phonons being examples of the later. Observe that $W_-(5, \infty) = \Gamma(6)\zeta(5) = 124.431$, where $\Gamma(n+1)$ is the gamma function and $\zeta(n)$ is the Riemann $\zeta$-function: they are 120 and 1.03693, respectively, for $n = 5$. The evaluation of $W_-(5, x)$ generally requires numerical methods for evaluation when its argument $x$ is of order unity, but the limiting cases are useful. In the small $x$ limit, the exponentials can be Taylor expanded to find

$$W_-(5, x \ll 1) \approx \int_0^x \frac{12s^3}{12 + s^2} ds = 6x^2 - 72 \ln\left(1 + \frac{x^2}{12}\right) \tag{7.60}$$

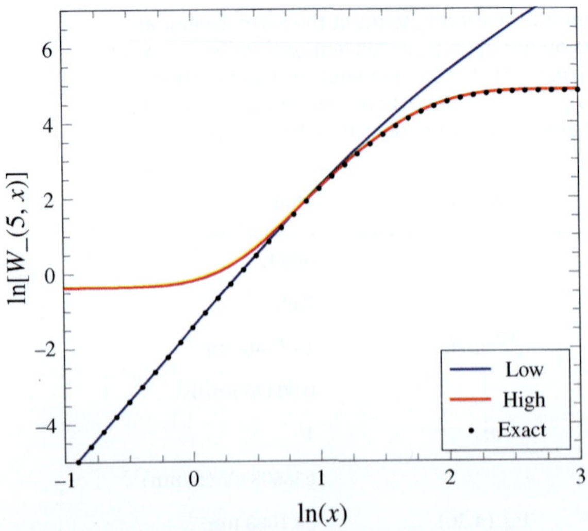


**Figure 7.9** Behavior of $W_{-}(5, x)$: Numerical evaluation of Eq. (7.59) (black solid dots) compared to its approximation by Eqs. (7.60) and (7.61). A good approximation to the former is to use the smaller of the later two approximations.

Conversely, for large argument, reconfigure the integral as $\int_0^x \to \int_0^\infty - \int_x^\infty$ and approximate $1/(e^s - 1)(1 - e^{-s}) \approx e^{-s} + 2e^{-2s}$ to find

$$W_{-}(5, x \gg 1) \approx \Gamma(6)\zeta(5) - w(x)e^{-x} - \frac{w(2x)}{2^5}e^{-2x}$$
$$w(x) = \sum_{j=0}^{5} \frac{\Gamma(6)}{\Gamma(j+1)} x^j \tag{7.61}$$

A not-entirely-crude approximation to $W_{-}(5, x)$ is to take it as the smaller of these two limiting cases, as shown in Figure 7.9 (see Algorithm B.2.11). When $T_D/T$ small, then $W_{-}(5, T_D/T) \approx (1/4)(T_D/T)^4$, and so Eq. (7.57) is recovered. At temperatures lower than $T_D$, though, $1/\tau_{ep} \propto T^5$.

Next, consider electron–electron scattering. For an electron above the Fermi energy $\mu$, then [149, 239, 241]

$$\tau_{ee}(E) \to \tau_o \frac{4\pi}{r_e} \frac{(K_s \beta E)^2}{[\beta(E - \mu)]^2 + \pi^2} \frac{1}{\gamma(2k/q_o)} \tag{7.62}$$

where $\tau_o = \hbar a_o / 2Q = 0.048378$ fs is a characteristic time scale, $\beta = 1/k_B T$, $q_o = \sqrt{4 r_e k_F / \pi K_s a_o}$, $r_e = m_n/m$ is the effective mass ratio with $m_0$ being the rest mass, and the other symbols are as found in Table 2.2. Lastly, the function $\gamma(x)$ is defined as

$$\gamma(x) = \frac{x^3}{4}\left[\tan^{-1}(x) + \frac{x}{1 + x^2} - \frac{\tan^{-1}(x\sqrt{x^2 + 2})}{\sqrt{x^2 + 2}}\right] \tag{7.63}$$

When compared to Eq. (7.54) for $A_{ee}$, then at the Fermi level [149, 272]

$$A_{ee} = \frac{k_B^2 A_o}{\hbar \mu}; \quad A_o = \frac{\pi r_e R_y}{4 \mu K_s^2} \gamma\left(\frac{2k_F}{q_o}\right) \tag{7.64}$$

and so, at room temperature, the ratio of $\tau_{ee}/\tau_{ep} > 300$: electron–phonon scattering dominates at the Fermi level.

Finally, consider ionized impurity scattering, which contributes at low temperatures [149, 243, 298]. Introduce $E_o = (2Qk_F/\pi r_e)$ (e.g. for $\mu = 1$ eV and $r_e = 1$, then $E_o = 3.10641$ eV) in terms of which

$$\frac{\tau_{ii}(E)}{\tau_0} = \frac{K_s^2}{4\pi r_e Z^2 N_i a_0^3}\left(\frac{E_o}{E}\right)^{1/2} H\left(\frac{E}{E_o}\right) \tag{7.65}$$

$$H(s) \equiv \frac{s^2(1+s)}{(1+s)\ln(1+s) - s} \tag{7.66}$$

where $N_i$ is the number density of the ionized impurities, $r_e = m/m_o$ is the effective mass ratio as before, and $Z$ is the charge on the impurity in units of $q$ (often unity). Even for $N_i \approx 10^{19}$ cm$^{-3}$, $N_i a_0^3 \approx 1.5 \times 10^{-6}$, and so the temperature has to be very low before the effects of $\tau_{ii}$ are seen to compare with $\tau_{ep}$ and $\tau_{ee}$ for copper-like metals.

## 7.5 Optical Properties

Light interacting with matter causes the flow of electrons: in metals, the flow readily happens, but in semiconductors and insulators, the motion is impeded by restoring forces, giving rise to dielectric behavior [78, 245]. The properties of most interest to photoemitters revolve around the reflectivity $R(\omega)$ and the laser penetration depth $\delta(\omega)$. Maxwell's equations relate the electric $\vec{\mathcal{E}}$, magnetic $\vec{H}$ fields, and current $\vec{J}$ with regards to their spatial and temporal behavior, for which

$$\vec{\nabla} \times \vec{\mathcal{E}} + \mu_o \partial_t \vec{H} = 0 \tag{7.67}$$

$$\vec{\nabla} \times \vec{H} - \varepsilon_o \partial_t \vec{\mathcal{E}} = \vec{J} \tag{7.68}$$

are the most important here, and are expressed in their conventional form, where $\varepsilon_0 = q^2/16\pi Q = 0.0552635$ [q$^2$/eV nm] is the permittivity and $\mu_0 = 1/\varepsilon_0 c^2 = 0.000201335$ [eV fs$^2$ / nm q$^2$] the permeability of free space. A wave equation is obtained by evaluating $\vec{\nabla} \times (\vec{\nabla} \times \vec{\mathcal{E}}) = \vec{\nabla}(\vec{\nabla} \cdot \vec{\mathcal{E}}) - \nabla^2 \vec{\mathcal{E}}$. If no free charge is present, then $\vec{\nabla} \cdot \vec{\mathcal{E}} = 0$ from Poisson's equation. Using Eq. (7.67) for $\vec{\nabla} \times \vec{H}$ results in

$$(\nabla^2 - \frac{1}{c^2}\partial_t^2 \vec{\mathcal{E}}) = \frac{16\pi Q}{q^2 c^2} \sigma \partial_t \vec{\mathcal{E}} \tag{7.69}$$

where the conductivity $\sigma$ is familiar. Let

$$\vec{\mathcal{E}} = \vec{\mathcal{E}}_o \exp(i\vec{\kappa} \cdot \vec{r} - i\omega t) \tag{7.70}$$

where $\vec{\kappa}$ is the propagation constant (called $\kappa$ here so as not to confuse matters with $k$ when it is taken to be a damping constant, another necessary accommodation to unfortunate overlaps in notation) and $\omega$ is the frequency, then

$$-\kappa^2 + \left(\frac{\omega}{c}\right)^2 = -i\frac{16\pi Q\sigma\omega}{q^2c^2} \to \kappa = \left(1 + i\frac{16\pi Q\sigma}{\omega q^2}\right)^{1/2}\left(\frac{\omega}{c}\right) \tag{7.71}$$

The coefficient of $\omega/c$ is the index of refraction

$$\hat{n} = \left(1 + i\frac{16\pi Q\sigma}{\omega q^2}\right)^{1/2} \equiv n - ik \tag{7.72}$$

where $n$ is the usual index of refraction, and $k$ is called the damping constant. The odd collection of parameters attending the conductivity term can be collected to define the plasma frequency $\omega_p$

$$\hbar\omega_p \equiv \hbar\sqrt{\frac{16\pi Q\rho_o}{m}} = \left[\frac{32\mu k_F Q}{3\pi}\right]^{1/2} \tag{7.73}$$

In the Drude model, finding $n$ and $k$ is then made possible by expanding $(n+ik)^2 = 1+\hat{\chi}$ and then equating the real and imaginary parts of the left- and right-hand sides. Drude theory predicts

$$\begin{aligned} n^2 - k^2 &= 1 - \frac{\omega_p^2\tau^2}{1+\omega^2\tau^2} = K_r \\ 2nk &= \frac{\omega_p^2\tau}{\omega(1+\omega^2\tau^2)} = K_i \end{aligned} \tag{7.74}$$

where the permittivity $\hat{\varepsilon} = \varepsilon_r + i\varepsilon_i = (K_r + iK_i)\varepsilon_0$ and $\varepsilon_0$ is the permittivity of free space familiar from Eq. (2.3). With a bit of effort, $n$ and $k$ can be expressed in terms of $K_r$ and $K_i$, or

$$2n^2 = \sqrt{K_r^2 + K_i^2} + K_r \tag{7.75}$$

$$2k^2 = \sqrt{K_r^2 + K_i^2} - K_r \tag{7.76}$$

from which $n(\omega)$ and $k(\omega)$ are readily isolated, then $R(\omega)$ and $\delta(\omega)$ are evaluated in turn, even when the contributions to $n$ and $k$ from Lorentzian terms are included below. For wavelengths up to infrared, these equations give a nice account of optical properties; into the visible, that cozy agreement degrades [141–143], and the radiation field must be handled through second quantization methods.

Confine now the analysis to one dimension, or $\vec{\mathcal{E}} = \mathcal{E}\hat{x}$ so that $\vec{\kappa}\cdot\vec{r} = \kappa x$, and resume the notation $F = q\mathcal{E}$. $F$ as a force accelerates an electron over a distance $l$ whereupon it acquires or loses an energy $dE = Fl = Fvdt$. The power density being exchanged with the electromagnetic wave is then $\rho Fv = J\mathcal{E} = \sigma\mathcal{E}^2$ after recalling that $J = qv\rho$). The intensity of light is proportional to $|\mathcal{E}|^2$, and it declines as light passes through a material specifies the decay parameters as a function of wavelength. The decay parameter for a function $y(x) = y_o e^{-x/a}$ is obtained from $-[\partial_x \ln(y(x))]^{-1}$. Thus, the decay parameter for light into a material is

$$\delta(\omega) = \left\{-\partial_x \ln\left[|\mathcal{E}(x)|^2\right]\right\}^{-1} = \frac{c}{2k\omega} \tag{7.77}$$

The absorption coefficient $T(\omega)$ is measured by finding the transmission through varying thicknesses of material, in which case $T(\omega)$ changes as [143]

$$T(\omega) = (1-R)^2 e^{-d/\delta}/(1 - R^2 e^{-2d/\delta}) \tag{7.78}$$

where $R$ is the reflectivity and $d$ is the thickness of material. When $d \gg \delta$, then the exponential decay goes as $e^{-x/\delta}$ into the material.

If the electric field is written as $\vec{\mathcal{E}} = \hat{x}\mathcal{E}_o \exp(i\kappa x - i\omega t)$, then the magnetic field is $\vec{H} = \hat{y}H_o \exp(i\kappa x - i\omega t)$ from Eq. (7.67), where $H_o = (c\kappa/\omega)\mathcal{E}_o$. Consider when a wave crosses a boundary at the origin. Some of the wave is reflected, some is transmitted. Introduce the subscripts $i$, $r$ and $t$ to denote incident, reflected, and transmitted waves, respectively. The matching of the waves and their first derivatives at $x = 0$ results in a matrix equation much like those encountered in

Section 6.6, and for good reason given the similarities between quantum mechanics and electromagnetics [299]. It is (where hats on $\hat{\kappa}$ denote complex quantity)

$$\begin{pmatrix} 1 & -1 \\ \kappa_i & \kappa_i \end{pmatrix} \begin{pmatrix} \mathcal{E}_o^i \\ \mathcal{E}_o^r \end{pmatrix} = \begin{pmatrix} 1 & -1 \\ \hat{\kappa}_t & \hat{\kappa}_t \end{pmatrix} \begin{pmatrix} \mathcal{E}_o^t \\ 0 \end{pmatrix} \tag{7.79}$$

Solving for the reflected component,

$$E_o^r = \frac{\hat{\kappa}_t - \kappa_i}{\hat{\kappa}_t + \kappa_i} E_o^i \tag{7.80}$$

Reflectivity $R$ is the ratio of the absolute square of the magnitudes of the reflected and incident wave coefficients, giving *Beer's equation* [134, 143]

$$R(\omega) = \frac{|\hat{\kappa}_t - \kappa_i|^2}{|\hat{\kappa}_t + \kappa_i|^2} = \frac{|\hat{n} - 1|^2}{|\hat{n} + 1|^2} = \frac{(n-1)^2 + k^2}{(n+1)^2 + k^2} \tag{7.81}$$

Both Drude and Lorentz provided models that described metals and insulators, respectively. The Lorentz–Drude model [281, 364] provides a way of understanding low and high frequency behavior. For the latter, unlike electrons in a metal, the electrons in an insulator are bound to atoms. The electrons being tugged by the electric field cause the atoms to become polarized, whereby the positive and negative centers of charge are separated, creating a dipole. The dipole moment $\vec{p}$ is proportional to the electric field $\vec{\mathcal{E}}$: numerous dipoles acting together produce a macroscopic polarization $\vec{P}$ defined by

$$\vec{P} = \varepsilon_0 \chi \vec{\mathcal{E}} \tag{7.82}$$

When dielectrics are present, Maxwell's equation $\vec{\nabla} \cdot \vec{\mathcal{E}} = q^2\rho/\varepsilon_0$ has to be modified to account for them by

$$\vec{\nabla} \cdot (\varepsilon_0 \vec{\mathcal{E}} + \vec{P}) = q^2 \rho \Rightarrow \vec{\nabla} \cdot \vec{D} = q^2 \rho \tag{7.83}$$

where $\rho$ is the density of free (not bound) charges and

$$\vec{D} = \varepsilon_0 \vec{\mathcal{E}} + \vec{P} = \varepsilon_0 (1 + \chi) \vec{\mathcal{E}} \equiv \varepsilon \vec{\mathcal{E}} \tag{7.84}$$

for static conditions. When conditions are not static, then $\vec{P}$ lags behind $\vec{\mathcal{E}}$, and $\chi$ takes over the role of a response function such that

$$\vec{P}(t) = \varepsilon_0 \int_{-\infty}^{t} \chi(t - t') \vec{\mathcal{E}}(t') dt' \tag{7.85}$$

Taking Fourier transforms of both sides introduces the frequency dependence $\omega$ and by the convolution theorem results in

$$\vec{P}(\omega) = \varepsilon_0 \hat{\chi}(\omega) \vec{\mathcal{E}}(\omega) \tag{7.86}$$

The connection between $\hat{n}$ and the susceptibility $\hat{\chi}(\omega)$ is then finally given by

$$\hat{n}(\omega)^2 = 1 + \hat{\chi}(\omega) \tag{7.87}$$

where the little hats over the symbols are to reinforce that they are complex quantities. Finding $\hat{n}(\omega)$ has therefore become the problem of finding $\hat{\chi}(\omega)$, for which two pedagogical approaches are available:

**i.** treat the bound electron as a damped harmonic oscillator, or
**ii.** use a distribution function approach (the quasi-classical Boltzmann Transport Equation or BTE).

In the second case, "quasi-classical" is important: in semiconductors like GaAs, the presence of a bandgap, the reduced electron density, and relaxation times that are not constant as for low frequencies conspire to cause the absorption coefficient to change from a $\lambda^2$ dependence predicted by the Drude model to a $\lambda^3$ or $\lambda^4$ dependence derivable from a full quantum mechanical theory [141–143]. For present purposes, though, showing why the reflectivity and penetration depth are frequency-dependent can be accomplished by finding their relationship to $\hat{\chi}(\omega)$.

A sinusoidal electric field causes an electron to oscillate harmonically (a driving force), but if bound to an atom it also experiences a damping term. Let $\vec{r}(t)$ be the time-dependent distance the electron is from its equilibrium position. The force equation for it is then [135, 154]

$$-\vec{F} = m\partial_t^2\vec{r}(t) + \frac{m}{\tau}\partial_t\vec{r}(t) + m\omega_o^2\vec{r}(t) \tag{7.88}$$

and is recognized as a damped harmonic oscillator and the three terms on the right are the acceleration, the dampening term, and the restoring force, respectively. The $\tau$ term is familiar from the Drude model (dissipation due to scattering events). Performing a Fourier transform on it to consider the frequency dependence gives

$$\vec{r}(\omega) = \frac{\vec{F}}{m}\left(\omega^2 - \omega_o^2 - i\frac{\omega}{\tau}\right)^{-1} \tag{7.89}$$

For metals, there is no restoring force because the electrons are not bound and so the $\omega_o \to 0$ limit recovers the Drude model. This equation is for one bound electron, but what is needed is a description of many of them. If the number density of electrons is $\rho_o$, then many dipoles agitating in tandem results in

$$\hat{\chi}(\omega) = \frac{16\pi Q\rho_o}{m}\left(\omega^2 - \omega_o^2 - i\frac{\omega}{\tau}\right)^{-1} \equiv \frac{\omega_p^2}{(\omega^2 - \omega_o^2 - i\omega\tau)} \tag{7.90}$$

where $\omega_p$ is the plasma frequency. Consider copper as a representative example. For it, $\mu = 7$ eV entails that $k_F = \sqrt{2m\mu}/\hbar = 13.5546\ \text{nm}^{-1}$ and so $\rho_o = k_F^3/3\pi^2 = 84.1088\ \text{nm}^{-3}$. Therefore $\omega_p = 16.3611\ \text{fs}^{-1}$ or $\lambda_p = 115.130$ nm (ultraviolet light). In the small frequency limit, the static dielectric constant $\varepsilon(0)$ is given by

$$\varepsilon(0) = \varepsilon_0\left(1 + \chi(0)\right) = \varepsilon_0\left(1 + \frac{\omega_p^2}{\omega_o^2}\right) \equiv K_0\varepsilon_0 \tag{7.91}$$

For example, the static dielectric constant of silicon is $K(0) = K_s = 11.9$.

Seemingly, $K_r$ and $K_i$ appear unrelated, but in fact they are connected by the Kramer–Kronig relations [203]

$$K_r(\omega) = \frac{1}{\pi}\mathbb{P}\int_{-\infty}^{\infty}\frac{K_i(\omega)}{\omega' - \omega}\,d\omega' \tag{7.92}$$

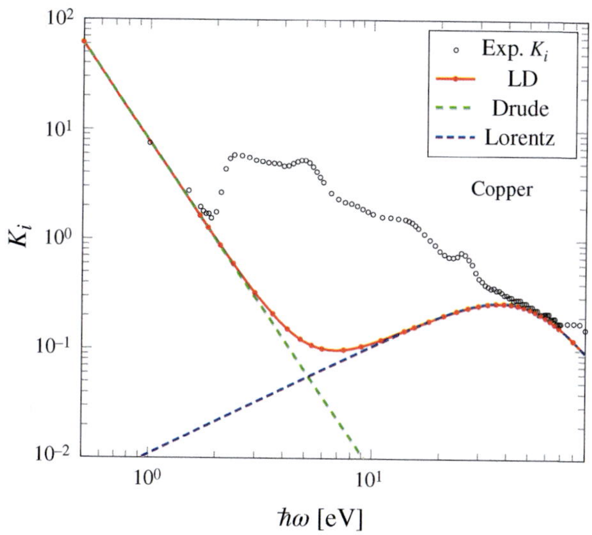


**Figure 7.10** Measured (black ∘) Dielectric term $K_i = \Im(\hat{K})$ compared to the Drude model (green dashed), Lorentz model (blue densely dashed) and their sum LD (red • and line). The Drude, Lorentz, and plasma frequency parameters are given in Table 7.4.

where $\mathbb{P}(\cdots)$ is the principal part of the integral. The same equation holds with $K_r \leftrightarrow -K_i$. Thus, in comparing to measured data, examining $K_i(\omega)$ is sufficient to determine both. A reasonable hunch that $K_i(\omega)$ can be found for all $\omega$ by finding its metal-like (Drude or small $\omega$) and insulator-like (Lorentz or large $\omega$) behaviors, and then extrapolating to middle $\omega$ frequencies is only partially adequate. Copper is a good test case, given that it is not of a silver color and therefore reflects wavelengths non-uniformly in the optical regime, portending difficulties there. The sum of just the Drude and Lorentz components is seen to miss important behavior in the visible region as shown in Figure 7.10, so more must be done. Before embarking on that, though, consider another approach to the Lorentz–Drude model based on a distribution function approach.

The Drude model came about through classical notions, and the Lorentz model from a harmonic oscillator formalism, and neither strays too far from familiar ideas. Electrons in a material, however, interact incessantly amongst themselves and with the vibrating atoms. In preparing for the quantum approach, then, putting the statistical approach on center stage is useful. Let the distribution function $f$ describing the electrons be written somewhat differently, that is, let it be the sum a distribution $f_0$ (Fermi–Dirac distribution) existing when fields are absent with another distribution $f_1$ that accounts for when the fields are turned on, so that $f = f_0 + \lambda f_1$. Here, $\lambda$ functions as a strength parameter: when it is 0, the field is off, and when it is 1, the field has been completely turned on. The trick is now to consider the time evolution of the distribution function in Eq. (7.44) in powers of $\lambda$ using a perturbation approach. The lowest order ($\lambda^0$) is

$$\partial_t f_0 + \frac{\hbar k}{m}\partial_x f_0 = 0 \tag{7.93}$$

The next order ($\lambda^1$) terms are

$$\partial_t f_1 + \frac{\hbar k}{m}\partial_x f_1 - \frac{F}{\hbar}\partial_k f_0 = -\frac{1}{\tau} f_1 \tag{7.94}$$

The Fermi–Dirac distribution depends on $E$, and so it is reasonable to expect that the difference between $f$ and $f_0$ can be expressed in a Taylor-like expansion

$$f - f_1 \approx \varphi(x, k, t)\,(-\partial_E f_0) \tag{7.95}$$

where $\varphi$ accounts for the variations with $x$ and $k$ because $F = q\mathcal{E}$ has spatial and temporal variation from $\mathcal{E}$ in Eq. (7.70), leading to the expectation that

$$\varphi(x, k, t) = \varphi_o(k) \exp(i\kappa x - i\omega t) \tag{7.96}$$

Let $\partial_k f = \partial_k E\, \partial_E f$. Eq. (7.44) becomes

$$\left[\left(-i\omega + i\kappa\frac{\hbar k}{m}\right)\varphi_o(k) - \frac{F_o}{\hbar}\partial_k E\right](-\partial_E f_o) = -\frac{1}{\tau}\varphi_o(k)\,(-\partial_E f_o) \tag{7.97}$$

From the usual parabolic relation $\partial_k E = \hbar^2 k/m$ and the light relation $\kappa = \hat{n}(\omega/c)$, followed by equating the coefficients of $(-\partial_E f_o)$ it follows for $\hbar k/mc \ll 1$ that

$$\varphi_o(k) = \frac{\tau F_o}{1 - i\omega\tau}\left(\frac{\hbar k}{m}\right) \tag{7.98}$$

The equation for DC or steady state conductivity $\sigma$ from Eq. (7.50) is then modified when the electric field varies using the approximation for $f_1$ in the low temperature limit. For convenience, it becomes

$$\begin{aligned}\sigma(\omega) = \frac{q}{F}J &= \frac{q^2}{2\pi F_o}\int_{-\infty}^{\infty}\frac{\tau F_o}{1 - i\omega\tau}\left(\frac{\hbar\, k}{m}\right)^2\left(\frac{m}{\pi\hbar^2}\right)\Theta(k_F^2 - k^2)dk \\ &= \frac{\tau q^2 k_F^3}{3\pi^2 m}(1 - i\omega\tau)^{-1} = \sigma(0)\frac{1 + i\omega\tau}{1 + \omega^2\tau^2}\end{aligned} \tag{7.99}$$

which recovers the Drude model for conductivity encountered in Eq. (7.74)

$$\hat{K}(\omega) = 1 - \frac{\omega_p^2}{\omega\,(\omega - (i/\tau))} = 1 - \frac{\omega_p^2}{\omega^2 - i\omega\varepsilon_0\frac{\omega_p^2}{\sigma(0)}} \tag{7.100}$$

where the relation

$$\tau = \frac{16\pi Q}{q^2\omega_p^2}\,\sigma(0) \tag{7.101}$$

can be used to express $\hat{K}(\omega)$ back in terms of the conductivity.

Historically, the small $\omega$ limit of $R(\omega)$ has been known as the Hagen–Rubens relation. When $\omega\tau$ is small, then so is $K_r$ compared to $K_i$, which becomes large as in Figure 7.10. The impact on reflectivity from Eq. (7.81) is then found from the leading order approximations to $n$ and $k$, and is

$$R(\omega \to 0) \approx 1 - \frac{2}{\omega_p}\sqrt{\frac{2\omega}{\tau}} \tag{7.102}$$

where Eq. (7.101) can express it in its more conventional form in terms of $\sigma(0)$. The reflectivity of aluminum illustrates the behavior nicely [154].

Returning now to Figure 7.10, it is time to consider the behavior of $K_i(\omega)$ in the visible to UV regime, and that requires a quantum mechanical approach. Importantly, remember that if a single Lorentzian like Eq. (7.90) satisfies the Kramer–Kronig relation of Eq. (7.92), then so does a sum of such terms. That will be seen to allow the peaks to be modeled by a sum of Lorentzians. A quantum approach [134, 135, 154, 281, 364] gives a justification for that. Let the initial state of an electron bound to an atom be designated as $|\phi_o\rangle$. When perturbed by a force $\vec{F} = F(t)\hat{x}$, then other eigenstates $|\phi_j\rangle$ become involved, leading to

$$|\psi(t)\rangle = c_o|\phi_o\rangle e^{-i\omega_o t} + \sum_{j\neq o} c_j|\phi_j\rangle e^{-i\omega_j t} \tag{7.103}$$

where $c_o$ is close to unity and the eigenstates $\langle\phi_i|\phi_j\rangle = \delta_{ij}$ are for when $F$ is absent. When $|\psi(t)\rangle$ is put into Schrödinger's equation, then

$$i\hbar\, \partial_t c_j(t) = \langle\phi_j|\, F(t)\hat{x}\, |\phi_o\rangle\, e^{i(\omega_j-\omega_o)t} \tag{7.104}$$

where the $|\phi_j\rangle$ are solutions to the Hamiltonian without the $F(t)\hat{x}$ contribution. Introducing the notation $\omega_{jo} \equiv \omega_j - \omega_o$, then

$$\begin{aligned} c_j(t) &= \frac{F_o}{i\hbar}\left\langle\phi_j\, |\hat{x}|\, \phi_o\right\rangle \int_0^t \cos\omega t\, e^{i\omega_{jo}t} dt \\ &= \frac{F_o}{2\hbar}\left\langle\phi_j\, |\hat{x}|\, \phi_o\right\rangle \left\{\frac{1-e^{i(\omega_{jo}+\omega)t}}{\omega_{jo}+\omega} + \frac{1-e^{i(\omega_{jo}-\omega)t}}{\omega_{jo}-\omega}\right\} \end{aligned} \tag{7.105}$$

Now use the fact that $c_o$ is close to unity and taken as constant, and all the $c_j(t)$ for $j \neq o$ are so small that terms of order $c_j^2$ can be neglected. Then

$$\langle\psi(t)|q\hat{x}|\psi(t)\rangle \approx \frac{2qF_o}{\hbar}\sum_j |x_{jo}|^2 \left(\frac{1}{\omega_{jo}-\omega} + \frac{1}{\omega_{jo}+\omega}\right)\cos(\omega t) \tag{7.106}$$

$$|x_{jo}|^2 \equiv \left|\langle\phi_j|\, \hat{x}\, |\phi_o\rangle\right|^2 \tag{7.107}$$

The dipole moment is therefore oscillating with the same frequency as the electric field, and so the coefficient of $\cos\omega t$ gives the strength of the dielectric function. From $(\omega_{jo}-\omega)^{-1} + (\omega_{jo}+\omega)^{-1} = 2\omega_{jo}/(\omega_{jo}^2-\omega^2)$ and the introduction of an "oscillator strength" factor

$$f_j = \left(\frac{2m}{\hbar^2}\right)\hbar\omega_j|x_{jo}|^2 \tag{7.108}$$

then the atomic polarizability is obtained. In order to account for the transition of one state into another, the initial state is taken to decay as $\exp(-t/\tau_o)$ and the other states $|\phi_j\rangle$ grow as $\exp(t/\tau_j)$, suggesting the replacement [281, 364]

$$\hbar\omega_{jo} \to \hbar\omega_{jo} + \frac{i\hbar}{2\tau_j} \tag{7.109}$$

which results in

$$\langle\psi(t)|q\hat{x}|\psi(t)\rangle = \frac{2qF_o}{\hbar}\sum_j |x_{jo}|^2\left\{\frac{2\omega_{jo}e^{i\omega t}}{\omega_{jo}^2-\omega^2+it\tau_j} + \frac{2\omega_{jo}e^{-i\omega t}}{\omega_{jo}^2-\omega^2-it\tau_j}\right\} \tag{7.110}$$

where $(\omega_{jo}\tau_j)^2 \gg 1$. Combine the terms in the curly brackets to get

$$\{\cdots\} = \frac{\omega_{jo}(\omega_{jo}^2-\omega^2)\cos\omega t}{(\omega_{jo}^2-\omega^2)^2+(t/\tau_j)^2} + \frac{(\omega_{jo}/\tau_j)\omega\sin\omega t}{(\omega_{jo}^2-\omega^2)^2+(t/\tau_j)^2} \tag{7.111}$$

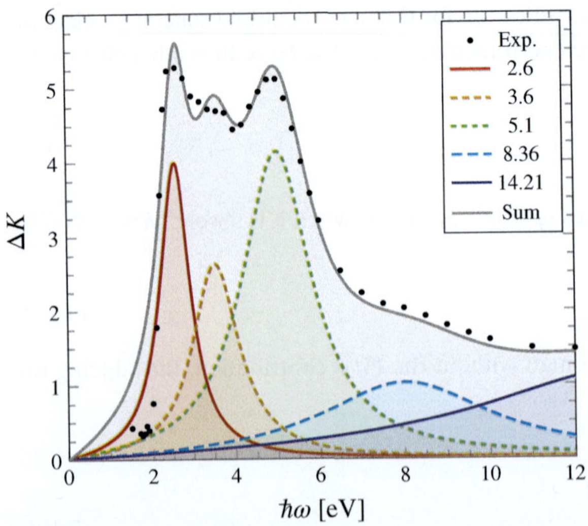


**Figure 7.11** Measured (black ∘) Dielectric term $\Delta K = K_i - K_i^D - K_i^L$, where the superscripts $D$ and $L$ refer to the Drude and Lorentz components, respectively, compared to an *ad hoc* sum of Lorenztians characterized by Table 7.4.

The first fraction oscillates with the electric field, and the second fraction is out of phase: the out-of-phase part leads to absorption and transition. It gives rise to the imaginary part of $\hat{K}$. Below, abandon the "$o$" subscript on $\omega_{jo}$ and let it simply be given by $\omega_j$. Introduce an oscillator strength factor $f_j$

$$f_j = \frac{2m\hbar^2}{\hbar}\omega_j|x_{jo}|^2 \tag{7.112}$$

Recalling that $(a - ib)/(a^2 + b^2) = 1/(a + ib)$ then

$$\chi(\omega) = \sum_j \frac{f_j\omega_p^2}{\omega_j^2 - \omega^2 + i(\omega/\tau_j)} \tag{7.113}$$

Thus, additional lorentzian terms are introduced due to the optical transitions, and these account for the structure shown in Figure 7.11. To refer to the enlarged model, these new terms will be called Resonant.

The purpose of the discourse is to justify a model useful for the prediction of $R(\omega)$ and $\delta(\omega)$, particularly, for subsequent photoemission models. For that purpose, *ad hoc* Lorentzians can be added to the Drude and Lorentz contributions to model the problematic visible-UV region and thereby recover correspondence over all frequencies between the theory and experiment lines of Figure 7.10 (see Algorithm B.2.12). To emphasize that a model is being presented that is more favorable to simulation needs, allow the notation to be slightly altered to match literature conventions with regards to how resonant terms are modeled by Lorentzians [154, 184, 284–286]. Instead of $1/\tau_j$, introduce $\Gamma_j$ as the broadening factor for the $j^{th}$ resonance term. Retain $f_j$ as the dimensionless weighting factor for each term. Lastly, speak of a Drude part, a Lorentz part, and a new sum of resonance terms, or Resonant, part. When all three are present, the model is now referred to as the Lorentz–Drude–Resonant or LDR model. The terms are [157]

$$K(\omega) = 1 + \chi_f(\omega) + \chi_b(\omega)$$
$$\chi_f(\omega) = -\frac{f_d\omega_p^2}{\omega(\omega + i\Gamma_d)} \tag{7.114}$$
$$\chi_b(\omega) = -\sum_{j=0}^{N_b} \frac{f_j\omega_p^2}{\omega^2 - \omega_j^2 + i\omega\Gamma_j}$$

where $\chi_f(\omega)$ is the Drude contribution ($f$ for "free"), the $j = 0$ part of $\chi_b(\omega)$ is due to the Lorentz contribution ($b$ for "bound"), and the new terms for $j \in (1, 2, \cdots N_b)$ in the summation in $\chi_b$ are here called the Resonant contributions. Insofar as resonant terms are used to model the optical to UV region of $K_i(\omega)$ in, for example, Figure 7.10, then the number of added terms $N_b$ can be made as large as needed to match experimental behavior as closely desired. As a model, then, there is nothing untoward about allowing the $f_j$ oscillator strength terms to be negative where needed [164] (the parameters are challenging to match even for an analytically known model [188]), although as an aesthetic matter, here they are taken as non-negative, and their values and those of the damping rates $\Gamma_j$ are adjusted until a match is good enough to get $R(\omega)$ and $\delta(\omega)$ where they are needed. They are quickly obtained by first evaluating the plasma frequency term $\hbar\omega$ of Eq. (7.73) and then calculating $n$ and $k$ via

$$2n^2 = \sqrt{K_r^2 + K_i^2} + K_r$$
$$2k^2 = \sqrt{K_r^2 + K_i^2} - K_r \tag{7.115}$$

where $K = K_r + iK_i$. Lastly, $R(\omega)$ and $\delta(\omega)$ are then given by

$$R(\omega) = \frac{(n-1)^2 + k^2}{(n+1)^2 + k^2}$$
$$\delta(\omega) = \frac{c}{2k\omega} \tag{7.116}$$

By way of example, consider copper, for which the LDR parameters are as listed in Table 7.4. If $\Delta K$ is $K_i$ with the Drude and Lorentz terms removed (as in Figure 7.10), then the sum of the five Lorentzians of Table 7.4 add to make up a theoretical approximation that is reasonably effective at modeling the range shown in Figure 7.10 (were the range extended, then a sixth lorentzian is needed to model another peak brought into range). Once the factors are found, the LDR model can be used in place of a data table to evaluate $R(\omega)$ and $\delta(\omega)$ quickly for arbitrary $\hbar\omega$. Even a small ($N_b = 5$) number of $f_j$ give reasonable ($K_i$, $K_r$) lines rapidly as shown in Figure 7.12 for copper.

**Table 7.4** Parameters for the modeling of the optical parameters of copper based on methods of [157] and [188]. "$D$" refers to Drude, $L$ to Lorentz, $p$ to plasma, and # to Resonant. $R$-parameters are determined by hand. Units $[\hbar\omega_j] = [\hbar\Gamma_j] = [\text{eV}]$; $f_j$ is dimensionless.

| $j$ | $p$ | $D$ | $L$ | 1 | 2 | 3 | 4 | 5 |
|---|---|---|---|---|---|---|---|---|
| $\hbar\omega_j$ | 10.77 | – | 54.18 | 2.60 | 3.60 | 5.10 | 8.36 | 14.21 |
| $\hbar\Gamma_j$ | – | 0.0912 | 87.85 | 0.90 | 1.36 | 2.00 | 4.90 | 9.00 |
| $f_j$ | – | 0.7360 | 8.972 | 0.08 | 0.11 | 0.36 | 0.35 | 1.28 |

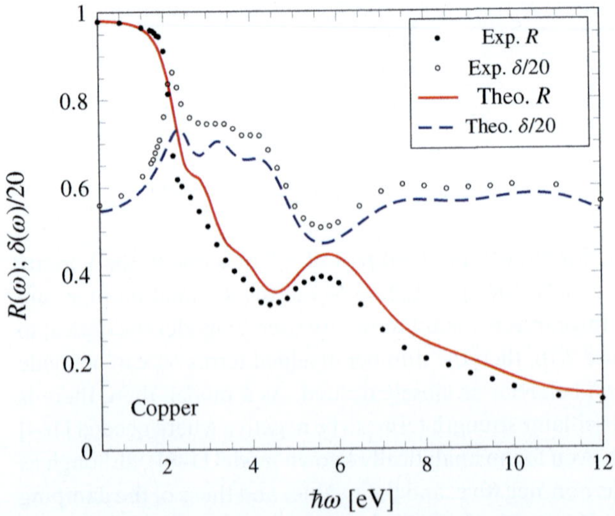


**Figure 7.12** Reflectivity ($R$, •) and penetration depth ($\delta$, ∘) as a function of photon energy evaluated from measured dielectric parameter $\hat{K}$. Thick red theory line is $R$ and thin blue line $\delta$ are evaluated using the LDR model, showing reasonable correspondence.

## 7.6 Exercises

**Exercise 7.1** A copper-like material with a Fermi level of $\mu = 7$ eV has a DC conductivity $\sigma_o = 5.92534 \times 10^7$ [1/Ω m]. What is the relaxation time $\tau$? What is the plasma frequency $\omega_p$? What is the complex conductivity $\hat{\sigma}(\omega)$ for a wavelength $\lambda = 1064$ nm (infrared)?

**Solution** $\tau = 25$ fs. $\hbar\omega_p = 10.7691$ eV. $\hat{\sigma}(\omega) = \sigma_o(1 + 14.088i)/1959.8$.

**Exercise 7.2** Evaluate the Weidemann–Franz constant $L_{wf}$ in both [nfeq] and [MKSA].

**Solution** $L_{wf} = 2.4430 \times 10^{-8}$ in both [Volt$^2$/K$^2$] and [eV$^2$/K$^2$ q$^2$].

**Exercise 7.3** Derive the Hagen–Rubens relation for reflectivity $R(\omega \to 0)$. Using it, find $R(\omega)$ for the parameters of Exercise 7.1 and $\lambda = 1064$ nm.

**Solution** Eq. (7.102). $R(\omega) = 95.4\%$.

**Exercise 7.4** Find the relaxation time $\tau$ in aluminum if the plasma frequency is $\hbar\omega = 14.98$ eV and $\sigma(0) = 3.5 \times 10^5$ [1/Ω m].

**Solution** $\tau = 7.632$ fs.

**Exercise 7.5** Confirm the form of $\hbar k$ in Eq. (7.41). What is $\beta$?

**Exercise 7.6** What is the thermal wavelength of an electron in an alpha semiconductor with a bandgap of $E_g = 2.177$ eV at $T = 300$ K?

**Solution** $\lambda_T = 10.76$ nm.

**Exercise 7.7** Show Eq. (7.21).

**Exercise 7.8** Find the middle integral of Eq. (7.21) in the limit of small $u$. Compare it to Eq. (7.25) and to $\delta f_r(u)$ for $u = 1$.

**Solution** In the small $u$ limit, the integral is $(2/3)u^{3/2}[(415/1024) - (43/320)u + (1/160)u^3]$ and equals 0.18477, compared to Eq. (7.25) which equals 8.00042. Compare to $(2/3)\delta f_r(1) = \sqrt{2}\ln(1 + e^{-1}) = 0.44301$.

**Exercise 7.9** Find when $\mu(T) = 0$ if $\mu_o = 0.1$ eV and $r_e = 1$ given $\zeta(3/2) = 2.61238$.

**Solution** $T = 1147.38$ K.

**Exercise 7.10** What is the reflectivity $R$, penetration depth $\delta$, and complex index of refraction $\hat{n}$ for $\lambda = 266$ nm light incident on copper if $\hat{K} = -0.487802 + 5.10421i$? Assume copper has the parameters of Ex. 7.1.

**Solution** $R = 33.589\%$. $\delta = 12.633$ nm. $\hat{n} = 1.5231 + 1.6756i$.

# Part III

# Current and Barriers

# 8

# Current Density Formalism

*And thus the native hue of resolution*
*Is sicklied o'er with the pale cast of thought,*
*And enterprises of great pith and moment*
*With this regard their currents turn awry ...*

W. Shakespeare, *Hamlet III.1*

## 8.1 Master Equation

Although the canonical equations are limiting cases of a general $J(F, T)$ applied to the Schottky–Nordheim (SN) barrier, a GTF emission equation can be developed that is actually more general: it can deal with barriers that tweak and modify the SN barrier, or even replace it with something more complex. To deal with such barriers beyond the SN barrier, first, a shape factor method is introduced, and second, a number of seemingly different barriers are recast into a formulation that can make use of the canonical equations, particularly the Richardson–Laue–Dushman (RLD) and Fowler–Norheim (FN) equations. What now must be done is to consider the transition between the asymptotic limits represented by the RLD and FN equations, and so to determine what a general thermal-field, or GTF, equation behaves like.

Recall that current density is the product of a charge $q$ with an integration of the product of velocity $\hbar k/m$ with a distribution of particles, as in Eq. (2.14). In the case of the iconic rectangular barrier, the region to the right of the barrier is flat and taken to be such that $V(x > L) = 0$. The distribution in this region is then independent of $x$, and can be reasonably well approximated by what comes at the barrier (the supply function $f(k)$ of Eq. (6.46)) modified by what tunnels through (the transmission probability $D(k)$ as given by the Kemble approximation, or Eq. (6.77) with $P(k) = 1$). As a result, one-dimensional current density past a general barrier is given by

$$
\begin{aligned}
J(F, T) &= \frac{q}{2\pi} \int_{-\infty}^{\infty} \frac{\hbar\, k_x}{m} D(k_x) f(k_x) dk_x \\
f(k_x) &= \frac{m}{\pi \beta_T \hbar^2} \ln\{1 + \exp[\beta_T(\mu - E(k_x))]\} \\
D(k_x) &= \frac{1}{1 + \exp[\theta(k_x)]}
\end{aligned}
\tag{8.1}
$$

*Fundamentals of Electron Emission Physics*, First Edition. Kevin L. Jensen.

where $\beta_T \equiv 1/k_B T$. In the Sommerfeld model of a metal from which $f(k)$ is developed, it is convenient to take the energy as parabolic in momentum $E(k_x) = \hbar^2 k_x^2/2m$ and switch to an integration over the normal energy $dE_x \to dE$ where the subscript can be dropped because heretofore the 1D model is understood to be under discussion. Letting $\theta(k) \to \theta[E(k)] \equiv \theta(E)$ then allows $J(F, T)$ to be written as

$$\begin{aligned} J(F,T) &= A_{RLD} T^2 \int_0^\infty \frac{\ln\{1+\exp[\beta_T(\mu - E)]\}}{1+\exp[\theta(E)]} \beta_T dE \\ &\equiv A_{RLD} T^2 N[\theta(E)] \end{aligned} \tag{8.2}$$

where $N[\theta(E)]$ is the dimensionless integral part: its numerator is drawn from the supply function, and its denominator from the Kemble approximation to $D[E(k)]$ with the coefficient $P(k) \to 1$ and $\theta_o(k) \to \theta[E(k)]$ in Eq. (6.77) for barriers more general than just the rectangular barrier case. Below, when there is cause to discuss the integrand of $J(F, T)$ because of a need to refer to the energy $E = E_m$ where the integrand is at its maximum, it is convenient to introduce the function $dN(E)$ defined by[1]

$$\begin{aligned} J(F,T) &= A_{RLD} T^2 \int_0^\infty dN(E) \\ dN(E) &= \frac{\ln\{1+\exp[\beta_T(\mu - E)]\}}{1+\exp[\theta(E)]} \beta_T dE \end{aligned} \tag{8.3}$$

The location of the maximum of the integrand, $E_m$, is then defined such that $dN(E)/dE|_{E=E_m} = 0$. It is important to emphasize, first, that $\theta(E)$ is given by Eq. (6.92) for general barriers, and second, that $N[\theta(E)]$ is an integral over energy with $\theta(E)$ as part of the integrand and so is a functional (in the sense of a function of a function). The $T^2$ in the coefficient is because one energy slope factor of $\beta_T$ arises from the coefficient of $f(k)$, and the other is to offset the $\beta_T dE$ in the integral put there to make the integral dimensionless. The coefficient $A_{RLD}$ (the odd name of which will become justified in Section 8.2) is a combination of fundamental constants (after $g = 2$ for spin) given by

$$A_{RLD} = \frac{qmk_B^2}{2\pi^2\hbar^3} \tag{8.4}$$

or 120.1732 [A/cm$^2$ K$^2$] in MKSA units (see Ex. 8.1).

There are two integrals used in Eq. (8.2), the obvious one over $dE$ and the less obvious one embedded in $\theta(E)$ for general barriers. The canonical emission equations treated crudely in Section (2.2) are put into their conventional forms by examining limiting cases of $J(F, T)$ that are analytic after a suitable approximation to $\theta(E)$ is employed. The canonical equations are not as accurate as the forms arising from the GTF form or developed using Transfer Matrix Approach (TMA) methods, but they are pedagogically valuable. The canonical forms are reliant on the approximation that the Gamow factor is linear in energy, or

1 In later sections, $dN(E)$ will be replaced by $dN(F, T; E)$ to point out that it also depends on $F$ and $T$, but here, that is a notational complication that need not be indulged.

$$
\begin{aligned}
\theta(E) &\approx \beta_F(E_m)(E_o - E) \\
E_o &\equiv \frac{\theta(E_m)}{\beta_F(E_m)} + E_m \\
\beta_F(E_m) &\equiv -\frac{d}{dE}\theta(E)\bigg|_{E=E_m}
\end{aligned}
\tag{8.5}
$$

where $E_m$ is the expansion point for a Taylor expansion to first order, and $\beta_F(E_m)$ is defined with a minus sign to make it a positive quantity, as $\theta(E)$ is a decreasing function as $E$ increases for the most common barriers. The notation of $\beta_F$ is designed to mimic that of $\beta_T = 1/k_B T$, but unlike the thermal energy slope factor $\beta_T$, the field energy slope factor $\beta_F$ is in general an involved function. With the linear approximation to the Gamow factor, it is then possible to define

$$
\begin{aligned}
z &\equiv \beta_F(E_m)(E_o - E) \\
\beta_T(\mu - E) &= n\beta_F[-(E_o - \mu) + (E_o - E)] \\
&\equiv n(-s + z)
\end{aligned}
\tag{8.6}
$$

where the important new functions $n$ and $s$ are defined by

$$
\begin{aligned}
n(F,T) &= \frac{\beta_T}{\beta_F(E_m)} = \frac{1}{k_B T \beta_F(E_m)} \\
s(F,T) &= \beta_F(E_m)(E_o - \mu) \\
&= \theta(E_m) + \beta_F(E_m)(E_m - \mu)
\end{aligned}
\tag{8.7}
$$

As a consequence, $N[\theta(E)]$ can be approximated by a simpler integral, and so

$$
\begin{aligned}
J(F,T) &\approx A_{RLD} T^2 N(n,s) \\
N(n,s) &\equiv n \int_{-\infty}^{\infty} \frac{\ln\left[1 + e^{n(z-s)}\right]}{1 + e^z} dz
\end{aligned}
\tag{8.8}
$$

under the approximation that $\beta_F(E_m)\mu \gg 1$ and can therefore be represented by $-\infty$ in the lower limit of the integral. Observe that $N(n,s)$ is a function of $n$ and $s$ rather than a functional like $N[\theta(E)]$, and it is formed by approximating $\theta(E)$ by a linear function. The limiting cases of Eq. (8.8) provide a rapid means of determining the canonical equations. In the cases which follow herein, keep in mind that the coefficient $P(k)$ in the Kemble approximation of Eq. (6.77) is always approximated by unity.

The form of $V(x)$ used to define $\theta(E)$ and $\beta_F(E)$ dictate $n$ and $s$, and so must be specified for general barriers and their associated current densities, but two cases are primary: thermal emission over a rectangular barrier, which does not depend on $\beta_F$; and field emission, which does require $\beta_F(\mu)$ and $\theta(\mu)$, but for which a triangular barrier provides simple and analytic representations. Their consideration, therefore, precedes an analysis using the SN barrier of Chapter 10.

## 8.2 Richardson–Laue–Dushman Equation

The factor $A_{RLD}$, as its name implies, was identified in the early derivations of the RLD equation for thermal emission [129, 154, 296]. The present preference to say "thermal" rather than "thermionic" is because the latter name originated when it was thought that electrons emitted from thermal processes ("thermions") were somehow different than those electrons liberated by high electric fields, insofar as the thermal current density was strongly dependent on temperature, whereas the field emission current density was not. The RLD equation makes the following approximations.

1. The expansion point $E_m \approx \mu + \Phi$, or the height of the barrier, for which $\theta(E_m) = 0$;
2. The ratio $n = \beta_T/\beta_F(E_m) \to 0$, because $\theta(E)$ is very large for thermal emission barriers and declines rapidly as $E \to E_m = \mu + \Phi$; similarly, $s = \beta_F(E_m)\Phi \gg 1$;
3. The Kemble form of $D(E)$ is therefore well approximated by the step function $D(E) \approx \Theta(E - \mu - \Phi)$, that is, the integrand of $N(n, s)$ is significant only when $z < 0$, for which the denominator of the integrand is simply unity;
4. $n(z - s)$ is therefore large and negative, so that $\ln\left[1 + e^{n(z-s)}\right] \approx e^{n(z-s)}$ in the numerator of the integrand.

With these approximations,

$$N_{RLD}(n, s) \approx n \int_{-\infty}^{0} e^{n(z-s)} dz = e^{-ns} \tag{8.9}$$

Because $ns = \beta_T\Phi$ in the present approximation, it follows

$$\begin{aligned} J_{RLD}(F, T) &= A_{RLD}T^2 \exp(-\beta_T\Phi) \\ &\equiv A_{RLD}T^2 \exp\left(-\frac{B_{RLD}}{T}\right) \end{aligned} \tag{8.10}$$

where $A_{RLD}$ is from Eq. (8.4), and $B_{RLD} = \Phi/k_B$. Eq. (8.10) is close to the canonical equation for thermal emission: accounting for SN lowering of the barrier (Schottky factor) due to applied field involves only the replacement $\Phi \to \phi = \Phi - \sqrt{4QF}$ in the definition of $E_m$, and $\theta(E_m) = 0$. As a consequence, Eq. (8.10) continues to hold with the replacement $\Phi \to \Phi - \sqrt{4QF} = \phi$ and renders Eq. (8.10) the canonical equation for thermal emission.

## 8.3 Fowler–Nordheim Equation

The first field emission equation for $J(F, T)$ by Fowler and Nordheim [95] assumed that the barrier is triangular and used the zero temperature limit of the supply function. Its extension to the SN barrier of Section 10 came later and a treatment of that is deferred to Section 10.2 and beyond. For the $T = 0$ and triangular approximations,

1. The expansion point $E_m = \mu$ is at the Fermi level because as $T \to 0$, $D(E)f(E)$ is largest near there;
2. At $E_m = \mu$, then $s = \theta(\mu)$ and so is the same as the lower bound of the integrand of $N(n, s)$

With these approximations,

$$N_{FN}(n,s) \approx n \int_s^{\infty} n(z-s)e^{-z}dz = n^2 e^{-s} \tag{8.11}$$

It follows

$$J_{FN}(F,T) = A_{RLD}T^2 n^2 e^{-s} = \frac{A_{RLD}}{[k_B\beta_F(\mu)]^2}\exp[-\theta(\mu)] \tag{8.12}$$

For a triangular barrier $V(x>0) = \mu + \Phi - Fx$. The zeros of $V(x) - E$ are then at $x_- = 0$ and $x_+ = (\mu + \Phi - E)/F$. The integration to find $\theta(E)$ from Eq. (6.92) is then simple, and results in

$$\begin{aligned}\theta_{FN}(E) &= \frac{4\sqrt{2m}}{3\hbar F}(\mu+\Phi-E)^{3/2}\\ \beta_F(E) &= \frac{2\sqrt{2m}}{\hbar F}(\mu+\Phi-E)^{1/2}\end{aligned} \tag{8.13}$$

from which $\theta(\mu) = 4\sqrt{2m\Phi^3}/3\hbar F$ and $\beta_F(\mu) = 3\theta(\mu)/2\Phi$. As a result, for the triangular barrier FN equation,

$$\begin{aligned}J_{FN}(F,T) &= \frac{A_{RLD}}{[k_B\beta_F(\mu)]^2}P(\mu)\exp\left(-\frac{4\sqrt{2m\Phi^3}}{3\hbar F}\right)\\ &\equiv \frac{A_o}{\Phi}P(\mu)F^2\exp\left(-\frac{B_o}{F}\right)\end{aligned} \tag{8.14}$$

where the "$o$" superscript signifies that image charge contributions are neglected, and where $P(\mu)$, previously approximated by unity, is re-inserted from the Kemble form and requires the Airy Transfer Matrix Approach for its evaluation [154]. Although often called the "Fowler Nordheim equation" in deference to its appearance in Fowler and Nordheim's seminal 1928 manuscript, Eq. (8.14) is not the canonical field emission equation because it does not as yet include Schottky barrier lowering. Explicitly, the $A_o$ and $B_o$ parameters are defined by

$$A_o = \frac{q}{16\pi^2\hbar};\quad B_o = \frac{4}{3\hbar}\sqrt{2m\Phi^3} \tag{8.15}$$

The similarity of the RLD and FN equations induced Millikan and Lauritsen [255] to hypothesize a general equation involving $X = T + cF$ rather than $T$ or $F$ as a "thermal-field" equation of the form $AX^2\exp(-B/X)$, but the reality of a GTF equation is more complex and relies on a form of $N(n,s)$ when $n \approx 1$ and the explicit consideration of the SN or image charge barrier, as shall be shown below.

## 8.4 Fowler–DuBridge Equation

A model of photoemission using the Sommerfeld model of a metal material was developed by Fowler [94] and subsequently modified by DuBridge [67]. Its derivation by means of a theory similar to the field emission derivation [67, 94] uses the Fowler-DuBridge function [108,161, 163] defined by

$$U(x) \equiv \int_{-\infty}^{x} \ln(1+e^s)\,ds \tag{8.16}$$

and satisfies the relation

$$U(x) + U(-x) = \frac{x^2}{2} + \frac{\pi^2}{6} \tag{8.17}$$

The form of Eq. (8.16) can be recast in terms of the dilog function $-\mathrm{Li}_2(1 + e^x)$. Belying its visual simplicity, Eq. (8.16) engenders computational challenges in simulations of photoemission, and it is therefore desirable to have an analytical approximation of reasonable accuracy if the dilog function is not available. One is possible [163] by first replacing the integrand by its Taylor expansion for small $e^{-x}$ and then performing a term-by-term integration to get the series

$$U(x) = \sum_{j=1}^{\infty} \frac{(-1)^{j-1}}{j^2} \exp(-jx) \tag{8.18}$$

and so $U(0) = \zeta(2)/2$ using Eq. (A.5). The approximation $U(-x) = e^x(1 - be^{ax})$ compared to Eq. (8.18) and its first derivative evaluated $x = 0$ allows for the determination of the approximation parameters

$$b \equiv 1 - \frac{\zeta(2)}{2}; \quad ab = 1 - b - \ln(2) \tag{8.19}$$

or $b = 0.177533$ and $a = 0.728427$. As a result,

$$U(x) \approx \begin{cases} e^x\,(1 - be^{ax}) & (x \leq 0) \\ \dfrac{x^2}{2} + \dfrac{\pi^2}{6} - e^{-x}\,(1 - be^{-ax}) & (x > 0) \end{cases} \tag{8.20}$$

the absolute error of which does not exceed 0.12%, and so makes the approximation useful for simulation.

A modified form of the FD equation for quantum efficiency then becomes [154, 161,176]

$$QE = [1 - R(\omega)]F_\lambda \frac{U[\beta(\hbar\omega - \phi)]}{U(\beta\mu)} \tag{8.21}$$

where $R(\omega)$ is the reflectivity and $F_\lambda$ accounts for losses due to scattering of photoexcited electrons on the way to the surface. The ratio of the $U$-functions is then seen to be the proportion of photoexcited electrons with sufficient energy to be emitted that have a normal energy $E_\perp$ greater than the barrier height.

In contrast, here the current density equation of Eq. (8.8) is the preferred approach to obtaining the FD approximation. It is extended to the inclusion of photons by the replacement $E \rightarrow E + \hbar\omega$, but the integral remains daunting. Consider a heuristic approach, namely, the inclusion of photon energy may change the sign of $s$ because the energy $E$ may be below $\mu$ but still allows $E + \hbar\omega > \mu + \Phi$. Here, however, treat $s$ as a positive number such that $(-s)$ results when $E \rightarrow E + \hbar\omega$. The current density on which QE depends will then be proportional to $N(n, -s)$ for $s$ positive. For the same $s$, $N(n, s)$ is exponentially smaller and so a profitable trick is to consider the sum $N(n, s) + N(n, -s)$, where the arguments of $s$ are rearranged so that it is positive when applied to emission, and at the conclusion of the calculation, observe that the sum is still dominated by $N(n, -s)$. Next, assume the tunneling contribution is so suppressed by low fields that contributions from $z > 0$ are small, and so replace the lower limit of the integrals by 0. Next, assume $n(z \pm s)$ is large, so that the logarithm terms can use the approximation $\ln(1 + e^x) \approx x$ for large and positive $x = n(z \pm s)$. Lastly, $n(z + s)$ will contribute to the $N(n, -s)$ integral for $-s < z < 0$ in a significant way, so the lower bound to that integral is $-s$. As a result, using $\ln(1 + e^x) \approx x$ for large $x$, then for $s$ large,

$$N(n,s)+N(n,-s)\approx n\left[\int_{0}^{\infty}\frac{n(z-s)}{1+e^{z}}dz+\int_{-s}^{\infty}\frac{n(z+s)}{1+e^{z}}dz\right] \tag{8.22}$$

The second integral can be separated into two integrals via $\int_{-s}^{\infty}\rightarrow\int_{-s}^{0}+\int_{0}^{\infty}$, the second term of which can be combined with the first integral of Eq. (8.22) to find

$$\begin{aligned} N(n,s)+N(n,-s) &\approx n\left\{2n\int_{0}^{\infty}\frac{z}{1+e^{z}}dz+\int_{-s}^{0}n(z+s)dz\right\} \\ &= \frac{(n\pi)^2}{6}+\frac{(ns)^2}{2} \end{aligned} \tag{8.23}$$

The absence of a denominator in the $\int_{-s}^{0}$ integral is from the approximation that $e^{z}$ is small for negative $z$ and therefore can be neglected.

Such a treatment is not rigorous, but a systematic series expansion of the integrands of $N(n,-s)+N(n,s)$ (see Section 12.2) and term-by-term integration, followed by a regrouping of terms in those series expansions [150, 154] does result in

$$N(n,s)+N(n,-s)=\frac{1}{2}(ns)^2+\zeta(2)\left(n^2+1\right) \tag{8.24}$$

where the Riemann zeta function $\zeta(2)=\pi^2/6$ as in Appendix A.2. As a consequence, when returning to Eq. (8.8), the term $k_B^2T^2(ns)^2=(\hbar\omega-\phi)^2$ arises as the dominant term, and the modified Fowler–DuBridge equation becomes

$$QE\propto\frac{1}{2}(\hbar\omega-\phi)^2+\zeta(2)\left(\beta_F^{-2}+\beta_T^{-2}\right) \tag{8.25}$$

and is taken to be the canonical equation of photoemission.

## 8.5 Exercises

**Exercise 8.1** Find $A_{RLD}$ in [efnq] units. How many electrons are emitted from a coated crystal face with an area of 3 $\mu$m$^2$ in one nanosecond for $T=1273$ K, if the effective work function is $\phi=2.1$ eV?

**Solution** $A_{RLD}=7.50063\times10^{-9}$ [q/fs nm$^2$ K$^2$]. Number of electrons emitted equals 177.

**Exercise 8.2** Dispenser cathodes are subject to grid potentials on the order of 1 kV for grid distances of 1 mm. What is the Schottky factor for a grid held at 1 kV and an AK gap spacing of 0.9 mm? What is the ratio of the current density under those conditions to the current density when the grid potential is zero, if $T=1373$ K?

**Solution** Schottky factor = 0.04 eV. Ratio = 1.402.

**Exercise 8.3** For the triangular barrier Fowler–Nordheim equation subject to a field corresponding to $F=6$ eV/nm, what is the Gamow factor $\theta(E)$ and the energy slope factor $\beta_F(E)$ for $E=\mu$, assuming tungsten-like parameters of $\Phi=4.5$ eV? What is $n=\beta_T/\beta_F$ if $T=1373$ K?

**Solution** $\theta(\mu)=10.868$. $\beta_F(\mu)=3.6226$ [1/eV]. $n=2.3331$.

**Exercise 8.4** For the triangular barrier with $F = 5$ eV/nm and $\Phi = 4.5$ eV, what is $A^o_{FN}$ and $B^o_{FN}$ if $P(\mu) = 1$?

**Solution** $A^o_{FN} = 0.0046931$ [q/fs eV$^2$]. $B^o_{FN} = 20.050$ [eV/nm].

**Exercise 8.5** Show that the Fowler–DuBridge function $U(x)$ of Eq. (8.16) satisfies Eq. (8.17).

**Solution** Use Appendix A.2 and $\zeta(2) = \pi^2/6$.

# 9

# Simple Barriers

*If there be any impediment, I pray you discover it.*

W. Shakespeare, *Much Ado About Nothing* III.2

## 9.1 Shape Factor Method

The tunneling barriers through which electrons must pass can have varied shapes, but all fit inside a square with dimensions of their maximum height (in energy) and maximum width (in position). The rectangular barrier is unique for being a barrier characterized by a constant height and a single width. Analogously, the Gamow $\theta(E)$ integrand $2\sqrt{V(x) - E}\, dx$ is bounded by an area $2\sqrt{V(x_o) - E}\, L \propto \theta_o(E)$ (compare Eq. 6.77). The shape factor is the dimensionless ratio of the area under the curve of $2\sqrt{V(x) - E}$ (i.e. the integral between the zeros of the integrand) to $\theta_o(E)$ and is defined by

$$\sigma(E) = \int_{x_-}^{x_+} \left[ \frac{V(x) - E}{V(x_o) - E} \right]^{1/2} \frac{dx}{x_+ - x_-} \tag{9.1}$$

For reasons that will become evident in due course, it is convenient to change the nomenclature from the outset. The following conventions shall be adopted:

- the maximum of $V(x)$ is at $x_o$, such that $dV/dx = 0$ at $x = x_o$;
- the zeros of the function $V(x) - E = 0$ are indexed by $j = 1, 2, \cdots$, with $x_-$ identified with the smallest root and $x_+$ the next smallest root (for the SN barrier, $x_- = x_1$ and $x_+ = x_2$ trivially, but for more complex barriers such as those of Section 11.3, the roots are no longer ordered by size and the distinction matters);
- a dimensionless factor $y(E)$, encountered in Eq. (2.12), is defined as the ratio of the extent to which the barrier is reduced by the application of an applied field $F$ to the height of the barrier above $E$, or

$$y(E) \equiv \frac{V_o - V(x_o)}{V_o - E} \equiv \frac{\Delta V}{V_o - E} \tag{9.2}$$

where $V_o$ is the maximum height of the barrier in the absence of a field, e.g. for the SN barrier of Eq. (2.9), with $V_o = \mu + \Phi$ and $\Delta V = \sqrt{4QF}$.

*Fundamentals of Electron Emission Physics*, First Edition. Kevin L. Jensen.

With the new nomenclature, then the formal definition of the Gamow factor $\theta(E)$ is, in the shape factor method

$$\begin{aligned}
\theta(E) &\equiv 2\sigma[y(E)]\kappa(E)L(E) \\
\kappa(E) &\equiv \frac{1}{\hbar}\sqrt{2m[V_o - E]} = \sqrt{k_o^2 - k^2} \\
L(E) &\equiv x_2(E) - x_1(E) \\
\sigma[y(E)] &= \int_{x_1}^{x_2} \left[\frac{V(x) - E}{V(x_o) - E}\right]^{1/2} \frac{dx}{L(E)}
\end{aligned} \tag{9.3}$$

where $y(E)$ relies on Eq. (9.2) for general barriers and $V(x_o)$ is the maximum height of the barrier, and for the SN barrier $x_2 = x_+$ and $x_1 = x_-$ in Eq. (9.1) (a convention that is modified for the polynomial barriers of Section 11.3). For shape factors, it is implicitly assumed that $k_o > k$ because the interest is in tunneling.

The shape factors of several simple barriers are analytic [165] and are also constant for all $E$

- The rectangular barrier is evidently a special case for which $\sigma_\square = 1$, easily shown by $x_1(E) = 0$ and $x_2(E) = L_o$ for all $E$, so that $L(E) = L_o$. $V(x) = V_o\Theta[(L_o/2)^2 - x^2]$ where $\Theta(x)$ is the Heaviside step function, for which $x_\pm = \pm L_o/2$, where $L_o$ is understood to be where $V(x)$ intersects the $x$ axis. It follows

$$\sigma_\square = \int_0^{L_o} \frac{dx}{L_o} = 1 \tag{9.4}$$

As a consequence, the Gamow factor for the rectangular barrier is $\theta_o \equiv 2\kappa(E)L_o$ where $\kappa^2 = k_o^2 - k^2$, or, in terms of energy $E = \hbar^2 k^2/2m$

$$\theta_o(E) = \frac{2L_o}{\hbar}\sqrt{2m(V_o - E)} \tag{9.5}$$

- The triangular barrier of Fowler and Nordheim [95] is the next simplest case, resulting in

$$\sigma_\triangleright = \int_0^{L(E)} \left[\frac{V_o - E - Fx}{V_o - E}\right]^{1/2} \frac{dx}{L(E)} = \frac{2}{3} \tag{9.6}$$

where $L(E) = (V_o - E)/F$. Consequently, $\theta_\triangleright = (4/3)\kappa(E)L(E)$ (compare Eq. 8.13), the special case of which is $\theta_\triangleright(\mu) = 4\sqrt{2m\Phi^3}/3\hbar F$.

- Symmetric barriers for which

$$V_n(x) = V_o\left[1 - \left(\frac{|x|}{L_o}\right)^{2/n}\right] \tag{9.7}$$

can be characterized by introducing $\lambda = V_o/(V_o - E) = k_o^2/\kappa(E)^2$ such that $x_2 = L_o\lambda^{-n/2}$. As a result,

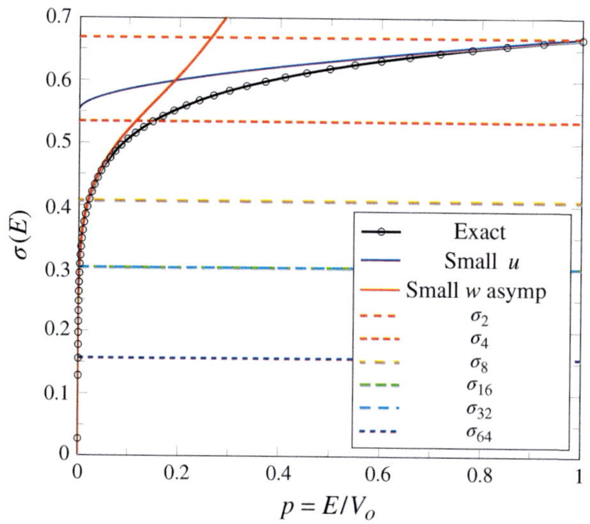


**Figure 9.1** Behavior of the shape factor for an exponentially declining $V(x)$. $E = pV_o$. Dashed lines: the values of $\sigma_n$ from Eq. (9.8). Symbols o: $\sigma_e(E)$ from Eq. (9.11). Solid red and blue lines: the asymptotic approximations to $\sigma_e(E)$ for small $u$ and $w = (\pi/2) - u$, respectively, the later of which shows $-1/\ln(p)$ behavior.

$$\sigma_n = \frac{1}{L(E)} \int_0^{L(E)} \left[1 - \lambda\left(\frac{x}{L(E)}\right)^{2/n}\right]^{1/2} dx \tag{9.8}$$

$$= \int_0^1 \left(1 - s^{2/n}\right)^{1/2} ds = \frac{\Gamma(3/2)\Gamma[(n/2)+1]}{\Gamma[(n+3)/2]}$$

for which $\sigma_0 = 1 = \sigma_\square$ is the rectangular barrier and $\sigma_2 = 2/3$ for the isosceles triangle is equivalent to $\sigma_\triangleright$. The parabolic barrier gives $\sigma_1 \equiv \sigma_\cap = \pi/4$, and $\sigma_3 = 3\pi/16$. As $n$ increases, $\sigma_n$ monotonically decreases [165] crudely as $1/\sqrt{n}$, as can be shown using Stirling's approximation for $\Gamma(x)$ (Section A.1), and indicated in Figure 9.1 (see Algorithm B.3.1).

## 9.2 Exponential Barrier

Although the simple barriers of the previous section gave rise to constant shape factors, more realistic barriers lead to shape factors that in fact vary in response to the energy of the tunneling particle. Of them, the Schottky–Nordheim (SN) barrier is the best known, and leads to modifications to the Richardson–Laue–Dushman (RLD) and Fowler–Norheim (FN) equations. There are simpler barriers that introduce the notion of a varying shape factor $\sigma(E)$ and so consider them first, beginning with an exponential barrier of the form

$$V(x) = V_o \exp\left(-\frac{x}{a}\right)\Theta(x) \tag{9.9}$$

where the Heaviside step function $\Theta(x)$ causes the barrier to vanish to the left of the origin. Therefore, $x_1 = 0$ for all $E$, but $x_2 = a\ln[V_o/E] \equiv -a\ln(p)$ where $p = E/V_o$. The shape factor $\sigma_e(E)$ is then

$$\sigma_e(E) = \frac{1}{x_2(E)} \int_0^{x_2(E)} \left[ \frac{e^{-x/a} - p}{1-p} \right]^{1/2} dx \tag{9.10}$$

The integral is analytic, but the solution is visually simplified by introducing a change of variables such that $p(u) = [\cos u]^2$ to get

$$\sigma_e(E) = \frac{u \cot(u) - 1}{\ln[\cos(u)]} \tag{9.11}$$

For $u$ and $w = (\pi/2) - u$ small, respectively, then the limiting forms are

$$\sigma_e(u) \approx \begin{cases} \dfrac{4(15+u^2)}{15(6+u^2)} & u \ll \dfrac{\pi}{2} \\ \dfrac{-1+[(\pi/2)-w]w}{\ln[\sin(w)]} & w \ll 1 \end{cases} \tag{9.12}$$

which are indicated by the blue (small $u$) and red (small $w = (\pi/2) - u$) lines in Figure 9.1, respectively. The small $u$ form shows that $\sigma_e(E)$ goes over to $\sigma_\triangleright$ limit, whereas the small $w$ form goes to 0 as $1/\ln(p)$, where the denominator is logarithmically singular as $E \to 0$.

## 9.3 Trapezoidal Barrier

Metal–insulator–metal (MIM) tunnel junctions [183, 312] and metal–oxide–semiconductor (MOS) barriers are other simple structures that have a varying shape factor $\sigma_\diamond(E)$. Such barriers are familiar from nano-electronic devices with nano- and subnano-scale junctions and contacts [12, 375]. The MIM barrier is a trapezoidal barrier that appears like a triangular barrier but has a constant $x_2$ for $E$ sufficiently below the maximum. The barrier is given by [157]

$$V_\diamond(x) \equiv (V_o - Fx)\,\Theta(x)\,\Theta(L_o - x) \tag{9.13}$$

where $L_o$ is the fixed width of the barrier. Observe that for $E > V_o - FL_o$, then the shape factor is the same as for the FN triangular barrier or $\sigma_\diamond = \sigma_\triangleright$. For energies below that transition point, though, $\sigma_\diamond(E)$ varies with energy. Introduce the dimensionless factor $\lambda(E)$ defined by

$$\lambda(E) = \frac{L_o F}{V_o - E} \tag{9.14}$$

in terms of which $\sigma_\diamond$ for $V_o > E$ is

$$\sigma_\diamond(\lambda) = \frac{2}{3\lambda}\left[1 - (1-\lambda)^{3/2}\right] \tag{9.15}$$

This is a consequence of replacing $x_2(E) = L(E) = (V_o - E)/F$ in the FN case with a constant $L_o$. For $\lambda \to 1$ ($E$ is just entering the triangular upper part of the barrier), then $\sigma_\diamond = \sigma_\triangleright$, as expected. Conversely, as $F \to 0$ and so $\lambda \to 0$, then $\sigma_\diamond \to 1$, meaning that the barrier resembles its rectangular barrier limit. It is seen, therefore, that the variation in $\sigma_\diamond(E)$ is not large, being $2/3 \le \sigma_\diamond(E) \le 1$.

## 9.4 Depletion Barrier

A semiconductor-metal contact has an interface barrier created by uniformly distributed charged dopants [185, 294, 295], the shape of which can be modeled by $V_\cup(x) = V_o[1 - (x/w)]^2$, where $w$ is the depletion width: the simple form enables an analytic $\sigma_\cup(E = 0)$ to be found, which elementary integration shows to be $\sigma_U(0) = 1/2$. For $E > 0$, however, it is more complex. In anticipation of the treatment of image-charge-modified barriers, particularly in Section 11.3, there is utility in reframing the depletion barrier to become instead

$$V_U(x) = \chi - Fx + \gamma x^2 \tag{9.16}$$

for which $x_- = 0$ and $w \to x_p = F/2\gamma$. The evaluation of $\sigma_U(E)$ is then straightforward, even if the integral is tricky

$$\begin{aligned}\sigma_U(E) &= \frac{1}{L(E)\sqrt{\chi - E}} \int_0^{L(E)} \left[\chi\left(1 - \frac{x}{x_p}\right)^2 - E\right]^{1/2} dx \\ L(E) &= x_p\left(1 - \sqrt{\frac{E}{\chi}}\right)\end{aligned} \tag{9.17}$$

where $L(E)$ is the width of the barrier when $E > V_U(x_p)$. The integral is analytic, and so

$$\sigma_U[u(E)] = \frac{1}{2(1-u)}\left[1 - \frac{u^2}{\sqrt{1-u^2}} \ln\left(\frac{1 + \sqrt{1-u^2}}{u}\right)\right] \tag{9.18}$$

where $u(E) \equiv \sqrt{E/\chi}$, for which $L[u(E)] = (1-u)x_p$. The limiting cases $\lim_{u\to 0} \sigma_U(u) = 1/2$ and $\lim_{u\to 1} \sigma_U(u) = \sigma_\triangleright = 2/3$ are associated with a parabolic and triangular barrier, respectively. The behavior is shown in Figure 9.2, showing that a simple polynomial of the form $\sigma_U[u(E)] \approx -0.156749u^4 + 0.462760u^3 - 0.565804u^2 + 0.426460u + 0.5$ is a very good approximation.

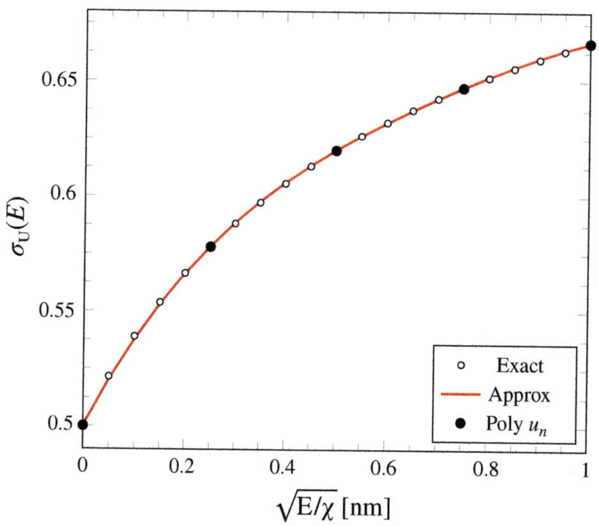


**Figure 9.2** Evaluation of Eq. (9.18) (∘) compared to a fourth order polynomial (red line) constructed from the five equispaced • exact evaluations, for which $\sigma_U(0) = 1/2$ and $\sigma_U(1) = 2/3 = \sigma_\triangleright$ are the parabolic and triangular limits.

## 9.5 Exercises

**Exercise 9.1** Confirm that $\sigma_{\triangleright}$ can be represented as

$$\sigma_{\triangleright} = \int_0^1 \sqrt{s}\, ds = \frac{2}{3}$$

**Exercise 9.2** Show that

$$\lim_{u \to 0} \sigma_{\cup}(u) = 1/2$$

$$\lim_{u \to 1} \sigma_{\cup}(u) = 2/3$$

using Eq. (9.18).

# 10

# Schottky–Nordheim Barrier

*Leap o'er the walls for refuge in the field ...*

W. Shakespeare, Henry 6.1 II.2

## 10.1 SN Shape Factors

The consideration of the common, or better, canonical emission equations of Sections 8.2–8.4 has revealed that all can be interpreted to rely on representations of the Gamow factor $\theta(E)$ and the energy slope factors $\beta_T$ and $\beta_F(E)$. In the canonical equations, these quantities have prescribed forms, but a generalization to other barriers can use the mechanics of Eqs. (8.7) and (8.8) from which to fashion equations for current density for other barriers and interfaces. Of the alternatives to the triangular barrier model, the previous and abbreviated treatment of the Schottky–Nordheim (SN) barrier $V_{sn}(x)$ is by far the most important of the tunneling barriers, and so an extended treatment of it is undertaken first. After its consideration, various ways to generalize it and even supplant it are considered.

The form of the SN barrier was originally encountered in Eq. (2.9) and first treated in Section 2.1.3 to treat the RLD and FN equations briefly. It, too, does not have a constant associated shape factor: $\sigma_{sn}(E)$ varies between the limiting cases of the triangular $\sigma_{\triangleright} = 2/3$ and parabolic $\sigma_1 = \pi/4$ barriers (recall that the power of $x$ in $V(x)$ is $2/n$ for the $\sigma_n$ shape factors of Eq. 9.8) as can be intuited from Figure 10.2 (see Algorithm B.3.2). That behavior can be anticipated by rewriting $V_{sn}(x)$ as

$$V_{sn}(x) - E = \frac{-Q + (V_o - E)x - Fx^2}{x} \tag{10.1}$$

An illuminating representation of $V(x)$ is then

$$V(x) - E = \frac{F}{x}\left[x - x_-(E)\right]\left[x_+(E) - x\right] \tag{10.2}$$

When the roots are far apart, then $Q$ is small compared to the other terms in the numerator for much of the barrier, and when the roots are close together, then the denominator does not depart much from $x_o$ and so is constant to a first approximation. The zeros $x_\pm$ defined by $V(x_\pm) - E = 0$ are

$$\begin{aligned} x_\pm(E) &= \frac{\Delta}{2F} \pm \frac{L}{2} \\ L(E) &= \frac{1}{F}\sqrt{\Delta^2 - 4QF} \\ \Delta(E) &\equiv V_o - E = \mu + \Phi - E \end{aligned} \tag{10.3}$$

*Fundamentals of Electron Emission Physics*, First Edition. Kevin L. Jensen.

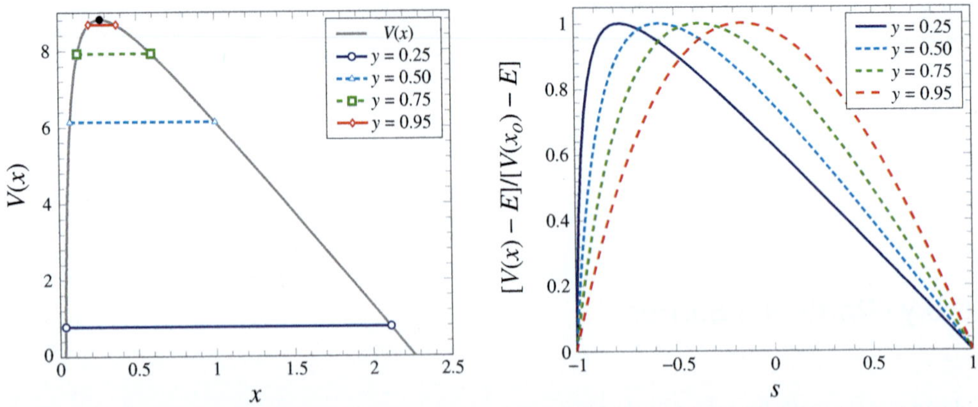


**Figure 10.1** (*left*) The Schottky–Nordheim barrier for copper-like $\mu = 7$ eV, $\Phi = 4.5$ eV, and $F = 5$ eV/nm. Horizontal lines are $E(y)$ as found from Eq. (2.12). (*right*) The Schottky–Nordheim barrier $V_{SN}(x) - E$ for energies shown on left, where $x(s) = (x_o/y)[1 + s\sqrt{1-y^2}]$ (see Eq. 10.4) as $s$ ranges from −1 to 1.

In terms of $y(E) = \sqrt{4QF}/\Delta$ (a rewriting of Eq. 2.12), the zeros are

$$x_\pm(E) = \frac{x_o}{y(E)}\left[1 \pm \sqrt{1 - y(E)^2}\right] \tag{10.4}$$

where $x_o = \sqrt{Q/F}$ is the location of the maximum of $V(x)$. Several relationships immediately follow: $(i)\, x_+x_- = x_o^2$ is independent of $E$, $(ii)\, y \to 1$ as $E \to \mu + \phi$, and $(iii)$ near $y \approx 1$, the roots $x_\pm \to x_o$ and the tunneling barrier becomes parabolic in $x$ to leading order because

$$\lim_{y\to1}\left[(x_+ - x)(x - x_-)\right] = \lim_{y\to1}\left[-x^2 - x_o^2 + \frac{2xx_o}{y}\right] = -(x - x_o)^2 \tag{10.5}$$

A representation of $V_{SN}(x)$ in dimensionless coordinates is shown for various energies $E$ in Figure 10.1. For $y \to 0$, the barrier increasingly resembles a triangular barrier because $x_-(E) \to 0$ in Eq. (10.2); conversely, for $y \to 1$, the barrier begins to appear parabolic in nature, as only the top rounded portion of the SN barrier is above $E$. When used to describe a metal, $y$ has a minimum value dictated by Eq. (2.12) of $y_{min} = \sqrt{4QF}/(\mu + \Phi) = 0.233326$ when $E = 0$ for the copper-like parameters of Figure 10.1 (see Algorithm B.3.2).

$V_{sn}(x)$ allows for the determination of the SN barrier reduction factor $\sqrt{4QF}$, and so affects the form of the Richardson–Laue–Dushman (RLD) equation by replacing $\Phi \to \phi = \Phi - \sqrt{4QF}$, where $V_{sn}(x)$ is profoundly important in the treatment of field emission. Since the present section is devoted to the SN barrier, it is convenient to forgo the "*sn*" subscript henceforth. The modifications the image charge term $Q/x$ makes to the triangular barrier Fowler–Nordheim (FN) equation of Eq. (8.3) give rise to the elliptical integral functions $v(y)$ and $t(y)$ (*aka* SN functions in the terminology of Forbes and Deane [91] of Eq. (8.15) and which are often used in the literature (but not in the shape factor method). Thus, before examining the SN barrier using the shape factor method, first describe it using conventional methods.

## 10.2 Standard Fowler–Nordheim

The genesis of the modern formulation of the FN equation is indebted to Murphy and Good [264], particularly for how they introduce the elliptical integral functions $v(y)$ and $t(y)$ that dress the $B_o$ and $A_o$ terms of Eq. (8.15), respectively. Approximations to those functions are often sought when

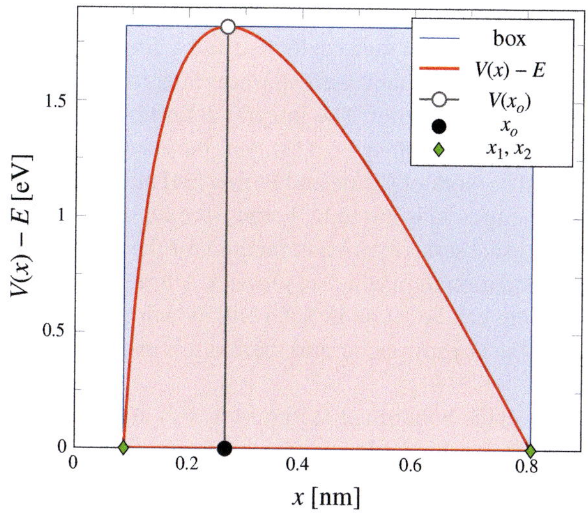


**Figure 10.2** Schottky–Nordheim barrier $V(x) - E = \mu + \Phi - Fx - Q/x - E$ for $\mu = E = 7$ eV, $\Phi = 4.5$ eV, and $F = 5$ eV/nm. Green diamonds are $x_\pm$ where $V(x_\pm) - E = 0$ and $L(E) = x_2 - x_1$. Black • is $x_o$, and open ∘ is $V(x_o)$. Blue box has an area of $(V(x_o) - E)L(E)$ and is related to $\theta_o(E)$.

comparing to field emission data, an early approximation being $v(y) \approx v_o - y^2$ and $t(y) = \sqrt{1.1}$ due to Spindt et al. [329], but by far the most significant approximation to these functions, and therefore to the very form of $J_{FN}(F)$, are due to the emendations of Forbes and Deane [91]. A modern formulation of the standard approach to the FN equation for field emission is given by Forbes [88, 90], with a bit more quirky rendition given here.

To account for image charge modifications to the triangular barrier model, Fowler and Nordheim advocated linearizing the Gamow factor $\theta(E)$ about the chemical potential $\mu$. As a result, they used the approximation

$$\theta(E) \approx \theta(\mu) + \beta_F(\mu)(\mu - E) \tag{10.6}$$

throughout their derivation of $J_{FN}(F)$ for any field, rather than searching for the value of $E_m$ for which the current density integrand was at its maximum, as done below. Consequently, the $y$ that appears as the argument of the elliptical integral functions $v(y)$ and $t(y)$ is actually $y(\mu)$ from Eq. (2.12), and so will be rendered as $y_o = y(\mu)$ below. The approximations they used are

$$\begin{aligned} \theta_{FN}(\mu) &= \frac{B_o}{F} v(y_o) \rightarrow \frac{B_{FN}}{F} \\ \beta_F(\mu) &= \frac{2\sqrt{2m\Phi}}{\hbar F} t(y_o) \rightarrow C_{FN} \end{aligned} \tag{10.7}$$

using Eq. (8.15) and result in the Murphy–Good form of the FN equation by amending Eq. (8.15) terms, resulting in

$$J_{MG}(F) \equiv \frac{A_o}{\Phi t(y_o)^2} F^2 \exp\left(-\frac{B_o}{F} v(y_o)\right) \tag{10.8}$$

with the underlying approximation that $P(\mu) = 1$. Eq. (10.8) is the basis of the canonical equation of field emission that will be given in Eq. (10.16) below.

The actual integral form of $v(y)$ (the related $t(y)$ requires its derivative) can be inferred from rendering the Gamow factor $\Theta(E)$ as the product of a dimensioned coefficient and a dimensionless integral, but even that simplified form lacks utility, and the tabulated elliptical integral functions from which $v(y)$ can be constructed are only marginally better. The integral definitions of these functions are in Murphy and Good [264]: $v(y)$ is given by their Eq. (13), and the relationship to the standard elliptical integrals by their Eq. (14). The work of Deane and Forbes [54] builds on and justifies enormously useful approximations to the functions found by Forbes (see Eq. 10.14), and Forbes and Deane have, moreover, developed an exact series expansion method for the evaluation of these functions and examined the differential equation from which $v(y)$ arises. Alternate integral and series expansions of the $v(y)$ and $t(y)$ functions can be found in Ref. [149], which approaches them from integrals arising in the evaluation of the Gamow factor and therefore is useful for isolating their asymptotic limits.

For pedagogical purposes, however, having a single unadorned integral for $v(y)$ allows for its rapid numerical calculation, against which various fashionable analytic approximations can be compared. The procedure here will be to consider a proposed analytic form for $v(y)$ and deduce from it a form for $t(y)$ making use of a relation highlighted by Murphy and Good [264]

$$t(y) \equiv v(y) - \frac{2}{3} y \frac{d}{dy} v(y) \tag{10.9}$$

For calculational simplicity, then, use the integral form of $v(y)$ given by [161]

$$v(y) \equiv (1-y) \int_0^{\pi/2} \frac{y + \sin^2 \xi}{\sqrt{2y + (1-y)\sin^2 \xi}} d\xi \tag{10.10}$$

from which $v(1) = 0$ and $v(0) = 1$ are readily shown. The integrand is designed to be smoothly varying and simple, so as to make numerical methods for its evaluation efficacious. Using Eq. 10.9 then gives

$$t(y) \equiv \int_0^{\pi/2} \frac{(1-y)(3-2y)\sin^4 x + y(9-5y)\sin^2 x + 4y^2}{3[2y + (1-y)\sin^2 x]^{3/2}} dx \tag{10.11}$$

The evaluation of these functions by numerical integration gives the values shown in Table 10.1. In particular, such integrals enable finding the functions at the endpoints analytically, e.g., $t(1) = \sqrt{2\pi}/4$ (see Ex. 10.7).

Consider two early and widely used approximations to $v(y)$ from the literature: $v_{sb}(y)$ is from the foundational work by Spindt et al. [329] and $v_c(y)$ is based on the cosine function. From them, the approximations to $t(y)$ are easily found

$$v_{sb}(y) = 0.95 - y^2; \quad t_{sb}(y) = 0.95 + \frac{y^2}{3} \tag{10.12}$$

$$v_c(y) = \cos(\pi y/2); \quad t_c(y) = \cos(\pi y/2) + \frac{\pi y}{3} \sin(\pi y/2) \tag{10.13}$$

They are reasonably accurate approximations and the former in particular allows for simplifications in Eq. (10.7). The errors of these approximations compared to the exact results of Table 10.1, are shown in Figure 10.3 (see Algorithm B.3.2). Before the routine availability of breathtaking computational power on hand-held electronics that eclipse the slide rulers of a previous generation, errors of a few percent were tolerable, especially when the approximations reduced the computational overhead endured in the tedious evaluations behind even simple plots.

**Table 10.1** Elliptical integral functions $v(y)$ and $t(y)$ numerically calculated from Eqs. (10.10) and (10.11).

| $y$ | $v(y)$ | $t(y)$ | $y$ | $v(y)$ | $t(y)$ |
|---|---|---|---|---|---|
| 0.00 | 1.000000 | 1.000000 | 0.55 | 0.635111 | 1.050177 |
| 0.05 | 0.994772 | 1.001118 | 0.60 | 0.576815 | 1.056588 |
| 0.10 | 0.981682 | 1.003618 | 0.65 | 0.515209 | 1.063113 |
| 0.15 | 0.962183 | 1.007030 | 0.70 | 0.450413 | 1.069734 |
| 0.20 | 0.937037 | 1.011124 | 0.75 | 0.382531 | 1.076435 |
| 0.25 | 0.906766 | 1.015750 | 0.80 | 0.311663 | 1.083201 |
| 0.30 | 0.871763 | 1.020801 | 0.85 | 0.237896 | 1.090023 |
| 0.35 | 0.832341 | 1.026197 | 0.90 | 0.161314 | 1.096889 |
| 0.40 | 0.788760 | 1.031877 | 0.95 | 0.081992 | 1.103791 |
| 0.45 | 0.741240 | 1.037792 | 1.00 | 0.000000 | 1.110721 |
| 0.50 | 0.689968 | 1.043903 | | | |

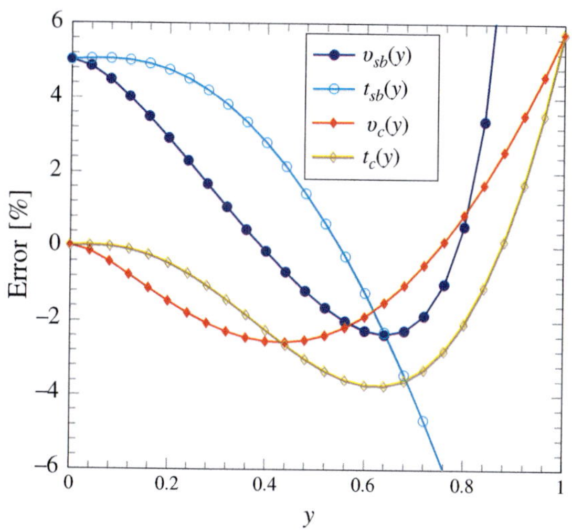


**Figure 10.3** Errors ($Err = 100 \times [1 - (f_a/f)]$ where $f_a$ is the approximation and $f$ the exact result) associated with the usage of Eqs. (10.12) for "sb" and (10.18) for "cos" compared to the exact results of Table 10.1.

For the present generation, far more is presumed. The approximations of Forbes and Deane [91], borne of an intuitive insight by Forbes [84], have emerged as the favored form of the SN functions: they proposed

$$\begin{aligned} v_{fd}(y) &\approx 1 - \frac{y^2}{3}[3 - \ln(y)] \\ t_{fd}(y) &\approx 1 + \frac{y^2}{9}[1 - \ln(y)] \end{aligned} \tag{10.14}$$

forms which have an absolute error of less than 0.4% as shown in Figure 10.4: the differences between them and the exact forms are close to indistinguishable to the naked eye when plotted.

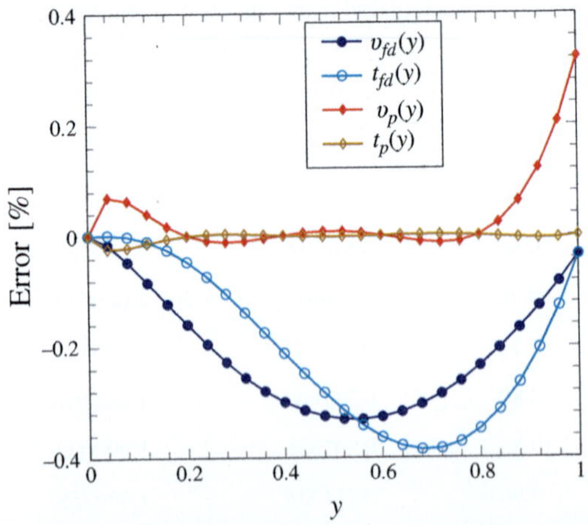


**Figure 10.4** Errors ($Err = 100 \times [1 - (f_a/f)]$ where $f_a$ is the approximation and $f$ the exact result) associated with the usage of Eqs. (10.14) for "fd" and (10.18) for "p" compared to the exact results of Table 10.1.

Through the use of additional terms, the accuracy of $v_{fd}(y)$ and $t_{fd}(y)$ can be improved further [54]. Although the shape factor method of Section 10.3 will become the preferred approach for the emission equations here, it is worth emphasizing a most agreeable feature of Eq. (10.14). Start with

$$\begin{aligned} \theta(\mu) &= \frac{4}{3\hbar F}\sqrt{2m\Phi^3}\, v(y_o) \\ \beta_F(\mu) &= \frac{2}{\hbar F}\sqrt{2m\Phi}\, t(y_o) \end{aligned} \tag{10.15}$$

where $y_o = y(\mu)$. Use of the Forbes–Deane approximations render the Murphy–Good form of the FN equation in a compact and eminently usable form that is hard to resist because of how it assists analysis. Using Eq. (10.14) in Eq. (10.8) results in the Forbes–Deane form of $J_{MG}(F)$ given by

$$\begin{aligned} J_{MG}(F) &\to A_{MG}\left(\frac{F}{B_o}\right)^{2-\nu} \exp\left(-\frac{B_o}{F}\right) \\ A_{MG} &\equiv \frac{A_o B_o^2}{\Phi t(y_o)^2}\left(\frac{\Phi^2 e^6}{4QB_o}\right)^{\nu} \end{aligned} \tag{10.16}$$

where $A_o$ and $B_o$ are familiar from Eq. (8.15) and $\nu = (8Q/9\hbar)\sqrt{2m/\Phi}$. As far as numerical evaluation goes, Eq. (10.16) [152, 183] is equivalent to Eq. (6.3) of Forbes and Deane [91], but it is designed so that the terms in parentheses are rendered dimensionless. A minor tweak to the equation is to replace the weakly varying $t(y_o)$ with a constant term $t_o$, where the value of $t_o = 1.06131$ is motivated by its connection to the familiar Spindt–Brodie approximation (see Ex. 10.6). In either case, it makes taking theoretical ratios of field emission current density values convenient, as those ratios do not involve any functions disagreeable to compute, and so intuitive comparisons are enabled to investigate evolution of the current density when, for example, the field is incrementally increased.

Importantly, Eq. (10.16) reveals a defect in the standard representation of field emission data. A "Fowler–Nordheim" plot shows $\ln[J_{FN}(F)/F^2]$ *vs.* $1/F$, and by virtue of Eq. (8.14) is often suggested to be a straight line. And yet, as rued by theorists of the discipline that came to be known as vacuum nanoelectronics, and as heard at past conferences, "everything looks linear on a log plot."[1] The issue here can be inferred from considering representative values for $\Phi = 4.5$ eV that give $B_o = 65.2073$ eV/nm, $\nu = 0.772808$ and $(\Phi^2 e^6/4QB_o) = 87.0050$, showing that $\nu$ in particular is of order unity. The form of Eq. (10.16) then shows instead that $\ln[J_{MG}(F)/F^{2-\nu}]$ *vs.* $1/F$ should be the straight line. That, taken with the fact that current and not current density is measured means that other features, such as emission area and statistical distribution of emitters, will further monkey with the power to which $F$ is raised. The presumption that an FN plot should use $\ln[J(F)/F^2]$ is therefore misguided, and in fact a Millikan plot of $\ln(J)$ *vs.* $1/F$ is preferable for many reasons [89], not the least of which is that it more quickly allows for the inference of what current density actually is. Still, FN plots are often used.

Although $J_{MG}(F)$ was derived for zero temperature, Murphy and Good [264] recognized that field emitters can run warm, and indeed, they were interested in the cross-over from field emission to thermal emission. To account for temperature effects, they appended a term that arose when $\beta_T$ was finite, which they expressed as

$$J_{MG}(F, T) = J_{MG}(F)\frac{\pi k_B T C_{FN}}{\sin(\pi k_B T C_{FN})} \tag{10.17}$$

and which shall be derived in Eq. (12.7). This is less useful than it appears, although it can account for room temperature or near-room-temperature conditions. First, the impact of temperature will be correctly accounted for by the general thermal-field equation below. Second, the MG correction is singular when $C_{FN} = 1/k_B T$ and so the temperature-dependent addition cannot track the entry of current density into the thermal-field regime from the field regime side, thereby setting an upper temperature for the utility of the formula. Practically, this means that for $F = 2$ eV/nm and $\Phi = 4.5$ eV, then temperature is restricted to being $T \leq 1/2k_B C_{FN} = 509$ K. Consequently, Eq. (10.17) is not considered further. Below, $J_{MG}(F)$ shall refer to $J_{MG}(F, T = 0)$ exclusively.

The final high-accuracy analytic approximation to $v(y)$ and $t(y)$ to consider will motivate a method to be taken up next (when the shape factor approach is applied to the SN barrier in Section 10.3): the polynomial fit. The functions $v(y)$ and $t(y)$ are approximately quadratic and linear in $y$, respectively – but not exactly. Following the methods of Appendix A.6, by selecting $N + 1$ data values from Table 10.1, a unique polynomial of degree $N$ may be constructed that exactly reproduces the $N + 1$ chosen values of $y$, which need not be equi-spaced but for convenience are. For a quartic fitting, $N = 4$, leading to

$$v_p(y) \equiv \sum_{n=0}^{N} C_n y^n; \quad t_p(y) \equiv \sum_{n=0}^{N} D_n y^n \tag{10.18}$$

The solution of such equations can be rapidly found using the conventional linear algebra methods of Appendix A.6 to obtain the $C_n$ (and similarly, the $D_n$) in Table 10.2. Observe that the table of coefficients is such that $\sum_{n=0}^{N} C_n = 0$ because $v(1) = 0$, and $\sum_{n=0}^{N} D_n = 1.110720$ because $t(1) = \sqrt{2}\pi/4$. The errors associated with the polynomial approximations $v_p(y)$ and $t_p(y)$ are also shown in Figure 10.4, where their performance compares favorably with (and for intermediate $y$ values, even better than) the much more aesthetic Forbes–Deane approximation. The error for $v_p(y)$

1 The comment was made by Paul Cutler in response to a claim that the linearity of data on an FN plot indicated tunneling. He made the response at the *Int'l Vacuum Microelectronics Conference* (IVMC) in Nagahama, Japan in 1991.

**Table 10.2** Polynomial approximation to $\mathbb{F}_n[p(y)]$ of Eq. (10.27). Note: $\mathbb{F}_n(0) = C_0 = \pi/2$ and $\mathbb{F}_n(1) = \sum C_j = 4\sqrt{2}/3$. *Err* is error associated with the polynomial approximation to $\mathbb{F}_n(p)$ for that $n$, the magnitude of which is less than the value shown.

| $y$ | $x(y)$ | $p(y)$ | $\mathbb{F}_n[p(y)]$ | $j$ | $C_j$ |
|---|---|---|---|---|---|
| n = 2 | | | | | Err < 0.109% |
| 1.000000 | 0.00 | 0.000000 | 1.570796 | 0 | 1.570796 |
| 0.333333 | 0.50 | 0.942809 | 1.794543 | 1 | 0.580164 |
| 0.000000 | 1.00 | 1.000000 | 1.885618 | 2 | −0.265342 |
| n = 4 | | | | | Err < 0.026% |
| 1.000000 | 0.00 | 0.000000 | 1.570796 | 0 | 1.570796 |
| 0.600000 | 0.25 | 0.800000 | 1.699456 | 1 | 0.591405 |
| 0.333333 | 0.50 | 0.942809 | 1.794543 | 2 | −0.335370 |
| 0.142857 | 0.75 | 0.989743 | 1.858163 | 3 | 0.131396 |
| 0.000000 | 1.00 | 1.000000 | 1.885618 | 4 | −0.072609 |
| n = 6 | | | | | Err < 0.008% |
| 1.000000 | 0.000000 | 0.000000 | 1.570796 | 0 | 1.570796 |
| 0.714286 | 0.166667 | 0.699854 | 1.660529 | 1 | 0.590150 |
| 0.500000 | 0.333333 | 0.866025 | 1.734689 | 2 | −0.334254 |
| 0.333333 | 0.500000 | 0.942809 | 1.794543 | 3 | 0.178278 |
| 0.200000 | 0.666667 | 0.979796 | 1.840515 | 4 | −0.250222 |
| 0.090909 | 0.833333 | 0.995859 | 1.871988 | 5 | 0.226839 |
| 0.000000 | 1.000000 | 1.000000 | 1.885618 | 6 | −0.095969 |

is somewhat deceptive: $v(1) = 0$ exactly, and so any small error in its approximation looks disproportionately large near $y \approx 1$. Still, that such an inelegant solution actually works suggests a moral: even models of great elegance may be computationally less useful than crude fits. This practical conclusion does not outweigh the beauty behind Eq. (10.16). Even so, such a moral has application to the next method to calculate the Gamow factor based on the shape factor method, where the end-points used in the polynomial fit are analytically known, leaving the intermediate points to be simply calculated by any number of accurate numerical means.

## 10.3 Shape Factor FN Model

The most important advantage of the shape factor method is that it is applicable to more general barriers, and in particular, SN barriers modified by space charge or curvature of the emitter. Nevertheless, to demonstrate the method, the standard SN barrier will be examined in detail so that the shape factor $\sigma_{sn}$ associated with it can suggest forms of $\Theta(\mu)$ and $\beta_F(\mu)$ that may be compared to their FN competitors $B_{FN}$ and $C_{FN}$ of Eq. (10.7). The numerical evaluation of $\sigma_{sn}$ is routine, and

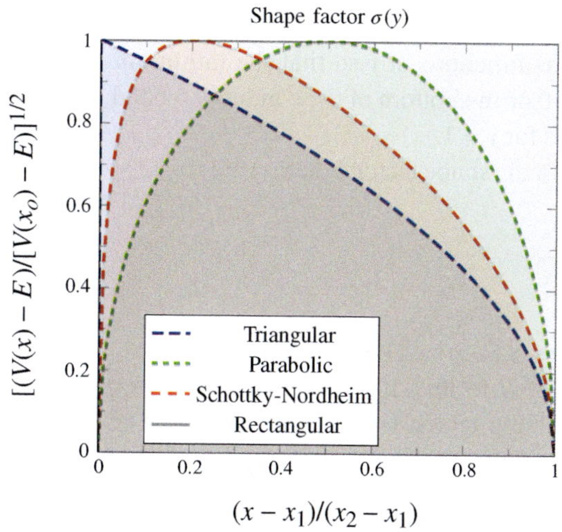


**Figure 10.5** Shape factor $\sigma(E)$ integrand for the SN barrier shown in Figure 10.2 for $\mu = 7$ eV, $\Phi = 4.5$ eV, $F = 4$ eV/nm, and $E = \mu - 0.5$ eV, compared to its limiting cases of triangular for $\sqrt{4QF}/\Phi \to 0$), and parabolic for $E \to \mu + \Phi - \sqrt{4QF}$). The blue shaded region is $\sigma_{\triangleright} = 2/3$, and the green shaded region is $\sigma_{\cap} = \pi/4$.

the behavior of its integrand in Figure 10.5 (see Algorithm B.3.2) clearly shows that $\sigma_{\triangleright} < \sigma_{sn} < \sigma_{\cap}$, that is, the SN shape factor lies above the triangular barrier and below the parabolic barrier shape factors.

Reconsider now the treatment of Section 3.1 where $y$ was treated as a function of energy $E$ and the roots of $V_{sn}(x) - E = 0$ found accordingly. It is now profitable to instead treat $E$ as a function of $y$, and therefore express both the Gamow factor and field tunneling factor as functions of $y$ as well. The numerator of Eq. (10.1) is a quadratic polynomial and has two real roots for $V(x_j) - E = 0$ when $V_o > E$. The maximum $x_o = \sqrt{Q/F}$ is independent of energy, and so $0 < x_1 < x_o$ and $x_o < x_2 < (V_o - E)/F$. Specifically, the width of the tunneling barrier for an electron at an energy $E$ is explicitly

$$\begin{aligned} x_j(E) &= \frac{\Delta}{2F} + \frac{(-1)^j}{2F}\sqrt{\Delta^2 - 4QF} \\ L(E) &= \frac{1}{F}\sqrt{\Delta^2 - 4QF} \\ \Delta(E) &= \mu + \Phi - E \end{aligned} \tag{10.19}$$

By virtue of the Lagrange interpolation formula (Eq. (A.44) Appendix A.6), the numerator can be represented by its zeros (compare Eq. 10.2)

$$V_{sn}(x) - E = \mu + \Phi - E - Fx - \frac{Q}{x} = \frac{F}{x}(x_2 - x)(x - x_1) \tag{10.20}$$

Consequently, Eq. (2.12) can be reconfigured as

$$x_{\pm}(y) = \frac{x_o}{y} \pm \frac{L}{2} \Rightarrow x_2 = x_+, \;\; x_1 = x_- \tag{10.21}$$

$$L(y) = x_+ - x_- = \frac{2x_o}{y}\sqrt{1 - y^2} \tag{10.22}$$

$$\kappa(y) = \frac{2}{\hbar}\left[\frac{mQ}{x_o}\frac{(1 - y)}{y}\right]^{1/2} \tag{10.23}$$

where $y(E) = \sqrt{4QF}/\Delta(E)$. Although all of these parameters are normally defined in terms of $E$, it is mathematically simpler to treat them here as functions of $y$ so that the minimum value of $y_{min} = \sqrt{4QF}/(\mu + \Phi)$ to be encountered is at $E = 0$ or the bottom of the conduction band, and the maximum $E$ that still constitutes tunneling occurs for $y = 1$.

In this revised outlook, the usage of Eq. (10.20) in the shape factor equation of Eq. (9.1) results in

$$\sigma_{sn}(y) = \int_{x_-}^{x_+} \left[ \frac{x_o(x_2 - x)(x - x_1)}{x(x_2 - x_o)(x_o - x_1)} \right]^{1/2} \frac{dx}{x_+ - x_-} \tag{10.24}$$

which is the form when $V(x) - E$ has two roots (see Section 11.3 for a generalization). Note the ordering of the roots $x_j$ and $x$ compared to $V(x) - E$ in Eq. (10.20): it matters for $V(x)$, but for the shape factor, the sign of $(x - x_j)$ in the tunneling region $(x_- \leq x \leq x_+)$ matches $(x_o - x_j)$ in the denominator of $\sigma$ in Eq. (10.24) so that $\sigma > 0$. Note also that $x_2 = x_+$ and $x_1 = x_-$. Making a change of variables $x \equiv [(x_2 + x_1) + (x_2 - x_1)\cos\varphi]/2$ in Eq. (10.24) results in

$$\sigma_{sn}(y) = \frac{\sqrt{1+y}}{2\sqrt{2}} \int_0^{\pi} \frac{\sin^2\varphi \, d\varphi}{\left[1 + \sqrt{1 - y^2}\cos\varphi\right]^{1/2}} \tag{10.25}$$

a form that can be used to analytically demonstrate that $\sigma_{sn}(0) = \sigma_{\triangleright} = 2/3$ (the triangular barrier) and $\sigma_{sn}(1) = \sigma_{\cap} = \pi/4$ (the parabolic barrier). The Gamow factor $\theta(E)$ can now be rewritten as a function of $y$ rather than a function of $E = V_o - \sqrt{4QF}/y$ by

$$\begin{aligned}
\theta_{sn}(y) &= \frac{2\sqrt{2mQx_o}}{\hbar} \left\{ \frac{(1 - y^2)}{y^{3/2}} \mathbb{F}\left[p(y)\right] \right\} \\
\mathbb{F}(p) &\equiv \int_0^{\pi} \frac{\sin^2\varphi}{\sqrt{1 + p\cos\varphi}} d\varphi \\
p(y) &\equiv \sqrt{1 - y^2}
\end{aligned} \tag{10.26}$$

where $F = Q/x_o^2$ to result in a cleaner equation. The boundary cases of $\mathbb{F}[p(y)]$ are analytic: $\mathbb{F}(1) = 4\sqrt{2}/3$, and $\mathbb{F}(0) = \pi/2$. This allows a very convenient simple approximation that is nevertheless quite accurate [156, 157]

$$\begin{aligned}
\sigma_{sn}(y) &\approx \frac{\sqrt{1+y}}{2\sqrt{2}} \mathbb{F}_n\left[p(y)\right] \\
\mathbb{F}_n\left[p(y)\right] &\equiv \sum_{j=0}^{n} C_j \left[\frac{1-y}{1+y}\right]^j \\
\mathbb{F}'_n\left[p(y)\right] &\equiv \sum_{j=1}^{n} j C_j \left[\frac{1-y}{1+y}\right]^{j-1}
\end{aligned} \tag{10.27}$$

The values of $C_j$ for several different $n$ are shown in Table 10.2, and where $\mathbb{F}'_n[p(y)]$ is defined for later usage, corresponding to the derivative $\mathbb{F}' = \partial_x \mathbb{F}_n$ with $x = (1 - y)/(1 + y)$. Even for the lowest $n = 2$ case the absolute accuracy is better than 0.11% so that the difference between $\mathbb{F}$ and $\mathbb{F}_n$ are not easily discernible: consequently, for standard numerical evaluations, $\mathbb{F}_2[p(y)]$ is sufficiently accurate and $\mathbb{F}_4[p(y)]$ is very accurate, making the shape factor method a hair's breadth away from "exact" for numerical purposes.

Although not readily apparent in Eq. (10.26), it can nevertheless be shown through the definitions of $\sigma_{sn}(y)$ combined with $\kappa(y)$ and $L(y)$ through Eqs. (10.22) and (10.23) that

$\theta_{sn}(y) = 2\sigma_{sn}(y)\kappa(y)L(y)$ as needed. The usage of the approximation $\mathbb{F}[p(y)] \approx \mathbb{F}_n[p(y)]$ makes the evaluation of $\theta_{sn}$ computationally simple, and therefore rapidly executed. One more term, however, is needed in order to use Eq. (8.12) for the evaluation of current density: the field energy slope factor $\beta_F(y)$. For that, the derivative with respect to $E$ needs to be re-expressed for the SN barrier as

$$\begin{aligned}\beta_F[E(y)] &= -\frac{d}{dE}\theta(E) \to -\frac{dy}{dE}\frac{d}{dy}\theta(y)\\ &= -\frac{y^2}{\sqrt{4QF}}\frac{d}{dy}\theta(y) \to \beta_F(y)\end{aligned} \tag{10.28}$$

with that, the derivative of $\theta_{sn}(y)$ becomes [157]

$$y^2\frac{d}{dy}\left[\frac{\sqrt{1-y^2}}{y^{3/2}}\right] = -\frac{3-y^2}{2\sqrt{y(1-y^2)}} \tag{10.29}$$

$$y^2\frac{d}{dy}\mathbb{F}_n[p(y)] = -\frac{2y^2}{(1+y)^2}\mathbb{F}'_n[p(y)] \tag{10.30}$$

where $\mathbb{F}'$ is defined in Eq. (10.27). Combining these results gives

$$\begin{aligned}\beta_F(y) &= \frac{1}{\hbar}\sqrt{\frac{mx_o^3}{2Qy}}\left[(y^2+3)\mathbb{F}_n[p(y)] + 4y\left(\frac{1-y}{1+y}\right)\mathbb{F}'_n[p(y)]\right]\\ &= \frac{1}{\hbar}\sqrt{\frac{mx_o^3}{2Q}}\left\{\frac{1}{\sqrt{y}}\sum_{j=0}^{n}(y^2+4jy+3)C_j\left[\frac{1-y}{1+y}\right]^j\right\}\end{aligned} \tag{10.31}$$

Clearly, $\beta_F(y)$ is nearly as easy to calculate as $\theta_{sn}(y)$, and both are freed from having to numerically evaluate the integral implicit in Eq. (6.92) for the Gamow factor or Eq. (10.26) for $\mathbb{F}[p(y)]$.

Although applied to the SN barrier, the method of expressing $\theta(y)$ in terms of $y$-dependent coefficients from $2\sigma\kappa L$ and a generally smoothly varying polynomial originating from $\sigma(y)$ is in fact a general method: variations on it can be used for more complex potentials characteristic of space charge, interface barriers, and curved trajectories from hyperbolic emitters [165, 183], all of which shall be investigated in time. For the SN barrier, Figure 10.5 makes clear that $\sigma_{\triangleright} \le \sigma_{sn}(y) \le \sigma_{\cap}$, and indeed, as direct numerical evaluation confirms, $\sigma_{sn}(y)$ smoothly transitions between them as shown in Figure 10.6 (see Algorithm B.3.2).

As a once-off calculation, the motivation to consider $\theta(y)$ and $\beta_F(y)$ instead of the direct numerical evaluation of Eq. (9.3) for $\theta(E)$ and from it a finite-difference numerical approximation to $\beta_F(E)$ hardly seems worth the effort, but then a moment's thought[2] brings home the appreciation that rarely is the Gamow factor or the field energy slope factor required in isolation. The two functions are repeatedly evaluated to determine the location of $E_m$ in the current density integrand, and then repeatedly used if $N[\theta(E)]$ is used instead of its approximation $N(n, s)$ to find $J(F, T)$. When evaluations are repetitively required, as they are in beam simulation codes, then "repetitively" means millions of places over a surface mesh describing the cathode and millions of times during the evolution of the beam that comes off of it. That computational overhead is exhausting and time-consuming, and awareness of it is behind all the incessant caterwauling about "computational utility" here. Continuing, recast $\theta(y)$ and $\beta_F(y)$ as

$$\theta(y) = \frac{2\sqrt{2mQx_o}}{\hbar}h_\theta(y); \quad \beta_F(y) = \frac{1}{\hbar}\sqrt{\frac{mx_o^3}{2Q}}h_\beta(y) \tag{10.32}$$

2 The English poet Alfred E. Housman once sarcastically quipped that " ... thought is irksome and three minutes is a long time" but physics requires greater fortitude.

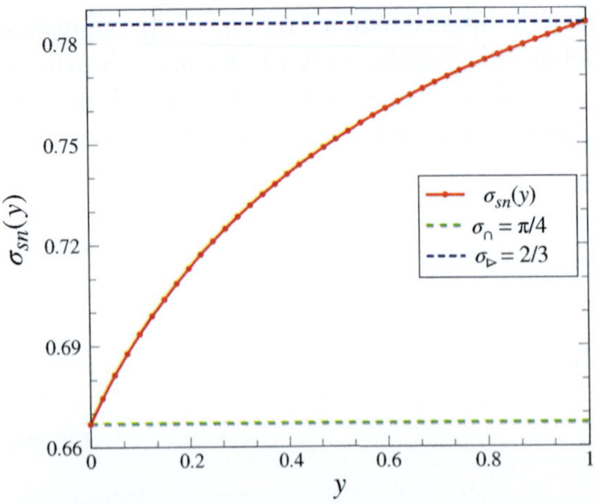


**Figure 10.6** The function $\sigma_{sn}(y)$ as evaluated using $\mathbb{F}_6[p(y)]$ (Table 10.2) and shown to be bounded by the triangular and parabolic shape factors $\sigma_{\triangleright}$ and $\sigma_{\cap}$.

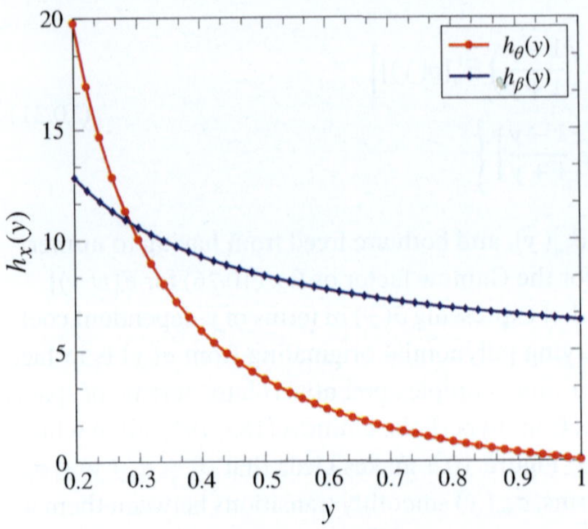


**Figure 10.7** The functions $h_\theta(y)$ and $h_\beta(y)$ used in Eq. (10.32).

is desirable, where the functions $h_\theta(y)$ and $h_\beta(y)$ are the factors in the curly brackets in Eqs. (10.26) and (10.31) respectively. Being smooth functions, they could even be easily interpolated or splined using a table of data for each, but that is not significantly more advantageous than rapidly calculating them from their series approximations. Their behavior is shown in Figure 10.7 (an analogous approach [173] gives essentially the same results, but differs in the integrals that require numerical evaluation). To relate them to the Gamow and field energy slope factor then only requires a straight-forward evaluation of the coefficients in Eq. (10.32), *for example*, if the field is such that $x_o = 0.25$ nm, then $(2/\hbar)\sqrt{2mQx_o} = 3.07386$ and $(1/\hbar)\sqrt{mx_o^3/2Q} = 0.533670$ [1/eV]. For a given field $F$, these coefficients do not change, and so variations in energy $E$ for the SN barrier become variations in $y$ via

$$E(y) = \mu + \Phi - \frac{\sqrt{4QF}}{y} \tag{10.33}$$

Finding $E_m$ numerically then amounts to finding $E(y_m)$ instead, which is now rapid because it no longer requires integrations of $\theta(E)$ as in Eq. (6.92) or $\sigma(E)$ as in Eq. (9.3) to determine its location. Rather, $\theta(y)$ and $\beta_F(y)$ are rapidly and accurately provided using Eqs. (10.26) and (10.31) with the approximations $\mathbb{F}_n[p(y)]$ and $\mathbb{F}'_n[p(y)]$.

## 10.4 Exercises

**Exercise 10.1** Consider an SN barrier $V_{SN}(x)$ (Eq. 2.9) and an electron at $E = \mu$. Letting $\Phi = 4.5$ and $F = 6$ eV/nm, find the following quantities: $x_o$, $\phi$, $\kappa(\mu)$, $L(\mu)$, and $x_\pm(\mu)$.

**Solution** $x_o = 0.26832$ nm; $\phi = 1.8168$ eV; $\kappa(\mu) = 6.90536$ [1/nm], $L(\mu) = 0.36125$ nm; $x_\pm(\mu) = (0.45 \pm 0.36125)$ nm.

**Exercise 10.2** For the same parameters as Ex. 10.1 find $y_o$, $v(y_o)$, and $t(y_o)$.

**Solution** $y_o = 0.596277$; $v(y_o) = 0.583175$; $t(y_o) = 1.05993$.

**Exercise 10.3** For the same parameters as Ex. 10.1 find $B^o_{FN}(\mu)$ and $A^o_{FN}$ (assuming $P(\mu) = 1$).

**Solution** $A^o_{FN} = 0.00213797[\text{q/fs eV}^2] = 3.42541 \times 10^{-7}$ A/eV$^2$; $B^o_{FN}(\mu) = 65.2073$ [eV/nm].

**Exercise 10.4** For the same parameters as Ex. 10.1 find the parameters of Eq. (10.16) of $A_{MG}$, $\nu$ and $\Phi^2 e^6/4QB_o$.

**Solution** $A_{MG} = 255.231$ [q/fs nm$^2$]. $\nu = 0.772808$. $\Phi^2 e^6/4QB_o = 87.0050$.

**Exercise 10.5** For the same parameters as Ex. 10.1 find $\sigma_{sn}(y_o)$. How well does $(B^o_{FN}/F)v(y_o)$ agree with $\theta(\mu) = 2\sigma(\mu)\kappa(\mu)L(\mu)$?

**Solution** Using Eq. (10.24), $\sigma_{sn}(y_o) = 0.75971$. The two approaches agree to within 0.35%.

**Exercise 10.6** Connect the Spindt–Brodie approximation of Eq. (10.12) to the Forbes–Deane approximation of Eq. (10.14) by writing the former as $v(y) = v_o - y^2$ and by Taylor expanding the later in $y^2$ about $y_c^2$ such that the coefficient of the $y^2$ term is $(-1)$. What is $y_c$, $v_o$ and $t(y_c)$?

**Solution** $y_c = 1/\sqrt{e}$. $v_o = 0.938687$. $t(y_c) = 1 + (1/6e) = 1.06131$.

**Exercise 10.7** From the definition of $v(y)$ in Eq. (10.10), find $v(1-\delta)$ as $\delta \to 0$. Show that this confirms $t(1) = \sqrt{2}\pi/4$. Compare to the same estimates using the Forbes–Deane approximation.

**Solution** $v(1-\delta) = (3\sqrt{2}\pi/8)\delta$. $v_{fd}(1-\delta) = 5\delta/3$. $t_{fd}(1) = 10/9$.

# 11

# Modified SN Barriers

*Hie thee hither,*
*That I may pour my spirits in thine ear,*
*And chastise with the valor of my tongue*
*All that impedes thee from the golden round ...*

W. Shakespeare, *Macbeth I.5*

## 11.1 Semiconductor Surface

The Schottky–Nordheim (SN) barrier of Eq. (2.9) was developed to account for the changes to a triangular barrier $V(x) = V_o - Fx$ wrought by the inclusion of an image charge modification to the barrier due to the departing (emitted) electron from a planar metallic conductor. Modifications to that barrier to account for material and non-planar conditions result in changes to $\theta[E(y)]$ and, therefore, to Eq. (8.12) as an estimate of current density. Four particular kinds of changes can be treated as extensions to the SN barrier formalism, with the more substantially different barriers waiting for their consideration below. The four changes are roughly characterized as changes due to

1. the surface being a semiconductor,
2. the image charge over a non-planar conductor departing from $Q/x$,
3. modifications due to surface curvature or space charge, and
4. the barrier being modified by quantum effects.

Consider each in turn.

Electrons in a metal are free to move in response to an electric field, and so are able to gather at the surface and shield out the effects of an external charge. It follows that there are no electric fields within the metal bulk: should one exist, electrons would accumulate somewhere until the field was eliminated. As a result, electric field lines terminate on, and are normal to, the surface of a metal. By comparison, electrons in insulators are bound, and so the action of an electric field is to pull them away from the positive atom to which they are attached, causing the atom to distort into a dipole. The dielectric that results is composed of an arrangement of little dipoles positioning themselves in response to an electric field. They reduce field penetration, but do not eliminate it.

*Fundamentals of Electron Emission Physics*, First Edition. Kevin L. Jensen.

Let the external or vacuum region outside the surface of a semiconductor be called Region 1, and the bulk dielectric material be in Region 2. The electric field $\vec{\mathcal{E}}_2$ in the material is related to the electric field $\vec{\mathcal{E}}_1$ in the vacuum by a dielectric constant $\varepsilon$ of the material by [78]

$$\left(\vec{F}_1 - \varepsilon \vec{F}_2\right) \cdot \hat{n} = 0 \tag{11.1}$$

where $\hat{n}$ is the normal to the surface and $\vec{F} = q\vec{\mathcal{E}}$. The tangential field components are continuous across the boundary.

A charge $q_s$ embedded in the dielectric will similarly cause the dipoles to reorient themselves in response, so that the potential energy associated with a test charge $q$ at a distance $r$ from $q_s$ satisfies

$$V(r) = \frac{qq_s}{4\pi K_s \varepsilon_0 r} = \frac{4Q}{r}\left(\frac{q_s}{q}\right) \tag{11.2}$$

where $K_s$ is such that the dielectric constant is $\varepsilon \equiv K_s \varepsilon_0$. To understand how the dipoles react to an external charge, remember how the free charges in a metal react: the charge distribution at the surface configures itself to resemble a mirror charge equidistant from the surface as the external charge, with an opposite sign. That no field exists inside the metal is visualized as an image charge of opposite sign but equal magnitude, located as far from the surface as the external charge but inside the material. When the metal is replaced by a semiconductor with a dielectric constant of $K_s$, the magnitude of the image charge changes from $q_s$ to $q_i < q_s$.

To find $q_i$, let a charge $q_s$ be at a distance $a$ outside the surface of a semiconductor characterized by $K_s$, and work in cylindrical coordinates. The distance from $q_s$ to an observation point at $(\rho, z)$ is $r_-$, where $r_\pm^2 = \rho^2 + (z \pm a)^2$. The potential energy is the sum of the effects of the external charge $q_s$ and its image $q_i$, and so

$$V(\rho, z > 0) = \frac{qq_s}{4\pi\varepsilon_0 r_-} - \frac{qq_i}{4\pi\varepsilon_0 r_+} \tag{11.3}$$

where $q_i$ is the image charge magnitude. It vanishes when $r_+ = r_-$ because $z = 0$. Inside the semiconductor

$$V(\rho, z < 0) = \frac{qq_s}{4\pi\varepsilon_0 r_-} - \frac{qq_i}{4\pi\varepsilon_0 r_-} \tag{11.4}$$

where $r_-$ is in the denominator of the second term because $q_i$ is in the vacuum. Now impose Eq. (11.1) to find

$$\varepsilon_0 \partial_z V(\rho, 0_-) = K_s \varepsilon_0 \partial_z V(\rho, 0_+) \tag{11.5}$$

Using $r_+ = r_-$ for $z = 0$, then collecting and simplifying, results in

$$q_s - q_i = K_s\left(q_s + q_i\right) \rightarrow q_i = \frac{K_s - 1}{K_s + 1} q_s \tag{11.6}$$

Writing down the image charge barrier for a semiconductor requires one last accommodation: semiconductors are better characterized by an electron affinity $\chi$ because (as shall be seen below) the electron density in the conduction band is generally so low that band bending occurs, and the Fermi level no longer serves as a good reference point. The bottom of the conduction band takes over as the reference, and so let $\chi$ be the distance from the bottom of the conduction band to the top of the barrier in the absence of fields or image charge effects. Thus, for a semiconductor, the dielectric constant $K_s$ modification of the image charge contribution $Q/x$ is required. To facilitate that generalization, let $Q \rightarrow Q_o \equiv \alpha\hbar c/4$, and let the semiconductor SN barrier be given by [333, 334]

$$V(x) \equiv \chi - Fx - \left(\frac{K_s - 1}{K_s + 1}\right)\frac{Q_o}{x} \equiv \chi - Fx - \frac{Q_s}{x} \tag{11.7}$$

which is seen to be equivalent to the SN barrier with the replacement $Q \rightarrow (K_s - 1)Q_o/(K_s + 1) \equiv Q_s$. The metal limit corresponds to $K_s \rightarrow \infty$ and $\chi \rightarrow \mu + \Phi$ accounts for the preference to discuss electron affinity rather than work function. Finite $K_s$, as Figure 11.1 shows, makes the image charge potential barrier somewhat higher than for metals.

The dielectric-modified image charge of Eq. (11.7) extends the SN barrier to bulk semiconductors, but it is often the case that semiconductors are applied as coatings that can be thin (down to the thickness of a few monolayers). The image charge model then becomes more difficult, as there are images of image charges that are needed to enforce boundary conditions (see the MIM barrier) below. A thin layer of atoms on a metal surface of thickness $l$ and dielectric constant $K_s$ results in [256]

$$V_{image}(x) \rightarrow \chi - Fx - \left(\frac{K_s - 1}{K_s + 1}\right)\frac{Q}{x} - \left(\frac{2}{K_s + 1}\right)\frac{Q}{x + l} \tag{11.8}$$

The last term is often small and its neglect is therefore common, but Modinos treats field emission from adsorbate-covered surfaces (Chapter 6, Ref. [256]) that will be seen to share features familiar from Gyftopoulos–Levine theory [118, 161] in treating how coatings like barium and cesium can reduce the work function at the surface of a metal.

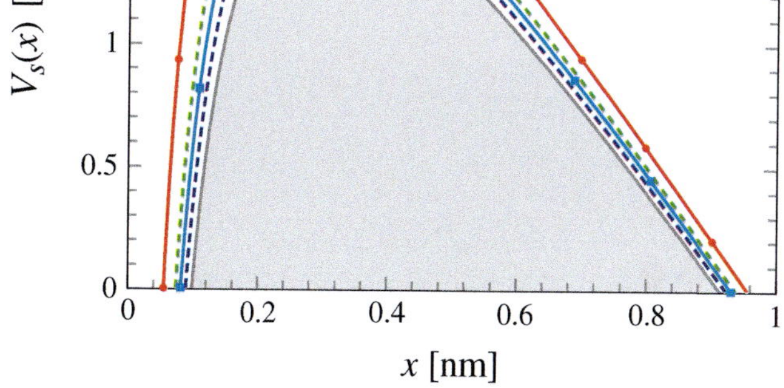


**Figure 11.1** Semiconductor image charge potential barrier based on Eq. (11.7) with $\chi = 4.05$ eV (after Silicon) and $F = 4$ eV/nm. Legend is $K_s$, with $K_s = \infty$ corresponding to a metal.

## 11.2 Non-planar Image Charge

The Method of Images is an often-used technique to readily find the potential associated with a charged particle outside a conductor surface of a convenient shape, the most common pedagogical case being that of a sphere, as the comparison of it to formalistically powerful approaches using expansion methods involving Legendre polynomials is very useful to highlight the power of each approach [78]. The solution for a charge outside a sphere is pedagogically aesthetic. The present account, based on Ref. [186], serves to introduce more challenging ellipsoidal and hyperbolic geometries.

The model assumes a metal sphere held at ground and positioned so its center is at the origin. Recall that in [efnq] units, the product of potentials with unit charges $q$ renders their combination as potential energy, so that speaking of potentials in units where $q = 1$ is directly analogous. Using the language of classical electrostatics is convenient for the moment, and so a distinction will not be drawn between the two. The sphere is held at zero potential, but the potential outside the sphere is due to both the external charge and the charge redistribution on the sphere itself. That redistribution serves to shield external fields from the interior of the sphere. Take the external charge to be $-q$ for simplicity, and let it be a distance $d$ from the origin, as shown in Figure 11.2. The potential energy between a test charge $q$ at the observation point $O$ due to the external charge $-q$ leads to $V(\vec{r}_2) = -4Q/r_2$, whereas the potential energy between a test charge $q$ and the image charge $+\lambda q$ is $V(\vec{r}_1) = 4\lambda Q/r_1$, where the axis of the polar coordinates is along the line joining the center of the sphere to the external charge, and where $\lambda$ is a dimensionless ratio of the image charge magnitude to the external charge magnitude. The surface of the sphere is defined by $|\vec{r}| = a$. On the surface, the sum of the potentials cancels. That means that, after the cancelation of common factors

$$\frac{1}{\sqrt{a^2 + d^2 - 2ad\cos\theta}} = \frac{\lambda}{\sqrt{a^2 + d'^2 - 2ad'\cos\theta}} \tag{11.9}$$

These two visually different sides have to be equal, and that specifies both $d'$ and $\lambda$. Two unknowns require two equations: as the equality holds for any $\theta$, it must hold for any two values, with $\theta = 0$ and $\theta = \pi/2$ being convenient. One then finds that

$$d' = \frac{a^2}{d}; \qquad \lambda = \left[\frac{a^2 + d'^2}{a^2 + d^2}\right]^{1/2} = \frac{a}{d} \tag{11.10}$$

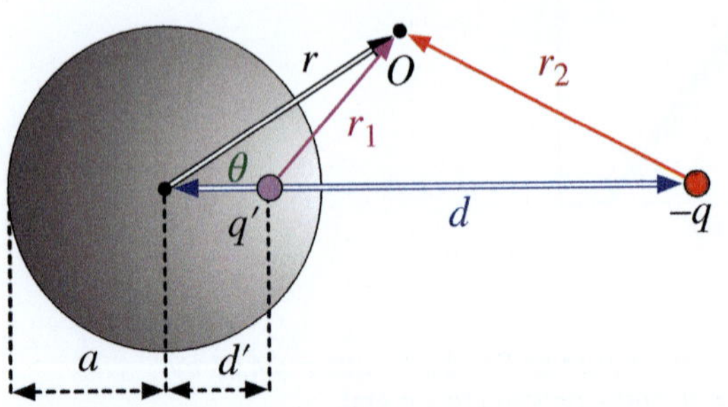


**Figure 11.2** Parameters for the determination of the image charge $q'$ of a charge $q$ outside a conducting sphere. $(r, \theta)$ are the spherical coordinates; the polar axis is along $d$. An observation point $O$ is at $\vec{r}_2$ from $-q$, and $\vec{r}_1$ from $q'$. Adapted from Ref. [186] Figure 1. See: https://doi.org/10.1116/1.4968007.

does the trick. Thus, along a line joining the charge to its image charge (that is, $\theta = 0$), the potential energy of interaction between the external charge and the image charge is $-4\lambda Q/[(d-a)+(a-d')]$, but because the observation point is now where the external charge is, then $d \to r$, and so the potential $V_i/q$ due to the image charge inside the sphere becomes

$$V_i(r) = -\frac{4aQ}{r^2 - a^2} \tag{11.11}$$

That is not convenient: emission gives a privileged role to the surface, so let $r = a + x$ where now $x$ is the radial distance from the surface (not a Cartesian axis). When that is done, then along the line connecting the center of the sphere to the external charge,

$$V_i(x) = -\frac{2aQ}{(a+x)^2 - a^2} = \frac{2aQ}{x(2a+x)} \approx -\frac{Q}{x}\left[1 - \frac{x}{2a} + \frac{x^2}{4a^2}\right] \tag{11.12}$$

an equation that is a rather fortuitous circumstance: as long as the radius of curvature $a$ is much larger than the width of the tunneling barrier (that is, $x_2(E) \ll a$ in Eq. 10.20), then Eq. (11.12) signifies that the field emission current density can be still represented by a Fowler–Nordheim-like result. This is seen by writing

$$\begin{aligned} V_{sn}^{sphere}(x) &= \mu + \Phi - Fx - \frac{2aQ}{x(2a+x)} \\ &\approx \mu + \left(\Phi + \frac{Q}{2a}\right) - \left(F + \frac{Q}{4a^2}\right)x - \frac{Q}{x} \end{aligned} \tag{11.13}$$

so that defining effective SN parameters $\Phi_{eff} = \Phi + Q/2a$ and $F_{eff} = F + Q/4a^2$ means that the SN formalism in both the FN equation and the shape factor method can be retained by letting $F \to F_{eff}$ and $\Phi \to \Phi_{eff}$.

And yet, emitters are not usually round, but rather wire-like, ellipsoidal, or hyperboloids. In the latter two instances, the image charge is no longer a point, but is better modeled as an image ring for an emitted ring of charge [186], as awkward as that sounds. If instead one desires to model a single electron being emitted from the sides of a prolate spheroidal emitter [51, 154, 383], then the curvature of the surface closest to that emitted charge has to be accounted for, and it is not a spherical radius but more like an ellipsoidal surface. Kyritsakis [210] has found that in such cases the mean radius of curvature that allows Eq. (11.13) to be used is obtained by letting the surface height be defined by a function $z_s = g(x, y)$, where the $z$-direction is normal to the surface at the $(x, y)$ origin, and $g_s = \partial_s g$, $g_{ss'} = \partial_s \partial_{s'} g$ for $(s, s') \in (x, y)$. The effective radius of curvature is then given by

$$\frac{1}{a_{eff}} = -\frac{(1+g_y^2)g_{xx} - 2g_x g_y g_{xy} + (1+g_x^2)g_{yy}}{2\left(1+g_x^2+g_y^2\right)^{3/2}} \tag{11.14}$$

which recovers the spherical limit [191, 211] of Eq. (11.12).

## 11.3 Polynomial Barriers

Field emitters are generally characterized by pointy tips or sharp ridges to enhance the local fields at the apex or edge. Because of that, the linear field approximation behind the SN barrier is in for challenges. Fields away from pointy artifacts drop off, a familiar macroscopic example being a lightning rod, but that means that there are additions to the SN barrier to slow the drop of $V_{sn}(x)$, and they can be modified by adding higher-order polynomials to the barrier equation. Such additions also can be used to model depletion regions at the interfaces of doped semiconductor contacts

[185, 190, 228, 257, 294, 295, 301] or the fall-off in field associated with prolate spheroidal geometries [154]. Such changes also go under the name of "quadratic" or "parabolic" models [53, 183, 212], and accounting for it in electron emission often relies on powerful numerical methods behind particle-in-cell codes simulating electron beams [279] because dynamic space charge effects cause a similar alteration of the tunneling barrier.

A tunneling problem shall be presumed. In the tunneling region, approximate a general barrier as

$$V_N(x) - E = \mu + \Phi - E - \frac{Q}{x} + \sum_{j=1}^{N} C_j x^j = \frac{C}{x}\prod_{j=1}^{N}(x_j - x) \tag{11.15}$$

where $C_1 \equiv -F$. The second form introduces the roots defined by $V_N(x_j) - E = 0$: What $C$ is will be returned to in a moment. The "linear" or familiar SN barrier corresponds to $V_1(x) = V_{sn}(x)$ of Section 10.20: it is so-called because in the absence of the image charge term $Q/x$, it is a triangular barrier with a linear field term such that the highest power of $x$ in $V(x)$ is $N = 1$. For $N > 2$ case, then the shape factor of Eq. (10.24) generalizes to [165, 173]

$$\sigma_N = \int_{x_-}^{x_+} \left[\frac{x_o}{s}\prod_{j=1}^{N}\frac{x_j - s}{x_j - x_o}\right]^{1/2} \frac{ds}{x_+ - x_-} \tag{11.16}$$

where the argument of $\sigma$, either $E(y)$ or $y(E)$ depending on need, reflects the most convenient manner of evaluating $x_j$ and $x_o$. $x_-$ is the smallest of the zeros $x_j$, and $x_+$ is that $x_j$ corresponding to where a particle of energy $E$ emerges from $V(x)$, a point returned to below. The term $x_o$ retains its privileged position of identifying where $V(x)$ is maximum.

Because tunneling barriers are so thin, it is often sufficient to add on a single term $\gamma x^2$ to $V_{sn}(x)$, that is, a "quadratic" model for which $N = 2$ and $C_2 \equiv \gamma$, and this is particularly important when adding image charge effects to the depletion barrier of Eq. (9.16) for metal-semiconductor contacts [294]. The $N = 2$ case is simpler to treat than the $N > 2$ cases, but the methods are generalizable. Therefore, the $N = 2$ barrier is treated exclusively. Using a "2" subscript on $V(x)$ and $\sigma(E)$ to designate its special functions is not very informative, and the ∪ subscript was already used to denote a parabolic barrier without an image charge modification. Thus, use a "$\gamma$" subscript to denote when $V(x)$ is a polynomial barrier instead, and so

$$\begin{aligned} V_\gamma(x) - E &\equiv \mu + \Phi - E - \frac{Q}{x} - Fx + \gamma x^2 \\ &= \Delta(E) - \frac{Q}{x} - Fx + \gamma x^2 \end{aligned} \tag{11.17}$$

where, as in Eq. (10.19), the height of the barrier above the tunneling electron is governed by $\Delta(E)$. In the limit $\gamma \to 0$, the quadratic model will reproduce the SN model. The roots of $V_\gamma(x) - E$ and the maximum of $V_\gamma(x)$ continue to be defined by

$$\begin{aligned} &V_\gamma(x_j) - E \equiv 0 \\ &\lim_{x \to x_o, x_p} \partial_x V_\gamma(x) \equiv 0 \end{aligned} \tag{11.18}$$

but the forms of the roots $x_j$ and the max/min points $(x_o, x_p)$ are different than those for the SN model, even though $x_o$ continues to lie between the two smallest roots, and $x_p$ between the two largest. Caution is required, however, because $x_j$ no longer have the same correspondence to $x_\pm$ as in the SN barrier case: the $x_j$ are not ordered by size with increasing index, and that must be kept in mind when the known analytic form of the cubic roots is used.

Cubic equations have a historical legacy extending back to the time of Shakespeare and Thomas Harriot (TH worked on them), where they caused a stir because they introduced the very unsettling idea of imaginary numbers: *cases irreducibilis* (irreducible case) cubic equations had three real roots even though their analytic expression involved imaginary numbers [297] because the roots are sums of complex conjugates. As a result, the roots $x_j, j \in (1, 2, 3)$ can be formulated in terms of trigonometric functions. In addition to the roots are two more points $x_o$ and $x_p$, which are the roots of the gradient of $V_\gamma(x)$. Using Eq. (11.18), these five points are specified by

$$\begin{aligned} &\gamma x_j^3 - F x_j^2 + H x_j - Q = 0 \quad [j \in (1, 2, 3)] \\ &2\gamma x_{o,p}^3 - F x_{o,p}^2 + Q = 0 \end{aligned} \tag{11.19}$$

where the "$o, p$" subscript notation denotes either/or. The relationship between the $x_j$ roots and the more familiar $(F, H, Q, \gamma)$ factors of $V_\gamma(x)$ is given by

$$\begin{aligned} F &= \gamma\,(x_1 + x_2 + x_3) \\ H &= \gamma\,(x_1x_2 + x_1x_3 + x_2x_3) \\ Q &= \gamma\,(x_1x_2x_3) \end{aligned} \tag{11.20}$$

In principle, one could simply specify the three $x_j$ and numerically vary them to obtain $(F, H, Q)$ for a given $\gamma$, but this runs afoul of the fact that $Q$ is a physical constant: it is preferable to calculate $x_3$ instead from the relation $x_3 = Q/\gamma x_1 x_2$, so that $Q$ retains its known value. The roots of a cubic equation with real coefficients are generally known to be

$$x_j = \frac{F}{3\gamma}\left\{1 + 2\omega\cos\left[\frac{1}{3}(\varphi - 2\pi j)\right]\right\} \tag{11.21}$$

$$x_o = \frac{F}{6\gamma}\left[1 + 2\cos\left(\chi - \frac{2\pi}{3}\right)\right] \tag{11.22}$$

$$x_p = \frac{F}{6\gamma}\left[1 + 2\cos(\chi)\right] \tag{11.23}$$

where $x_2 < x_1 < x_3$ because of how Eq. (11.21) depends on $j$. Thus, $x_- = x_2$ and $x_+ = x_1$. From the definitions of the roots $x_j$ and trigonometric relations such as $\cos(a \pm b) = \cos a \cos b \mp \sin a \sin b$, it follows that the angles $(\chi, \varphi)$ in Eq. (11.21) are defined by

$$\cos\varphi \equiv \frac{2(3\omega^2 - 1) + \delta}{4\omega^3}, \quad \cos(3\chi) \equiv 1 - \delta \tag{11.24}$$

and the factors $\delta$ and $\omega$ are defined as

$$\delta = \frac{54\gamma^2 Q}{F^3}, \quad \omega = \sqrt{1 - \frac{3\gamma H}{F^2}} \tag{11.25}$$

A comparison of $V_2(x) = V_\gamma(x)$ with $V_1(x) = V_{sn}(x) = \lim_{\gamma\to 0} V_2(x)$ is shown in Figure 11.3 (see Algorithm B.3.2). As $E$ gets smaller, the roots $x_1$ and $x_3$ converge on $x_p$: an alternate way of seeing this is in Figure 11.4 (see Algorithm B.3.2) which shows the change of the roots $x_j(E)$ with energy, which is seen to be the previous figure lying on its side. For the roots to be real, $\omega$ has to be real, and this implies that for a given $F$, then the energy $E$ must be larger than that value for which $H = V(x_o) - E$ satisfies $F^2 > 3\gamma H$, or

$$E \geq V(x_o) - \frac{F^2}{2\gamma} \tag{11.26}$$

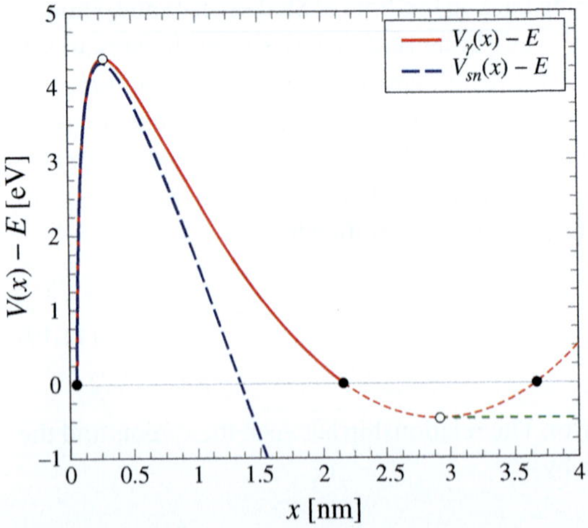


**Figure 11.3** $V_\gamma(x)$ for $\gamma = 0.85$ eV/nm$^2$, $H = 7$ eV and $F = 5$ eV/nm. Also shown is $V_{sn}(x)$, which is the $\gamma = 0$ case of $V_\gamma(x)$. The roots $x_j$ are shown by a black •, and the max/min $x_{o,p}$ are shown by ∘. The dashed red line is the extension of $V_\gamma(x)$ past $x_+ = x_1$. The dashed green line is the extension of a depletion barrier for $x > x_p$.

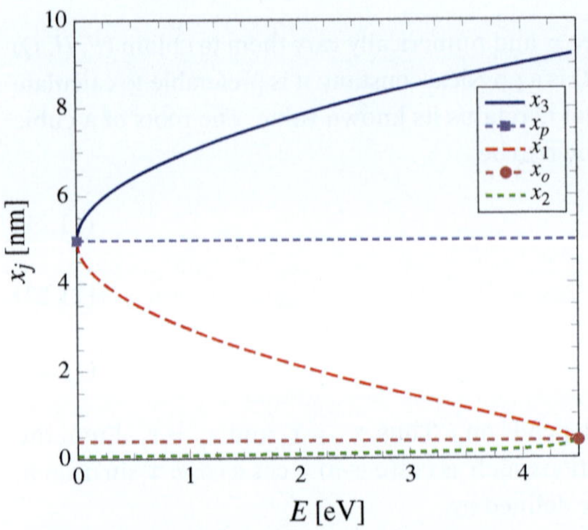


**Figure 11.4** The roots $x_j(E)$ for $F = 2.5$ eV/nm, $\gamma = 0.25$ eV/nm$^2$, $V_o = 6.3222$ eV, and $K_s \to \infty$ such that $V(x_p) = 0$. Also shown are the locations of $x_o$ and $x_p$. As $E \to V(x_o)$, then $x_1$ and $x_2$ converge; similarly, as $E \to 0$, then $x_1$ and $x_3$ converge.

For $E$ below that, the barrier is taken to be infinitely thick, and transmission through it comes to a stop. The connection to the SN barrier is seen to be the $\gamma \to 0$ limit: for small $\gamma$, $\chi \approx 2\gamma\sqrt{3Q/F^3}$ from which it follows

$$\begin{aligned}\lim_{\gamma\to 0} x_o &= \sqrt{\frac{Q}{F}} + \gamma\frac{Q}{F^2}\\ \lim_{\gamma\to 0} x_p &= \frac{F}{2\gamma} - 2\gamma\frac{Q}{F^2}\end{aligned} \tag{11.27}$$

that is, $\lim_{\gamma\to 0} x_o = \sqrt{Q/F}$ as known for the Schottky barrier, and $x_p \to \infty$, as expected.

Both $\kappa(E)$ and $L(E)$ can be evaluated analytically from $V_\gamma(x_o)$ and $x_1(E) - x_2(E)$, respectively, but the final component required to evaluate the Gamow factor $\theta = 2\sigma\kappa L$ is the shape factor term $\sigma(E)$. Returning to Eq. (11.16) for $N = 2$ and make the change of variables $x \to x_2 + L\sin^2 s$ [186] (remember that for the present barrier, $x_- \leftrightarrow x_2$ and $x_+ \leftrightarrow x_1$). Introduce $H_o \equiv V_\gamma(x_o) - E$ and recall the generalized $Q$ from Eq. (11.7) is $Q_s = (K_s - 1)Q_o/(K_s + 1)$ and that $Q_s = \gamma x_1 x_2 x_3$ from Eq. (11.20). Doing all that shows

$$\sigma_\gamma(E) = \left(\frac{LQ_s}{4H_o x_1 x_2}\right)^{1/2} \int_0^{\pi/2} h(s)ds \tag{11.28}$$

$$h(s) = \left[\frac{1}{\sin^2 s + \epsilon} - \rho\right]^{1/2} \sin^2(2s) \tag{11.29}$$

where $\epsilon = x_2/L = x_2/(x_1 - x_2)$, and $\rho \equiv L/x_3 = (x_1 - x_2)/x_3$. It is seen that $K_s \to 1$ recovers the triangular barrier $\sigma_\triangleright$ for $\gamma \to 0$ or the depletion barrier $\sigma_\cup(0)$ when $\gamma > 0$ and $V(x_p) = 0$, or that as $E \to V(x_o)$ then the top resembles a concave down parabola and so recovers $\sigma_\cap$. For general conditions, then, $\sigma_\gamma(E)$ lies between $\sigma_\triangleright$ and $\sigma_\cup$ suggesting that like $\sigma_{sn}(E)$, it may be approximated by a low order polynomial in $E$ much like $\sigma_{sn}[E(y)]$ was in Eq. (10.27), but clearly Eq. (11.28) requires numerical evaluation to generate that polynomial. It is a tedious integral, although the behavior of the integrand, for which the pleasingly behaved $h(s)$ and its derivatives vanish at the limits, indicates that the accurate Euler–MacLaurin series for approximating integrals becomes the simple summation of a series, making

$$\int_0^{\pi/2} h(s)ds \approx \frac{\pi}{2(n+1)}\sum_{j=1}^{n} h(s_j) \tag{11.30}$$

where $s_j = j\pi/2(n+1)$, numerically a good approximation even for small $n$, having an accuracy much better than normally attributed to Simpson's rule. By way of example, for the $x_j$ constructed from $H = 7$ eV, $F = 5$ eV/nm and $\gamma = 0.5$ eV/nm$^2$, resulting in $(x_1, x_2, x_3) = (1.61691, 0.053458, 8.32963)$ [nm], then the exact value of the left-hand side of Eq. (11.30) is 0.589725, whereas the right-hand side for $n = (1, 2)$ is (0.605176, 0.591733), respectively, indicating quite good agreement. The $n = 1$ case in particular suggests a reasonable approximation is

$$\int_0^{\pi/2} h(s)ds \approx \frac{\pi}{4} h\left(\frac{\pi}{4}\right) = \frac{\pi}{4}\left[\frac{1 + 2\epsilon}{2 - (1 + 2\epsilon)\rho}\right]^{1/2} \tag{11.31}$$

The devil, however, is always in the details. As long as the barrier looks more or less like an SN barrier, then $\sigma_\gamma$ is between the limits $\sigma_{sn}$ confines itself to, but near the base of $V_\gamma(x)$, $E \to 0$, and so $\rho = x_3/L$ can become large, as shown in Figure 11.5, making $\sigma_\gamma(E)$ depart from its triangular 2/3 limit to a smaller parabolic 1/2 limit [173], thereby frustrating a polynomial approximation. Even so, a polynomial approximation for $\theta_\gamma(E)$ may not be frustrated because the behavior of $2\kappa(E)L(E)$ can offset the problem cases.

## 11.4 Metal–Insulator–Metal Barrier

The metal–insulator–metal (MIM) barrier in its simplest form is the trapezoidal barrier of Section 9.3, for which the shape factor $\sigma_\Diamond(E)$ is given by Eq. (9.15) for $E < V_o - FL_o$. For energies above that limit, then the shape factor is simply given by the triangular barrier result, and so [183]

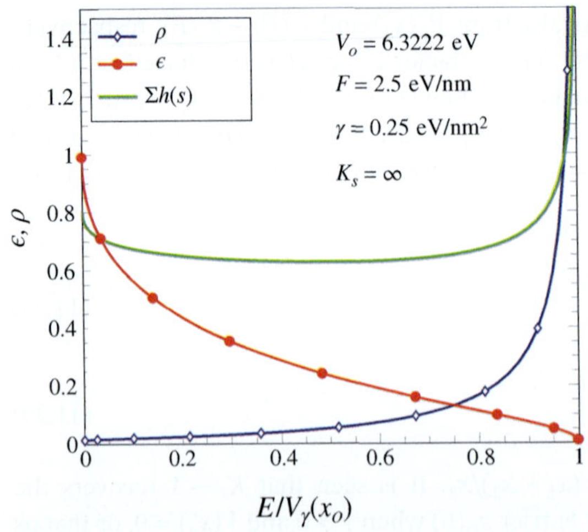


**Figure 11.5** Behavior of $\epsilon = x_2/(x_1 - x_2)$, and $\rho = (x_1 - x_2)/x_3$ in Eq. (11.29) for the $(V_o, F, \gamma, K_s)$ parameters of Figure 11.4. Green line labeled $\sum h(s)$ refers to the approximation of Eq. (11.30).

$$\sigma_\diamond(\lambda) = \begin{cases} \dfrac{2}{3\lambda}\left[1 - (1-\lambda)^{3/2}\right] & (\lambda \le 1) \\ \sigma_\triangleright = 2/3 & (\lambda > 1) \end{cases} \tag{11.32}$$

for $\lambda(E) = L_o F/(V_o - E)$ as in Eq. (9.14).

The sharp edges of the trapezoid are a consequence of the neglect of image charge modifications. Their inclusion, one for each metal contact on the left and right sides, in the interface barrier between metal and semiconductor/insulator causes unexpectedly rich complications investigated by Simmons [312] and later by others [12, 183, 375]. Notably, a Green's function method [136] used to find the barrier between the metal (or semiconductor) plates results in MIM-like barriers of a similar shape to those found in the present analysis, with the virtue of avoiding difficulties with the image charge contributions at the boundaries plunging to $-\infty$, although achieving that result requires methods outside the present purview. The simplest classical model of an MIM barrier adds two image charges and so results in a barrier resembling

$$\begin{aligned} V_M(x) &= V_o - F_s x - V_m(x) \\ V_m(x) &\approx \frac{Q_s}{x} + \frac{Q_s}{W - x} \end{aligned} \tag{11.33}$$

where $F_s = V_b/(K_s W)$ with $V_b$ arising from the potential bias across the insulator and $V_m(x)$ replacing the standard image charge term $V_i(x) = Q_s/x$, with $Q_s$ being the semiconductor form of Eq. (11.7). However, that form neglects the image charge of the image charges, and then the image charge of those image charges and so on in an infinite cascade of terms. When all of them are added up, Simmons found that $V_m(x)$ should be instead[1]

$$V_m(s) = \frac{2Q_s}{W}\left\{\frac{1}{2s} + \sum_{j=1}^{\infty} \frac{s^2}{j(j^2 - s^2)}\right\} \tag{11.34}$$

---

1 Simmons took the image charge to be $q^2/8\pi\varepsilon_0 x = 2Q/x$ instead of $Q/x$ in his Eq. (33), which has been corrected here.

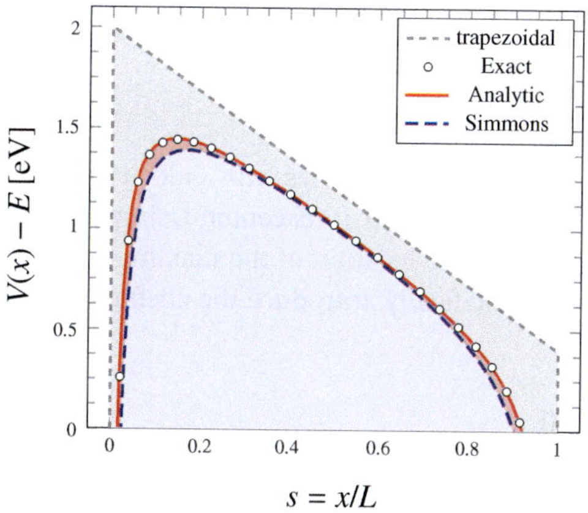


**Figure 11.6** An MIM barrier constructed using $V_o - E = 2$ eV, $F = 0.4$ eV/nm, $W = 4$ nm, and dielectric constant $K_s = 8$. Gray background is the trapezoidal barrier without image charge effects (e.g. $Q_s = 0$). The exact (∘) is from using Eq. (11.34); the red line is the approximate form of Eq. (11.38). The blue line is Eq. (11.35) developed. Adapted from [312].

where $s = x/W$, although the equation here is in a simpler but nevertheless equivalent form to Eq. (31) of Simmons [312]. Due to his finding the infinite series awkward, Simmons approximated $V_m(s)$ by

$$V_m(s) \approx 0.6931 \frac{Q_s}{Ws(1-s)} \tag{11.35}$$

where the numerical coefficient corrects a small error in Simmons' formulation. The resulting MIM barrier is shown by the blue line in Figure 11.6 (see Algorithm B.3.3).

The infinite series in Eq. (11.34), however, need not be off-putting if Riemann's zeta function (Appendix A.2) is employed [183]: if the $1/(j^2 - s^2)$ term is Taylor expanded and Eq. (A.5) is used, then

$$\sum_{j=1}^{\infty} \frac{1}{j(j^2 - s^2)} = \frac{1}{1-s^2} + \sum_{k=0}^{\infty} [\zeta(2k+3) - 1]\, s^{2k} \tag{11.36}$$

where the first term on the right-hand side is the $j = 1$ term and is pulled out so that the small coefficient $\zeta(2k+3) - 1$ can be used, and lastly, the relation $1/(1-\delta) = \sum_{k=0}^{\infty} \delta^k$ for $\delta = (s/j)^2$ is used. As the argument of $\zeta(p)$ becomes ever larger, then $\zeta(p)$ becomes ever closer to unity by Eq. (A.7), and so the coefficients $\zeta(2k+3) - 1$ drop off rather quickly as $2^{-(2k+3)}$. That indicates only the first few terms of the series need to be retained. A superior approximation to Eq. (11.36) than that obtained by Simmons is then

$$\begin{aligned} \sum_{j=1}^{\infty} \frac{1}{j(j^2 - s^2)} &= \frac{1}{1-s^2} + \sum_{n=0}^{\infty} [\zeta(2n+1) - 1] s^{2n} \\ &\approx \frac{s^2}{1-s^2} + \zeta(3) + [\zeta(5) - 1]s^2 + [\zeta(7) - 1]s^4 \end{aligned} \tag{11.37}$$

Using $\zeta(3) = 1.20206$, $a \equiv \zeta(5) - 1 = 0.0369278$ and $b \equiv \zeta(7) - 1 = 0.00834928$ then $V_m(x)$ can be rewritten for $x = sW$ as

$$V_m(s) \approx \frac{Q_s}{Ws(1-s)} v(s)$$
$$v(s) = \frac{1-s}{2} + \frac{s^3}{1+s} + s(1-s)[\zeta(3) + as^2 + bs^4] \tag{11.38}$$

where Simmons approximated $v(s) \approx 0.6931$ as a constant. Instead, it varies with $s$ albeit weakly, ranging from $v(0) = v(1) = 1/2$ at its boundaries to $v(1/2) = 0.636280$ at its center. Using a cubic fit better approximates the shape of $v(s)$, and thereby allows the location of the maximum to be determined from the quadratic equation $dv/ds|_{s=s_o} = 0$. Specifically, introduce the coefficients $A$ and $B$ such that for $s_a = 1/3$ and $s_b = 2/3$,

$$-(A+B)\,s_a^3 + As_a^2 + Bs_a + \frac{1}{2} = v(1/3) = 0.629170$$
$$-(A+B)\,s_b^3 + As_b^2 + Bs_b + \frac{1}{2} = v(2/3) = 0.615582$$

where the first coefficient $(A+B)$ is a consequence of the boundary value $v(1) = 1/2$. The value $s_o$ which maximizes $v(s_o)$ is then

$$s_o \approx \frac{A + \sqrt{A^2 + 3BA + 3B^2}}{3\,(A+B)} = 0.459195 \tag{11.39}$$

from which Eq. (11.38) returns $v(s_o) \approx 0.637297$, very close to the actual value of 0.637304, which serves as the better coefficient to Eq. (11.35).

The analytic form of Eq. (11.38) makes possible the usage of numerical techniques to find the zeros of Eq. (11.33), that is, of $V_M(x_\pm) - E = 0$ with Eq. (11.38) used for $V_m(x)$, as well as the maximum $V_M(x_o)$. It is clear, however, that the complexity of the problem must be addressed numerically, and so a discussion of it here is deferred: numerical means to examine the shape factor $\sigma_M(y)$ and the related Gamow factor $\theta_M(E)$ are treated elsewhere [183].

## 11.5 Quantum Modifications

The wave nature of the electron entails that waves incident on the surface of a metal are completely reflected if the surface barrier is treated as an infinite barrier, as it was in Section 6.4. That, though, creates something of a paradox: the electrons in a Sommerfeld metal are free to roam over the charged metal ions that constitute the background charge, and in an electrically neutral metal, the background positive charge is matched by the electron's negative charge. But if, as is often done, the background positive charge is taken as uniform and equal to $+q\rho_o$, then there is an excess of it compared to the negative charge taken as $-q\rho_o\xi(2k_F x)$ as in Eq. (6.51): the positive charge seemingly is constant right up to $x = 0$, but the negative charge is pushed back from the origin a bit as in Figure 6.3. That implies a surfeit of positive charge and contradicts the notion of global charge neutrality.

A way to correct the imbalance is to take the origin of the background positive charge as being different from the origin of the electron density, as the location of the electrical surface [85] need not be the same as the origin used in the electron density or field emission calculations. If the electron density origin is defined to be at $x_e = 0$ and the ion density is at $x_i < 0$, then the metal will have no net charge as long as

$$\int_{-\infty}^{0} \{\rho_e(x) - \rho_o\Theta(|x_i| - |x|)\}\, dx = 0 \tag{11.40}$$

where the Heaviside step function $\Theta$ term is attached to the background positive charge term, and the origin of $x$ is chosen such that $x_e \equiv 0$. Invoking Eq. (6.51), then the relation is equivalent to

$$\int_{-\infty}^{0} [\xi(s) - 1]\, ds = 2k_F x_i \tag{11.41}$$

where the substitution $s = 2k_F x$ is made. There are two ways to evaluate the left-hand side which result in the same outcome, but rely on different illustrative approaches. Consider both[2] after observing that the integrand is an even function of $s$, so that Eq. (11.41) is unchanged if $\int_{-\infty}^{0} \cdots ds \to \int_{0}^{\infty} \cdots ds$ instead, which makes discussing the integral easier.

The first approach considers alternate representations of $[\xi(s) - 1]$. Using the form

$$\xi(s) - 1 = \frac{3}{2}\frac{d}{ds}\left(-\frac{\cos(s)}{s} + \frac{\sin(s)}{s^2} - \mathrm{Si}(s)\right) \tag{11.42}$$

where the Sine Integral is defined as

$$\mathrm{Si}(s) \equiv \int_{0}^{s} \frac{\sin(x)}{x} dx \tag{11.43}$$

which then entails that

$$\int_{2\pi n}^{2\pi(n+1)} [\xi(s) - 1] ds = \frac{3}{2}\left[\mathrm{Si}(2\pi n) - \mathrm{Si}[2\pi(n+1)] + \frac{1}{2\pi n(n+1)}\right] \tag{11.44}$$

for (nonzero) integer $n$. Clearly, the same equation holds when the lower limit of the integral is extended downward to $j$ and the upper limit to $\infty$. With some care, the limits $j = 0$ (using Eq. 11.42 directly) and $n \to \infty$ limit can be taken to get

$$\int_{0}^{\infty} [\xi(s) - 1] ds = \frac{3}{2}\,[\mathrm{Si}(0) - \mathrm{Si}(\infty)] = -\frac{3\pi}{4} \tag{11.45}$$

where for $\delta = 2\pi j \to 0$, then $-3\delta\cos(\delta) + 3\sin(\delta) \to \delta^3$ so that its ratio with $2\delta^2$ vanishes as $j$ vanishes, and $\mathrm{Si}(\infty) = \pi/2$ can be found from either look-up tables or methods considered next.

The second approach uses the Calculus of Residues, which also handles the integrals behind $\mathrm{Si}(\infty) = \pi/2$ by extending the integration region using the even parity of the integrand

$$\mathrm{Si}(\infty) = \frac{1}{2}\int_{-\infty}^{\infty} \frac{\sin(s)}{s} ds = \frac{1}{2}\Im\left\{\int_{-\infty}^{\infty} \frac{e^{is}}{s} ds\right\} = \frac{\pi}{2} \tag{11.46}$$

where $e^{is} = 1$ at the pole $s = 0$, and so the second integral is $i\pi$ because the pole is on the real axis. Applying the same methods to $(\xi(s) - 1)$ is more nuanced because the order of the pole is larger (of order 3), so that

$$\begin{aligned} \frac{1}{2}\int_{-\infty}^{\infty} (\xi(s) - 1) ds &= \frac{3}{2}\Re\left\{\int_{-\infty}^{\infty} \frac{e^{is}(i+s)}{s^3} ds\right\} \\ &= \frac{3}{2}\Re\left\{\frac{\pi i}{3!}\left(\frac{d}{ds}\right)^2 [e^{is}(s+i)]\Big|_{s=0}\right\} = -\frac{3\pi}{4} \end{aligned} \tag{11.47}$$

2 The approaches here contrast with the two approaches to the same problem in Ref. [154], with the present methods being more streamline.

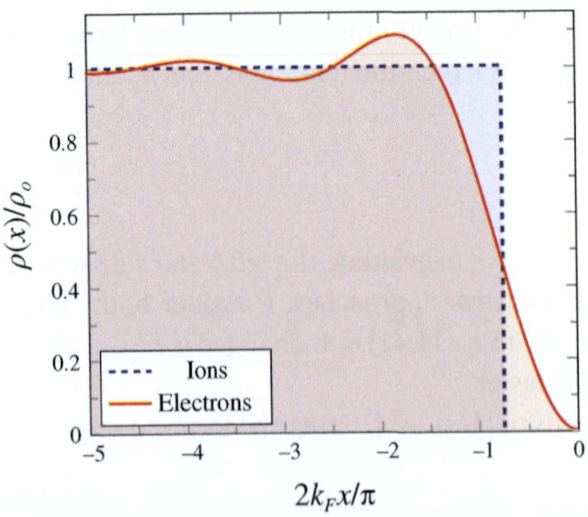


**Figure 11.7** Electron density (red) atop the background positive charge due to ions (blue) for an infinite barrier at $x = 0$; ion density begins at $2k_F x_i = -3\pi/4$.

Either approach used in the evaluation of Eq. (11.41) results in finding the location of the background positive charge boundary to be

$$x_i\,[\text{nm}] = -\frac{3\pi}{8k_F} = -\frac{1.1781}{k_F} \tag{11.48}$$

relative to the infinite barrier confining the electrons at $x = 0$. The constant background positive charge is overlaid on the electron density variation in Figure 11.7. The shape of the electron density looks vaguely familiar to the hyperbolic tangent (tanh) function, and that approximation has been used in examining many-body (exchange-correlation) effects [5, 145, 154, 193] on the nature of the surface barrier (see Ex. 11.6).

In looking at such a figure, it is natural to infer that the "surface" for the electrons is where the background positive charge ends, but that is not so: for the electrons, the origin is at $x = 0$. But the inference that an excess of electrons outside the metal and a deficit inside right near $x_i$ creates a dipole contribution to the work function [218] is warranted, and because the size of that dipole is dependent on the height of the barrier, and therefore the field at the surface because of the Schottky barrier lowering factor $\sqrt{4QF}$, there are grounds to conclude that the work function has a field-dependent component to it [145, 154].

The calculation of $x_i$ is pleasing as an analytic model, but its simplicity is indebted to the approximation of an infinite barrier at $x = 0$ for the electron gas. Real materials do not have infinite barriers, and the barrier an electron at the Fermi level experiences for a metal is governed by the work function $\Phi$ and the application of an electric field specified by $F$. What happens when $\Phi$ is finite, and more importantly, what happens when $\phi = \Phi - \sqrt{4QF}$ changes as the result of variations in field? As the electron density pushes into the barrier via quantum mechanical tunneling and penetration, the shifting charge density causes variations in the work function that are a consequence of how the barrier height $V_o = \mu + \Phi - \sqrt{4QF}$ changes. A simple model can be used to assess what the consequences are.

To that end, consider the rectangular barrier of Section 6.6 again, where the reflection coefficient $r(k)$ for $k < k_o$ is found from Eq. (6.72) by the prescription $\kappa \to i\kappa$, where $\kappa^2 = k_o^2 - k^2$, or

$$
\begin{aligned}
r(k) &= \frac{-i(k^2 - (i\kappa)^2)\sin(i\kappa L)}{2ik\kappa\cos(i\kappa L) - i(k^2 + (i\kappa)^2)\sin(i\kappa L)} \\
&= \frac{(k^2 + \kappa^2)\tanh(\kappa L)}{2ik\kappa + (k^2 - \kappa^2)\tanh(\kappa L)}
\end{aligned}
\tag{11.49}
$$

When $\kappa L$ becomes large, either through a high or wide (or both) barrier, then $\tanh(\kappa L) \approx 1$, and so

$$
r(k) \approx \frac{(k^2 + \kappa^2)}{2ik\kappa + (k^2 - \kappa^2)} = \frac{k - i\kappa}{k + i\kappa} \equiv e^{2i\varphi} \tag{11.50}
$$

that is, if the incident wave is $\psi_k^+(x) \propto e^{ikx}$, then the reflected wave is $\psi_k^-(x) \propto e^{-ikx+i\varphi}$, a result that applies to triangular and other kinds of barriers under similar conditions [146, 152, 154]. The phase is

$$
\begin{aligned}
\varphi &= \frac{1}{2}\arcsin\left\{\frac{1}{2i}\left[\frac{k - i\kappa}{k + i\kappa} - \frac{k + i\kappa}{k - i\kappa}\right]\right\} \\
&= -\frac{1}{2}\arcsin\left[\frac{2k\kappa}{k_o^2}\right] \approx -\frac{k}{k_o}\left(1 + \frac{k^2}{6k_o^2}\right)
\end{aligned}
\tag{11.51}
$$

to leading order, therefore, $\varphi \approx -kx_o$, where here, $x_o = 1/k_o$ and is not to be confused with the same term in the SN barrier analysis. As long as $k_F^2/6k_o^2$ is sufficiently small, then all of the wave functions that make up $\rho(x)$ are affected by the same displacement of $x_o$, and so whereas for the infinite barrier, $|\psi_k^+(x) + \psi_k^-(x)|^2 \propto |e^{ikx} - e^{-ikx}|^2 = \sin^2(kx)$ as in Eq. (6.51), now it resembles

$$
\begin{aligned}
\left|e^{ikx} - e^{-ik(x-2x_o)}\right|^2 &= \left|e^{ikx_o}\left(e^{ik(x-x_o)} - e^{-ik(x-x_o)}\right)\right|^2 \\
&= \sin^2[k(x - x_o)]
\end{aligned}
\tag{11.52}
$$

In other words, the density has been shifted by a roughly constant amount $x_o \approx 1/k_o$ dependent on the barrier height, a result mirrored by sufficiently tall triangular barriers as well as well [152, 154]. That displacement results in a shift of the electron density compared to the background positive charge, and in a one-dimensional problem, a thin sheet of charge $\rho(x)dx$ produces a field independent of distance from it given by $F = 8\pi Q\sigma = q^2\sigma/2\varepsilon_0$ (recall Section 2.2.5 and see Section 15.3) where $\sigma(x) \equiv \rho(x)dx$. Adding up all of the fields from all of the sheets is therefore an integration over density, and results in a change in barrier height $\Delta\Phi$ that can be tacked on to the work function $\Phi$, where, on using integration by parts,

$$
\begin{aligned}
\Delta\Phi &= -\int_{-\infty}^{0} F(x)dx = \int_{-\infty}^{0} x\frac{dF}{dx}dx \\
&= 4\pi Q\int_{-\infty}^{0} x[\rho_e(x) - \rho_i(x)]dx
\end{aligned}
\tag{11.53}
$$

where Poisson's equation has been used to relate $dF/dx$ to total charge density, namely the excess or deficit of electrons compared to the background positive charge. But from the analysis following Eq. (11.40) and observing that another way of representing $[\xi(s) - 1]$ is as

$$
\xi(s) - 1 = \frac{3}{s}\frac{d}{ds}\left(\frac{\sin s}{s}\right) \tag{11.54}
$$

then it follows upon switching the integration variable to $s = 2k_F x$ that

$$\Delta\Phi = 4\pi Q\rho_o \left[ \frac{3}{4k_F^2} \int_{-\infty}^{0} \frac{d}{ds}\left(\frac{\sin s}{s}\right) ds - \frac{x_i^2}{2} \right] \tag{11.55}$$

The integration is trivial and only requires $(\sin s)/s$ at the integration limits or $\lim_{s\to\infty}(\sin s)/s = 0$ and $\lim_{s\to 0}(\sin s)/s = 1$, resulting in a simple analytic expression

$$\Delta\Phi = \frac{Qk_F}{\pi}\left[1 - \frac{3\pi^2}{32}\right] = 0.161810\, k_F a_o \text{ [eV]} \tag{11.56}$$

As an example, for copper, $\mu = 7$ eV so that $k_F = 13.5546$ [1/nm], and so $\Delta\Phi = 0.116063$ [eV]. This calculation makes an assumption about where $x_i$ is, but as the barrier height changes, $x_i$ changes in turn, and so another field-dependent wrinkle is added because the field should originate on the metal surface, not necessarily the $x = 0$ point for calculating electron density.

These arguments suggest three modifications to the SN barrier, modifications that seem independent, but which are tied together. First is that the image charge term $Q/x$ has more to do with where the surface of the metal $x_s$ is than where the electron origin is, and so the image charge term should be modified by something like

$$\frac{Q}{x} \to \frac{Q}{x + x_s} \tag{11.57}$$

where $x_s$ is field-dependent and the origin of the background positive charge has to be determined by the condition of global charge neutrality. Second, the onset of the applied field should modify the $Fx$ term to something more like $F(x - x_s')$, where room is left for the possibility that $x_s$ and $x_s'$ differ. That they do is a consequence of the global charge neutrality argument behind Eq. (11.55). Third, and by analogy with the second argument regarding how field affects electron penetration into the barrier, as the temperature of the electron gas in the metal increases, higher energy states are populated in the Boltzmann tail of the Fermi–Dirac distribution function $f_{FD}(E)$ and those higher states also penetrate further into the barrier than when conditions are cooler. As a result, the work function $\Phi$ portion of the barrier must be viewed as both temperature- and field dependent [145]. These arguments are based on the simple and aesthetic Sommerfeld model of an electron gas: real metals are horribly different, making the simple arguments require modification. Even so, in the case of thermal emission and the approximation $\Phi(T) = \Phi_o + \tilde{\alpha}T$, experimental values of $\tilde{\alpha}$ can be tabulated, and the magnitude of the $\tilde{\alpha}$'s so found [119] are comparable to the simple theoretical values [154].

More on that pesky $\tilde{\alpha}$ factor: the temperature dependence of the work function is behind a subtle matter that must be accounted for in interpreting Richardson plots of $\ln(J/T^2)$ *vs.* $1/T$, namely that the slope of such plots is a measure of $\Phi/k_B$ and the intercept a measure of $\ln(A_{RLD})$, as pointed out by Shelton [308]. He observed specifically that if a simple linear dependence of the form $\Phi(T) = \Phi_o + \tilde{\alpha}T$ is assumed (the tilde is to distinguish $\tilde{\alpha}$ from the fine structure constant), then the $A_{RLD}$ coefficient on a Richardson plot deduced from the intercept is mangled because

$$\begin{aligned} \ln\left(\frac{J}{T^2}\right) &= \ln(A_{RLD}) - \frac{\Phi(T)}{k_B T} \\ &= \ln\left(A_{RLD} e^{\tilde{\alpha}/k_B}\right) - \frac{\Phi_o}{k_B T} \end{aligned} \tag{11.58}$$

where $\Phi_o$ is the zero temperature value most commonly associated with "work function." Shelton's analysis demonstrated two different measurement means of determining slope and intercept on

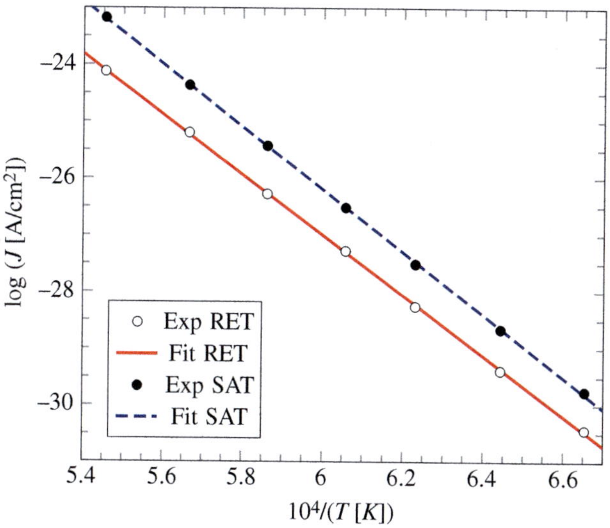


**Figure 11.8** Retarding potential (RET) and saturation current (SAT) data of thermionic emission from Tantalum digitally extracted from Figure 6, Adapted from [308]. $y_{RET} = -5.31 \times 10^4 + 4.83/T$; $y_{SAT}(T) = -5.51 \times 10^4 + 6.85/T \Rightarrow A_{ret} = 125$ A/cm$^2$, $A_{sat} = 943$ A/cm$^2$, and $\tilde{\alpha} = 2.02k_B$.

Richardson plots: the “retarding potential (RET)” measurements (negative voltage applied to the collector) measure the thermally independent parameters by taking into account the shifting of the contact potential, whereas the “saturation current (SAT)” measurements require that the intercept be reduced by $\tilde{\alpha}/k_B$ to get $A_{RLD}$. Shelton’s measurements, reproduced in Figure 11.8, confirm this: a linear fit of both RET and SAT data sets gives a measure of $A_{RLD}$ and $\tilde{\alpha}/k_B$ that are comparable to accepted values.

And so, on a simplistic level suggested by experimental measurements, $\Phi(T) = \Phi_o + \tilde{\alpha}T$ and where $\tilde{\alpha}/k_B$ is generally on the order of unity for metals and is not restricted to positive numbers (see Table 28.1 of Ref. [154]). As a result, the SN Barrier should be closer to

$$V_{sn}(x) = \mu + \Phi_o + \tilde{\alpha}T + \Delta\Phi - F(x - x_s') - \frac{Q}{x + x_s} \tag{11.59}$$

a form that is put on better footing than the arguments here by accounting for the properties of the materials using the Thomas–Fermi approximation to the dielectric function of the metal [136, 137].

The form of Eq. (11.59) seemingly undermines the elegant simplicity of the canonical equations. And yet, all is not lost: consider a shifting of the coordinate system $x' = x + x_s$ and a collection and renaming of terms to write

$$\begin{aligned} V_{sn}(x') &= \mu + \Phi_{eff} - Fx' - \frac{Q}{x'} \\ \Phi_{eff} &\equiv \Phi_o + \tilde{\alpha}T + F(x_s + x_s') \end{aligned} \tag{11.60}$$

where $V_{sn}(x')$ once again has all the trappings of the SN barrier and so allows the retention of the methods of evaluating current density through it. The cost of doing so is having to be careful to include modifications due to temperature and field. It is sufficient, therefore, to keep that caveat in mind and go back to speaking of $\Phi$ and $Q/x$ as though quantum effects did not trouble the discussion of them.

## 11.6 Exercises

**Exercise 11.1** Silicon is characterized by $K_s = 11.9$. How do the size of the Schottky factor and the location of the barrier maximum for silicon compare to that for a metal?

**Solution** $\Delta\Phi_s/\Delta\Phi = 0.91922 = x_o(K_s)/x_o(\infty)$.

**Exercise 11.2** Show Eq. (11.10)

**Exercise 11.3** Show Eq. (11.20).

**Exercise 11.4** What is the ratio of the thermal emission current density on axis for a sphere at $T = 1373$ K and in a background field of 10 MV/m compared to that for a flat surface for a sphere radius of $5 \times 10^n$ nm for $n \in (0, 1, 2, 3)$?

**Solution** For $J_n(a)$ equal to the RLD equation using effective parameters, then $J_n(a)/J_{RLD}(T) = (0.7432, 0.9701, 0.9970, 0.9997)$ for $n = (0, 1, 2, 3)$.

**Exercise 11.5** What is the ratio of the field emission current density on axis for a tungsten sphere for $\Phi = 4.5$ eV in a background field of $F = 6$ eV/nm compared to that for a flat surface for a sphere radius of $5 \times 10^n$ nm for $n \in (0, 1, 2, 3)$? Neglect thermal effects. Assume $t(y_o) \approx 1.06131$.

**Solution** For $J_n(a)$ equal to the FN equation using effective parameters, then $J_n(a)/J_{FN}(F) = (0.9003, 0.9890, 0.9989, 0.9999)$ for $n = (0, 1, 2, 3)$.

**Exercise 11.6** Approximate the electron density by $\rho_e(x)$ by

$$\rho_a(x) \equiv \frac{\rho_o}{\exp[\lambda k_F(x - x_i)] + 1}$$

where $\rho_o \equiv k_F^3/(3\pi^2)$ and $\rho_i(x) = \rho_o$ for $x < x_i$ (remember that $x_i$ is negative). What is $\lambda$ if Eq. (11.56) holds for copper so that $\Delta\Phi = 0.11606$ eV? Hint: global charge neutrality applies.

**Solution** $\Delta\Phi = 2\pi Q k_F/9\lambda^2$ so $\lambda = 5.418$.

# Part IV

# Emission

# 12

# Thermal-field Emission

*They shall stand in fire up to the nav'l, and in ice up to th' heart, and there th' offending part burns, and the deceiving part freezes ...*

W. Shakespeare, *Two Noble Kinsmen* IV.3

## 12.1 The Thermal-field Regime

The Richardson–Laue–Dushman (RLD) equation of Section 8.2 and the Fowler–Nordheim (FN) equation of Section 8.3 described the limiting cases of $J(F, T)$ as governed by the behavior of $N(n, s)$ in Eqs. (8.9) and (8.11) as $n$ became very small or large, respectively, where the thermal and field limits caused $N(n, s)$ to take the different forms of $e^{-ns}$ and $n^2 e^{-s}$, respectively. Now explore the physics of why that is so. To begin, first observe that both the RLD and FN equations are field dependent, and so consider what happens when the field is progressively raised, say for a typical metallic work function of $\Phi = 4.5$ eV as in Figure 12.1. There, several instances of the RLD equation (Eq. 10.16) at various fixed temperatures are shown alongside on zero-temperature field emission evaluation (Eq. 10.16). The (relatively weak) field dependence of the RLD equation is due to the factor $\exp[-(\Phi - \sqrt{4QF})/k_B T]$; the (much stronger) dependence of the Murphy–Good form of the FN equation is because the factor $\exp[-B_o/F]$ of Eq. (10.16) is far more sensitive. The location where the two equations meet, called a "nexus" [102, 217, 235], indicates where the physics is changing. What is changing in the present case is that thermal emission is giving way to field emission as the barrier progressively thins when $F$ increases.

The behavior evinced in Figure 12.1 (see Algorithm B.4.1) has been observed in experimental data taken from Ba-dispenser and scandate cathodes operated at high enough anode voltages to reveal the transition from thermal emission to field emission. The data of Geittner et al. [105] are shown in Figure 12.2 and compared to curves constructed from Eqs. (8.10) and (10.16) under the approximation that $I(V) \approx A \exp(B\sqrt{V})$ for thermal emission, and $I(V) = AV^p \exp(-B/V)$ for field emission, where the $(A, B)$ constants are determined from the pair of orange (RLD) and cyan (FN) data points in Figure 12.2, and $p \approx 1.2272$ for consistency with $J_{MG}$ although a range of $p$ values near unity works comparably well.

What is changing is the relative strength of the exponential decline of the supply function $f(E)$ compared to the exponential rise of the transmission probability $D(E)$ in Eq. (8.1) [60], a behavior which has parallels in experimentally measured energy distributions of emitted electrons [97, 373]. Comparisons are made simply and analytically for the triangular barrier of Eq. (8.3) so

*Fundamentals of Electron Emission Physics*, First Edition. Kevin L. Jensen.

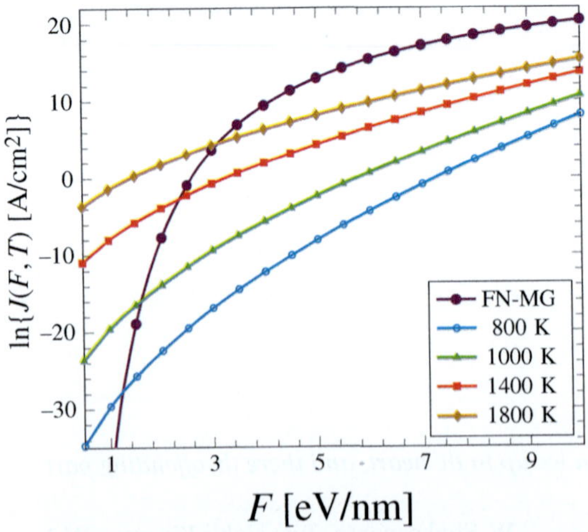


**Figure 12.1** Comparison of $J_{MG}(F, 0)$ evaluated using Eq. (10.16) (open ∘) to $J_{RLD}(F, T)$ (Eq. (8.10)) evaluated at various temperatures. Metal-like parameters of $\Phi = 4.5$ eV assumed.

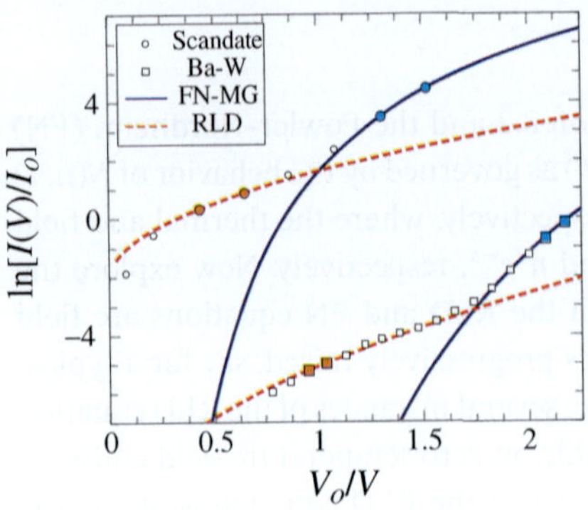


**Figure 12.2** Current from a standard (W-I) cathode compared to a scandate ("Sc"/Re-I) cathode for data digitally extracted from of Figure 2 of Geittner et al., [105] / AIP Publishing for $V_o = 1$ kV and $I_o = 1\ \mu$A, compared to theoretical field emission (Eq. (10.16)) and thermal emission (Eq. (8.10) with $\Phi \to \phi$) curves (solid and dashed lines) using the orange and cyan data points to parameterize.

consider it for what follows. Plotting the integrand $dN(F, T; E)$ of Eq. (8.3) shows the transition for a logarithmic $y$-axis in Figure 12.3 (see Algorithm B.4.2), where a smaller $\Phi = 4$ eV is used to compensate for a triangular barrier being much higher than an SN barrier. For a given temperature of $T = 1500$ K, the effect of progressively higher $F$ is to move the peak of $dN(E)$ from near $\mu + \Phi$ down to near $\mu$, as suggested by three instances in Figure 12.4. In that figure, the white circle identifies where $E_m$ is: if the green curves are normalized by $dN(E_m)$ and stacked together with energy $E$ on the $x$-axis and $F$ on the $y$-axis, then a contour of the evolution of $E_m$ with increasing $F$ can be charted, as in Figure 12.3 (see Algorithm B.4.2). The behavior of the resulting contour plot indicates just how $E_m$ varies with $F$: for a triangular barrier, analytic methods thankfully suffice, but numerical methods work just as well for more complex barriers.

When the maximum of the current density integrand $dN(E)$ lies between the Fermi energy and the maximum of the barrier, that is, when $\mu < E_m < \mu + \Phi$ for the triangular barrier (in contrast to the Schottky–Nordheim, or SN, barrier, for which $\Phi \to \phi = \Phi - \sqrt{4QF}$), then $E_m$ is between the

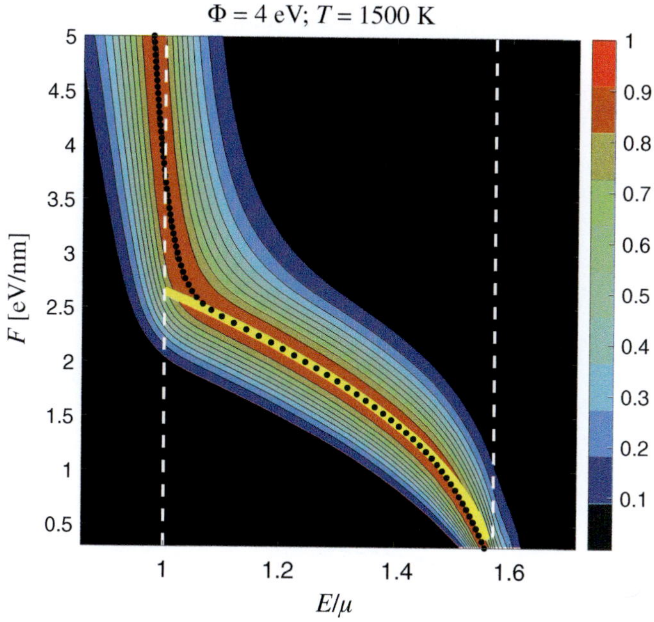


**Figure 12.3** Contours of $dN(E)/dN(E_m)$ for a triangular barrier with $\mu = 7$ eV, $\Phi = 4$ eV and $T = 1500$ K. Yellow line is Eq. (12.3) for $E_m$. Black dots over yellow line are $E_m$ determined using quartic method of Appendix A.8. Source: After Ref. [157] / AIP Publishing/CC BY 4.0.

field and thermal regimes, and is therefore said to be in the thermal-field, or TF, regime. There, $\beta_T(\mu - E_m)$ is large and negative so that $f(E) \propto \exp[\beta_T(\mu - E_m)]$. So too is $\theta(E_m)$ large so that $D(E) \approx \exp[-\theta(E)]$. As a consequence, in the TF regime and to leading order

$$dN(\mu < E < \mu + \Phi) \approx \exp[-\beta_T(\mu - E) + \theta(E)] \tag{12.1}$$

The location of $E_m$ in the TF regime is then dictated by

$$\frac{1}{dN(E)} \frac{d}{dE}[D(E)f(E)]\bigg|_{E=E_m} = 0 \Rightarrow \beta_T - \beta_F(E_m) \approx 0 \tag{12.2}$$

In the case of the triangular barrier, where $\theta_{FN}(E)$ and $\beta_F(E)$ are both known from Eq. (8.13), then the solution of Eq. (12.2) is

$$E_m \approx \mu + \Phi - \frac{1}{2m}\left[\frac{\hbar F}{2k_B T}\right]^2 \tag{12.3}$$

If the definition of the TF regime is enlarged a bit to include where temperature starts making a difference to field emission, or when fields start making a difference to thermal emission apart from the Schottky factor, then the TF regime can be defined by the behavior of

$$n(F, T) \equiv \frac{\beta_T}{\beta_F(E_m)} \approx 1 \tag{12.4}$$

or, more specifically, the thermal regime corresponds to $n \lesssim 0.95$, the field regime to $n \gtrsim 1.05$, and so the TF regime corresponds to $0.95 < n < 1.05$). On reflection, it is clear that $n \approx 1$ for the TF regime for all barriers, not just the triangular barrier one, because of how $E_m$ is defined in Eq. (12.2).

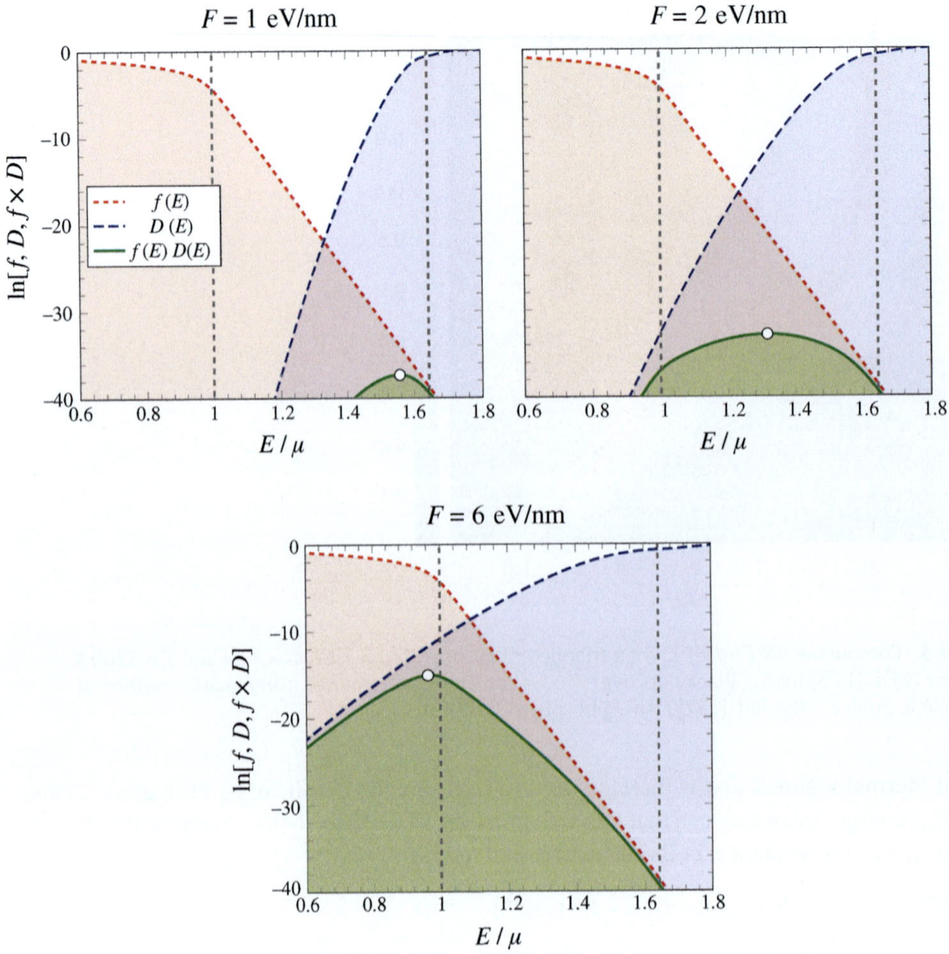


**Figure 12.4** Normalized supply function $f(E)/f(0)$ (red fill), transmission probability $D(E)$ (blue fill), and their product $dN(E) \propto D(E)f(E)/f(0)$ (green fill) for metal parameters of $\mu = 7$ eV, $\Phi = 4$ eV, a triangular barrier, and a temperature $T = 1500$ K for the fields $F$ shown. Left vertical dashed line shows the location of $\mu$, right vertical dashed line the location of $\mu + \Phi$. White ∘ denotes location of the maximum point $dN(E_m)$.

Consider two complementary cases: first, hold the field fixed at a representative value and increase the applied temperature as in Figure 12.5*a*, and second, hold the temperature fixed and increase the applied field as in Figure 12.5*b* (see Algorithm B.4.2). In both cases, parameters are set so that the energy maximum $\mu + \Phi \geq E_m \geq \mu$, that is, between the thermal and field regimes. Although on a log-plot, it is evident that the broadness of $D(E)f(E)$ is substantially larger in the TF regime than in either the thermal or field regimes.

The same methodology holds for more complex barriers like the SN barrier: all that is required is a means to calculate both $\theta(E_m)$ and $\beta_F(E_m)$ exists, analytically if possible, numerically if not. The shape factor approach to evaluate $\theta(E) = 2\sigma[y(E)]\kappa(E)L(E)$ was simple in the case of the triangular barrier because $\sigma$ was constant, and both $\kappa(E)$ and $L(E)$ were analytic, making $\beta_F(E)$ analytic as well. Direct evaluation then shows that the calculation of $n(F, T) = \beta_T/\beta_F(E_m) \approx 1$ in the TF regime to a reasonable approximation provides a fast method to determine the Gamow factor and its derivatives [183].

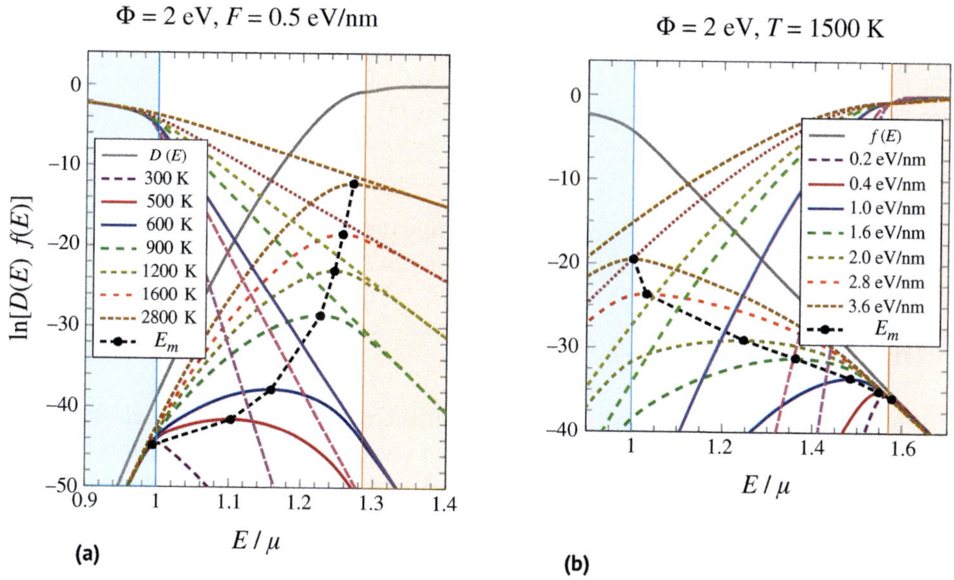


**Figure 12.5** (a) Gray line: triangular barrier $D(E)$ for $\mu = 7$ eV, $\Phi = 2$ eV, $F = 0.5$ eV/nm. Thin color lines: $f(E)$; Thick color lines: $dN(E)$. Legend labeled by $T$ [K]. Black bullet (•) is $dN(E_m)$. Blue region is for $E < \mu$; orange region is for $E > \mu + \Phi$. (b) Gray line: $f(E)$ for $\mu = 7$ eV, $\Phi = 4$ eV, $T = 1500$ K. Thin color lines: triangular barrier $D(E)$; Thick lines: $dN(E)$. Legend labeled by $F$ [eV/nm]. Black bullet (•) is $dN(E_m)$. Blue region is for $E < \mu$; orange region is for $E > \mu + \Phi$. Source: [157]/AIP Publishing / CC BY 4.0.

The movement of $E_m$ into the TF regime when conditions are ripe for it has spawned numerous efforts to understand how current from field emitters changes when the temperature rises, and so to provide temperature-dependent correction factors to the FN equation [60, 88, 97, 373]. The study by Murphy and Good [264] provides an early and widely used equation: although it will be eclipsed by the general thermal-field (GTF) equation below, consider it. They observed that using a simple exponential form $\exp[-\theta(E)]$ for the transmission probability resulted in Eq. (8.2) becoming

$$J(F,T) \approx A_{RLD}T^2 \int_0^{\infty} e^{-\theta(E)} \ln\left[1 + e^{\beta_T(\mu - E)}\right] \beta_T dE \tag{12.5}$$

Let $\theta(E) = \theta(\mu) - \beta_F(\mu)(\mu - E) = \theta(\mu) - p\beta_T(\mu - E)$, which results in

$$J(F,T) \approx A_{RLD}T^2 e^{-\theta(\mu)} \int_0^{\infty} z^{p-1} \ln(1 + z) dz \tag{12.6}$$

where $z = e^{\beta_T(\mu - E)}$ and $p = 1/n = \beta_F(\mu)/\beta_T$. The integral is analytic and known to be equal to $[\pi/p \sin(\pi p)] = p^{-2}[\pi p/\sin(\pi p)]$. The coefficient $p^{-2}$ is taken over into $J_{MG}(F,0)$ giving the temperature-independent form attributed to Murphy and Good for the FN equation. The term in square brackets is then identified as the temperature-dependent multiplier and now expressed as

$$J_{MG}(F,T) \approx J_{MG}(F,0) \frac{\pi k_B T \beta_F(\mu)}{\sin[\pi k_B T \beta_F(\mu)]} \tag{12.7}$$

where $\beta_F(\mu)$ takes over the role of $C_{FN}$ in Eq. (10.7). Eq. (12.7) is commonly offered as the thermal modification to field emission, but it has limitations. As the temperature increases, then

$J(F, T)$ must monotonically increase because the supply function $f(E)$ acquires an increasingly pronounced thermal tail, yet the thermal factor written as $\eta/\sin(\eta)$ shows that as $\eta$ passes $\pi$, the thermal factor increases rapidly until $\eta = \pi$ whereupon the ratio becomes infinite and for larger $\eta$, it switches to a negative value. This occurs at a temperature of

$$T_{max} = \frac{\hbar F}{4k_B\sqrt{2m\Phi}} \tag{12.8}$$

That relation suggests that for $F = 2\,\text{eV/nm}$ and a work function of $\Phi = 4.5\,\text{eV}$, then the prediction of the thermal factor should be looked at with askance for $T > T_{max}/2 = 534$ K.

## 12.2 The *N(n,s)* Function

The $N(n, s)$ function (Eq. (8.8) of Section 8) was used to find thermal modifications to field emission and field modifications to thermal emission. The function was explored in the TF regime defined by $n \approx 1$ where both become comparable in contribution. An accurate approximation to $N(1, s)$ was shown. Rapid computational methods for the evaluation of $J(F, T)$ that do not rely on numerical integration were developed. The first and faster original GTF (*o*GTF) approach uses linear approximations to $\theta(E)$ using the Fermi level $\mu$ or the barrier maximum $\mu + \phi$ as the expansion points. The second and more accurate reformulated (*r*GTF) approach uses the maximum of the current density energy integrand as the expansion point. Extending $N(n, s)$ to $N(n, -s)$ inserts thermal and field effects into the Fowler–DuBridge formula for photoemission.

Given the size of typical values of $s$ in the $N(n, s)$ function for the SN barrier, the integral definition of $N(n, s)$ in Eq. (8.8) is not an easy integral to calculate numerically for generally encountered values of $n$, particularly for low temperatures or low fields (or both). A tractable approach begins by recasting the integrand into various analytic series expansions that allow for a collection and simplification of aggregates of terms. The expansions are made possible through the standard series expansions for a general term $x < 1$

$$\frac{1}{1+x} = \sum_{j=0}^{\infty}(-1)^j x^j \tag{12.9}$$

$$\ln(1+x) = \sum_{j=1}^{\infty}\frac{(-1)^j}{j}x^j \tag{12.10}$$

The difficulty is that setting $x$ to either of the terms $e^z$ and $e^{n(s-z)}$ is undercut by the range of values over the integration region that exist where these terms exceed unity, and in places substantially so. Consequently, the $N(n, s)$ integral of Eq. (8.8) becomes four such integrals that do allow the usage of the series expansions, or $N(n, s) = \sum_{j=1}^{4} N_j(n, s)$, where

$$N_1(n, s) = n\int_s^{\infty}\frac{n(z-s)}{e^z+1}dz \tag{12.11}$$

$$N_2(n, s) = n\int_s^{\infty}\frac{\ln\left[1+e^{-n(z-s)}\right]}{e^z+1}dz \tag{12.12}$$

$$N_3(n, s) = n\int_{-\infty}^{0}\frac{\ln\left[1+e^{n(z-s)}\right]}{e^z+1}dz \tag{12.13}$$

$$N_4(n,s) = n\int_0^s \frac{\ln\left[1+e^{n(z-s)}\right]}{e^z+1}dz \tag{12.14}$$

where $N_1$ and $N_2$ are "field-dominated," and $N_3$ and $N_4$ are "thermal-dominated" [150]. Each of the integrals is set up so that the series expansions of Eqs. (12.9) and (12.10) can be used, which then allows the integrals to be over exponential functions and therefore simple. Of them, $N_1$ can be expressed in terms of easier functions to work with, namely, the dilogarithm function

$$\mathrm{Li}_2(x) \equiv -\int_0^x \frac{\ln(1-u)}{u}du = \sum_{j=1}^{\infty}\frac{x^j}{j^2} \tag{12.15}$$

which satisfies the relation $\mathrm{Li}_2(x) + \mathrm{Li}_2(-x) = (\pi^2/6) + (x^2/2)$ (compare Eq. (8.17)) and has the specific value $\mathrm{Li}_2(1) = \zeta(2) = \pi^2/6$. In terms of it,

$$N_1(n,s) = \frac{(\pi n)^2}{6} + \frac{(ns)^2}{2} + n^2\mathrm{Li}_2(s) \tag{12.16}$$

In spite of that elegant result, do not use it. $N_1$ and the other $N_j(n,s), j \in (2,3,4)$ integrals are better expressed for purposes here as infinite series resulting from the application of Eq. (12.10) and a term-by-term integration. The resulting series are then

$$N_1(n,s) = n^2\sum_{j=1}^{\infty}(-1)^{j+1}\frac{e^{-js}}{j^2} \tag{12.17}$$

$$N_2(n,s) = n\sum_{j=1}^{\infty}\sum_{k=1}^{\infty}\frac{(-1)^{j+k}e^{-js}}{k(j+kn)} \tag{12.18}$$

$$N_3(n,s) = n\sum_{j=0}^{\infty}\sum_{k=1}^{\infty}\frac{(-1)^{j+k+1}e^{-kns}}{k(j+kn)} \tag{12.19}$$

$$N_4(n,s) = n\sum_{k=1}^{\infty}\sum_{j=1}^{\infty}\frac{(-1)^{j+k+1}}{k(j-kn)}\left(e^{-js} - e^{-kns}\right) \tag{12.20}$$

where it is important to remember that $j$ and $k$ are dummy indices that can be swapped if needed, and whether the $j$ summation begins at $j = 1$ or $j = 0$ is important. Notice now that the sums contain two types of terms: those with a $e^{-js}$ and those with a $e^{-kns}$. Regrouping into those two types give

$$\begin{aligned} N(n,s) = n^2 &\sum_{j=1}^{\infty}\frac{(-1)^{j+1}e^{-js}}{j^2}\left(1 + 2\sum_{k=1}^{\infty}\frac{(-1)^k j^2}{j^2 - k^2n^2}\right) \\ &+ \sum_{j=1}^{\infty}\frac{(-1)^{j+1}e^{-jns}}{j^2}\left(1 + 2\sum_{k=1}^{\infty}\frac{(-1)^k j^2}{j^2 - \frac{k^2}{n^2}}\right) \end{aligned} \tag{12.21}$$

where the terms with $e^{-js}$ are *field-like* and those with $e^{-jns}$ are *thermal-like*. Now make use of the fact that even though $n$ can be smaller than unity for the thermal-like terms or larger than unity for the field-like terms, $s$ is of such a size that both $s$ and $ns$ are large with $ns > s$ for field-like and $s > ns$ for thermal-like. Either way, the sum over $j$ can be truncated at the first term, or

$$N(n,s) \approx n^2e^{-s}\left[1 + 2\sum_{k=1}^{\infty}\frac{(-1)^k 1}{1-k^2n^2}\right] + e^{-ns}\left[1 + 2\sum_{k=1}^{\infty}\frac{(-1)^k}{1-\frac{k^2}{n^2}}\right] \tag{12.22}$$

The quantities inside the square brackets are independent of $s$, and so constitute a new function $\Sigma(1/n)$ for the field-like term, and $\Sigma(n)$ for the thermal-like term, where for $x < 1$

$$\Sigma(x) \equiv 1 + 2x^2 \sum_{k=1}^{\infty} \frac{(-1)^k}{x^2 - k^2} \tag{12.23}$$

so that the GTF approximation to Eq. (8.8) becomes

$$N(n, s) \approx n^2 e^{-s}\, \Sigma\left(\frac{1}{n}\right) + e^{-ns}\, \Sigma\,(n) \tag{12.24}$$

Therefore, quite naturally, the full solution of $N(n, s)$ separates itself into a thermal part hearkening to Eq. (8.9) and a field part hearkening to Eq. (8.11). It follows, therefore, that for the SN barrier, the relationship of $J(F, T)$ from Eq. (8.8) to the canonical equations reliant upon Eq. (8.9) for $N_{RLD}(n, s)$ and Eq. (8.11) for $N_{FN}$ result in the identification that

$$\begin{aligned} &\lim_{n\to\infty} A_{RLD} T^2 N(n, s) \approx J_{FN}(F) \\ &\lim_{n\to 0} A_{RLD} T^2 N(n, s) \approx J_{RLD}(T) \end{aligned} \tag{12.25}$$

where both $s$ and $ns$ are large but one is much larger than the other. The relations are only approximate because $\Sigma(x)$ is only close but not equal to unity for large and small $n$, and $E_m$ is generally different from $\mu$ for field emission or $\mu + \phi$ for thermal emission. A pragmatic boundary of $0.95 \geq n$ separates the thermal from the TF regime, and $n \geq 1.05$ separates the TF from the field regime.

The evolution of $n(F, T)$ from the thermal to the field regime can be appreciated using the triangular barrier model, which has the benefit of being analytic: it gives qualitatively similar results to the SN barrier, but at a fraction of the latter's complexity. Therefore, allow $\theta(E)$ and $\beta_F(E)$ for the present inquiry to be drawn from Eq. (8.13) for $dN(E)$ as evaluated using Eq. (8.3), as previously done for Figure 12.4, but now consider a range of $F$ that run the gamut from the thermal to field regimes. To find $E_m$, an extension of the transmission probability above the barrier maximum is required, and is most readily met by the approximation that for $E = \mu + \Phi - x$, then $\theta(-x) = -\theta(x)$, that is, the Gamow factor is antisymmetric across the barrier maximum. The maximum of the triangular barrier $dJ_{\triangleright}(E)$ is then found quickly using the methods of Appendix A.8, from which the ratio $n(F, T) = \beta_T/\beta_F(E_m)$ is straightforwardly evaluated and the evolution of $n(F, T)$ as a function of $F$ charted: Figure 12.6 (see Algorithm B.4.2) shows how it changes for various temperatures and its relation to the various regimes. The dashed lines there correspond to when $\beta_F(E_m)$ corresponds to $\beta_F(\mu)$ or $\beta_F(\mu + \Phi)$, particular cases, which are of central importance in the development of the oGTF model of Section 12.3.

Clearly, the interesting behavior is near $n = 1$, which is where $\Sigma(n)$ or $\Sigma(1/n)$ make a difference: note that $\Sigma(1/2) = \pi/2 = 1.57080$, showing that the canonical equations are in need of adjustment well before the TF region (as even Eq. (12.7) suggested). Therein lies a difficulty: Eq. (12.23) is numerically unpleasant to work with inside the TF regime, requiring, as it does, the summation of many terms before acceptable convergence is achieved, and so an approximation is desired intently. Fortunately, quickly convergent series are possible by making use of another series expansion involving $x/(1-x)$ for $x$ small, where $x = n/k$ for thermal and $x = 1/(kn)$ for field, to get

$$\Sigma(x) = 1 + \sum_{j=1}^{\infty} (1 - 2^{1-2j})\zeta(2j) x^{2j} \tag{12.26}$$

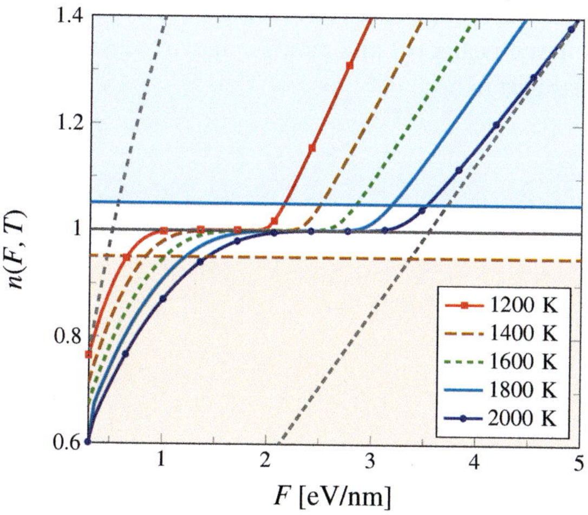


**Figure 12.6** Evolution of the ratio factor $n(F, T) = \beta_T/\beta_F(E_m)$ where $E_m$ is the maximum of $dJ_{\triangleright}(E)$, for a triangular barrier for which $\mu = 7$ eV and $\Phi = 4$ eV. Shaded regimes: Bluish = field; reddish = thermal; white = TF.

While this series rapidly converges for small $x$, because $(1 - 2^{1-2j})\zeta(2j)$ is close to unity for large $j$, the series does not rapidly converge for $x$ closer to unity. It can be made to do so by rewriting the series as

$$\begin{aligned}\Sigma(x) &= 1 + 2\sum_{j=1}^{\infty} x^{2j} + 2\sum_{j=1}^{\infty}[(1 - 2^{1-2j})\zeta(2j) - 1]x^{2j} \\ &= \frac{1+x^2}{1-x^2} + 2\sum_{j=1}^{\infty}[(1 - 2^{1-2j})\zeta(2j) - 1]\, x^{2j}\end{aligned} \tag{12.27}$$

Now the series is rapidly convergent for all $x \leq 1$, but with a term that is singular at $x = 1$. Another difficulty is that both $\Sigma(1/n)$ and $\Sigma(n)$ are required, and only one at a time has an argument smaller than unity. A suitable approximation is to truncate the $\Sigma(x)$ series to a small number of terms and allow it to extend beyond $x = 1$ boundary, relying on both $e^{-s}$ and $e^{-ns}$ to reduce the downside of that approximation. As a result,

$$\begin{aligned}\Sigma(x) &\approx \frac{1+x^2}{1-x^2} + [\zeta(2) - 2]x^2 + \frac{1}{4}[7\zeta(4) - 8]x^4 \\ &= \frac{1+x^2}{1-x^2} - 0.355066x^2 - 0.105934x^4\end{aligned} \tag{12.28}$$

This approximation removes the singularity in Eq. (12.24) because the thermal and field term singularities cancel when $e^{-ns} = e^{-s}$ at $n = 1$. Use Eq. (12.28) to show this. Letting $\Sigma_o(x) = (1 + x^2)/(1 - x^2)$ be the singular part, then

$$\begin{aligned}\lim_{n\to 1} e^{-ns}\Sigma_o(n) + n^2 e^{-s}\Sigma_o\left(\frac{1}{n}\right) &= \lim_{n\to 1}\frac{(n^2+1)[n^2e^{-s} - e^{-ns}]}{n^2 - 1} \\ &= (s+2)e^{-s}\end{aligned} \tag{12.29}$$

Including the additional terms, or $\Sigma(x) - \Sigma_o(x)$, from Eq. (12.28) lowers that result to $(s + 1.078)e^{-s}$: as more terms in the $\zeta$-function form of Eq. (12.27) are included, the closer the coefficient of $e^{-s}$ approaches $(s+1)$. This indicates that returning to Eq. (12.23) to investigate the $n \to 1$ limit is desirable. In that limit, part of $\Sigma(x)$ is well-behaved near $x = 1$ and part is singular: separate out the singular part and let $x = 1$ in the non-singular part to find

$$\Sigma(x \approx 1) = \frac{1+x^2}{1-x^2} + 2\sum_{k=2}^{\infty} \frac{(-1)^{k+1}}{k^2-1} \tag{12.30}$$

The summation part is surprisingly simple

$$\begin{aligned} 2\sum_{k=2}^{\infty} \frac{(-1)^{k+1}}{k^2-1} &= \sum_{k=2}^{\infty} \frac{(-1)^{k+1}}{k+1} + \sum_{k=2}^{\infty} \frac{(-1)^{k+1}}{k-1} \\ &= \sum_{k=3}^{\infty} \frac{(-1)^{k}}{k} - \sum_{k=1}^{\infty} \frac{(-1)^{k+2}}{k} = -\frac{1}{2} \end{aligned} \tag{12.31}$$

As a result, when joined with Eq. (12.29),

$$N(1,s) = (s+1)e^{-s} \tag{12.32}$$

An alternate and elegant analysis by Chernin [44] finds the exact result for $N(n,s)$ (Eq. (8.8)) in an aesthetic summation form that is free of singularities that otherwise require caution in the present analysis. It gives for $N(1,s)$

$$N(1,s) = -s\ln(1-e^{-s}) + \mathrm{Li}_2(e^{-s}) \tag{12.33}$$

which in the limit of large $s$ reproduces $N(1,s) \approx (s+1)e^{-s}$ as found above, and confirms the accuracy of the series-based expansion method.

The special case $N(1,s)$ is important because it identifies the TF regime separating the field regime $N(n>1,s)$ from the thermal regime $N(n<1,s)$. As in Figure 12.5, it occurs when $E_m$ lies between the Fermi level energy and the top of the barrier energy. As a result, $n = 1$ is a dividing line, a separation starkly revealed in a contour plot of $dN(E)$ on the axes of energy $E$ and ratio $n$ for a fixed $s$, as in Figure 12.7 (see Algorithm B.4.3): there, along $n = 1$, a distinct boundary is encountered. Importantly, although the GTF equation finds its most prominent application in the SN emission barrier characteristic of metal and semiconductor surfaces abutting vacuum, the definitions of $n(F,T)$ and $s(F,T)$ in Eq. (8.7) do not require this, only that $\Theta(E)$ and $\beta_F(E)$ can be constructed, and that a maximum of $dN(F,T;E)$ be found. Therefore, a fixed $s$ is instructive, but not typical of realistic barriers. In the representation shown in Figure 12.4, the contours were for the $(F,E)$ axes, but given the relation between $F$ and $n$ due to the definition of $\beta_F(E)$ in Eq. (8.13), Figure 12.4 can be recast instead as contour lines along the $n$ and $E$ axes, as shown in Figure 12.8 (see Algorithm B.4.3). Now the very manner in which $dJ(E)$ spreads out in the TF regime is visible, and the figure gives a better feel for why $|n-1| < 0.05$ is a useful definition of that regime: it is where $E_m$ enters the energy gap between $\mu$ and the barrier maximum.

Compared to the numerical efforts required to evaluate $N[\Theta(E)]$ in Eq. (8.2), approximating $\Theta(E)$ as a linear function in $E$ has resulted in a reasonably accurate and analytic form of $N(n,s)$ as given in Eq. (12.24) when the approximation to $\Sigma(x)$ of Eq. (12.28) is used: this sequence of approximations shall be known hereafter as the $L\Theta$ approximation for the SN barrier in particular, but to be useful, the $L\Theta$ approximation further requires that finding $E_m$ be simplified, which is somewhat daunting given that the shape factor for the SN barrier is not constant. A computationally rapid method gives rise to the $o$GTF equation, whereas a more accurate but therefore more nuanced approach gives the $r$GTF equation. Consider each in turn.

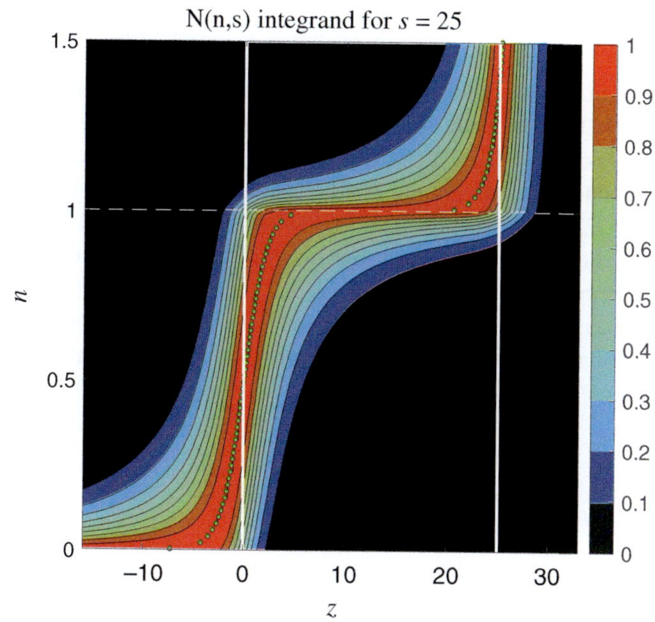


**Figure 12.7** Integrand of $N(n, s)$ of Eq. (8.8) normalized to unity for each value of $n$ for a fixed parameter $s = 25$. Green dots show location of the maximum of the integrand. White lines frame the regimes: below the dashed line at $n = 1$ is the thermal regime and above is the field regime, whereas the TF regime is along the $n = 1$ dashed line.

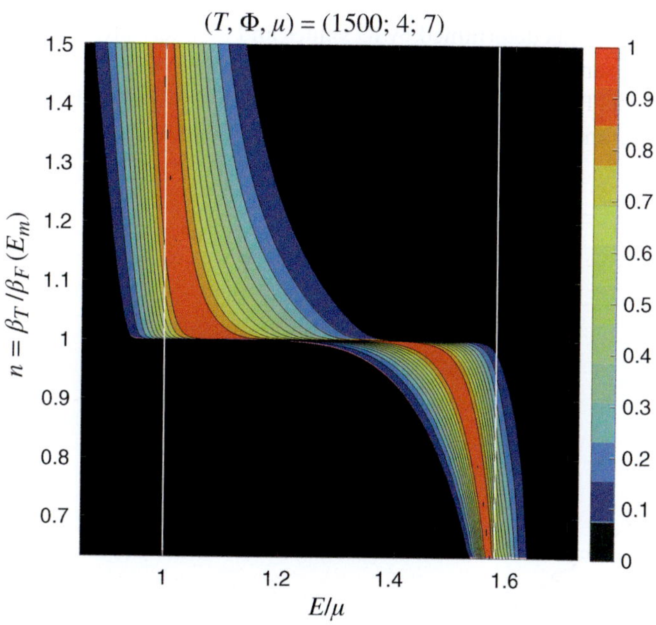


**Figure 12.8** The representation of Figure 12.4 but with the vertical axis replaced by $n(F, T) = \beta_T/\beta_F(E_m)$, where $\beta_F(E_m)$ is based on Eq. (8.13), for a triangular barrier with $\mu = 7$ eV, $\Phi = 4$ eV and $T = 1500$ K. $E_m$ is approximately given by Eq. (12.3) in the TF regime where $n \approx 1$. After: [157] / AIP Publishing / CC BY 4.0.

## 12.3 Original GTF Equation

The *o*GTF equation [149, 150, 160] was an effort to transparently recover the MG form of the FN equation in the field regime, and the RLD equation in the thermal regime. It took its cue from the fact that Fowler and Nordheim approximated $E_m \approx \mu$ in their derivation of the canonical FN equation, as this choice was eminently practical and resulted in an elegant form of $J_{FN}(F)$. In the same spirit, it is also practical to choose $E_m = \mu + \phi$ (where $\phi = \Phi - \sqrt{4QF}$) in the thermal limit, and use the fact that for energies near the barrier maximum, the barrier is parabolic to a good approximation. The requirement to so approximate $E_m$ is because of the mild convexity of $\theta(E)$, which causes $\beta_F(E)$ to change between the thermal and field limits, as in Figure 12.9 (see Algorithm B.4.3). Alongside $\theta(E)$ are the conventional (linear) lines associated with the FN approximation and the parabolic approximation. For the FN line, (recall Eq. (10.15))

$$\begin{aligned} \theta_{FN}(E) &= \theta_{FN}(\mu) - \beta_F(\mu)(E - \mu) \\ \theta_{FN}(\mu) &= \frac{4}{3\hbar F}\sqrt{2m\Phi^3}\, v[y(\mu)]; \quad \beta_F(\mu) = \frac{2}{\hbar F}\sqrt{2m\Phi}\, t[y(\mu)] \end{aligned} \tag{12.34}$$

where $y(\mu) = \sqrt{4QF}/\Phi$. For the parabolic line, Eq. 9.5 with $V_o = \mu + \phi$ is used to show that

$$\begin{aligned} \theta_p(E) &= \theta_p(\mu + \phi) - \beta_F(\mu)(E - \mu - \phi) \\ \theta_p(\mu + \phi) &= 0; \quad \beta_F(\mu + \phi) = \frac{\pi}{\hbar F}\sqrt{m\Phi}\sqrt{y(\mu)} \end{aligned} \tag{12.35}$$

Although $\theta(E)$ appears reasonably straight, its comparison with its tangent lines $\theta_{FN}(E)$ at $E = \mu$ and $\theta_p(E)$ at $E = \mu + \phi$ show that it is not. The non-linearity of $\theta(E)$, which is weak, can approximated by a cubic polynomial with coefficients determined by its values and first derivatives at its boundaries, or $\theta(\mu)$, $\theta(\mu + \phi) = 0$, $\beta_F(\mu)$ and $\beta_F(\mu + \phi)$. If energy is written as $E(p) = \mu + p\phi$ with $0 \le p \le 1$, then $p$ spans the range of the TF region. Demanding that the cubic polynomial gives those boundary terms results in $\theta[E(p)]$ and $\beta_F[E(p)] = -d\theta/dE$ are, after regrouping,

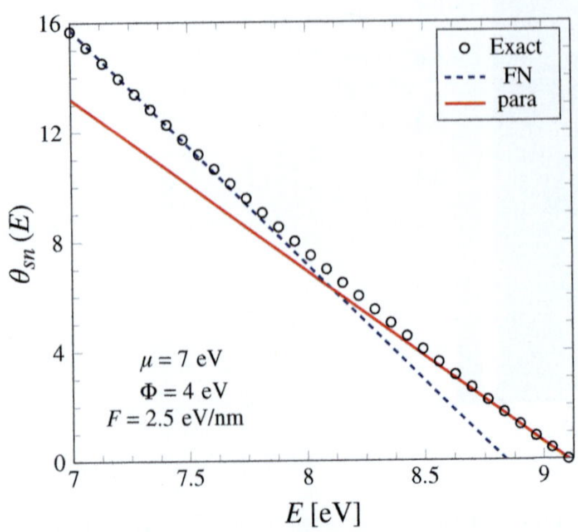


**Figure 12.9** Comparison of $\theta_{sn}(E)$ (∘, Eq. 10.26) to $\theta_{FN}(E)$ (blue line, Eq. (12.34)) and $\theta_p(E)$ (red line, Eq. 12.35).

$$\begin{aligned} \theta[E(p)] &= (1-p)^2(2p+1)\theta(\mu) \\ &\quad - \phi p(1-p)\left[(1-p)\beta_F(\mu) - p\beta_F(\mu+\phi)\right] \end{aligned} \tag{12.36}$$

$$\begin{aligned} \beta_F[E(p)] &= 6p(1-p)\frac{\theta(\mu)}{\phi} \\ &\quad + (1-p)(1-3p)\beta_F(\mu) + p(3p-2)\beta_F(\mu+\phi) \end{aligned} \tag{12.37}$$

When $F$ is fixed and $T$ is varying, the calculation of current density in the $L\theta$ approximation for a given $F$ follows a recipe given by [156, 157]

1. Define $T_{min}$ and $T_{max}$ as the smallest and largest $T$ for which $n(F,T) = 1$
2. Assign $E_m$ according to
   - for $T < T_{min} = 1/k_B\beta_F(\mu)$, set $E_m = \mu$;
   - for $T > T_{max} = 1/k_B\beta_F(\mu+\phi)$, set $E_m = \mu + \phi$; and
   - for $T_{min} < T < T_{max}$, determine $E_m$ from $\beta_F(E_m) = \beta_T$.
3. Evaluate $n(F,T) = \beta_T/\beta_F(E_m)$ and $s(F,T) = \theta(E_m) + \beta_F(E_m)[E_m - \mu]$
4. Approximate $N(n,s)$ using Eq. (12.24)
5. Evaluate $J(F,T) = A_{RLD}T^2N(n,s)$.

The same algorithm works if $T$ is given and $F$ varies, except that the determination of $F_{max}$ and $F_{min}$ from Eq. (12.34) and (12.35), respectively, is required. $F_{min}$ is straightforward from its definition

$$F_{min} = \left[\frac{\pi k_B T}{\hbar}\sqrt{2m}\, Q^{1/4}\right]^{4/3} \tag{12.38}$$

whereas $F_{max}$ is best solved using an iterative approach (Appendix A.10). Let the "guess" field be indexed by $j$, with $F_0 = \Phi^2/4Q$, which makes $y(\mu) = 1$. Then perform

$$F_{j+1} = \frac{2k_BT\sqrt{2m\Phi}}{\hbar}\, t\left[\sqrt{4QF_j}/\Phi\right] \tag{12.39}$$

where $F_{max} = \lim_{j\to\infty} F_j$. As an example, for $\Phi = 4.5$ eV and $T = 1500$ K, then $F_{min} = 1.88930$ eV/nm, and for $j = 0$ to 4, $[F_j] = (14.0628, 3.1217, 2.9310, 2.9256, 2.9255)$ eV/nm, with no further changes to the fourth decimal for $j > 4$.

It is instructive to see how the recipe for evaluating oGTF current density works in practice for representative parameters of $\mu = 7$ eV, $\Phi = 4.5$ eV, $T = 2300$ K, and $F = 4$ eV/nm. Proceeding in steps

1. $\beta_F(\mu+\phi) = 4.40774\ \text{eV}^{-1}$ and $\beta_F(\mu) = 5.71363\ \text{eV}^{-1}$. It follows $T_{max} = 2632.76$ K and $T_{min} =$ 2031.02 K. The temperature $T$ for the example is therefore within the TF regime.
2. $\phi = \Phi - \sqrt{4QF} = 2.10003$ eV, $\beta_T = 5.04544$ eV, $y(\mu) = 0.533327$.
3. $\theta_{FN}(\mu) = 10.6933$. Setting $\beta_T = \beta_F[E(p)]$ gives

   $$-0.187915p^2 - 1.11797p + 0.668193 = 0$$

   for which the roots are $p = (0.54733, -6.49667)$. The root between 0 and 1 is the one of interest. Therefore, $E_m = \mu + p\phi = 8.14941$ eV.
4. From Eq. (12.36), then $\theta(E_m) = 4.49929$. For TF, $\beta_F(E_m) = \beta_T$ and Eq. (8.7), $s(E_m) = \theta[E(p)] + \beta_T(E_m - \mu) = 10.2986$.
5. $N(1,s) = (s+1)e^{-s} = 3.80551 \times 10^{-4}$.
6. $J = A_{RLD}T^2N(1,s) = 2.41923 \times 10^5\ \text{A/cm}^2$.

Such a procedure adapts well to programming languages, for which scripts are available [154], and it meshes well with the canonical forms of the thermal and field emission equations. The transition to the thermal or RLD equation occurs by virtue of $ns = \beta_T\phi$ for $n < 1$ and $\Sigma(n \ll 1) \approx 1$. The transition to field emission is aided by using the SN functions $v(y)$ and $t(y)$ in the definition of $\theta(\mu)$ and $\beta_F(\mu)$ in Eq. (12.34), thereby mimicking the derivation of the FN equation as done by Murphy and Good. Most importantly, the *o*GTF recipe for the evaluation of $J(F, T)$, being computationally rapid, is very well suited for simulation codes.

If the *o*GTF recipe is applied repeatedly for a range of either $T$ or $F$, then the behavior of $J(F, T)$ can be represented and contrasted to the canonical $J_{RLD}$ and $J_{FN}$ equations. Consider two representative cases for metal-like parameters

1. For $\mu = 7$ eV, $\Phi = 4.5$ eV and $T = 1500$ K, Figure 12.10(a) results. The thermal regime ends at $F_{min} = 1.8893$ eV/nm; the field regime begins at $F_{max} = 2.9255$ eV/nm.
2. For $\mu = 7$ eV, $\Phi = 4.5$ eV and $F = 2.0$ eV/nm, Figure 12.10(b) results. The field regime ends at $T_{min} = 1035.5$ K; the thermal regime begins at $T_{max} = 1565.4$ K.

It is clear, then, what the advantages of the *o*GTF approach are:

**(i)** it connects the two canonical equations of field emission (FN) and thermal emission (RLD) directly;
**(ii)** it provides temperature modifications to FN and field modifications to RLD (apart from the field dependence of $\phi$) straightforwardly through the introduction of $\Sigma(1/n)$ and $\Sigma(n)$, respectively;
**(iii)** it provides a TF equation for the first time in the intermediate regime of warm field emission and cool thermal emission; and finally
**(iv)** it is a fast algorithm.

Its speed and ability to connect FN to RLD are a consequence of its reliance on the elliptical integral functions $v(y)$ and $t(y)$. Where the approximation is less satisfactory is, first, because the reliance on the elliptical integral functions makes the method not extensible to other kinds of barriers. A second downside of *o*GTF is revealed in the kinks in the $J(F, T)$ curves as the $J(F, T)$ line passes from the thermal to TF regime, and from TF to the field regime. These kinks are not

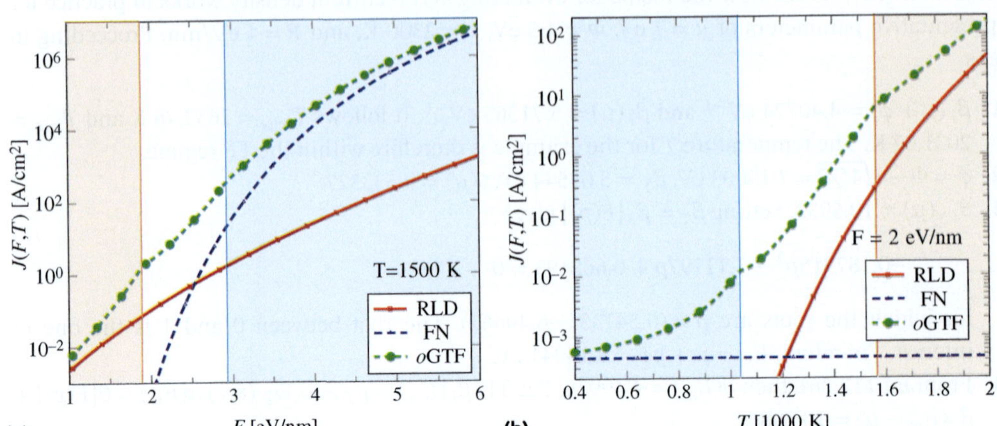


**Figure 12.10** Variation of current density with (a) field for a constant temperature and (b) temperature for a constant field. Comparison of $J(F, T)$ evaluated by the *o*GTF, RLD, and FN methods for metal-like parameters of $\mu = 7$ eV and $\Phi = 4.5$ eV. Observe the kinks at the transition from one regime to another.

plotting artifacts but reflect the actual behavior of $J(F, T)$ in the $o$GTF model. As found in the analysis by Kyritsakis and Xanthakis [214], the kinks persist for other barriers, not just the SN barrier behind the $o$GTF as Kyritsakis and Xanthakis found in their generalization of the GTF equation. The cause of the kinks is anticipated in Figure 12.3, where the analytic line tracking $E_m$ departed from the maximum of the $dJ$ integrand, signifying that $\beta_F(E_m)$ is departing from $\beta_T$, and then again in Figure 12.6, which showed that even for triangular barriers (as will be true for more general barriers), the approach of the $n$ vs $F$ line does not have a constant slope as $n = 1$ is approached, the main approximation for both the thermal and field regimes in Eqs. (12.34) and (12.35): convenient as it is to assume for $n$ away from unity that $\beta_F(E_m)$ takes on a constant value of $\beta_F(\mu)$ or $\beta_F(\mu + \phi)$, the small errors in doing so are magnified by their appearance in an exponential factor. Lastly, Figure 12.9 shows that differences exist between $\theta(E)$ and its approximations by $\theta_{FN}(E)$ and $\theta_p(E)$, particularly for $\mu < E < \mu + \phi$, that affect the accuracy of $s(F, T)$ in Eq. (8.7). Compared to a numerical integration of $N[\theta(E)]$, compared to $N(n, s)$ can be up to a factor of 3× larger than the former [156] when the $o$GTF methods are used. However, before a factor of 3× seems insufferable, consider Figure 12.10: it is a $\log_{10}$ plot, and even though the departures of $J_{GTF}$ from either $J_{FN}$ or $J_{RLD}$ seem visually small, the latter lines are in fact quite off: for the (a) plot, at $F = 2.54$ eV/nm (approximately where the RLD and FN lines cross), $J_{GTF}/J_{RLD} \approx 56$, whereas for the (b) plot, at $T = 1220$ K (as before, about where the RLD and FN lines cross), $J_{GTF}/J_{FN} \approx 124$. Those discrepancies are quite large by comparison to 3. In light of such departures, a factor of 3 is tolerable.

## 12.4 Reformulated GTF Equation

The $r$GTF methodology corrects the downsides of the $o$GTF equation, but at the cost of having to actually find $E_m$ instead of making assumptions about it. The $r$GTF approach therefore relies on the existence of a high accuracy solution to $\theta(E)$ such as Eqs. (10.26) and (10.27) (see Algorithm B.4.3) for the SN barrier or the generalizations of $\sigma(y)$ like Eq. (11.16) to treat generalized barriers using the shape factor method of Section 10.3. It is therefore assumed at the outset that a rapid computational method to evaluate $\theta(E)$ exists. The question is then what to do with it to evaluate $N(n, s)$? Finding what is in need of repair is foremost: consider the behavior of Figure 12.11 (see Algorithm B.4.3) that shows $dJ_x(E)/dJ(E_m)$, where the $x$ subscript can denote the FN approximation for $\theta(E)$, the parabolic (para) approximation, or when no subscript is present the exact $\theta(E)$, and where $E_m$ is the location of the integrand maximum for the exact $dJ(E)$. Three temperatures are shown, $T_{min}$, $T_{max}$, and their average $(T_{min} + T_{max})/2$, showing that the FN and para approximations overestimate the integrand because finding tangent lines to $\theta(E)$ at $E = \mu$ and $E = \mu + \phi$ have slight errors in the TF regime that get magnified when exponentials of $\theta(E)$ are taken in the denominator of $dJ(E)$. The ratio of the shaded regions, or $\int dJ_{ns}/\int dJ = J_{ns}/J$ (where the numerator uses the $N(n, s)$ approximation and the denominator uses $N[\theta(E)]$) can therefore be as large as $1 < J_{ns}/J \lesssim 3$ depending on parameters. The reformulated approach therefore demands that the tangent line of $\theta(E)$ be used where $E_m$ is found from the maximum of the exact $dJ(E_m)$, or

$$
\begin{aligned}
&\theta(E) \approx \theta(E_m) - \beta_F(E_m)(E - \mu) \\
&0 = \frac{d}{dE}\left\{\frac{\ln\left[1 + e^{\beta_T(\mu - E)}\right]}{1 + e^{\theta(E)}}\right\}\Bigg|_{E=E_m}
\end{aligned}
\tag{12.40}
$$

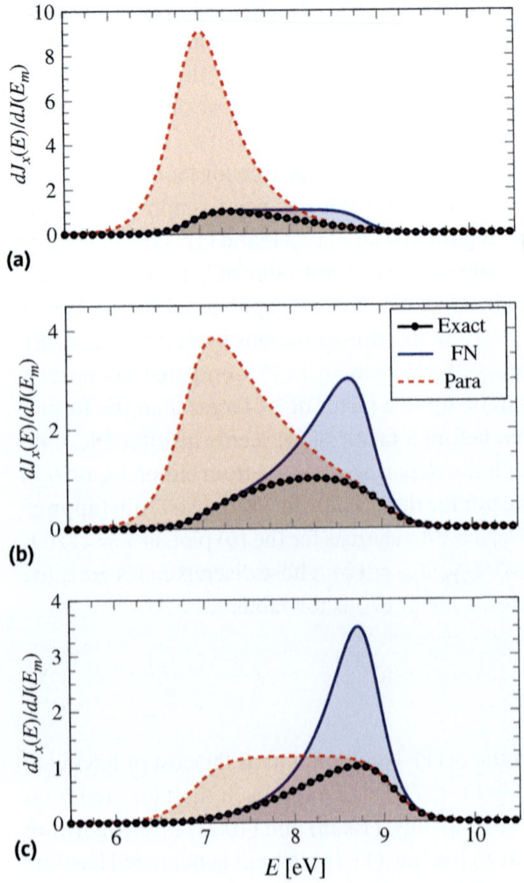


**Figure 12.11** Comparison of $dJ(E)$ to its FN and parabolic $\theta$ and $\beta_F$ approximations, all normalized to $dJ(E_m)$ for the exact case. Parameters are $\mu = 7$ eV, $\Phi = 4$ eV, and $F = 2.5$ eV/nm. (*a*) $T = 1356$ K; (*b*) $T = 1604$ K; (*c*) $T = 1851$ K

which updates the $L\theta$ approximation. This leads to the question, if one has $\theta(E)$ exactly, why approximate it with a linear function? The answer, of course, is to make use of $N(n, s)$ as an approximation to $N[\theta(E)]$ in the evaluation of $J(F, T)$ so as to be spared evaluating $J = \int dJ$ numerically. The first consequence is $dJ(E)$ and the $L\theta$ approximation to it have the same maximum, both in magnitude and location, thereby resolving the large differences in Figure 12.11*b*: whatever differences exist between $N[\theta(E)]$ and $N(n, s)$ are thus traceable to the quadratic term in a Taylor expansion of $\theta(E)$, which is small. The pay-off of the reformulated approach is that superior estimates of $n(F, T)$ and $s(F, T)$ are obtained, making the $L\theta$ approximation behind $N(n, s)$ more accurate.

The *r*GTF approach [156, 174] is designed to be a computationally expedient method to find $\theta(E)$ [183] for general barriers, albeit not as fast as the *o*GTF method. The refinements improve accuracy and are extensible to MIM, depletion, prolate spheroidal, space-charge modified, and other kinds of barriers. However, given that the machinery for the SN barrier of Section 10 is in place, plus the centrality of the SN barrier to emission physics, rooting the discussion in its terms is advantageous, with its generalization left as an exercise. The *r*GTF enhanced method is computationally implemented as follows [157]

1. Using Eq. (10.26) and $E(y) = \mu + \Phi - \sqrt{4QF}/y$, evaluate four equispaced values in $y$ between $y_{min} = \sqrt{4QF}/(\mu + \phi)$ and $y_{max} = 1$;
2. From those points, approximate $\theta(y)$ by a cubic equation in $\mu + \phi - E(y)$ by finding three coefficients $C_j$. The coefficient $C_0$ is by definition zero because $\theta[E(1)] = 0$;
3. $\theta(E)$ and $\beta_F(E)$ for $E(y) < \mu + \phi$ are then defined by

$$\begin{aligned} \theta[E(y)] &\Leftarrow \sum_{j=1}^{3} C_j[\mu + \phi - E(y)]^j \\ \beta_F[E(y)] &\Leftarrow \sum_{j=1}^{3} jC_j[\mu + \phi - E(y)]^{j-1} \end{aligned} \tag{12.41}$$

   For $E > \mu + \phi$, set $\theta(\mu + \phi - E) = -\theta(|\mu + \phi - E|)$ (antisymmetric).
4. Evaluate $N(n, s)$ using Eq. (12.24) then $J(F, T)$ using Eq. (8.8)

Observe that Step 1 makes use of $\theta(\mu)$ and $\theta(\mu + \phi) = 0$ just as *o*GTF did, but does not use the elliptical integral functions of Eq. (12.34) to do so, but rather by Eq. (9.3), or better yet, Eq. (10.26).

It may seem contradictory that $L\theta$ relies on a linear approximation to $\theta(E)$ whereas the *r*GTF approach models $\theta$ as a cubic in $(\mu + \phi - E(y))$, but it is not. Both $n(F, T)$ and $s(F, T)$ depend $\theta(E)$ and $\beta_F(E)$ evaluated at a single point $E = E_m$ to obtain $N(n, s)$. Nevertheless, outside the thermal-field region, the *r*GTF method replicates the FN ($n \to \infty$) and RLD ($n \to 0$) behavior in keeping with Eq. (12.2) and as directly indicated in Figure 12.12. In the absence of the $\Sigma(n)$ and $\Sigma(1/n)$ factors in Eq. (12.2), $J_{MG}(F, 0)$ and $J_{RTD}(F, T)$ are reproduced, but it is seen that the neglect of the $\Sigma$-factors is consequential by the separation between the ∘-data points and the (◊, □)-data points. More importantly, notice also that the discrepancy between $J(F, T)$ and either $J_{FN}$ or $J_{RLD}$ is larger. On top of the neglect of the $\Sigma(n^{\pm 1})$ factors, the choice of either $E_m = \mu$ or $E_m = \mu + \phi$, respectively, does not eliminate the error associated with $\theta(E_m)$ generally being different than $\theta(\mu)$ and $\theta(\mu + \phi)$, respectively. Now, as Figure 12.13 demonstrates, the error is much reduced.

The maximum error of approximately 70% is a significant improvement over the *o*GTF model and a substantial improvement over the FN and RLD equations in the TF regime shown in Figure 12.12, but there are means to reduce it further by a function that accounts for the peaky nature of Figure 12.13. Consider a correction [157, 174] of the form

$$\begin{aligned} \frac{dJ_{ns}}{dJ} &\approx 1 + h_1\mathbb{L}_1(T/T_o) + h_2\mathbb{L}_2(T/T_o) \\ \mathbb{L}_j(x) &\equiv \frac{1}{1 + p_j(x - \epsilon_j)^2} \end{aligned} \tag{12.42}$$

The parameters $p_j$ and $\epsilon_j$ may be found by finding the locations of the peak maxima, using, for example, the techniques of Appendix A.8, then finding their local curvature to approximate those parameters. Even crude estimates formed by eye can work reasonably well. After $p_j$ and $\epsilon_j$ are found by whatever means, then the weights $h_j$ have to be found from the matrix equation

$$\begin{bmatrix} 1 & \mathbb{L}_2(\epsilon_1) \\ \mathbb{L}_1(\epsilon_2) & 1 \end{bmatrix} \begin{bmatrix} h_1 \\ h_2 \end{bmatrix} = \begin{bmatrix} H(\epsilon_1) \\ H(\epsilon_2) \end{bmatrix} \tag{12.43}$$

where $\mathbb{L}_j(\epsilon_j) \equiv 1$ and $H(\epsilon_j) = [J_{ns}(F_j, T_j)/J(F_j, T_j)] - 1$. In the end, though, such fidelity to accuracy may be deemed ignorable as a practical matter, given that on a log plot, the visual differences are not significant. For beam optics simulations, e.g. particle-in-cell codes like MICHELLE [279, 280],

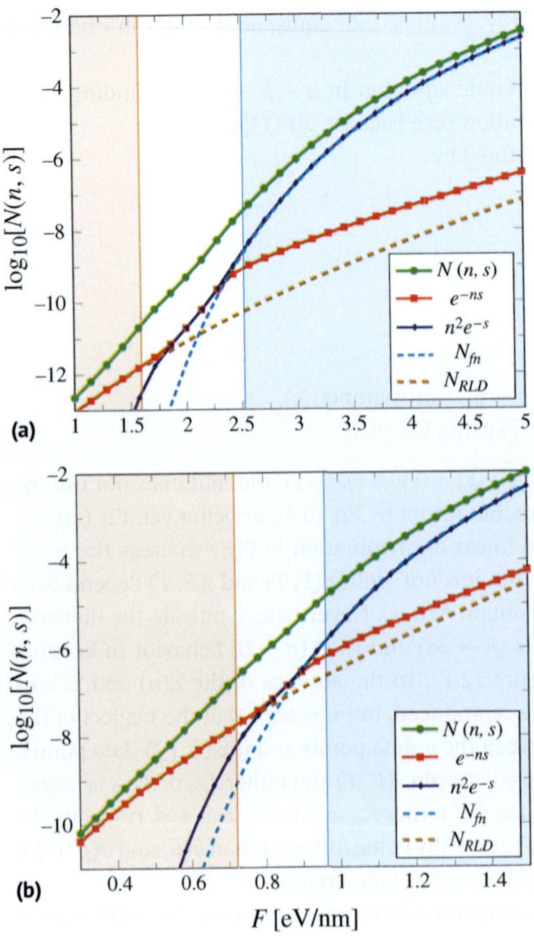


**Figure 12.12** Comparison of $N(n, s)$ (green •) to $e^{-ns}$ (red □) and $n^2e^{-s}$ (blue ◊) to show consequences of neglecting $\Sigma(n)$ and $\Sigma(1/n)$ in Eq. (12.2). Terms $n$ and $s$ evaluated for the SN barrier, with $\theta(E)$ and $\beta_F(E)$ evaluated using the shape factor method. Orange region is thermal regime, blue region is field regime, and white region is TF regime. To show convergence of the latter two to the canonical equations, $N_{fn} = J_{MG}/A_{RLD}T^2$ and $N_{rld} = J_{RLD}/A_{RLD}T^2$ are shown as dashed lines (cyan and orange, respectively). (*a*) Metal-like parameters with $\mu = 7$ eV, $\Phi = 4.5$ eV, and $T = 1273$ K. (*b*) Same, but now for $\Phi = 2.1$ eV and $T = 700$ K. Source: [157] / AIP Publishing / CC BY 4.0.

what matters more is the time taken to evaluate the current density on millions (or more) of surface elements forming an emission surface and to do so for millions (or more) time steps. That puts a premium on computational rapidity, given that actual surfaces have a range of differing parameters related to field, temperature, work function, and barrier shapes. In light of the uncertainties already present on a realistic emitter, the improvements provided by Eq. (12.42) may not warrant the investment of effort, but having options is always desirable.

How *r*GTF compares to actual data requires attention. Although field emission structures are sharpened needles, micro-fabricated cones, edges, or nanotubes, and emit current $I$ rather than current density $J$, if the not-quite-accurate assumption that relates current to current density is $I = JA$ where $A$ is the notional emission area, then the effects of temperature predicted by *r*GTF can be compared to experiment, a useful one being the data taken from heated tungsten needles by

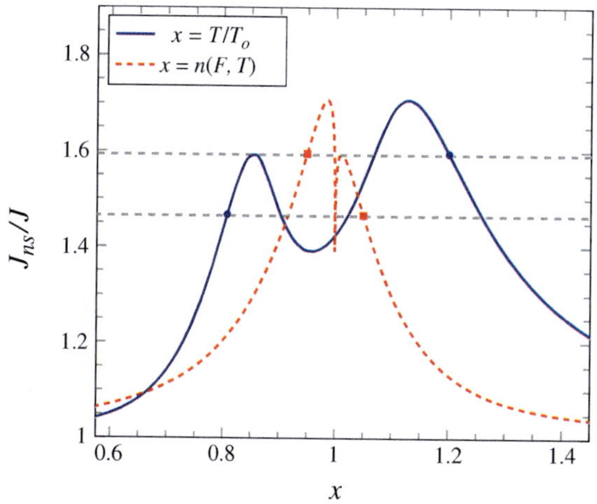


**Figure 12.13** The ratio $J_{sn}/J$ as a function of $T/T_o$ (where $T_o = 1250$ K) and $n(F, T)$ for the $x$-axis. Symbols (•, ■) occur at the same $J_{sn}/J$ in each $x$-coordinate. Source: [157] / AIP Publishing / CC BY 4.0.

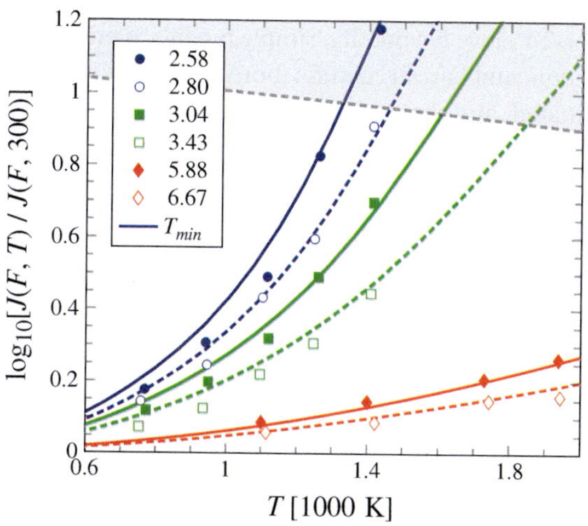


**Figure 12.14** Thermally assisted field emission from a sharpened tungsten needle: experimental data (symbols) extracted from Fig. (4) of Ref. [71]; theory lines use *r*GTF. Legend refers to the value of $F_d$ in [eV/nm] quoted by Dyke et al. The values of $F$ used in *r*GTF are $F = 1.2559F_d - 0.5027$. Gray band at top is approximately the thermal-field regime. Source: [157]/AIP Publishing/CC BY 4.0.

Dyke et al. [71]. The total emitted current from needles at a field high enough to give field emission was taken at progressively higher temperatures. A ratio of the current to its value at room temperature $T_o = 300$ K is then approximately given by $I(T)/I(T_o) \approx J(F, T)/J(F, T_o)$, where $F$ is the field at the apex of the tungsten emitter. The comparison is shown in Figure 12.14 (see Algorithm B.4.4), where the symbols are data digitally extracted from the figures in Dyke et al., and the lines are from the *r*GTF analysis (without a Lorentzian correction). Also shown is the entry into the TF regime (gray area), the boundary of which is predicted using the simpler *o*GTF approach for finding $T_{min}$ with the additional approximation that $t(y) \approx \sqrt{1.1}$. The labeling of the curves uses the approximate values of $F_d$ given by Dyke et al., but their model of the emitter makes assumptions about the relationship between the apex field $F_d$ and the potential difference of the experimental setup that creates those fields. It is therefore assumed that there is a linear relationship between their hypothetical fields and the "actual" fields at the apex given by $F = aF_d + b$. The values of $(a, b)$ are inferred by adjusting $F$ in $J(F, T)$ for each data set until the theoretical curve matches the experimental one, then finding the average of $a$ and $b$ from those matches. Doing so gives $a = 1.2559$ and $b = -5.027$. Such an analysis demonstrates that the usage of $J_{MG}(F, 0)$ alone can be in error by more than an order of magnitude when temperature effects are neglected. The complications from neglecting the thermal contributions to current from the sides of the needle using conical and prolate spheroidal models of the emitter can be significant [173, 174].

Some final observations are in order. The correspondence in Figure 12.14 is good even though there are several factors that undermine the analysis. First, the apex of a tungsten needle has several crystal faces, and each face has a different work function [74, 154]. Second, the decline in field from the apex value moving down the sides of a tungsten needle does so in a manner suggested by the prolate spheroidal models of Section 3.6. Third, because of the field fall-off, different regions of the tungsten needle will be in different regimes (field, TF, and thermal) and so the total current is the sum of all, but the contribution from each changes as the anode voltage changes because the effective emission area changes with field, and so makes the extraction of parameters such as effective work function or field enhancement complicated [172]. Even so, when field emission begins to contribute, it also begins to dominate, and so the correspondence between simple theory and experimental data shows analogous behavior. To move beyond the simple models, a means of evaluating surface field, temperature, work function, and barrier modifications from each surface element making up the emitter needs to be evaluated, and their separate contributions summed. The means to do so are best done in particle-in-cell [59, 279] or molecular dynamics codes [315, 342] to account for the multidimensional nature of the emitter surface, whatever the emitter protrudes from, and whatever structures (anodes, gates) are present that give rise to the field under which they operate.

## 12.5 Exercises

**Exercise 12.1** Show that knowing that the maximum of the integrand of $N(n, s)$ (Eq. 8.8) occurs for $s > z_m > 0$, coupled with $e^z$ and $e^s$ being large, entails that $n = 1$ (i.e. $\beta_T = \beta_F$) to leading order.

**Solution** $(d/dz)\exp[(n-1)z - ns] = 0$.

**Exercise 12.2** Let a cubic polynomial $f(x) = \sum_{n=0}^{3} C_n x^n$ for $0 \le x \le 1$ be such that $f(0) = f_a$, $df(0)/dx = g_a$, $f(1) = f_b$ and $df(1)/dx = g_b$. Find $C_n$.

**Solution** $C_0 = f_a$; $C_1 = g_a$; $C_2 = -3f_a - 2g_a + 3f_b - g_b$; $C_3 = 2f_a + g_a - 2f_b + g_b$.

**Exercise 12.3** Let $0 \le x \le 1$. Using the notation of Ex. 12.2, first show a linear approximation to $f(x)$ can be written

$$f(x) \approx f_a(1-x) + f_b x$$

Next, show that a cubic approximation can be written as

$$f(x) \approx f_a(1-x)^3 + Ax(1-x)^2 - Bx^2(1-x) + f_b x^3$$

What is $A$ and $B$?

**Solution** $A = g_a + 3f_a$; $B = g_b - 3f_b$.

**Exercise 12.4** Use Exs. 12.2 and 12.3 to show Eqs. (12.36) and (12.37).

**Exercise 12.5** Using the Forbes–Deane approximations to $v(y)$ and $t(y)$, first calculate $T_{min}$ for the values of $\Phi = 4.5$ eV and $F = 4$ eV/nm and second, calculate $T_{max}$ for the same values.

**Solution** $T_{min} = 2135.6$ K; $T_{max} = 2632.8$ K.

**Exercise 12.6** For the conditions of Figure 12.10b, the GTF equation is shown to be approximately 150× the RLD equation, and 100× the FN equation, at $T = 1200$ K and $F = 2$ eV/nm. Compare these ratios to the ratio of Eq. (12.2) with Eq. (12.32), specifically, the size of $(s + 1)$.

**Solution** Algorithm B.4.4 gives ratios of 140× and 98×, respectively.

# 13

# Photoemission

*Give me some light! Away!*

W. Shakespeare, *Hamlet* III.2

## 13.1 Simple Quantum Efficiency

A one-dimensional Fowler–DuBridge equation of quantum efficiency $QE$ of Eq. (8.25) successfully used the $N(n, \pm s)$ function to find the contributions of field and thermal effects to $QE$. The fields put on photocathodes in the laboratory ($\mathcal{E} \lesssim 1$ kV/m equivalent to $F \lesssim 10^{-6}$ eV/nm) are paltry compared to those in photo-injectors ($F \sim 10^{-4} - 10^{-3}$ eV/nm), but those fields in turn are much less than those needed for field emission. Therefore, a reasonably good approximation is to neglect the $\beta_F$ contribution to Eq. (8.25) and express QE as

$$\frac{QE(\omega)}{QE(\omega_o)} \approx \frac{(\hbar\omega - \phi)^2 + 2\zeta(2)(k_B T)^2}{(\hbar\omega_o - \phi)^2 + 2\zeta(2)(k_B T)^2} \tag{13.1}$$

where $\hbar\omega_o$ is a conveniently chosen photon energy larger than, but presumably near the value of $\phi = \Phi - \sqrt{4QF}$, given that the FD approximation is a threshold approximation. To assess the theory, copper is a well-understood and convenient metal photocathode material [65]. For measurements made on a copper photocathode by Dowell et al. [64], the account of the behavior of $QE$ using Eq. (13.1) is shown in Figure 13.1 (see Algorithm B.4.5), for which it is seen that the value of $\phi =$ 4.36 eV gives the optimal low photon energy, in keeping with FD theory being a threshold emission model. At higher photon energies, the FD theory curves depart from the data for the optimal $\phi$ there, but one has the impression that the disagreement is not terrible. In its more modern usage, it can be extended in various ways to treat emission from metals exhibiting multiphoton or field-assisted photoemission [2, 108, 162, 265], or in being able to describe changes wrought by additional physics. As an example, for the common photocathode material GaAs to generate spin-polarized electron beams [47], Rashba spin-orbit coupling (RSOC) causes photoexcited electrons to separate into different spin chirality populations, so that replacing the standard parabolic electron dispersion with a Rashba spintronic material (RSM) population in the FD framework gives [341]

$$QE \propto \frac{1}{2}(\hbar\omega - \phi)^2 + E_R(\hbar\omega - \phi) - \frac{E_R^3}{6} \tag{13.2}$$

where $E_R$ (a fraction of an eV) affects the apparent work function through the RSOC effect.

*Fundamentals of Electron Emission Physics*, First Edition. Kevin L. Jensen.

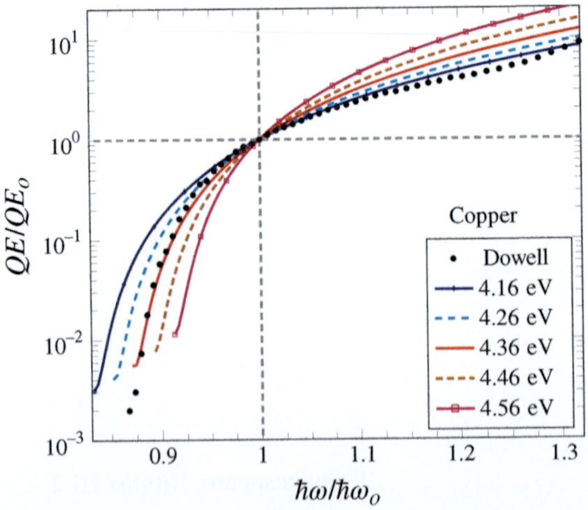


**Figure 13.1** The quantum efficiency of an H-ion beam cleaned copper surface [64] (•, data courtesy of D. Dowell, SLAC). Field at surface was essentially zero, so $1/\beta_F^2 \approx 0$, and measurements were done at room temperature. The reference point is $\hbar\omega_o = 4.999$ eV and $QE(\hbar\omega_o) = 1.79 \times 10^{-4}$. Theory lines (Eq. 13.1) shown for a range of values of $\phi$. The theory lines terminate at $\hbar\omega \approx \phi$. After [157] / AIP Publishing LLC/ CC BY 4.0.

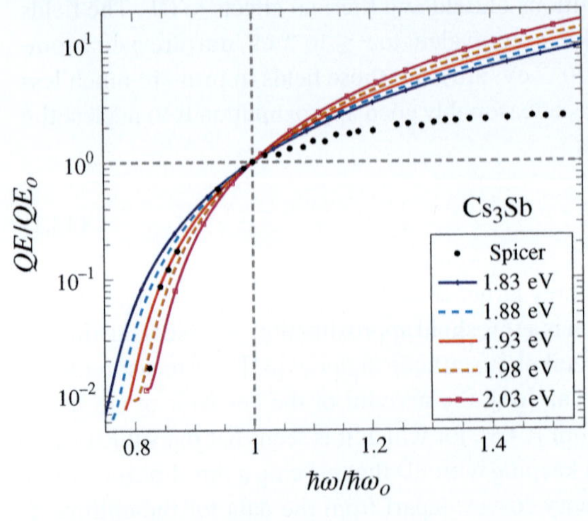


**Figure 13.2** The quantum efficiency of $Cs_3Sb$ [324] (•, data digitally extracted from Figure 8 of Spicer [324]. Field at surface was essentially zero, so $1/\beta_F^2 \approx 0$, and measurements were done at room temperature. The reference point is $\hbar\omega_o = 2.475$ eV and $QE(\hbar\omega_o) = 0.0580$. Theory lines (Eq. 13.1) shown for a range of values of $\phi$. The theory lines terminate at $\hbar\omega \approx \phi$. After [157] / AIP Publishing LLC/CC BY 4.0.

For QE data taken from another common semiconductor photoemissive material, cesium antimonide or $Cs_3Sb$ [322, 324, 325], the pretense of agreement is lost. Such data are shown in Figure 13.2 for data taken by Spicer, again alongside lines based on the FD equation (Eq. 13.1). Spicer's response was to develop a Three-Step Model (TSM) [22, 324, 365] that continues to enjoy widespread usage in understanding QE from metals [65, 319, 328] but it requires multidimensional notions from the outset and has to account for particular differences between metals and semiconductors.

Semiconductors and metals differ in key ways. The existence of a bandgap between the conduction and valence bands [203] has several important consequences. The first consequence is that the Fermi level of a semiconductor can change as a consequence of doping (Section 7.2), although even very high doping levels (on the order of $10^{20}$ cm$^{-3}$ = 0.1 nm$^{-3}$) are pitiful compared to a metal like copper (electron density of 84.11 nm$^{-3}$). Consequently, a more useful measure of the barrier height at the surface is given by the electron affinity $E_a$, which is the height of the vacuum level above the conduction band minimum. Some materials, like diamond, can have exceptionally large bandgaps, so that if the surface is treated (hydrogenated in the case of diamond [11, 52, 300, 370]), the vacuum level is below the conduction band minimum $E_c$, making diamond a negative electron affinity (NEA) material. The second consequence is the impact on scattering.

If photoexcited electrons scatter with other electrons from the valence band, then both daughters have to have energies above the conduction band minimum: final states within the bandgap are disallowed. As a result, for photoexcited electrons near threshold, that is, with an energy not much higher than the conduction band minimum, electron–electron scattering is forbidden. In metals, for electron–electron scattering, the "parents" (the photoexcited electron $p$, and an electron typically near the Fermi level $p'$) scatter and distribute their energy, resulting in the daughter electrons $d$ and $d'$, or $p + p' \rightarrow d + d'$ as in the dashed line of Figure 13.3. In metals, if the photon energy is not much more than $\phi$, the sharing of that energy generally results in both electrons having insufficient energy to surmount the surface barrier, thereby generally killing the chances to escape to vacuum of one or both of them. It also entails that, more often than not, both the primary electrons and the daughters $(p, p')$ and $(d, d')$ in Figure 13.3 are near the Fermi level, as the final states cannot be occupied, and so conservation of energy and momentum restricts what the final states can be. In semiconductors, electron–electron scattering is prevented if the final states result in daughters that are in the bandgap (band bending due to external field is ignored for the semiconductor side of Figure 13.3), and so a "magic window" [328] is said to exist where only electron–phonon scattering is allowed. Because *ep*-scattering (Section 7.4) has a longer relaxation time $\tau$, electrons can travel further without suffering a scattering, and even after scattering, their energy need not be reduced so precipitously as when they collide with one of their own. Since a much smaller exchange of energy occurs for *ep*-scattering (technically, acoustic phonon scattering – optical phonons are higher energy but not as prevalent), electrons in semiconductors can scatter multiple times without losing so much energy that they are emission ineligible [175, 179], which is in contrast to metals for which electron–electron scattering dominates the photoexcited electron relaxation dynamics.

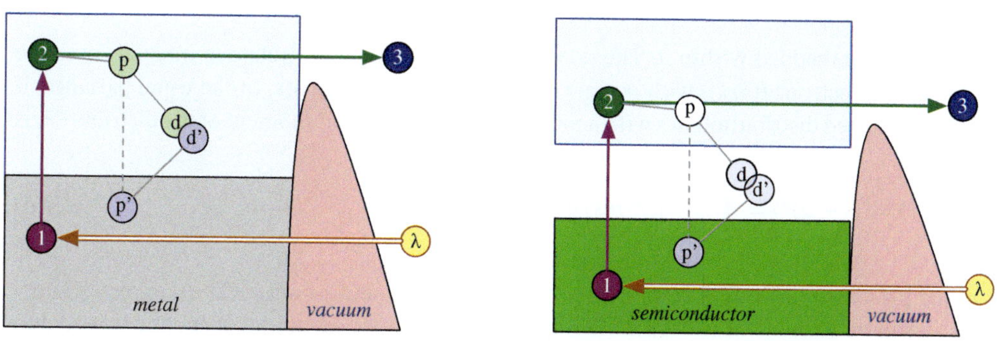


**Figure 13.3** Photoexcitation in metals (*left*) and semiconductors (*right*) in the Three Step Model: (1) electron absorbs incident photon $\lambda$; (2) electron energy near Fermi level $\mu$ increased by photon energy $\hbar\omega$ and electron transports to surface where (3) it is emitted if $E > \mu + \phi$. After [155] / IEEE.

## 13.2 Simple Moments Model

The model of photoemission due to Spicer [22, 325, 327, 328, 365] is commonly called the TSM [64, 154, 319, 378]. Unlike the FD model, it takes into explicit account the three-dimensional nature of electron excitation and transport. Spicer treated photoemission as a sequence of three steps [65, 208, 328] by taking the "steps" to be:

1. *Absorption* ($\mathbb{A}$): incident light (typically a laser in photoinjectors) illuminates the surface of a photocathode, where some of the light is reflected ($R(\omega)$). The absorbed light is attenuated as it passes into the photocathode material ($\delta(\omega)$). The absorbed light intensity $I_o = [1 - R(\omega)]I_\omega$ is attenuated as a function of distance as $I_o e^{-z/\delta(\omega)}$.
2. *Transport* ($\mathbb{T}$): an electron absorbs the photon energy $\hbar\omega$, and the excited electron moves in a random direction ($\theta$) which may be directed at the surface, where an electron is lost on average every mean free path $\ell = (\hbar k/m)\tau(E)$, where $\tau$ is the relaxation time, in proportion to its path length $r = z/\cos\theta$ to the surface, or a factor proportional to $e^{-z/\ell \cos\theta}$.
3. *Emission* ($\mathbb{E}$): upon reaching the surface, the electron escapes with a probability of emission $D(E_z)$, where $E_z = E\cos^2\theta$ is the normal, or "into the surface" energy. The total energy is the sum of the transverse and normal components, or $E = E_z + E_\perp$.

Berglund and Spicer made five simplifying approximations in pursuit of an analytical model. Those approximations are

i. the photoexcited electron distribution is isotropic;
ii. all scattering events are inelastic;
iii. the inelastic scattering events occur with a mean free path given by the product of velocity with a scattering time, or $\ell = \hbar k(E)\tau(E)/m$ that depends only on the electron's total energy $E$;
iv. the scattering is isotropic as well; and
v. the normal energy $E_x = E\cos^2\theta$ must be greater than the surface barrier to be emitted.

The Simple Moments Model [171] uses the same three-step structure and the same simplifying approximations, but works in a momentum-based rather than energy-based framework to evaluate an integral for current density

$$J(\hbar\omega) = \frac{2}{(2\pi)^3} \iiint \left(\frac{\hbar k_z}{m}\right) \mathbb{A} \times \mathbb{T} \times \mathbb{E}\, dk_z d\vec{k}_\perp \tag{13.3}$$

and the factors embedded within it. The nomenclature of "Moments" refers to integrals over the distribution of electrons $f(\vec{x}, \vec{k})$ with powers of momentum components. In the usual parlance, if $p(s)$ is a normalized distribution in $s$ with a mean $\langle s \rangle = \mu$, then the $n^{th}$ moment of the distribution is

$$M_n(s) = \langle (s-\mu)^n \rangle = \int_{-\infty}^{\infty} (s-\mu)^n\, p(s)\, ds \tag{13.4}$$

but in the present context, for calculating currents, densities, and (soon) mean transverse energy (MTE), it is more convenient to drop including the mean $\mu$ in the definition of $M_n$ and extend it to multi-dimensions. Therefore, "moments" for the Moments Model will be defined according to

$$M_n(k_j) \equiv \frac{g}{(2\pi)^3} \int k_j^n f(\vec{k}) d^3k \tag{13.5}$$

where $d^3k = dk_x dk_y dk_z$ in Cartesian coordinates or $k^2 \sin\theta \, dk d\theta d\phi$ in polar, $k_j$ is a Cartesian component of $\vec{k}$, e.g. $k_z = k\cos\theta$, a factor of $g = 2$ in the numerator of the coefficient accounts for electron spin, and $f(\vec{k})$ is the distribution function over which moments are evaluated. Note that so defined, $M_n(k_j)$ is not normalized, and so the evaluation of moments in the present usage will require a denominator to do so, an integral over $f(\vec{k})$ that assumes an infinite mean free path $\ell(E)$ (the loss factor $f_\lambda = 1$, see Eq. 13.16), a unitary transmission probability ($D(E_z) = 1$), and a recognition that electrons can go in all directions ($0 \le \theta \le \pi$) rather than just in the forward direction ($0 \le \theta \le \pi/2$). Finally, note that the coefficient of 2 that accounts for electron spin is thereby technically unnecessary in the new definition of $M_n$, as are the factors of $(2\pi)^3$, but they are included for clearness (even though taking ratios eliminates them anyway) because doing so retains a connection to other emission equation formalisms. The general form of $M_n(k_j)$ (where $k_j$ can denote the longitudinal $k_z$ or transverse $k_\perp$ momentum) therefore takes the form for metals [178] (with semiconductors having a different, but analogous, form)

$$M_n(k_j) = \frac{1}{(2\pi)^3}\int d^3k \quad k_j^n D[(E+\hbar\omega)\cos^2\theta] f_\lambda(\cos\theta, E) \\ \times f_{FD}(E)[1 - f_{FD}(E+\hbar\omega)] \tag{13.6}$$

where $E = \hbar^2 k^2/2m$ and the $E$ in $f_\lambda(x, E) = x/(x+p)$ means that the term $p(\tilde{E})$ is evaluated at the energy $\tilde{E} = E_{initial} + \hbar\omega$ of the photoexcited electron (an important observation because the equation for metals will differ from semiconductors in how $E$ will be interpreted in the integrand below).

In the Moments model, the number density is proportional to the $0^{th}$ moment, and the current density to the product of the $1^{st}$ moment of $k_z$ with $(\hbar/m)$. Quantum efficiency, being dependent on the current density, then becomes

$$QE \equiv [1 - R(\omega)] \frac{M_1(k_z)}{2\, M_1(k_z)|_{D=1, f_\lambda=1}} \tag{13.7}$$

where the 2 in the denominator takes into account that half the electrons are traveling away from the surface rather than towards it, thereby allowing the $\theta$ integration to be only to a maximum of $\pi/2$ on both numerator and denominator. In the limit that $\hbar\omega \to 0$, the Moments model would seemingly resemble Eq. (8.1), but a substantial modification arises for photoemission: the probability that an electron has a total energy $E$ before the absorption of a photon is still governed by the Fermi–Dirac distribution $f_{FD}(E)$, but now the probability that its final state is available must be taken into account by a factor of $[1 - f_{FD}(E+\hbar\omega)]$, for which the low-temperature limit results in the introduction of Heaviside step functions $\Theta(s)$ (0 for negative argument $s$, 1 for positive argument). Therefore, in place of a single $f_{FD}(E)$, the product

$$f_{FD}(E)[1 - f_{FD}(E+\hbar\omega)] = \Theta(\mu - E)\Theta(E - \mu + \hbar\omega) \tag{13.8}$$

appears and places limits on the $k$ integrations.

Compared to the puny fields of laboratory conditions, in photoinjectors, fields at the surface of the photocathode are often in the 50 MV/m (or $F = 0.05$ eV/nm) range [250], but can go as high as 500 MV/m (or $F = 0.5$ eV/nm) [304] (although exploiting field enhancement from sharpened tips for field-assisted photoemission [100, 138] can get to near the GV/m range). For such fields, which are still comparatively low by field emission standards, the transmission probability $D(k_z)$ can be treated as a step function that only allows electrons to escape if their energy after photoexcitation is $\tilde{E} = \hbar^2 k_z^2/2m < \mu + \phi$, where $\phi = \Phi - \sqrt{4QF}$. That step function is changed for semiconductors with a positive electron affinity $E_a$ and bandgap $E_g$. Switching to polar coordinates, $\hbar k_z/m = (\hbar k/m)\cos\theta$, and so a maximum angle $\theta_m$ exists for which emission is allowed (the "escape cone" argument [48, 319] according to

$$\begin{aligned} (E+\hbar\omega)\cos^2\theta_m &= \mu+\phi \qquad \text{(metals)} \\ \tilde{E}\cos^2\theta_m &= E_a \qquad \text{(semiconductors)} \end{aligned} \tag{13.9}$$

which can be re-expressed as

$$\cos\theta_m|_{metal} = \left[\frac{\mu+\phi}{E+\hbar\omega}\right]^{1/2} \tag{13.10}$$

$$\cos\theta_m|_{semi} = \left[\frac{E_a}{\tilde{E}}\right]^{1/2} \tag{13.11}$$

but with the important caveat that energy is being treated differently for metals and semiconductors in this formulation: in Eq. (13.10), $E$ is the energy of the electron before excitation, whereas in Eq. (13.11), as subsequent exposition will show, it is more convenient to let $\tilde{E}$ be the energy after excitation (hence the tilde on it). The distinction reflects how the integrals in $QE$ are defined in the simple Moments Model (SMM). Consequently, although the metal and semiconductor versions of the various equations appear to use the same $E$, they do not in the SMM formulation, and the calculation must take that into account [179]. Where possible, a tilde on the $\tilde{E}$ will serve as a reminder of this, but when $E$ is part of an integral, being an integration variable, the tilde will not appear so as to keep the integrals uncluttered: simply remember semiconductors and metals differ.

After setting up the integrals, it is convenient to transform from a momentum integration to an energy integration using the parabolic correspondence $E = \hbar^2k^2/2m$ and $dE = (\hbar^2 k/m)dk$. It then follows [157]

$$QE_{metal} = \frac{(1-R(\omega))}{\hbar\omega(2\mu-\hbar\omega)} \int_{\Delta_m}^{\mu} EdE\, F_\lambda(x_m, E) \tag{13.12}$$

$$QE_{semi} = \frac{(1-R(\omega))}{(\hbar\omega-E_g)^2} \int_{E_a}^{\Delta_s} EdE\, F_\lambda(x_m, E) \tag{13.13}$$

$$F_\lambda(x_m, E) \equiv \int_{x_m}^{1} D(Ex^2) f_\lambda(x, E)\, x dx \tag{13.14}$$

where $\Delta_m = \mu+\phi-\hbar\omega$ for metals and $\Delta_s = \hbar\omega - E_g$ for semiconductors, $x_m = \cos\theta_m$, $x = \cos\theta$, and the denominators of the coefficients are the integrals make the replacements $D(\tilde{E}x^2) \to 1$ and $x_m \to 0$, so that

$$\begin{aligned} 2\int_{\mu-\hbar\omega}^{\mu} EdE \int_0^{\pi/2} \sin\theta\, d\theta &= \hbar\omega(2\mu-\hbar\omega) \\ 2\int_0^{\hbar\omega-E_g} EdE \int_0^{\pi/2} \sin\theta\, d\theta &= (\hbar\omega-E_g)^2 \end{aligned} \tag{13.15}$$

The loss function $f_\lambda(\cos\theta, E)$ is proportional to the integral of the product of the attenuation of light into the material ($e^{-z/\delta}$) with the fraction of electrons lost getting back to the surface due to scattering ($e^{-z/\ell\cos\theta}$), and is normalized so that for $\ell \to \infty$ then $f_\lambda \to 1$, giving

$$\begin{aligned} f_\lambda(\cos\theta, E) &\equiv \frac{1}{\delta}\int_0^\infty \exp\left[-\frac{z}{\delta}\left(1+\frac{p}{\cos\theta}\right)\right] dz \\ &= \frac{\cos\theta}{\cos\theta + p(E)} \end{aligned} \tag{13.16}$$

where $p(E) = \delta(\omega)/\ell(\tilde{E})$, with $\delta(\omega)$ as per Eq. (7.116), $\ell(\tilde{E}) = (\hbar k(\tilde{E})/m)\tau(\tilde{E})$, and $\tau(E)$ as per Eq. (7.53). The evaluation of the laser penetration depth $\delta(\omega)$ requires the imaginary part of the index of refraction $\Im(\hat{n})$, values of which for materials commonly used in photoemission studies are given in Table 13.1. Two cautionary points: first, remember that the $k$ of $\hat{n} = n + ik$ is not the same as the wavenumber $k$ in the velocity term ($\hbar k/m$); second, remember that the evaluation of the mean free path $\ell(\tilde{E})$ requires the velocity of the photoexcited electron, calculated from a kinetic energy that is always the sum of the initial energy of the electron plus the photon energy that it absorbs, or $\tilde{E} = E_{initial} + \hbar\omega$. The approximation behind Eq. (13.16) is the same as assuming that any scattering event renders the daughter electrons with an energy insufficient to overcome the surface barrier, and so scattering is fatal to emission. Hence, Eq. (13.16) is referred to as the "Fatal Approximation" for metals. For semiconductors, scattering is not nearly as perilous, and so electrons that have scattered with phonons may do so a number of times before they are

**Table 13.1** Real and imaginary parts of $\hat{n}$ for typical wavelengths of used on photocathode materials. Adapted from [183] / IEEE.

| $\lambda$ [nm] | 1064 | 800 | 650 | 532 | 405 | 355 | 266 | 206 |
|---|---|---|---|---|---|---|---|---|
| $n$(Ag) | 0.405 | 0.469 | 0.571 | 0.74 | 1.10 | 1.28 | 1.51 | 1.56 |
| $k$(Ag) | 7.36 | 5.28 | 4.06 | 3.08 | 2.11 | 1.81 | 1.44 | 1.23 |
| $n$(Au) | 0.118 | 0.144 | 0.216 | 0.406 | 1.34 | 1.51 | 1.37 | 1.16 |
| $k$(Au) | 6.73 | 4.75 | 3.53 | 2.43 | 1.89 | 1.49 | 1.64 | 1.06 |
| $n$(Cu) | 0.414 | 0.328 | 0.365 | 0.582 | 1.26 | 1.25 | 1.54 | 1.13 |
| $k$(Cu) | 7.44 | 5.31 | 4.01 | 2.87 | 2.20 | 1.93 | 1.77 | 1.66 |
| $n$(Pb) | 2.97 | 2.19 | 2.76 | 2.79 | 1.56 | 1.23 | 1.05 | 0.708 |
| $k$(Pb) | 6.50 | 4.48 | 3.72 | 4.05 | 3.76 | 3.25 | 2.50 | 2.16 |
| $n(Cs_3Sb)$ | 3.61 | 3.73 | 3.81 | 2.96 | 1.98 | 1.37 | 1.01 | 0.378 |
| $k(Cs_3Sb)$ | 0.0434 | 0.512 | 1.26 | 2.12 | 2.24 | 1.98 | 1.70 | 1.16 |
| $n(CsK_2Sb)$ | 3.19 | 3.17 | 3.11 | 3.21 | 2.34 | 1.68 | 1.23 | 0.717 |
| $k(CsK_2Sb)$ | 0.123 | 0.313 | 0.564 | 0.992 | 2.06 | 1.84 | 1.55 | 1.47 |
| $n$(GaAs) | - | 3.68 | 3.83 | 4.13 | 4.42 | 3.54 | 3.66 | 1.26 |
| $k$(GaAs) | - | 0.09 | 0.18 | 0.34 | 2.07 | 2.02 | 3.34 | 2.47 |

emission-ineligible [175], which accounts for the long tails on emission pulses that depart from the Gaussian drive lasers which create them [18, 179].

Equation (13.14) is analytic if $p(E)$ is replaced by a constant (that is, $\ell$ does not vary much, an approximation that is increasingly reasonable the closer $\hbar\omega$ gets to $\phi$): since most of the photoexcited electrons for a metal originate from near the Fermi level, then letting $p(E) \rightarrow p(\mu)$ is a reasonable approximation (with a similar approximation for semiconductors), and so

$$F_{\lambda}(x,p) = \frac{1}{2}(1-x)(1+x-2p) - p^2 \ln\left(\frac{x+p}{1+p}\right) \tag{13.17}$$

This is an unpleasant function to have in an integrand, apart from whether or not $p(\mu)$ is a good approximation, and its presence makes the pursuit of a clean analytic representation of the integrals behind QE awkward. Pursuit of a reasonable analytic form that allows for a simple form of QE is desirable, and an approximation to $F_{\lambda}(x,p)$ that makes it so is therefore alluring. Observe that when the energy $E$ is at its minimum value of $\Delta$, then the electron has to be directed straight at the surface, or $\theta_m = 0$, and so $x_m = 1$. But $F_{\lambda}(1,p) = 0$ identically. A plot of $F_{\lambda}(x,p)$ for a variety of $p$ approximately constant shows that it has a mild curvature, and so considering a linear approximation is suggestive, and is simply

$$F_{\lambda}[x_m(E), p(E)] \approx F_{\lambda}[x_m(\mu), p(\mu)] \left[\frac{E-\Delta}{\mu-\Delta}\right] \tag{13.18}$$

The final integration over $E$ is then analytic, and so for metals,

$$QE_{metal} \approx [1-R(\omega)]F_{\lambda}[x_m(\mu), p(\mu)] \left[\frac{(\hbar\omega-\phi)(3\mu+\phi-\hbar\omega)}{6\hbar\omega(2\mu-\hbar\omega)}\right] \tag{13.19}$$

a form which conveniently separates the optical parameters part $[1-R]F_{\lambda}$ from the material parameters part dependent on $\phi$ and $\mu$, and so which constitutes the desirable $QE$ approximation being sought for metals. Its performance is shown in Figure 13.4.

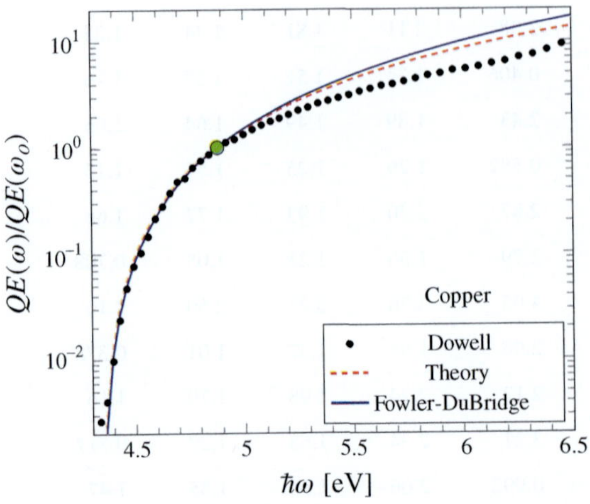


**Figure 13.4** Measurements of the QE of copper (data courtesy of D. Dowell (SLAC) and reported Ref. [64] for copper. Theory line corresponds to Eq. (13.19). Both theory lines assume that $\phi = 4.34$ eV, and that $R(\omega)$ and $p(\mu)$ do not change with $\hbar\omega$ and therefore factor out in the ratio. After [157] / AIP Publishing LLC/CC BY 4.0.

The form of Eq. (13.19) is simple, but perceiving how the Fowler–DuBridge equation lurks behind it is difficult. To tease out that behavior, take a step back and consider $F_\lambda$ with an albeit worse (but more transparent) approximation. Observe that (remembering that $x = \cos\theta$)

$$f_\lambda(x, E) = \frac{x}{x + p(E)} \approx \frac{x}{1 + p(E)} \tag{13.20}$$

near threshold. With that approximation, next look at $F_\lambda(x_m, E)$: for metals $D(Ex^2)$ is approximated by a Heaviside step function (0 when the electron is below the SN barrier maximum $V(x_o)$, unity when it is above), but by design the bounds of the integrals are such that $D(Ex^2)$ is always unity in Eq. (13.14). As a result,

$$\begin{aligned} F_\lambda(x_m, E) &\approx \int_{x_m}^{1} \frac{x^2\,dx}{1 + p(E)} = \frac{1 - x_m(E)^3}{3[1 + p(E)]} \\ &\approx \frac{1 - x_m(\mu)^3}{3[1 + p(\mu)]} \left( \frac{E - \Delta_m}{\mu - \Delta_m} \right) \end{aligned} \tag{13.21}$$

where the second line uses the logic seen in the linear approximation behind Eq. (13.18) again. Lastly, assume again that $\hbar\omega - \phi \equiv \delta$ is small. The pieces are then

$$\begin{aligned} \frac{1 - R(\omega)}{\hbar\omega(2\mu - \hbar\omega)} &= \frac{1 - R(\omega)}{\hbar\omega(2\mu - \phi - \delta)} \\ \frac{1 - x_m(\mu)^3}{3[1 + p(\mu)]} &\approx \delta \frac{[4(\mu + \phi) - 5\delta]}{8[1 + p(\mu)](\mu + \phi)^2} \\ \int_{\Delta_m}^{\mu} EdE \left( \frac{E - \Delta_m}{\mu - \Delta_m} \right) &= \delta \frac{(3\mu - \delta)}{6} \end{aligned} \tag{13.22}$$

Observe that lines 2 and 3 have a $\delta$ out front: therefore, to order $\delta^2$ the product of these three lines will have a $\delta^2 = (\hbar\omega - \phi)^2$ out front, so $\delta$ can be replaced by zero wherever it appears thereafter to get the leading order approximation. Doing so

$$QE \approx \frac{1 - R(\omega)}{4[1 + p(\mu)]} \frac{\mu(\hbar\omega - \phi)^2}{\phi(\mu + \phi)(2\mu - \phi)} \tag{13.23}$$

which is clearly exhibits the $(\hbar\omega - \phi)^2$ that the FD relation argues for.[1] Being a threshold approximation, though, the approximation degrades the further $\hbar\omega$ moves away from $\phi$.

Before taking leave of metals, observe that the Moments approach can also be used to calculate the mean transverse energy (MTE) of emitted electrons [17, 49, 83]. That is, just as $M_1(k_z)$ is used in the evaluation of $QE$, so too is $M_2(k_\perp^2)$ used in the calculation of MTE, where $k_\perp = k \sin\theta$. That relation can be used to show that the critical parameter of emittance, a parameter that gives a measure of how a beam of electrons spreads [31] can also be calculated in the Moments approach. This matters because the "brightness" of a beam is technologically very important, with a bright beam being one in which all the current $I$ from an area $A$ is in a narrow cone angle $\Omega$, or $B = I/A\Omega$, a relation that is often recast as $2I/\varepsilon^2$, where $\varepsilon_{n,rms}$ is the emittance [31, 65, 178]. By analogy, light sources with very bright beams are called "lasers."

1 Eq. (7) of Ref. [183] uses $\hbar\omega$ rather than $\phi$ in the denominator, a difference of small consequence if $\delta = \hbar\omega - \phi$ is small: here $QE$ is to leading order in $\delta$ only and avoids a singularity when $\hbar\omega$ is large.

Emittance is one of those concepts for which the definition is deceptively easy but difficult in practice to quantify [154]. It is defined in terms of distributions behind the calculation of average quantities, and by virtue of relying on distributions, it is best understood by visualizing it as a creature of phase space. The spread of a beam means that as the beam moves in the forward ($\hat{z}$) direction, it widens in the transverse direction, which, for the sake of argument, shall be called the $\hat{x}$ direction.[2] So, spread is measured by the ratio of the change in the size of the transverse direction (call it $\Delta x$) given how far the beam has moved in the forward direction (call it $\Delta z$), and that ratio is conventionally called $x' = \Delta x/\Delta z$ (the prime does not mean derivative here). And so, the difficulties start: phase spaces are not two-dimensional, but six-dimensional (three position dimensions, three spread dimensions). Take as a pair a position coordinate (e.g. $x$) and a spread coordinate (e.g. $x'$), or $(x, x')$, so that there are three such pairs, and focus on "emittance" as applied to one of them, chosen here to be $\varepsilon_x$. So, given a distribution of velocities and positions, every particle in a beam can have its $x'$ location plotted on the vertical axis, and its $x$ location plotted on the horizontal axis. Being particles with random velocities and positions, all these points in phase space are going to be spread out, but taking a pencil and drawing an oval that encloses all those points is possible, and the size of that oval is taken as a measure of emittance. Mostly. What is actually important is the root-mean-square size of the enclosing area, with its size along the position coordinate given by $\sqrt{\langle x^2 \rangle}$ and along the spread coordinate by $\sqrt{\langle x'^2 \rangle}$. If the enclosing area is oddly shaped, then a factor $\langle xx' \rangle^2$ corrects for its asymmetries. With that, (tracer) emittance is defined as $\varepsilon = \sqrt{\langle x^2 \rangle \langle x'^2 \rangle - \langle xx'^2 \rangle}$ [293]. Now, for that emittance, the spread of the beam in the $x$-coordinate can be related to the velocity of the beam in that coordinate via $\Delta x = v_x \Delta t$ compared to how far the beam moves in the forward (or $z$) direction, or $\Delta z = v_z \Delta t$, as in Figure 13.5, making the ratio $x' = v_x/v_z$, and therefore introducing an emittance more in keeping with phase spaces being position-velocity relations. Further modifications are made to describe electron beams, which are generally accelerated to very high velocities to do what they do (create microwaves, slam into metals to generate X-rays, act as the beams for electron microscopes, be used in particle accelerators, and so on). That means that the total velocity magnitude is dominated by the $v_z$ component, or $v_z \approx v$. Recalling Section 5.2, the relativistic notation $v = \beta c$ and $\gamma = \sqrt{1-\beta^2}$ means the momentum in the $z$ direction is approximately $mv_z = \beta\gamma mc$ as per Eq. (5.18). The "normalized" emittance divides out the $\beta\gamma$ factor, or $\beta\gamma\varepsilon \equiv \varepsilon_{n,rms}$. Writing velocity as $\hbar k/m$, then the normalized emittance becomes [31, 226, 293]

$$\varepsilon_{n,rms} \equiv \frac{\hbar}{mc}\sqrt{\langle x^2 \rangle \langle k_x^2 \rangle - \langle xk_x \rangle^2} \tag{13.24}$$

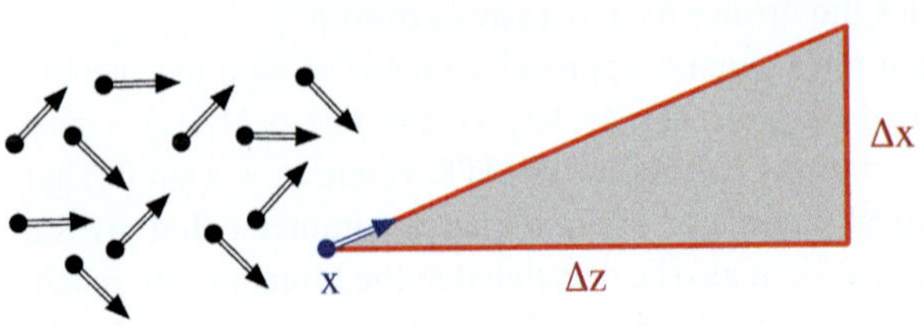


**Figure 13.5** Position $x$ and spread $x' = \Delta x/\Delta z$ for a group of particles randomly positioned and pointing in random directions, but all moving over time to the right (beam direction).

2 Alas, the notation follows convention at the risk of confusion: $x$ here is a coordinate normal to the $\hat{z}$ axis, whereas $x = \cos\theta$ in the discussion of $f_\lambda(x, E)$. Context and the alertness of the reader help in knowing which is which.

Keep in mind that the averages are over the entire distribution: for a Gaussian distribution in both momentum and radius (e.g. $f(\vec{r},\vec{k}) = \exp[-(r/a)^2 - \beta\hbar^2k^2/2m)]$, then $\langle x^2\rangle$ and $\langle k_x^2\rangle$ look like their one-dimensional forms

$$\langle x^2\rangle = \frac{\int_{-\infty}^{\infty} x^2 e^{-(x/a)^2} dx}{\int_{-\infty}^{\infty} e^{-(x/a)^2} dx} \tag{13.25}$$

with a similar equation for $\langle k_x^2\rangle$, but not all distributions are Gaussian and the ones that characterize intense space charge dominated beams in accelerators [23, 200] are often not friendly. So, as always, the details matter.

For axisymmetric beams (ones where the cross-section of the beam is ellipsoidal or circular), the second factor $\langle xk_x\rangle^2$ vanishes, and the first factor has a $\langle k_x^2\rangle$ in it: but $E_\perp = E_x + E_y = \hbar^2(k_x^2 + k_y^2)/2m = \hbar^2k_\perp^2/2m$, and so emittance is often said to be proportional to the MTE of the beam. Letting $\rho_c = \sqrt{\langle x^2\rangle}$ be the radius of the beam, then [178]

$$\varepsilon_{n,rms} = \frac{\rho_c}{2}\frac{\hbar}{mc}\sqrt{\frac{M_2(k_\perp)}{2M_0(k_\perp)}} \tag{13.26}$$

The calculation of the moments is a bit more subtle: attention must be restricted to electrons that actually get out of the metal, and that makes a factor of $f_\lambda$ appear in the denominator of the moments ratio seen in Eq. (13.7), but apart from that subtlety, the integrals can be approximated in a manner reminiscent of that used in treating $QE$ (see in particular Ref. [178], although hand-waving here gets to the same point. From $\varepsilon \propto \sqrt{MTE}$, the electrons that are photoemitted have an energy that goes as $\hbar\omega - \phi$, and part of that energy is in the transverse direction. If that is not obvious, think of an ideal gas which has a total energy of $E = (3/2)k_BT$ which, by the equipartition theorem, is distributed such that $E_x = E_y = E_z = (1/2)k_BT = E/3$, so that the transverse energy is $k_BT$ and therefore goes as $E$ does, apart from a different numerical coefficient. Thus, it is expected (and with effort shown) that $\varepsilon_{n,rms} \propto \sqrt{\hbar\omega - \phi}$. Dowell and Schmerge [65] pointed out that this implied

$$\left.\begin{aligned} QE &\propto (\hbar\omega - \phi)^2 \\ \varepsilon &\propto (\hbar\omega - \phi)^{1/2} \end{aligned}\right\} \Rightarrow QE \propto \varepsilon^4 \tag{13.27}$$

where the factor $\sqrt{\hbar\omega - \phi}$ takes over the role of $\sqrt{k_BT}$ in the thermal emittance relation (see Ex. 13.6). This crucial equation has a dispiriting consequence: in the pursuit of high brightness Free Electron Lasers (FELs) driven by photocathodes [63, 206, 230, 259, 269], to enable ever higher power performance, the $QE$ had to be increased at the same time as the emittance was minimized, and the Dowell–Schmerge (DS) relation shows the irony of that aspiration: improving one is at the expense of the other.

Now consider approximations to Eq. (13.13) for semiconductors. Let $\Delta_s \to \Delta = \hbar\omega - E_g$ and $x_m = \sqrt{E_a/E}$. However, there is an added peculiarity to the semiconductor version of $F_\lambda(x, E)$ from Eq. (13.14), more clearly rendered as

$$F_\lambda(x, E) = \int_{x_m}^{1} \frac{x^2}{x+p} D(Ex^2) dx \tag{13.28}$$

and that is the nature of $D(E)$ for surface barriers that photocathode materials possess. Many high QE semiconductor materials used cesium [82, 325]. Examples of recent interest are $Cs_3Sb$ and $K_2CsSb$ [17, 49, 195, 259, 332], CsTe [138, 282], and GaAs with a surface coating of Cs [47, 50, 128]. In the works of Spicer and separately Fisher, to explain a reduced QE of their measurements compared to their theoretical predictions, they postulated a sharp triangular or rectangular barrier at the surface that tended to bring down the theoretical estimates (see Figure 13.6). Intriguingly, quantum mechanical wave effects over a potential well can mimic the transmission probability of a triangular barrier [164], which recovers the step barrier of Eq. (6.80) in the limit of small field. Therefore, $D(E)$ is something more difficult in the case of the cesiated semiconductors, but the step barrier limit can prove useful to provide a qualitative account [259]. A second complication is that the resulting double integrals over angle and energy are unpleasant in the best of times, and so simplifying approximations are desired. One that is recurrently useful for some smoothly varying function $g(x)$ that is reasonably linear over the integration region is the trapezoidal approximation

$$\int_a^b g(x)dx \approx \frac{1}{2}[g(a) + g(b)](b - a) \tag{13.29}$$

It is very useful when the integrand $g(x)$ vanishes at one of the limits, as occurs in both the energy and angle integrals behind Eqs. (13.13) and (13.14). Use it to approximate $F_\lambda(x_m, E)$ by

$$D(E) \approx D_{step}[k(E)] = \frac{4\sqrt{E(E - E_a)}}{[\sqrt{E - E_a} + \sqrt{E}]^2} \tag{13.30}$$

After replacing $E \to Ex^2$ to get $D(Ex^2)$, then it follows

$$F_\lambda(x_m, E) = \frac{\sqrt{E} - \sqrt{E_a}}{2(1 + p)\sqrt{E}} D_{step}(E) \tag{13.31}$$

where $x_m = \sqrt{E_a/E}$ causes the integrand at the lower limit to vanish. Note that $p = \delta(\omega)/\ell(E)$ and so is evaluated at the upper limit $E$.

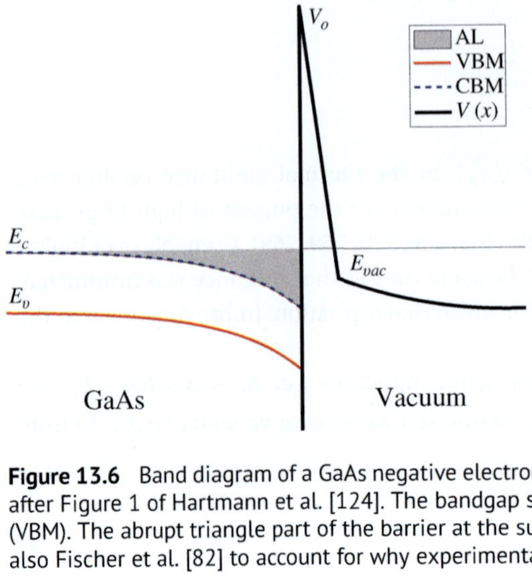


**Figure 13.6** Band diagram of a GaAs negative electron affinity (NEA) photoemitter ($E_a = E_{vac} - E_c(-\infty) < 0$), after Figure 1 of Hartmann et al. [124]. The bandgap separates the conduction from the valence band (VBM). The abrupt triangle part of the barrier at the surface was a hypothetical addition by Spicer [325] and also Fischer et al. [82] to account for why experimental yields were smaller than theoretical yields.

The next integral for the numerator of Eq. (13.13) using Eq. (13.28) may have analytic components, but using the trapezoidal approximation instead is again easier. It gives

$$\int_{E_a}^{\Delta} EF_{\lambda}(x_m, E)dE \approx \frac{\Delta^2}{(1+p)}\left[\frac{s^5}{u^2(1+u)(s+u)^2}\right] \tag{13.32}$$

where $\Delta = (1+s^2)E_a$ or $s = \sqrt{(\hbar\omega - E_g - E_a)/E_a}$, which is a useful dimensionless term, and $u \equiv \sqrt{1+s^2}$. This is fortuitous, because like the metal case before it, quantum efficiency is the product of an optical constants factor with a materials factor, and is represented as[3] (using $2\int_0^{\Delta} EdE \int_0^1 dx = \Delta^2$)

$$\begin{aligned} QE_{semi} &\approx \left[\frac{1-R(\omega)}{1+p}\right] h(s) \\ h(s) &\equiv \frac{s^5}{u^2(1+u)(s+u)^2} \end{aligned} \tag{13.33}$$

In his study of the alkali antimonides, Spicer proffered a simpler form for threshold emission conditions that is useful to compare to. His formula (Eq. 6 of Ref. [324]) is equivalent to

$$QE_{spicer} = G(\hbar\omega)\left(\frac{s^3}{s^3+\gamma}\right) \tag{13.34}$$

where $G(\hbar\omega)$ corresponds to Spicer's coefficient $G(h\nu)$, and $\gamma$ is a dimensionless ratio made of Spicer's $\gamma$ with $E_a^{3/2}$ (a more contemporary account of Spicer's 3-Step formula for semiconductors is in [232]). Eq. (13.34) suggests that Spicer's formula requires the approximation to $h(s)$ to be

$$h_{sp}(s) = G_o \frac{s^3}{4(s^3+\gamma)} \tag{13.35}$$

and so unlike Eq. (13.33), two additional "fitting parameters" $G_o$ and $\gamma$ are required in modeling data, apart from $E_a$ and $E_g$. By linking all four and making them dependent upon a least squares fit to a range of experimental data, small discrepancies in the parameters used by different authors can cause the values of $E_a$ and $E_g$ to differ even for the same material: for example, when $Cs_3Sb$ measured by Spicer [324] and separately by Taft and Philipp [338] are fitted by Eq. (13.34) represented as the Spicer-like $QE \approx B/[1 + g(\hbar\omega - V_g)^{\nu}]$ where $V_g = E_a + E_g$, the fits give differing values. The best fit assuming $\nu = 3/2$ for Spicer's data suggests $(B, g, V_g) = (0.186, 0.699, 2.038)$, whereas for Taft and Philipp's data, $(B, g, V_g) = (0.397, 1.701, 1.876)$. Part of the discrepancy is because different samples have material conditions that lead to different fit parameters, but observe how the fits imply differing values of $E_a + E_g$. In contrast, the usage of Eq. (13.33) to fit experimental data has less freedom: $h(s)$ has to be the same for both measurements, and assuming $E_a = 0.4$ eV and $E_g = 1.6$ eV for the Spicer and Taft & Philipp sets of data, it is only necessary to consider the ratio $QE(\omega)/QE(\omega_o)$ for each data. By taking the ratio, the coefficient depending on the optical parameters cancels, leaving $h(s)$ alone. For comparison here, let the reference point be $\hbar\omega_o = 2.6264$ eV. Doing so, the data of both Spicer and Taft & Philipp line up well with each other and with the function $h(s)$ as shown in Figure 13.7 (see Algorithm B.4.5). The hypothesis of why the fit to Spicer's data differed from that of Taft & Philipp conforms with observing their $\hbar\omega$ ranges was different and

3 Typographical errors, namely a neglected power of 2 on $(s+u)^2$ in the denominator, in prior accounts [154, 183, 259] are corrected in Eq. (13.33). The present account follows [157].

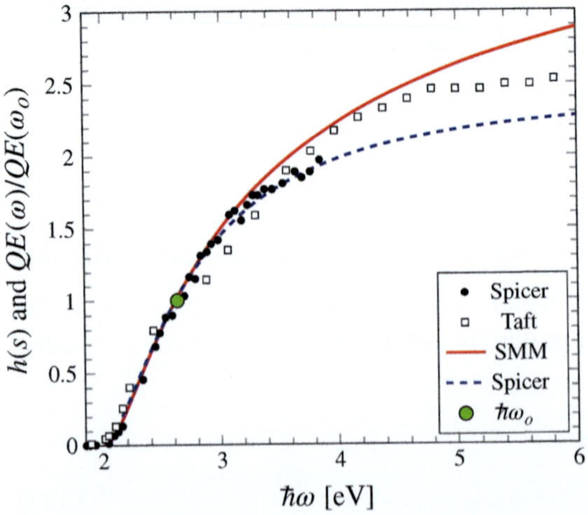


**Figure 13.7** Measurements of the QE of $Cs_3Sb$ from the data of Spicer [324] and Taft and Philipp [338] for $\hbar\omega_o = 2.6264$ eV, where $QE_{spicer}(\omega_o) = 0.0747$ and $QE_{taft}(\omega_o) = 0.1193$. Theory (SMM) line based on Eq. (13.33). After [157] / AIP Publishing LLC/CC BY 4.0.

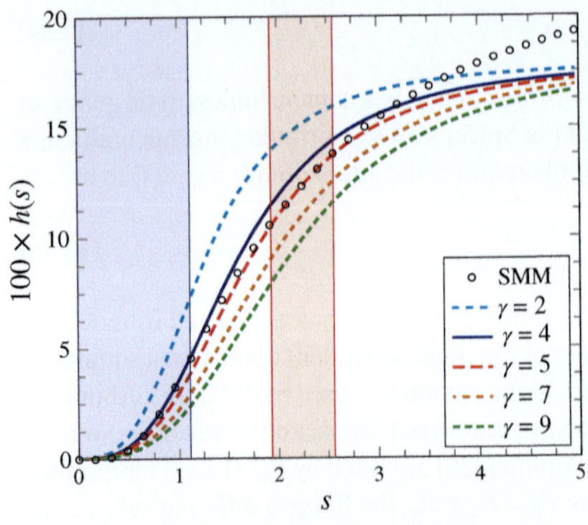


**Figure 13.8** Comparison of $h(s)$ (Eq. 13.33) to $h_{sp}(s)$ (Eq. 13.35) for $G_o = 4h(4)$. Observe that $h_{sp}(s)$ transfers from one $\gamma$ line to another depending on whether $s$ is small or large. The light blue region is for small $s$, the light red region for larger $s$. After [157] / AIP Publishing LLC/CC BY 4.0.

so a least squares analysis of data emphasized different regions in the $h(s)$ curve, as suggested by Figure 13.8 Taft and Philipp measured QE for higher $\hbar\omega$, and those QE estimates at higher photon energy measurements altered the best fit of a three parameter $(B, g, V_g)$ model.

In contrast, the SMM approach separates the determination of $E_g$ and $E_a$, determined by the shape of $h(s)$, from the determination of $R(\omega)$ and $p(\omega)$, which are specified by optical and transport models, and can be used to construct the coefficient $(1 - R)/(1 + p)$ separately [157]: in the SMM, they are assumed to not vary greatly, so that the combination can be well approximated by evaluating $p \approx \delta(\omega)/\ell(\mu)$, with subsequent variation in $p$ due to variations in $\delta(\omega)$ alone, and therefore only a single experimental $QE$ data point is required to fix $p(\mu)$. The accuracy of the optical model behind the SMM model was limited in the present example to a small number of resonant terms in the LDR model of Section 7.5. The number can be enlarged to achieve greater fidelity in modeling photoemission from, e.g. alkali antimonides like $Cs_3Sb$ or perovskites [233], but the procedure is much the same [183] as outlined herein.

## 13.3 Enhanced Moments Model

The approximation connecting SMM to FD in Eq. (13.23) picked the reference energy at which $p(E) = \delta(\omega)/\ell(E)$ was evaluated to be $E = \mu$ for a number of pragmatic reasons. First, the relaxation times near the Fermi level have well-known behaviors for $\tau_{ep}$ and $\tau_{ee}$ evaluated there. Second, the velocity $\hbar k(E)/m$ is increasing but $\tau_{ee}(E)$ decreases as $E$ moves away from $\mu$ [150, 176, 240], and a reasonable hope is that the effects counter each other. Thus, it is opportunistic to evaluate $\tau$ where it has a reasonable chance of being understood. Although $\tau_{ee}$ is small by comparison to $\tau_{ep}$ at the Fermi level, an electron with an energy well above the Fermi level scatters readily, and that approximation is no longer appropriate. An improvement on SMM is therefore to better approximate the behavior of $p(E)$ when evaluated at one of the limits of the energy integration in the trapezoidal approximation to $QE$. Further improvements in the exact evaluation of the angular and energy integrals require numerical methods, and so while they are "enhanced" they are not considered further.

The key quantity to the enhanced Moments Model is the proper consideration of the mean free path term $\ell(E) = (\hbar k(E)/m)\tau(E)$. The velocity term is easily rendered as $\hbar k(E)/m = \sqrt{2E/m}$. Focus, therefore, on the scattering term $\tau(E)$. From Eq. (7.53), several processes clearly contribute to it, the most important being the electron–phonon $\tau_{ep}$ and the electron–electron $\tau_{ee}$ contributions, and so revisit each to investigate their behavior at energies greater than $\mu$. For pedagogical purposes, it is desirable to ground the discussion in a particular relevant metal used in photoinjectors, for which copper is a well-examined choice [65]. The parameters used to examine copper shall be taken from Table 7.3 for analyses of $\tau_{ep}$ and $\tau_{ee}$ as examined in the literature based on methods of Refs. [104, 150, 161, 176, 239, 298, 350] and references therein.

The electron–phonon $\tau_{ep}$ and electron–electron $\tau_{ee}(E)$ terms were encountered in Eqs. (7.55) and (7.62), respectively, where for the sake of simplicity, ignore $\tau_{ii}$ as being generally much larger than the others for the metals of interest here. By Matthiessen's Rule (Eq. 7.53), the relaxation times add inversely, but near the Fermi level $\mu$, the contribution of $\tau_{ep}(\mu)$ dominates that of $\tau_{ee}(\mu)$ for metals like copper. Photoexcited electrons, however, are not near the Fermi level: for wavelengths of common usage in photoinjectors, they can be several eV higher, e.g. commonly used Nd:YAG lasers used for photoexcitation operate at wavelengths characterized by $(1064/n)$ [nm] (other lasers give other convenient wavelengths [62], where $n$ is an integer up to 5 [101, 161, 227], with the last two required to achieve appreciable photoemission from metals with work functions comparable to 4.5 eV. The largeness of $\beta = 1/k_B T$ in Eq. (7.62) then causes $\tau_{ee}(E)$ to become rather indifferent to temperature and show a rather steep decline as a function of energy $E \approx \mu + \hbar\omega$. The theoretical models are complex and the existence of the term $K_s$ introduces a new parameter. For copper using Table 7.3 parameters, Eq. (7.62) is compared to simulations [215, 239] in Figure 13.9 (see Algorithm B.4.6), showing that $K_s \approx 2.3$ is a reasonable value. If $\tau_{ee}$ is approximated as a power law in $E - \mu$, then the red theory line behaves to a good approximation as

$$\tau_{ee}(E) \propto \tau_{ee}(E_o)\left(\frac{E_o}{E-\mu}\right)^{1.93} \tag{13.36}$$

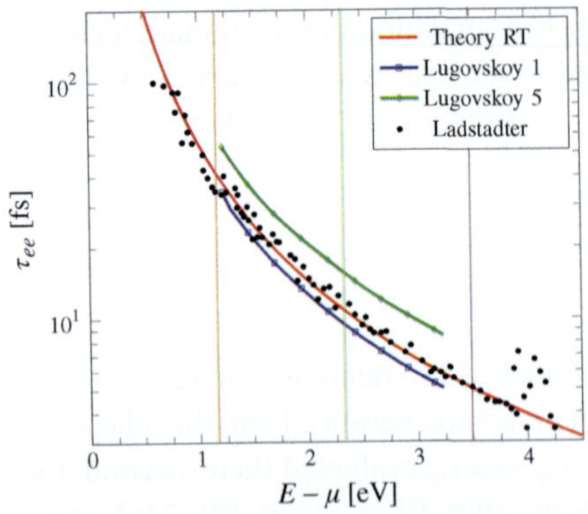


**Figure 13.9** Electron–electron scattering relaxation time (Eq. 7.62) for copper parameters of Table 7.3. Vertical lines are $E \sim \mu + \hbar\omega_n$, where $\hbar\omega_n$ are photon energies associated with an Nd:YAG laser $2\pi\hbar cn/\lambda_1 = (1.17, 2.33, 3.50, 4.66, 5.82)$ eV for $\lambda_1 = 1064$ nm and $n$ from 1 to 5 (only three smallest are shown).

where $E_o = 1$ eV, $\tau_{ee}(E_o) = 55.30$ fs, and $E$ is the energy of the photoexcited electron (if excited from the Fermi level, then $E - \mu = \hbar\omega$) making the power law approximation in what follows convenient.

A power-law behavior, however, entails that by the third harmonic of an Nd:YAG laser ($\hbar\omega_3 = 3.50$ eV) the electron–electron scattering rate for $E = \mu + \hbar\omega_3$ has dropped to 4.95 fs for copper, well below $\tau_{ep}(\mu + \hbar\omega_3) \approx 20$ fs. In fact, $\tau_{ep}$ at the higher energy is not much different than its value at the Fermi level of 24.5 fs. The rapid drop of $\tau_{ee}$ therefore, makes the approximation $\ell(E) \approx \ell(\mu)$ in $p(E) = \delta(\omega)/\ell(E)$ suspect and causes it to over-predict QE for large photon energies.

The hypothesis that more frequent scatterings change QE at higher photon energies can be vetted by again considering the QE data for copper from Figure 13.4, but now incorporating the photon energy into the scattering terms: the weaker dependence of $\tau_{ep}(E)$ from Eq. (7.55) further modifies $\tau(E)$, but given the smallness of $\tau_{ee}(E)$ for $\hbar\omega > 4.5$ eV, the impact of $\tau_{ep}$ on $\tau(E)$ is hard to discern and could be ignored to a first approximation. Now the theoretical QE estimate becomes (compare Eq. 13.19)

$$QE_{metal} \approx [1 - R]F_\lambda[x_m, p]\left[\frac{(\hbar\omega - \phi)(3\mu + \phi - \hbar\omega)}{6\hbar\omega(2\mu - \hbar\omega)}\right](1 + f_o) \tag{13.37}$$

where now $p \to p(\mu + \hbar\omega) = \delta(\omega)/\ell(\mu + \hbar\omega)$, and the last term $f_o$ takes care of any putative thermal effects that gave rise to $\zeta(2)/\beta_T^2$ in the Fowler–DuBridge model of Eq. (13.1).

All the parameters behind the theoretical model change with regards to their best-fit values, depending on the equation used to model QE. The data were characterized by [64] as satisfying $\Phi = 4.31$ eV, acknowledged to be smaller than standard cited values for the work function of copper, but nevertheless within "experimental uncertainty," and $F = 0$ eV/nm. For the theory model, the value of $\tau_{ee}(E_o)$ might be expected to be different for photoemitter-grade copper than values derived from the conductivity of copper wire or optical properties in the handbooks [10, 125, 357]. Therefore, small differences in parameters from what has come before are expected. The following values were used to achieve agreement between theory and the data of Dowell et al.,: $\Phi = 4.34$ eV, $\mu = 7$ eV, $T = 300$ K, $\tau_{ep}(E) = 24.5\sqrt{\mu/E}$ in fs, $\tau_{ee}(E) = 200/[(E/\mu) - 1]^{1.93}$ in fs, $f_o = (1/6)[\pi k_B T/(\hbar\omega - \phi)]^2$, and the optical constants unchanged from the SMM model. With these modifications, agreement at higher $\hbar\omega$ is improved, as shown in Figure 13.10.

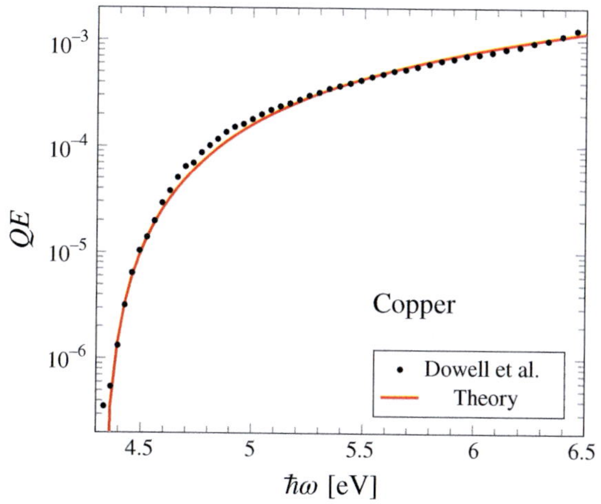


**Figure 13.10** Comparison of the QE of copper (adapted from [64] data as shown in Figure 13.4) to the enhanced Moments Model. Theory line corresponds to Eq. (13.19). See text for discussion. Compare to Figure 13.4.

## 13.4 Delayed Emission

The essence of the Fatal Approximation of Eq. (13.16) is that scattering is inimical to the probability of emission because it saps the energy of a photoexcited electron, rendering it unable to surmount the surface barrier. In the case of metals, where scattering is primarily between energetic photoexcited electrons and electrons at the Fermi level such that the daughter electrons redistribute the kinetic energy involved in the collision. As both conservation of momentum and energy have to be satisfied, and as the colliding particles are of the same mass, then consider two electrons denoted $a$ and $b$ in the frame of reference for which the second electron is at rest, that is, their initial velocity components, in the notation $\langle v| = [v_x, v_y]$, are $[V, 0]$ and $[0, 0]$ in coordinates for which the pre-scattering (parent) and post-scattering (daughter) electrons are in the same plane, where $V$ is the magnitude of the initial primary electron, and where the velocity of the target electron is zero. After the collision, they require energy and momentum to be conserved. Observing that there is no net momentum in the $\hat{y}$ direction, then the $y$-components of the daughter electrons are equal and opposite, and so the final velocities can be expressed as $[u, \delta]$ and $[v, -\delta]$. The remaining $\hat{x}$-component of momentum and conservation of energy then require

$$\begin{aligned} V &= u + v \\ V^2 &= u^2 + v^2 + 2\delta^2 \end{aligned} \tag{13.38}$$

Photoexcited electrons generally scatter with electrons at the Fermi level, and the final states in the lab frame must be unoccupied for the scattering to occur: this provides a restriction that the kinetic energy of both daughter electrons must be above the Fermi level where the states are generally unoccupied (their occupation probability is $[1 - f_{FD}(E)]$) [154]. This caveat can be set aside if the investigation of emission focuses on photon energies well past threshold. Solving for $u$ and $v$ gives

$$u = \frac{1}{2}\left(V - \sqrt{V^2 - 4\delta^2}\right)$$
$$v = \frac{1}{2}\left(V + \sqrt{V^2 - 4\delta^2}\right) \tag{13.39}$$

It follows that the vectors $\langle a| = [u, \delta]$ and $\langle b| = [v, -\delta]$ are orthogonal ($\langle a|b\rangle = uv - \delta^2 = 0$). The electrons share the kinetic energy (on average equally), such that for photon energies $\hbar\omega$ comparable to $\phi$, the daughter electrons lack the energy to overcome the work function barrier. For photoemission conditions near threshold, therefore, the Fatal Approximation is good, but well past threshold, one (or maybe both if the photon energy is high enough and the electrons are directed at the surface) of the electrons may have sufficient energy to contribute to the emission.

The problem is then to find the population of electrons that can still be emitted even after losing energy from a scattering event, and that requires accounting for the electrons not contained in $f_\lambda(\cos\theta, E)$ in Eq. (13.16). Profitable use can be made of the identity [153]

$$1 = \frac{\cos\theta}{\cos\theta + p} + \frac{p}{1+p} + \frac{2p\sin^2(\theta/2)}{(\cos\theta + p)(1+p)} \tag{13.40}$$

The first term on the right-hand side (RHS) is simply $f_\lambda(\cos\theta, E)$ for $p = p(E)$. Because the largest angle $\theta_m$ is such that $\cos\theta_m = x_m \approx 1$ near threshold, the last fraction will in general be small because of the sine term and because $p(E) = \delta(\omega)/\ell(\tilde{E})$ is generally large for metals due to the smallness of the relaxation time $\tau(\tilde{E})$, where $\tilde{E}$ is the energy of the excited electron (remember that for metals, $E$ without a tilde denotes the energy of the electron prior to the absorption of a photon, and $\tilde{E}$ with a tilde is the energy of the excited electron). That leaves $p/(1+p)$ as the term on the RHS accounting for the population of electrons that have scattered and which would thereby be ignored by the Fatal Approximation. A few simplifying assumptions help lead to an account. First assume that the excited electron originated from the Fermi level. Before scattering (0 collisions) that electron has an energy $\tilde{E}_0 = \mu + \hbar\omega$. If after a collision it shares its energy with an electron at the Fermi level, then its energy drops to $\tilde{E}_1 = \mu + \hbar\omega/2$, and after $n$ collisions its energy has dropped to $\tilde{E}_n = \mu + \hbar\omega/2^n$ (neglecting aforementioned momentum-energy restrictions) because it always shares its energy with an electron at the Fermi level. An upper limit on $n$ is established through the requirement that $\tilde{E}_{n+1} < \mu + \phi$. Second, assume that on average, only one of the two daughters is directed at the surface and therefore contributes to emission. Because of the second approximation, let the third approximation be that the spread of excited electrons in the material still mimics an $e^{-x/\delta(\omega)}$ behavior so that a new loss factor $f(\cos\theta, E)$ has the same form as the unscattered loss factor (a coarse but useful assumption). To distinguish them, let $f_n(\cos\theta, E) = \cos\theta/[\cos\theta + p_n(E)]$ for $n = 1$ describes the contribution to QE of electrons that have scattered but once, and $n = 0$ for those that have not scattered at all. And so, augment $F_\lambda(x_m, E)$ from Eq. (13.14) by including electrons that have scattered 0 or 1 times. A change of notation is now needed to keep track of the scatterings, and so sacrifice the $\lambda$ subscript on $F_\lambda$ to make it: introduce $F_n(x_m, E)$ where $n$ modifies the value of $\tilde{E}_n$ such that for metals, Eq. (13.10) can be rewritten as $x_m(\tilde{E}_n) = [(\mu + \phi)/\tilde{E}_n]^{1/2}$. And so

$$\begin{aligned} {}_\lambda(x_m, E) &\rightarrow F_0(x_m, E) + \frac{p_0(E)}{1 + p_0(E)} F_1(x_m, E) \\ F_n(x_m, E) &= \int_{x_m}^{1} D(Ex^2) f_n(x, E)\, x dx \end{aligned} \tag{13.41}$$

in Eq. (13.14), and where $p_n$ in $f_n(x, E)$ is understood as $p_n(E) = \delta(\omega)/\ell(\tilde{E}_n)$. The process can be extended to include electrons that have scattered twice and can still contribute to emission. The arguments that led to Eq. (13.41) can be iterated to obtain

$$F_\lambda(x_m, E) \to F_0(x_m, E) + \sum_{j=1}^{n} \left[ \prod_{k=0}^{j-1} \frac{p_k(E)}{1 + p_k(E)} \right] F_j(x_m, E) \tag{13.42}$$

Such an expression appears to be off-putting, but a suggestion of what it entails is possible if one is not overly zealous about accuracy. If the approximations behind Eq. (13.23) are incautiously made using the trapezoidal approximation, then at the lower limit, the energy integrand vanishes, and at the upper, $E \to \mu$ in the integrand. Then $F_n[x_m(\tilde{E}_n), E]$ at $E = \mu$ is what mostly dictates the contribution of the scattered electrons. If Eq. (13.19) is renamed as $QE_0$ or the quantum efficiency for the unscattered (or Fatal Approximation)] electrons, then the QE when scattered electrons are included approximately behaves like

$$\begin{aligned} QE &\sim QE_0 \left[ 1 + \sum_{j=1}^{n} C_j(\mu) \frac{F_j[x_m(\tilde{\mu}_j), \mu]}{F_0[x_m(\tilde{\mu}), \mu]} \right] \\ C_j(\mu) &= \prod_{k=0}^{j-1} \frac{p_k(\mu)}{1 + p_k(\mu)} = \prod_{k=0}^{j-1} \frac{\delta(\omega)}{\delta(\omega) + \ell(\tilde{\mu}_j)} \end{aligned} \tag{13.43}$$

where $\tilde{\mu}_j = \mu + \hbar\omega/j$, $x_m(\tilde{\mu}_j) = [(\mu + \phi)/\tilde{\mu}_j]^{1/2}$ and $x_m(\tilde{\mu}) = [(\mu + \phi)/(\mu + \hbar\omega)]^{1/2}$ as before. The value of $n$ is the largest $j$ for which $x_m(\tilde{\mu}_j) < 1$.

Seeing the calculation of terms in action is helpful. For the sake of illustration, assume copper-like values $\mu = 7$ eV and $\tau_{ep} = 10$ fs. Let the baseline simulation use $\delta = 12$ nm and $\phi = 4$ eV, then consider changes of $\delta$ to 100 nm or $\phi$ to 1.8 eV. If the unscattered photoexcited electron is $\hbar\omega$ above the Fermi level $\mu$, then because the scattering electrons equally share their kinetic energy, the excited electron after $n$ scatterings will be $\hbar\omega/2^n$ above $\mu$. Let $\tau_{ee} = A/(E_n - \mu)^2 = 25/2^{2n}$. For such conditions, both $F_j/F_0$ and $C_j$ drop quickly, and their product drops faster. The sum of their product plus unity, as in Eq. (13.43), gives the $y$-axis of Figure 13.11 (see Algorithm B.4.7). For this simple model, the initial execution must be twice the barrier height $\phi$ above the Fermi level before any scattered electrons can contribute to $QE$. Although *ad hoc*, the model provides a justification for the Fatal approximation for metals: at typical photon energies characteristic of drive lasers in photoinjectors, for which $\hbar\omega$ is generally less than 6 eV because generating higher harmonics of

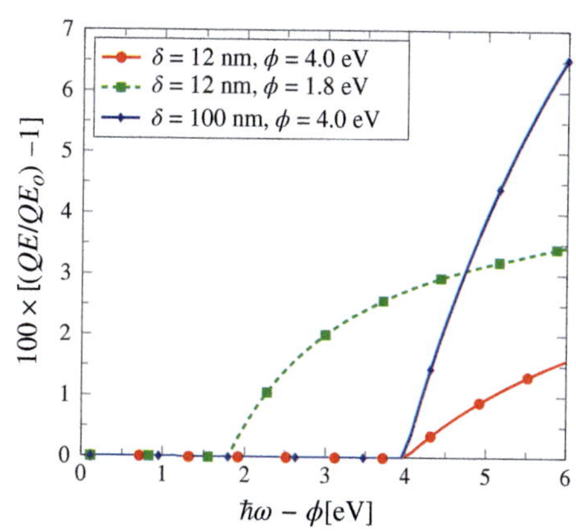


**Figure 13.11** Evaluation of Eq. (13.43) for three combinations of $\phi$ and $\delta$ as labeled in legend for *ad hoc* fixed penetration depth and $\tau_{ep} = 10$ fs and $\tau_{ee} = 25/2^{2j}$. Energy of excited electron after $n$ scattering events is taken to be $\mu + \hbar\omega/2^n$.

a fundamental laser frequency quickly becomes inefficient, a single scattering event in this simple model is enough to curtail emission, and even when it is not, the contribution of the scattered electrons to the emitted current is meager.

### 13.4.1 Simple Monte Carlo Model

The proper way to demonstrate how much scattered electrons contribute to photoemission from metals is best understood by using Monte Carlo methods, whereby a population of electrons is generated and then allowed to disperse. Scattering events can be calculated and the electrons re-directed, with those falling below the surface barrier removed from the simulation. Those that cross the surface then can be counted and used to calculate current across the surface plane. When this is done, the percentage of electrons that are emitted is starkly revealed: an example of the losses in copper is shown in Figure 13.12 for two cases: first, bare copper for which $\Phi = 4.5$ eV, and second, for copper with a layer of cesium on the surface for which $\Phi = 1.6$ eV; electrons that scatter off other electrons can still contribute for high enough photon energy [175].

Semiconductors change that understanding in several ways. First, the penetration depth of the incident laser light is often larger than for metals and so electrons can be excited much deeper into the bulk material. Second, scatterings between a photoexcited electron and a valence electron cannot occur if the final state is in the bandgap as schematically illustrated in Figure 13.3. Third, electron–phonon collisions decrement the energy of the excited electron by a small amount even for optical phonons, on the order of 10's of meV, and so an excited electron can actually enjoy many scattering events before it becomes emission ineligible, as suggested in Figure 13.13. Lastly, the barriers of semiconductors are often very small, especially if cesium is part of their constitution (e.g. alkali antimonides and tellurides like $Cs_3Sb$ and CsTe) or not present for negative electron affinity photocathodes like GaAs or GaN [62, 176, 179, 128, 322, 328]. These processes reduce the rapidity with which $C_j$ declines and changes the calculation of $F_j$, thereby giving rise to a larger scattered electron population that can be emitted, and which therefore shows a larger impact even in models of photoemission that use simple models for the optical parameters [169]. Studies involving the proper simulation of electron transport using Monte Carlo methods [47, 197, 239, 263, 283, 335, 365] reveal a great deal that the simple models miss, but also demonstrate that the simple approximations can suggest important effects. When scattered electrons contribute non-trivially to

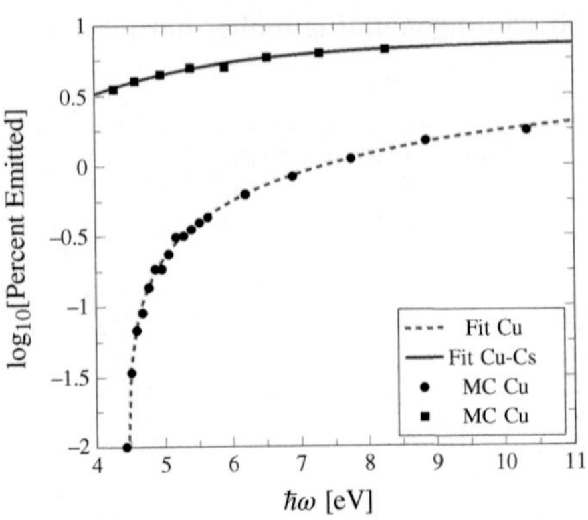


**Figure 13.12** Monte Carlo simulation of percentage of photoexcited electrons that are emitted as a function of photon energy and surface. Bare copper (Cu) has a work function of $\Phi = 4.5$ eV, cesiated copper (Cu-Cs) of $\Phi = 1.6$ eV. Lines are fits for the eye; symbols are MC evaluations. After Figure 6, adapted from [175].

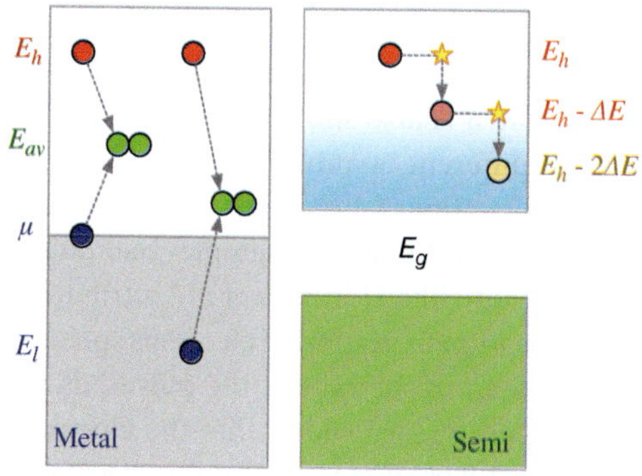


**Figure 13.13** Energy loss mechanism differences between metals and semiconductors: for metals, $e$–$e$ dominates and excited electrons lose energy quickly, whereas for semiconductors, $e$–$p$ dominates and the energy loss is slower. Adapted from [179].

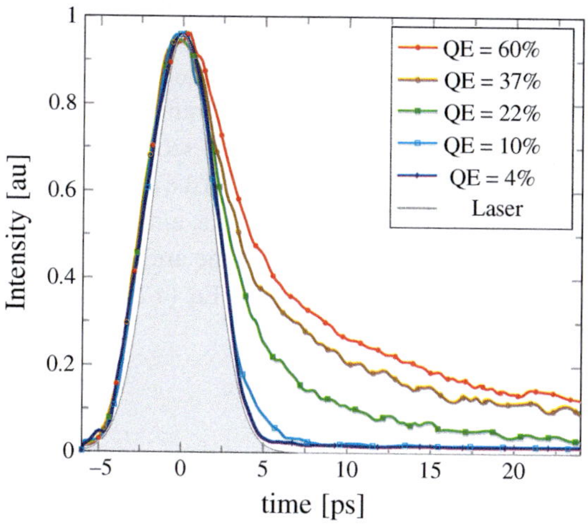


**Figure 13.14** Electron emission profiles from GaAsP for various values of quantum efficiency $QE$ after exposure to a 1 ps rms 520 nm laser pulse. Data digitally extracted from Figure 9 of Bazarov et al. [18]. Shaded gray area is the function $I_o \exp(-t^2/2t_o^2)$ for $t_o = 1.8$ ps and $I_o$ being the mean of the QE data maxima.

the quantum efficiency, then two consequential behaviors emerge: first, photoemission bunches acquire tails due to delayed emission, and second, the response time of the photocathode is affected. Some simple models can illuminate what is meant by this.

Consider delayed emission first. Its presence becomes apparent in measurements of emitted bunches of electrons when the shape of the emitted bunch departs from that of the means by which they are generated [18, 321, 366], often brought about by Gaussian laser pulses. A clear example of this is from Figure 9 of Ref. [18], recreated in Figure 13.14, where the temporal response from the negative electron affinity (NEA) photocathode material GaAsP was measured after excitation by a $\lambda = 520$ nm (or $\hbar\omega = 2.3843$ eV) laser pulse. An "effective" NEA material is one where an applied field induces the accumulation of electrons at the surface, effectively pulling the conduction band $E_c$ down such that the vacuum level $E_{vac}$ is below the conduction band minimum (CBM) in bulk as in Figure 13.6 for gallium arsenide (GaAs). This causes electrons to aggregate at the surface in the shaded gray region. Effective NEA is made possible by a surface coating of cesium that brings the vacuum level $E_{vac}$ to below the conduction band minimum (CBM) in bulk. In order to bring theoretical QE estimates into agreement with experiment, Spicer [325] and Fisher et al. [82] added a hypothetical thin, high surface barrier $V_o$ at the surface, which is also shown. The NEA material studied by [18] was similar, and as [124] also reported, different quantum efficiencies were associated with different samples due to differences originating in their fabrication. On examining the shape of the emitted pulses of electrons, Bazarov et al. showed higher QE samples had pulse shapes that departed from the shape of the laser pulse, and they (like Hartman et al.) attributed this to a diffusive contribution to the emitted current. Diffusion implies scattering electrons spreading out. For NEA photocathodes, their contribution can be seen in the tails of the pulses, as in Figure 13.14, but the overlay with the laser pulse (gray shaded region) shows that the electron bunches are already somewhat wider than the laser pulse ($t_o = 1.8$ ps) due to electron propagation effects after photoexcitation.

The consequences of a primitive diffusion model compared to a ballistic model using a graphical approach can give a feel for why a portion of the electron pulse follows the laser and a portion seems to trickle out afterwards. Let two populations of $N$ electrons each be excited at a fixed distance $\delta$ from the surface with an energy equal to $E = \mu + \hbar\omega$. At each time increment $\tau$, let the first population continue to travel ballistically (no scattering), and the second population be scattered such that at the $j^{th}$ scattering, its energy changes to $\mu + \hbar\omega/2^j$ and its direction is randomized.

By virtue of collisions which randomize direction, the second population will seem to expand more slowly than the first ballistic population: Figure 13.15 (see Algorithm B.4.8) shows this, plotting the *rms* radius of the no-scattering and scattering populations over time, alongside lines that are linear in both $t/\tau$ and $\sqrt{t/\tau}$. Radii that expand as the square root of time are a feature of diffusive behavior because of how the density distribution of particles $\rho(x, t)$ in Eq. (4.26) generalizes to three dimensions: letting $\rho(\vec{r}, t) = \rho(x, t)\rho(y, t)\rho(z, t)$, then

$$\rho(\vec{r}, t) = (4\pi Dt)^{-3/2} \exp\left(-\frac{r^2}{4Dt}\right) \tag{13.44}$$

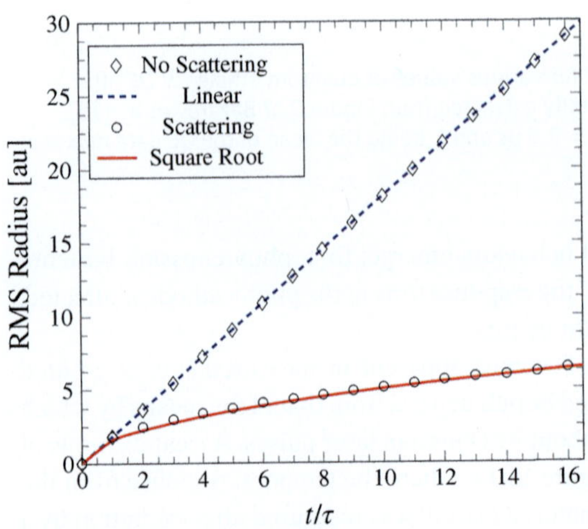


**Figure 13.15** Symbols: the root-mean-square (*rms*) radius of the ballistic and diffusive populations of electrons in Figure 13.16 evolving using Monte Carlo model. Lines: analytic linear ($t/\tau$) and square root ($\sqrt{t/\tau}$) behavior.

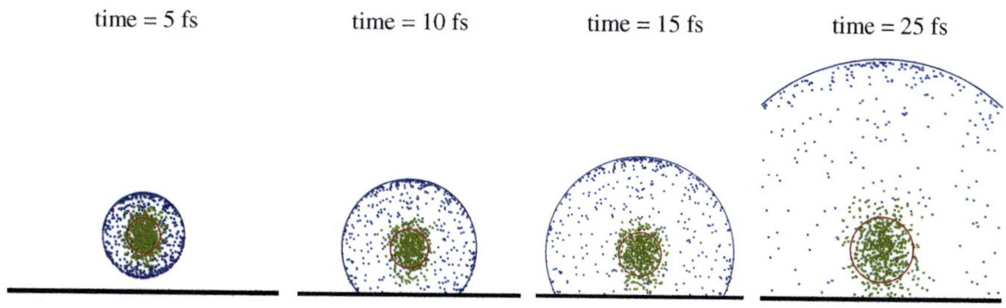


**Figure 13.16** Time evolution of two electron populations, one that does not undergo scattering (shell, blue) and one that does at each time step $\tau$. Copper parameters and $\lambda = 532$ nm. At $t = j\tau$, the energy of each electron is $\mu + \hbar\omega/2^j$ ($e$–$e$ scattering model) but the direction of its velocity has been randomized. Thin dark blue line is a circle with the root-mean-square (rms) radius of the blue population; similarly, thin dark red line is the rms radius of the red population. The surface is the thick black line at the bottom: electrons crossing it are taken to be "emitted."

where $D$ is the diffusion constant encountered in Eq. (4.21). The absence of an angular dependence means that the root mean square (*rms*) radius is straightforward to calculate

$$[r_{rms}(t)]^2 = \langle r^2 \rangle = \frac{\int_0^\infty r^4 e^{-r^2/4Dt} dr}{\int_0^\infty r^2 e^{-r^2/4Dt} dr} = 4Dt \frac{\int_0^\infty s^4 e^{-s^2} ds}{\int_0^\infty s^2 e^{-s^2} ds} \tag{13.45}$$

The numerator integral on the right-hand side is simply $2\Gamma(5/2)$ and the denominator is $2\Gamma(3/2)$. As a result, $r_{rms}(t) \propto \sqrt{t}$, and this behavior is evident in Figure 13.15. Therefore, the two populations of electrons can be referred to as "no scattering / ballistic" and "scattering / diffusive." The ballistic electrons lie on an expanding shell of charge; the diffusive electrons are a cloud of particles where the density $\rho(r, t)$ increases towards the center ($r \to 0$) but tapers off exponentially with distance from that center. Electrons which pass a plane representing the surface can be counted to find the total charge $\Delta Q(t)$ in vacuum, from which the current past that surface plane is taken to be the time rate of change of $\Delta Q(t)$. An example of the simple simulation is shown in Figure 13.16 (see Algorithm B.4.8): clearly, the diffusing population emerges after the ballistic population gets a head start. The figure is for one excitation site, but there is a whole line of excitation sites into the crystal that taper off with the intensity of the laser light, or as $e^{-z/\delta}$ if the $\hat{z}$-axis is normal to the surface. A sum over many such excitation sites then shows how current passing the surface changes.

### 13.4.2 Shell and Sphere Model

A single-parameter analytical model is possible by assuming that the relation of the emitted electron current density $J_\tau(t)$ to the laser intensity $I_\lambda(t)$ is

$$J_\tau(t) = \frac{q}{\hbar\omega} \left( \frac{QE}{\tau} \right) \int_{-\infty}^{t} I_\lambda(s) e^{-(t-s)/\tau} ds \tag{13.46}$$

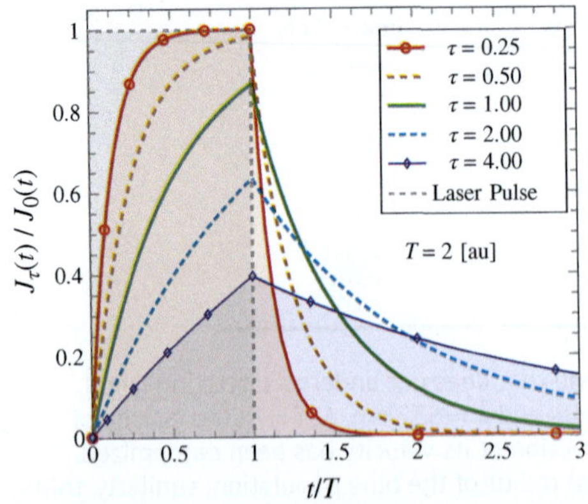


**Figure 13.17** Simple delay model of Eq. (13.47) for $T = 2$ [au] and $\tau$ as shown in legend. $I_0(t)$ is the $\tau \to 0$ limit, corresponding to an electron pulse that exactly follows the laser pulse $I_\lambda(t)/I_o$ (shaded gray region). The areas of the largest and smallest $\tau$ are shaded blue and red, respectively. Adapted from [179].

For a flat-top laser profile $I_\lambda(t) = I_o\Theta(t)\Theta(T - t)$ where $T$ is the pulse duration and $\Theta(t)$ is the Heaviside step function, then the integration is simple, and results in [179]

$$\frac{J_\tau(t)}{J_0(t)} = \begin{cases} 1 - e^{-t/\tau} & (0 \le t \le T) \\ \left(e^{T/\tau} - 1\right) e^{-t/\tau} & (t > T) \end{cases} \tag{13.47}$$

where $J_0$ is the $\tau \to 0$ limit of Eq. (13.47) for which $e^{-(t-s)/\tau}/\tau$ approximates a Dirac delta function $\delta(t - s)$. As the characteristic time $\tau$ approaches and exceeds the laser pulse width $T$, the effects of delayed emission alter the pulse shape more profoundly as in Figure 13.17 in a manner reminiscent of the tails in the experimental data of Figure 13.14.

A more instructive model of the experimental behavior is to consider the behavior of the expanding shell (ballistic) compared to the expanding sphere of charge (diffusive). The shell is made up of electrons that have escaped without scattering as assumed by the Fatal Approximation, and the sphere has undergone scattering processes and so meanders (diffuses) on its way to the surface as it is buffeted by collisions with other electrons (metals) or phonons (semiconductors). In the case of NEA photocathodes, phonon scattering dominates, and the absence of a barrier at the surface (or an exceptionally thin barrier as hypothesized by Spicer [325] or Fisher et al. [82], corresponding to the abrupt triangle in Figure 13.6) allows even multiply scattered electrons to contribute to the current density and therefore quantum efficiency. No scattering and fully diffusive behavior are opposing limits, but the actual behavior will exist between them. Using the schematic of Figure 13.15 as a guide, the problem is to find out how much charge is in the fraction of the shell cap in vacuum, in contrast to the spherical dome for diffusive expansion, as shown in Figure 13.15.

Let the charge in the thin blue shell in vacuum for the left side of Figure 13.18 be designated as $Q_s(t)$ where the "$s$" signifies "shell," given by the portion of a shell past $z = \delta$. The charge is found from infinitesimal strips of charge making up the shell cap. Let $Q_{so}$ is the total charge in the complete spherical shell, and let the radius of the sphere at time $t$ be given by $r(t) = v_o(t + t_o)$, where the kinetic energy of the electrons is assumed to be the same and given by $(1/2)mv_o^2 = \mu + \hbar\omega$ if the electrons are all excited from the Fermi level. Lastly, $t_o$ is how much time must pass before the front edge of the shell gets to the surface (that is, how long the shell takes to expand to

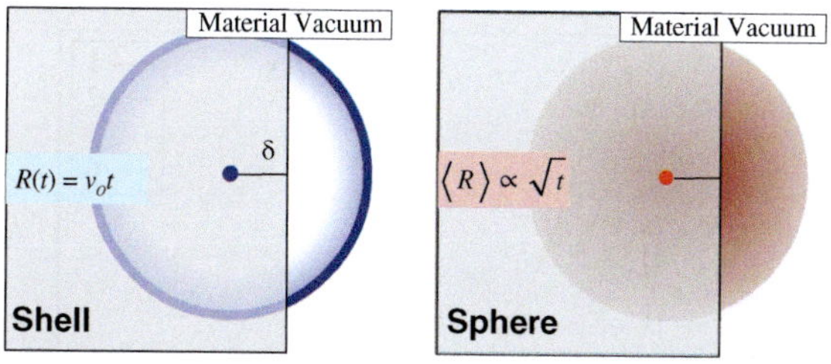


**Figure 13.18** An expanding shell of charge (left) compared to a diffusively expanding sphere of charge (right). The time rate of change of the charge in vacuum is the contribution of each to emitted current.

a radius of $\delta$). Loops of charge in the form of thin ribbons making up the cap have an area given by $dA = (2\pi r \sin\theta)(rd\theta) = 2\pi r^2 \sin\theta d\theta$. The fraction of the shell in vacuum is then

$$\frac{Q_s(t)}{Q_{so}} = \frac{\int_0^{\theta_m} 2\pi r(t)^2 \sin\theta d\theta}{\int_0^{\pi} 2\pi r(t)^2 \sin\theta d\theta} = \frac{t}{2(t+t_o)} \tag{13.48}$$

where $\cos\theta_m = r(0)/r(t)$.

Performing the same kind of calculation for the diffusive sphere is a bit more challenging. Again, let the total charge in the diffusive sphere be $Q_{do}$ (it will not equal $Q_{so}$), where the "$d$" signifies "diffusion." Rather than an area integration, the spherical dome is a volume integral using the density $\rho(r,t)$ from Eq. (13.44) subject to $Q_{do} = \int \rho(0)d^3r$. After each scattering event, the average energy of the diffusing electrons declines, but for the sake of argument, focus on semiconductors and let $\Delta E$ be small by comparison to the initial kinetic energy of the electrons: this will create an upper-bound estimate of $Q_d(t)$. Now use the relation $r^2 = x^2 + y^2 + z^2$ and observe that only diffusion in $\hat{z}$ matters and introduce $t_d \equiv \delta^2/4D$ to get

$$\frac{Q_d(t)}{Q_{do}} = \frac{\int_{\delta}^{\infty} e^{-z^2/4Dt} dz}{\int_{-\infty}^{\infty} e^{-z^2/4Dt} dz} = \frac{1}{2}\left[1 - \text{Erf}\left(\sqrt{t_d/t}\right)\right] \tag{13.49}$$

where Erf($x$) is defined in Appendix A.3. The total charge in vacuum is then $Q(t) = Q_s(t) + Q_d(t)$, but it is the derivative $I(t) = dQ/dt$ that is used to find the current, and so

$$I(t) = Q_{so}\frac{t_o}{2(t+t_o)^2} + Q_{do}\left(\frac{t_d}{4\pi t^3}\right)^{1/2} e^{-t_d/t} \tag{13.50}$$

The maximum of the first term occurs at $t = 0$, and that of the second term at $t = 2t_d/3$: as a result excitation sites close to the surface contribute differently than those further in, as indicated in Figure 13.19 (see Algorithm B.4.8) for representative but *ad hoc* parameters. The total current is the sum over all the sites where photons excite electrons progressively deeper into the photocathode material, in addition to whether there is an appreciable surface barrier. As a result, a range of behaviors are possible depending on how energy is lost due to repeated scatterings, how far into the material excitation occurs, and the nature of the electron affinity (positive or negative) at the

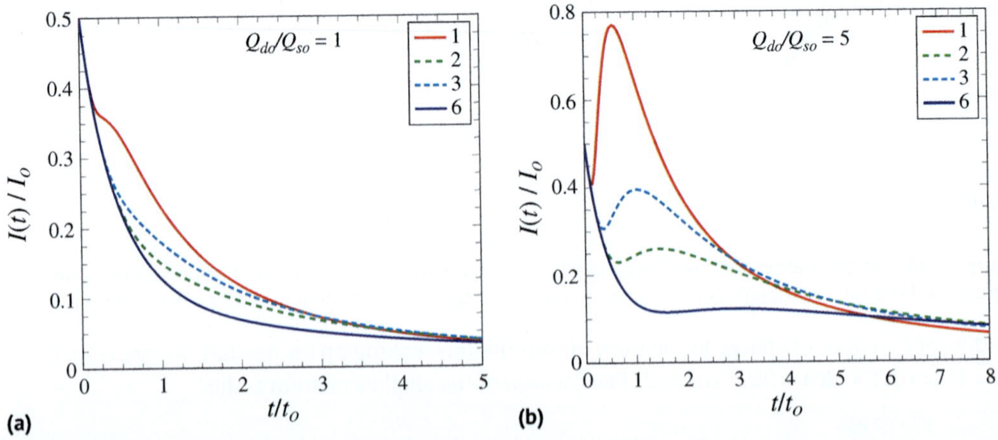


**Figure 13.19** Shell and Sphere current contributions (Eq. 13.50) for NEA conditions and a single excitation site at $\delta$: (*a*) close to surface where shell dominates (small $Q_{do}/Q_{so}$); (*b*) further in where sphere may dominate (large $Q_{do}/Q_{so}$). Legend is *ad hoc* ratio of $t_d/t_o$.

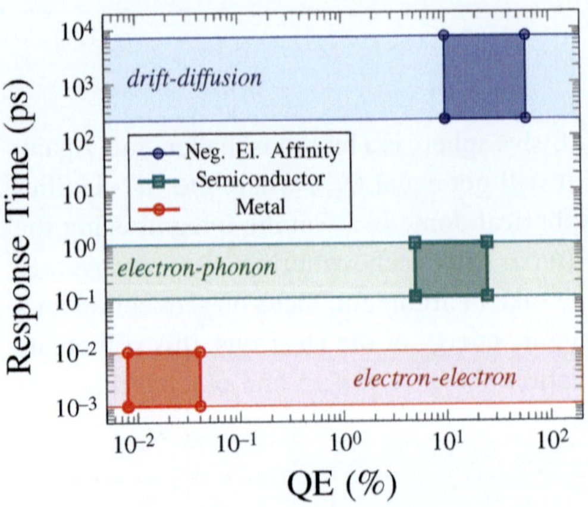


**Figure 13.20** Correlation between of response time and quantum efficiency for classes of photocathode materials (based on Figure 14 of Ref. [328]), showing the connection to the dominant scattering mechanisms. Boxes show the range of *QE* and response time for metal, semiconductor, and NEA (low to high).

surface [175], as the contrasting behaviors reported by Hartmann et al. [124] and Bazarov et al. [18] demonstrate. The full complexity of the current pulse as a theoretical exercise, however, requires keeping track of all the delays associated with all the excitation sites, and is a computationally and numerically intensive task [179]. From the above treatment, however, it is clear that quantum efficiency and emission delay are linked, as recognized in Figures 13.20 and 13.21.

As aesthetically simple as the shell-sphere model is, it neglects taking into account that electrons are not photoexcited at a single site at $z = \delta$. Instead, they are excited at multiple sites into the photocathode material. There are many expanding shells and spheres, not just two, and it is their collective emission which causes the shape of Figure 13.14 to emerge in a theoretical simulation [179]. The charge content of each shell and sphere declines the further into the material the excitation occurs as per the attenuation factor $e^{-z/\delta}$ encountered in Eq. (13.16), and monitoring the content of the numerous shell/spheres that are excited becomes a complex numerical problem [179] without an easy analytical representation. Still, the simple models here give a means to understand their plausible behavior.

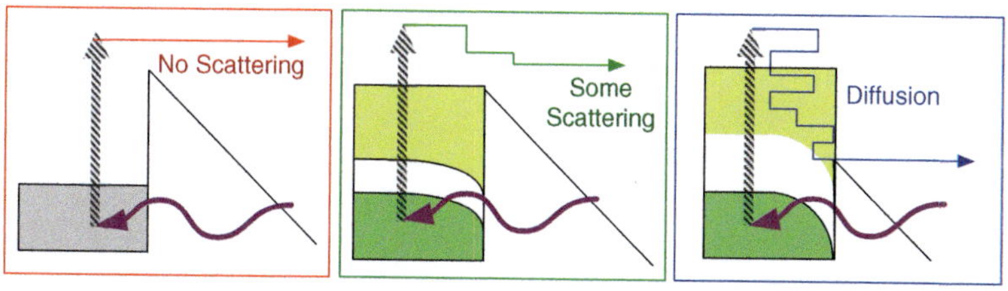


**Figure 13.21** Three types of photocathodes and their response time/QE domains in relation to the scattering processes that dominate their behavior for (left to right) metal, semiconductor, and NEA photocathodes. Adapted from [154].

## 13.5 Laser Heating

A final behavior of importance that, like delayed emission, resists simple analytical models is the behavior of photoemission when the laser intensity is high enough to cause thermal excursions in the electron distribution due to electron–electron scattering and the exchange of heat between the electrons and the lattice ions via electron–phonon scattering. Because the *e–e* and *e–p* scattering times differ, the temperature of the electrons $T_e$ evolves differently from that of the lattice $T_i$, and their departure from each other becomes evident when the laser intensities are extreme and the pulse widths very narrow [19, 161, 250, 272], further complicated by the fact that for very high laser intensities, the absorption of more than one photon by an excited electron gives rise to a multiphoton Fowler–DuBridge [20, 108, 109] model that qualitatively accounts for the emitted current.

If the temperature of the lattice is ignored, then it is reasonable to assume that the electron temperature satisfies a heat diffusion equation similar to Fick's law of diffusion in Eq. (4.21). The new behavior that changes the simplicity of Eq. (4.26) is that the processes–scattering–that give rise to energy exchange between diffusing electron populations are in fact temperature-dependent, and so the simple derivation of Eq. (4.21) needs emendation to account for that: $D$ is no longer constant. The methods of elementary kinetic theory for particles with a distribution in energy [289] suggest a revised equation. Recall that in the derivation of Fick's second law of Eq. (4.21), the diffusion constant was found to be $D \equiv \Delta x^2/4\Delta t$ if the jump probability $p = 1/2$. Consider this now from a scattering perspective. If the factor $\Delta t \approx \tau$, or the mean time between collisions, then $\langle v\rangle = \Delta x/\Delta t \approx \delta/\tau$ is the average velocity. Therefore, in the one-dimensional model, following the individual particles as they migrate ("tracer diffusion") suggests that $D = (\Delta x/\delta t)\Delta x/4 = \langle v\rangle\delta/4$. Consider this now from the perspective of particles that can scatter. In polar coordinates $(r, \theta, \phi)$ with the center being at a surface element $dS$ (taken to be on the $z = 0$ plane), then particles coming from a volume element $dV$ where they have just suffered their last collision that get to $dS$ without scattering again is proportional to $e^{-r/\delta}$. The fraction of particles from $dV$ that make the trip per unit time is then

$$\frac{1}{\rho}\frac{d\rho}{dt} = \frac{\langle v\rangle}{\delta}\frac{dV}{4\pi r^2}|\cos\theta|dS \tag{13.51}$$

where the $|cos\theta|$ factor is required because the particles coming from $dV$ are coming towards $dS$ at an angle $\theta$ with respect to the normal to $dS$ as in Figure 13.22. An integral over all space for those particles coming from above $(0 < \theta < \pi/2)$ gives

$$\int \frac{d\rho}{dt}dV = \frac{\langle v\rangle}{4\pi\delta}\int_0^{\infty}\int_0^{\pi/2} 2\pi r^2 \sin\theta dr d\theta\, \frac{\cos\theta}{r^2}e^{-r/\delta}\rho(r,\theta) \tag{13.52}$$

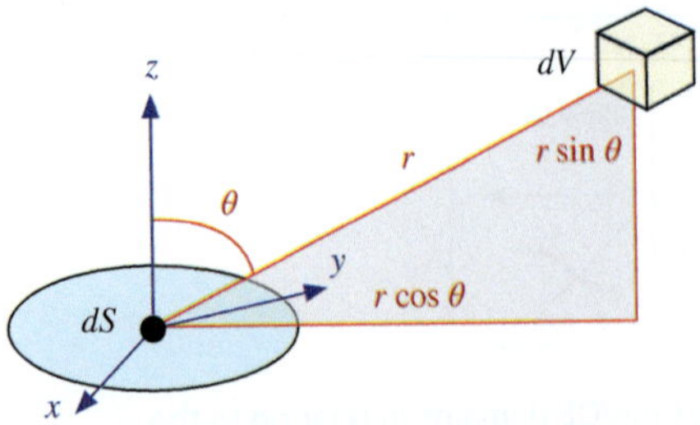


**Figure 13.22** Passage of particles from volume element $dV$ that pass through surface element $dS$ that is a distance $r$ away on $xy$-plane. $\theta$ is angle between $\vec{r}$ and $\hat{z}$, $\phi$ is the angle between the projection of $\vec{r}$ onto the $xy$-plane and $\hat{x}$.

If $\rho = \rho_o$ is uniform, then the double integration is analytic and results in $\rho_o\delta/2$ and so $D = \langle v\rangle\delta/4$ is recovered when the difference is taken for those particles coming from above with those coming from below. The passage of particles across a plane in a unit time and area is a current, and the difference in those respective currents that go as $\rho\langle v\rangle$ is then $J_+ - J_-$, with $\rho_+ - \rho_-$ begin replaced by $\partial_z\rho$. Fick's first law $J_D(z) = -D\partial_z\rho(z)$ results.

If instead $\rho(\vec{r})$ varies weakly in the $z$-direction, then it is expected that $\rho \approx \rho(0) + z\left(\partial_z\rho(0)\right) + (z^2/2)\left(\partial_z^2\rho(0)\right)$. The difference in currents for the first term $\rho(0)$ is identically zero. The next term introduces a $z = r\cos\theta$ into the integrals, for which $\int_0^\infty re^{-r/\delta}dr = \delta^2$ and $\int_0^{\pi/2}\cos^2\theta\ \sin\theta d\theta = 1/3$, resulting in $D = \langle v\rangle\delta/3$. Generalizing to three dimensions then introduces the gradient, and so Fick's first law becomes

$$J_D(\vec{r}) = -D\nabla\rho(\vec{r}) \tag{13.53}$$

Fick's second law is obtained by taking the difference of currents across a plane

$$\frac{J_D(z+\Delta z, t) - J_D(z-\Delta z, t)}{2\Delta z} \rightarrow \partial_z J_D = -\partial_z\left(D\partial_z\rho\right) \tag{13.54}$$

Consequently, when the factor $D$ can itself change (depends on $z$), then it is no longer outside the gradient operator.

Speaking of temperature now instead of density, the consideration of electrons as the carriers of that kinetic energy which defines the temperature requires some additional nomenclature based on the Sommerfeld model of a metal encountered in Section 6.4. For both simplicity and generalization, work with the density of states $D(E)$ from Eq. (7.15), and return to the convention that $x$ (not $z$) refers to the length parameter over which temperature varies. The electron heat capacity $C_e(T)$, or a measure of the heat $\Delta Q$ required to cause a change in temperature, goes as $\Delta Q = (d\langle E\rangle/dT)\Delta T = C_e\Delta T$, and so

$$C_e = \frac{d}{dT}\int ED(E)f_{FD}(E)dE \tag{13.55}$$

This can be modified for metals by observing that $\mu$ does not change much with temperature (Eq. 7.29), and so neither does $\rho$. As a result, the factor $\mu\rho$ can be subtracted from the integral on which $d/dT$ acts, allowing for a more useful form

$$C_e(T) = \frac{d}{dT}\int (E-\mu)D(E)f_{FD}(E)dE \tag{13.56}$$

Now the tricks used in treating conductivity in Section 7.4 can be used, where $f_{FD}(E)[1 - f_{FD}(E)]$ mimics the behavior of a delta function. As a result, $\mathcal{D}(E)$ can be approximated by its value at $E = \mu$ and removed from the integral, giving

$$C_e(T) \approx k_B^2 \mathcal{D}(\mu) T \int_{-\infty}^{\infty} s^2 \delta_l(s) ds = \frac{\pi^2}{3} k_B^2 \mathcal{D}(\mu) T \tag{13.57}$$

Such an approximation is justification for the oft-used $C_e(T) = \gamma T$, but clearly higher order terms participate. It can be shown that approximations which nominally account for those higher order factors is [160]

$$C_e(T) \approx \frac{\gamma T}{1 + \frac{7}{40}\left(\frac{\pi k_B T}{\mu}\right)^2} \tag{13.58}$$

$$C_i(T) \approx \frac{3NK_b}{1 + \frac{1}{20}\left(\frac{T_D}{T}\right)^2} \tag{13.59}$$

where $N$ is the number density (#/cm$^3$) of the crystal and $T_D$ is the Debye temperature. Analogous to electrical conductivity $\sigma$ which relates current to a gradient in potential, thermal conductivity $\kappa$ relates current to a gradient in temperature as in Eq. (7.8), and so

$$\kappa(T) = \frac{2\mu}{3m} \tau(\mu) C_e(T) \tag{13.60}$$

Analogous to Fick's first law, $J_T = \kappa(dT/dx)$, and so now a heat diffusion equation is possible to infer, where in addition to the self-diffusion of heat, a source term $G(x, t)$ can be added to write [19]

$$C_e(T)\partial_t T = \partial_x [\kappa(T)\partial_x T] + G(x, t)$$

but the heat diffusion for electrons requires a modification by accounting for the energy lost (or gained) from the lattice, and which therefore requires keeping track of the electron temperature $T_e$ as well as the lattice temperature $T_i$. What energy the electrons lose to or gain from the lattice is reflected in what the lattice gains from or loses to the electrons: designate this by a function $U(T_e, T_i)$. Now, two separate but coupled equations for the electron and lattice temperature arise, and are [108, 150, 161, 250, 272]

$$\begin{aligned} C_e(T_e)\frac{\partial}{\partial t} T_e &= \frac{\partial}{\partial x}\left(\kappa(T_e)\frac{\partial}{\partial x} T_e\right) - U(T_e, T_i) + G(x, t) \\ C_i(T_i)\frac{\partial}{\partial t} T_i &= U(T_e, T_i) \end{aligned} \tag{13.61}$$

Finding $U(T_e, T_i)$ is outside the scope of the present work, but is treated by various authors [108, 163, 194, 252]: the form used here are from complex derivations described in Refs. [150, 161]. The source function $G(x, t)$ is the amount of laser power absorbed (and not taken away by photoexcited electrons). They are defined by

$$U(T_e, T_i) = \frac{\pi^2}{6} \frac{m v_s^2 \rho}{\tau_{ep}(T_i)} \left(\frac{T_e - T_i}{T_i}\right) \tag{13.62}$$

$$G(x, t) = [1 - R(\omega)] I_\lambda(t) \left(\frac{e^{-x/\delta}}{\delta}\right) \left[1 - \frac{U[\beta(\hbar\omega - \phi)]}{U(\beta\mu)}\right] \tag{13.63}$$

where the FD function $U(x)$ is from Eq. (8.16) and the ratio of them represents the faction of the total number of electrons in the Fowler–DuBridge model with sufficient normal energy to be photoemitted that get to the surface without scattering and so take their energy with them when they depart. Note that the energy exchange $U(T_e, T_i)$ and the Fowler–DuBridge $U(x)$ are not the same functions, their symbols not withstanding: the conventions of the references are retained in spite of possible confusion.

The evaluation of the source-driven heat diffusion equation can make use of a sum of point source distributions based on Eq. (4.26) as Bechtel cleverly did [19], or use a computationally involved finite-difference, coupled matrix, and predictor-corrector methods [150, 161] that are required when the temperature dependence of $\kappa(T_i)$ is respected and which used analytical models for the relaxation times, thermal conductivities and heat capacities of copper and tungsten. As a result of the reliance on those models, the later "Numerical" approach made more extensive use of material parameters as shown in Table 13.2 (based on Table 13 of Jensen et al. [161]) than did the model of Bechtel. Although both are beyond an introductory treatment, the behaviors both can be intuited as being related to the manner in which a sum of heat sources with differing weights diffuse (akin to superposition methods in quantum mechanics), which is a simple computational model. On that basis, it is clear that a laser pulse in the shape of a Gaussian will give rise to a Gaussian-like temperature rise at the surface of the metal, but which will be wider than the laser pulse and which will evolve differently because thermal diffusion only occurs into the metal. For a very strong laser pulse, the electron temperature will rise significantly, potentially giving rise to a thermal emission component in addition to photoemission. The differences in the relaxation times $\tau_{ep}(T)$ and $\tau_{ee}(T)$ especially for electrons excited to well above the Fermi energy will cause the electron population to heat rapidly and the lattice to heat more gradually. Eventually, the electrons will cool, and the lattice will return energy to the electron population. Such expectations are confirmed in a

**Table 13.2** Parameters used for the heating of tungsten using a high intensity Gaussian laser pulse (After Table 13 of Adapted from [160]. Rows in Bechtel containing dashes signify that parameter is not used by him. Values shown for $C_e$, $C_i$, and $\kappa$ are evaluated at Room Temperature (300 K).

| Parameter | Symbol | Unit | Numerical | Bechtel |
|---|---|---|---|---|
| Wavelength | $\lambda$ | nm | 1064 | 1060 |
| Reflectivity | $R(\omega)$ | % | 60.3184 | 60 |
| Laser Penetration | $\delta$ | nm | 22.3654 | 25 |
| Laser Source | $G$ | GW / K cm$^3$ | 22.3654 | 25 |
| Specific Heat | $C_e$ | J / K cm$^3$ | 0.04094 | – |
| Specific Heat | $C_i$ | J / K cm$^3$ | 2.39981 | – |
| Number density | $N$ | mole/cm$^3$ | 0.1050 | – |
| Acoustic Phonon Param. | $\lambda_o$ | – | 2.238 | – |
| Sound velocity | $v_s$ | m/s | 5174 | – |
| Dielectric Const. | $K_s$ | – | 18.0396 | – |
| Debye Temp | $T_D$ | K | 400 | – |
| Fermi Level | $\mu$ | eV | 18.08 | – |
| Relax time | $\tau$ | fs | 1.37942 | – |

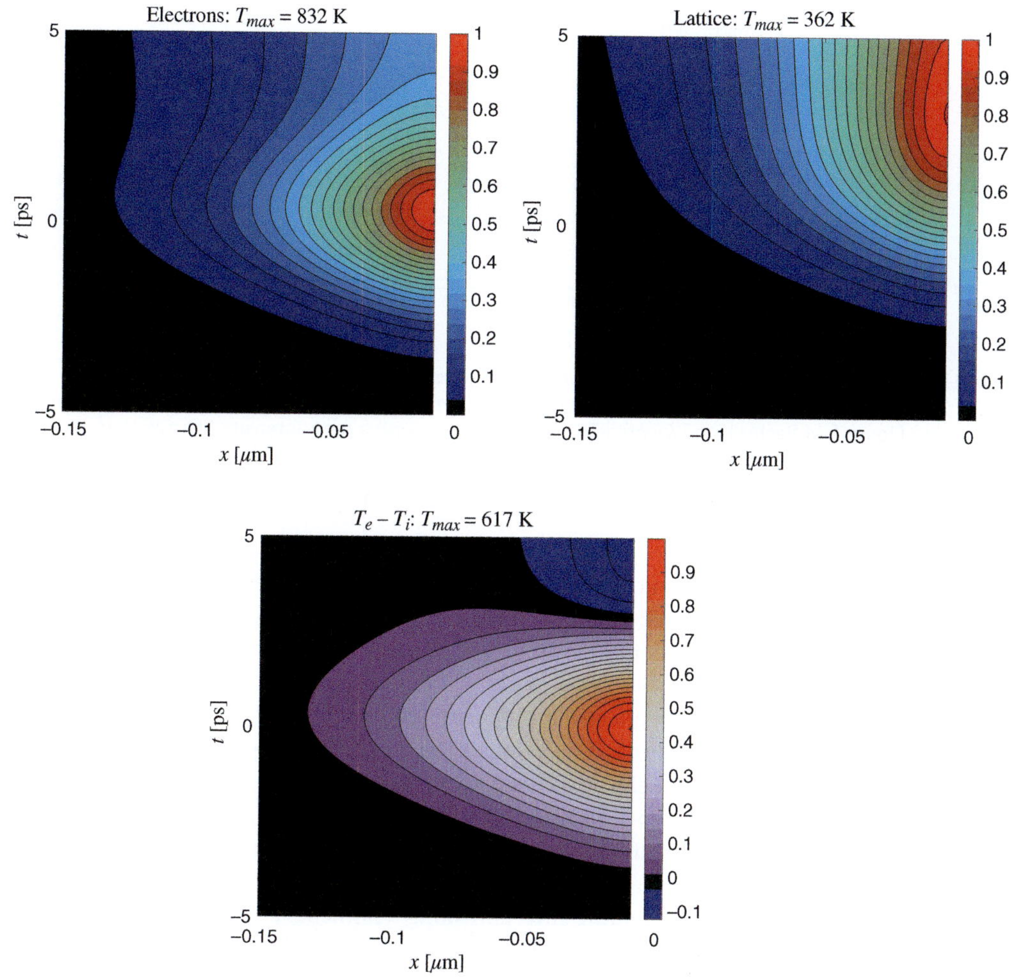


**Figure 13.23** Contour plots of $T_e/T_{max}$ (electrons) and $T_i/T_{max}$ (lattice), and their difference, as a function of position from the surface and time for laser heating on copper, for the $T_{max}$ shown in the title. Initial laser pulse is Gaussian $I_\lambda(t) = I_o \exp[-(t/\delta t)^2]$ with $\delta t = 2$ ps (the time origin is the center of the laser pulse, and the spatial origin is at the surface). Parameters. Laser parameters analogous to those of Papadogianis et al., [272] of $I_o = 3$ GW/cm$^2$, $F = 1$ MV/m, $\lambda = 266$ nm. Adapted from Figure 84c–e of [149] / with permission of Elsevier.

simulation of laser heating of a copper surface in Figure 13.23. The Numerical approach, however, shows the impact of the subtle temperature dependence of $C_e$, $C_i$, and $\kappa$ in Figure 13.2 for two different laser intensities and pulse lengths.

The surface temperature variation shown in Figure 13.24 is suggested by the temperature variation in bulk shown in Figure 13.23. Initially, the front edge of the Gaussian laser pulse heats the bulk material, but after a short delay, the heat is partially transferred from the electrons over to the lattice. The lattice stays warm after the electron population cools off, and then the electrons are warmed by the lattice returning energy to the electron population. For that reason, long tails are evident in the surface temperature. Given the substantial temperature rise associated with high-intensity laser pulses, the anticipation is that thermal-field emission of electrons accompanies their

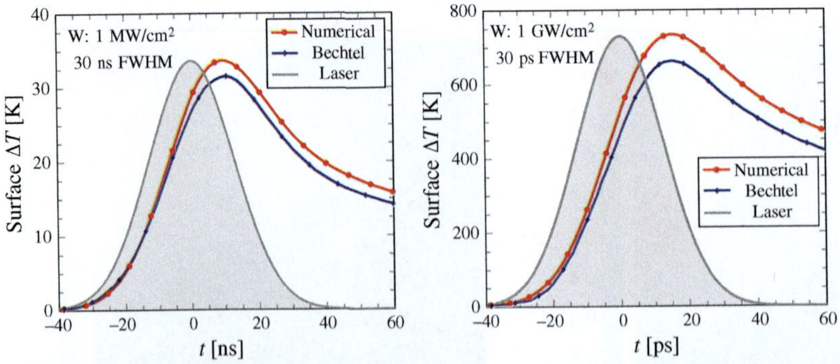


**Figure 13.24** Comparison of finite difference / predictor-corrector Eq. (13.61) for $T_e$ with theory of Bechtel [19] for *(left)* a 30 ns full-width half-max (FWHM) Gaussian laser pulse with a peak intensity of 1 MW/cm$^2$ compared to Figure 5 of Bechtel, and *(right)* a 30 ps FWHM pulse with a peak intensity of 1 GW/cm$^2$ compared to Figure 7 of Bechtel. Grey area indicates laser pulse profile normalized to "Numerical" peak. After Figures 7 and 8 of Ref. [161] / AIP Publishing LLC.

photoemission, giving rise to dark current that can persist even as the photoexcited electrons fade away. Whereas photoemission can be treated as more or less commensurate with the laser pulse, the dark current lags behind the laser pulse, and so can complicate the formation of a pulse of electrons and alter its shape.

A final matter of importance is the effect of multiple pulses. Modern applications (particle accelerators and Free Electron lasers being notable examples) require numerous pulses: high-power FEL program goals, for example, sought electron bunches of a duration of 10–50 ps, generated about once a nanosecond and containing up to a nanocoulomb of charge [161]. The temperature changes on the surface of a metal photocathode during a pulse are thereby complicated by rising temperature conditions associated with how long a train of pulses has hit a target. The expectation is that a metal photocathode would not long endure such punishing treatment, a conclusion supported by expectations drawn from Figure 13.24 and estimates of cumulative temperature changes after even a moderate number of pulses. By way of estimation, consider a crude calculation of how much the temperature may rise at the surface of a metal that absorbs a laser pulse. A common and reasonably good approximation is that $C_e(T) = \gamma T$, with $\gamma$ constant, a representative value being $\gamma = 134.6$ J/K$^2$ m$^3$ for tungsten (see Table I of Ref. [161], upon which the present treatment is based). The deposited energy $\Delta E$ can be estimated by integrating Eq. (13.61) over all space and over the duration of the pulse (from $-\Delta t/2$ to $\Delta t/2$) and observing that the spatial integral over $\partial_x(\kappa\partial_x T_e)$ vanishes because $\partial_x Te$ vanishes at both boundaries (diffusing heat is reflected from the surface). That means

$$
\begin{aligned}
\Delta E &\approx \int_{-\infty}^{0} dx \int_{-\Delta t/2}^{\Delta t/2} dt \frac{\partial}{\partial t}\left(\frac{1}{2}\gamma T_e^2 + C_i T_i\right) \\
&= \int_{-\infty}^{0} dx \left[\frac{1}{2}\gamma(T_e^2 - T_o^2) + C_i(T_i - T_o)\right]
\end{aligned}
\tag{13.64}
$$

where $\Delta E$ units are [eV/nm$^2$] and $T_o$ is the temperature of both the electron gas and the ion lattice at $t = -\Delta t/2$. As inferred from Figure 13.24(a), even lasers of MW/cm$^2$ intensity are not expected to produce large $\Delta T$ if the pulse length is on the order of 10 ps, so the approximations that both $T_e$ and

$T_i$ are comparable to $T_o$ make the approximations $T_i \approx T_e \to T$ and $T_e + T_i \approx 2T_o$ very convenient. Using them, then

$$\Delta E \approx \int_{-\infty}^{0} (T - T_o)(\gamma T_o + C_i) \tag{13.65}$$

This deposited energy diffuses into the metal according to

$$\begin{aligned} T &= T_o + 2\Delta T \sqrt{D_o \Delta t}\, u(x, \Delta t/2) \\ u(x,t) &= \frac{1}{\sqrt{4\pi D_o t}} \exp\left[-\frac{x^2}{4D_o t}\right] \end{aligned} \tag{13.66}$$

where $D_o \equiv \kappa / C_e$. That approximation allows the remaining spatial integration to be performed, resulting in

$$\Delta T = \frac{\Delta E}{\sqrt{D_o \Delta t}\,(\gamma T_o + C_i)} \tag{13.67}$$

$\gamma$ may be approximated using Eq. (7.15) by

$$\gamma = \frac{1}{3}(\pi k_B)^2 \eta D(\mu) \tag{13.68}$$

where $\eta$ is a dimensionless thermal effective mass term of order unity (e.g. $\eta = 1.375$ for copper). Similarly, approximate $C_i = 3Nk_b$, where $N$ is number density (moles/cm$^3$) of the metal.

For typical parameters, $\Delta T$ is on the order of a degree Kelvin (see Ex. 13.9). After the deposition of the energy of each laser pulse into the metal surface, heat diffusion into the bulk immediately begins, so that simply adding up the number of temperature increments $\Delta T$ is an obvious overestimate, but even so, one can infer that the diffusion is slow enough that the temperature rise will still be profound even for several thousand pulses at the surface, near the melting points of metals like copper [64] and lead [318] that are often the metals of choice. Even tens of thousands of pulses are hardly enough to meet the needs of a high-power FEL if it needs to dwell on a target for durations of a millisecond or more. Alternately, although increasing the intensity of drive lasers may seemingly meet the needs of other applications like X-ray FEL's, that avenue also risks heat-induced damage and back ion bombardment: QE is the primary cathode metric of several that must be improved to meet future technological demands [259]. Hence, high quantum efficiency photocathodes such as the multi-alkali antimonies ($Cs_3Sb$, Cs $K_2$ Sb), cesium telluride (Cs Te), or other exotic candidates which operate at visible wavelengths and thereby also reduce harsh demands on the drive lasers [62, 355] have been and are actively sought.

## 13.6 Exercises

**Exercise 13.1** From the definition of quantum efficiency (QE) from Eq. (2.22), $QE$ relates the number of electrons $N_q$ emitted to the number of photons $N_\omega$ incident on a surface. It can therefore be used to write

$$QE = \frac{\lambda_c}{\lambda}\frac{N_q}{N_\omega}$$

where $\lambda$ is the wavelength of the drive laser. What is $\lambda_c$?

**Solution** $\lambda_c = 1239.8$ nm.

**Exercise 13.2** In Eq. (13.16), what is $p(\tilde{E})$ for copper if the photon wavelength is $\lambda = 266$ nm? Assume $\tau$ is dominated by electron–electron scattering for electrons excited from the Fermi level. See Table 13.1.

**Solution** $p(\tilde{E}) = 2.083$.

**Exercise 13.3** In Eq. (13.16), what is $p(\tilde{E})$ for $Cs_3Sb$ if the effective mass is assumed to be $m/m_o = 0.12$ when the bandgap is $E_g = 1.6$ eV as suggested by Section 7.3, if the photon wavelength is $\lambda = 266$ nm? Use Table 13.1 values for other needed parameters. Assume the electron is excited from the top of the valence band across the bandgap. Assume as per Spicer [326] that $\ell = 25$ nm. What is the relaxation time $\tau$?

**Solution** $p(\tilde{E}) = 0.546$. $\tau = 8.35$ fs.

**Exercise 13.4** Depending on the purity of gallium arsenide (GaAs), a material useful for making spin-polarized electrons [47], the mean free path can range from 300 nm to a micron (or more). If the effective mass of the electron is $m_n/m = 0.063$ and the bandgap $E_g = 1.42$ eV, then what is the maximum value of $p(\tilde{E})$ if $\lambda = 532$ nm?

**Solution** $p(\tilde{E}) = 0.0423$.

**Exercise 13.5** For a distribution $f(\vec{r}, \vec{k})$ that is Gaussian in both position and momentum (see Eq. (13.25)), show that $\langle xk \rangle = 0$. Find $\langle x^2 \rangle$.

**Solution** $\langle x^2 \rangle = a^2/2$.

**Exercise 13.6** Assume a beam with a uniform radial distribution of particles that have a Maxwell–Boltzmann distribution in energy and a Gaussian in position such that $f(x, k) = \exp[-(x/\rho_c)^2 - E(k)/k_B T]$ where $E(k) = \hbar^2 k^2/2m$. What is the normalized emittance?

**Solution** $\varepsilon_{n,rms} = (\rho_c/2)\sqrt{k_B T/2mc^2}$. This is thermal emittance of a beam produced by a thermionic cathode.

**Exercise 13.7** Consider a photocathode with a cesiated metal surface ($\mu = 11.5$ eV, $\Phi = 1.6$ eV [258] that has a zero field $QE$ of 0.4% at a wavelength of $\lambda = 354.67$ nm ($n = 3$ for an Nd:YAG laser). If it were operated in a photoinjector with an electric field at the cathode of 10 MV/m, what is the approximate $QE$ for the other wavelengths of that laser? Assume temperature effects can be neglected, and that the reflectivity $R$ and $p(\tilde{E})$ are constant in Eq. (13.23). Compare to the predictions of the Fowler–DuBridge model.

**Solution** For $n \in (1, 2, 3, 4, 5)$, then $QE(n)[\%] = (0, 0.104, 0.400, 0.766, 1.176)$. Compare to $QE_{FD}(n)[\%] = (0, 0.071, 0.400, 0.996, 1.856)$.

**Exercise 13.8** Show Eq. (13.50). What is $t_o$ assuming Ex. 13.3 parameters? What is the diffusion term $D$ if $t_d/t_o = 6$ (an ad hoc number)?

**Solution** $t_o = 4.56$ fs. $D = 1.70$ nm$^2$/fs.

**Exercise 13.9** Approximate the rise in temperature at the surface of standard copper absorbing the energy of an Nd:YAG laser operating in the fourth harmonic ($n = 4$). Assume the following operational and material parameters: Table 13.1; starting temperature $T_o = 573$ Kelvin; incident laser intensity $I_\omega = 1$ MW/cm$^2$; pulse duration $\Delta t = 1$ ns; thermal mass factor $\eta = 1.375$; relaxation time $\tau(\mu) = 8.4$ fs; copper number density $N = 0.141$ [moles/cm$^3$]; copper Fermi level $\mu = 7$ eV. For simplicity, assume $C_e(T) \approx \gamma T$ and $C_i(T) \approx 3Nk_B$. Numerically find the intermediate quantities $(R, \gamma, \kappa)$ and express them in their conventional units (e.g. Table 13.1, Ref. [161]).

**Solution** $R(\omega) = 35.73\%$; $\gamma = 97.00$ [J/K$^2$ m$^2$]; $\kappa = 383.2$ [J/K m s]. Conclude $\Delta T = 0.6851$ [K].

# 14

# Secondary Emission

*Scatter'd by winds and high tempestuous gusts?*
*O, let me teach you how to knit again*
*This scattered corn into one mutual sheaf,*
*These broken limbs again into one body...*

W. Shakespeare, *Titus Andronicus* V.3

## 14.1 Simple Model Revisited

The brief model of secondary emission in Section 2.2.4 obscured important details about which electrons contributing to the yield are important. Several types of electrons are emitted after a material is bombarded by high-energy primaries.

- The primary electron may change its direction back toward the surface after experiencing a collision. Because the primary is scattering with another electron, it loses roughly half of its initial energy. If conditions are such that it is emitted, it is a back-scattered primary, and it contributes a high-energy peak at about half of the incident primary energy on a distribution curve;
- A primary electron that has undergone several scatterings but still retains much of its incident energy, or a secondary that scatters with another primary, is an intermediate primary; although it still has sizable energy, its kind is poorly represented in the distribution curve;
- Lastly, low-energy secondaries are those electrons that have been generated by the primary electrons as they shed energy through repeated collisions; these electrons have a low energy comparable to 20 eV.

The last group are generally called "true" secondaries [121], and they are the ones that diffuse through the bulk material until they get to the surface and are emitted, as in Figure 14.1 (see Algorithm B.4.9). Those secondaries are generated by a "three-step" process of their own with the steps of generation of the secondaries, their transport to the surface, and their emission [367], much like the photoemission model of Section 13.1. They are commonly said to go through a three-step model [15] of their own. In a confluence of physics, Krolikowski and Spicer considered high enough photon energies that secondary electron generation was part of their photoemission studies [208]. A description of secondary emission methods by Haas and Thomas [120] sounds exceedingly familiar to those who know Spicer's model.

*Fundamentals of Electron Emission Physics*, First Edition. Kevin L. Jensen.

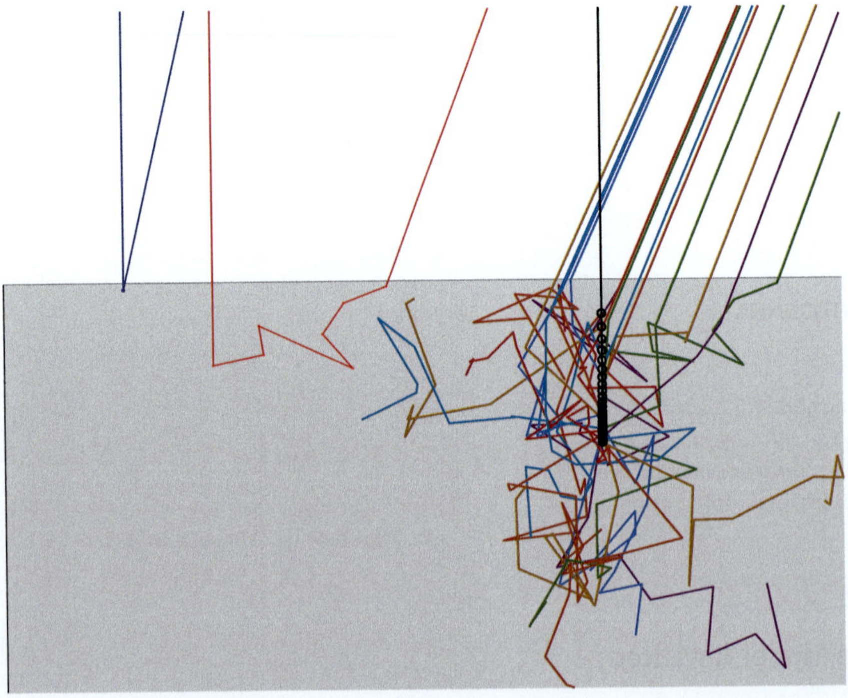

Figure 14.1 A high energy primary electron (vertical line in vacuum) can cause the emission of several types of electrons: (*left*) back-scattered primaries, (*middle*) intermediate primaries; and (*right*) low-energy or "true" secondaries. White/Gray area is vacuum/metal. Secondaries that cross surface boundary are emitted.

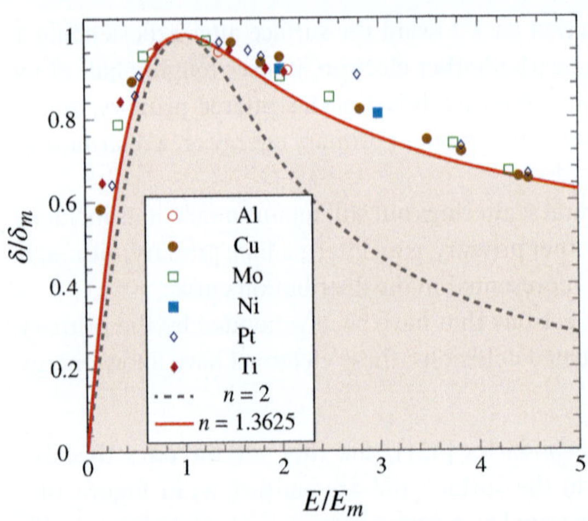


Figure 14.2 Ratio of SEY $\delta(E)$ to maximum value $\delta_m$ for six metals examined in Figure 1 of [15]. Curves are based on Eq. (2.27) with the approximation of Eq. (14.2). After [15, 157].

True secondaries are generated as the primary electron bores into the material, where repeated collisions with the atoms cause electrons to be liberated. Generally, their initial velocity is perpendicular to the primary electron's direction [263]. Metals have a small secondary electron yield (SEY) because of their typically large work functions. Baroody [15] observed that if $\delta_m$ is the maximum SEY of a material being struck by primary electrons of energy $E_m$ (with $E_o$ and $\delta$ being the primary energy and the resultant yield), then the yield ratio $\delta(E)/\delta(E_m)$ as a function of $E/E_m$ tends to follow a universal curve for metals, as in Figure 14.2 (see Algorithm B.4.9). Semiconductors with their higher yield are often preferred as electron sources. In those cases, the energy of the created secondary is approximately $2.5E_g$, where $E_g$ is the bandgap of the semiconductor [175, 185] entailing that for the ultra-wide bandgap emitters such as diamond, cubic boron nitride (c-BN), and aluminum nitride (AlN) [367], the secondaries that are generated can have considerable energy.

Baroody [15] was the first to observe that if $\delta_m$ is the maximum secondary emission yield of a material being struck by primary electrons of energy $E_m$ (with $E$ and $\delta$ being the primary energy and the resultant yield), then the yield ratio $\delta(E)/\delta(E_m)$ as a function of $E/E_m$ tends to follow a universal curve for metals. He also saw that there was a connection between the work function of a metal and its value of $\delta(E_m)$, as shown in Figure 14.3 (see Algorithm B.4.9).

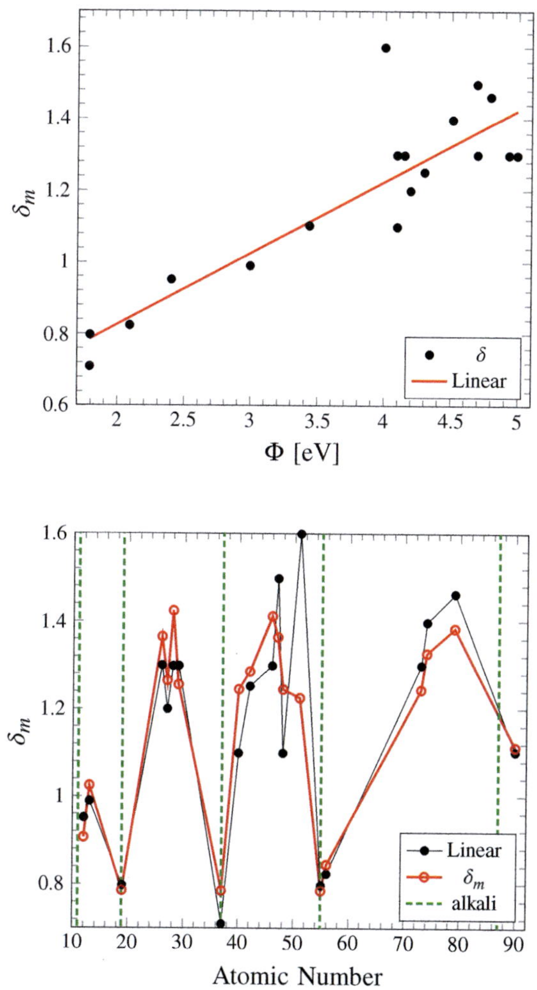


**Figure 14.3** (*top*) Maximum yield $\delta_m$ (black •) compared to least squares linear fit of $\delta_m$ as a function of $\Phi$ given by $y(\Phi) = 0.4264 + 0.1995\Phi$ for $[\Phi] = \text{eV}$. Data digitally extracted from Figure 2 of Ref. [15]. (*bottom*) Comparison of measured $\delta_m$ to linear fit $y(\Phi)$ relation. Vertical green dashed lines are at the AN of the alkali metals. *Source:* After [157] / AIP Publishing LLC/CC BY 4.0.

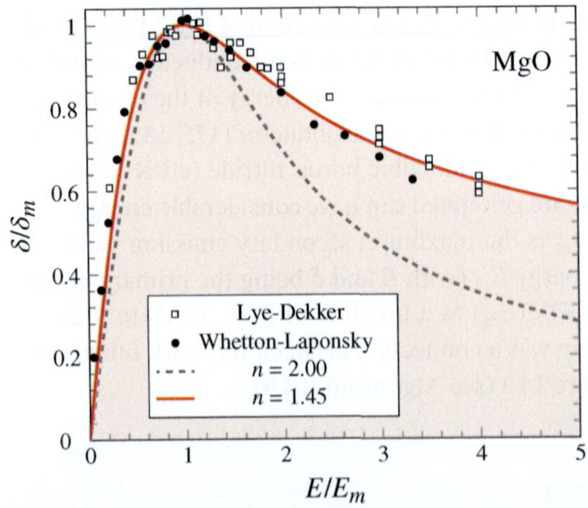


**Figure 14.4** Ratio of SEY $\delta(E)$ to maximum value $\delta_m$ for MgO as examined by Lye and Dekker and Wheton and Laponsky: Data digitally extracted from Figure 5 of Ref. [244]. Curves are based on Eq. (2.27) with the approximation of Eq. (14.2). Adapted from [154].

Semiconductors, too, tend to follow a universal curve, an example being MgO (diamond will be shown to give a similarly shaped yield curve). In the elementary treatment of Section 2.2.4, a power-law relation in Eq. (2.26) led to a range formula

$$R(E_o) = \frac{E_o^n}{nA} \tag{14.1}$$

for how far a primary electron would penetrate a material, such as a metal, before coming to a stop, where $n - 1$ is the power (not necessarily an integer) that the primary energy $E_o$ is raised to in the denominator of Eq. (2.25). That range equation was based on assuming $dE/dx \propto 1/E^{n-1}$. [15] had considered the value $n = 2$ in his study of the yield of various metals, a choice which he showed causes an underestimation of the yield curve for metals (Figure 14.2, see Algorithm B.4.9). Baroody [254] showed the same behavior for semiconductors (Figure 14.4). Still, Baroody's model captured an essential feature of the yield curve, namely, that $E_o/E_m$ vs. $\delta(E_o)/\delta(E_m)$ tended to fall along the same general curve described by Eq. (2.27).

## 14.2 Other Models

Different approximations to $dE/dx$ result in different approximations to the universal yield curve. Measurements by Young [371] for the range of primaries in aluminum of Figure 14.5 (see Algorithm B.4.9) suggest that the range equation for $E_o \lesssim 10$ keV is governed by $n \approx 1.36$: by comparison, $n = 2$ caused the theoretical $R(E_o)$ to visibly depart from the measured behavior. The optimal value of $n$ depends both on the maximum primary energy examined and the material. Lye and Dekker offered an analytic form for $G_n(\lambda)$ of Eq. (2.27) that can be found by using the approximation $s^n \approx 1 - n(1 - s)$ in the exponential of the integrand, resulting in

$$G_n(\lambda) \approx \frac{1 - e^{-\lambda n}}{\lambda n} \tag{14.2}$$

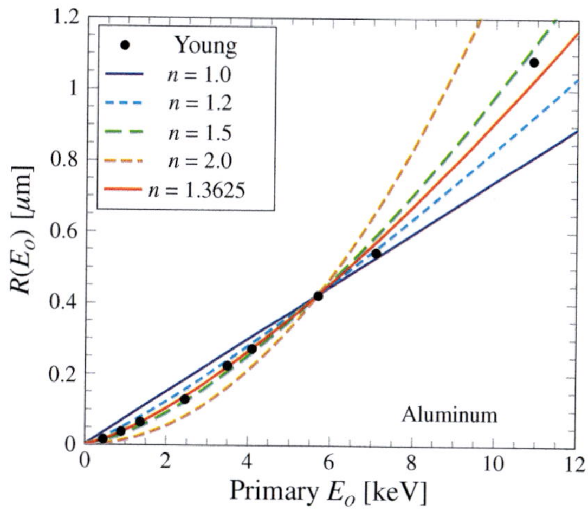


**Figure 14.5** Maximum range of primary electron penetration into Aluminum. Symbols (•) are data of Young (digitally extracted from Figure 5, Ref. [371]). Lines are $R(E) = R_o(E/E_j)^n$ with $R_o = 50.993$ nm, $E_j = 5.6987$ keV, and $n$ in the legend. Thick red line is from linear fit to $\ln[R(E_o)]$ for $1.3 < E_o < 7.1$ [keV].

Letting $\xi = E_o/E_m$ and $\lambda = \lambda_m \xi^n$ then results in [154]

$$\frac{\delta(E_o)}{\delta(E_m)} = \frac{1}{\xi^{n-1}} \left[ \frac{1 - e^{-\lambda_m n \xi^n}}{1 - e^{-\lambda_m n}} \right] \tag{14.3}$$

Kanaya and Okayama [196] developed a relatively simple and semi-empirical model of the range equation that is often used [367]. They assumed that the potential energy between electrons and atoms was governed by $V(r) = (ZQ/sa)(a/r)^s$, where $Z$ is the atomic number and $a$ is the effective screened radius of the atom $a = C_{tf}\, a_o/Z^{1/3}$, where $a_o$ is the Bohr radius and $C_{tf}$ is the Thomas-Fermi constant $C_{tf} \equiv (1/8)(6\pi)^{2/3} = 0.88532$. The parameter $s$ is such that $s = 1$ corresponds to Rutherford scattering and $s = 2$ results in $dE/dx$ equal to a constant. They deduce that $dE/dx$ is then given by

$$\left.\frac{dE}{dx}\right|_{ko} = \lambda_s \frac{3\pi 2^{5/3} a^{1/3} (2\sqrt{Q})^{10/3} \mathcal{N} Z}{E_o^{2/3}} \tag{14.4}$$

in their notation, where $\mathcal{N} = N_A\, \rho/\mathcal{A}$ is the number of atoms per unit volume in the target material, $N_A$ is Avogadro's number (a mole), $\rho$ is the mass density of the material, $\mathcal{A}$ is the atomic weight, $Z$ is the atomic number, $E_o$ is the incident primary energy, $\lambda_s$ is an empirically determined constant of order unity which they take to be $\lambda_s = 0.182$, and in [cgs] units, Kanaya and Okayama's $q_o = 2\sqrt{Q}$. From that, they find (or would find if they used [efnq]-based quantities) the range $R$ is given by

$$R(E_o) = \int_0^{E_o} \frac{dE}{dE/dx} = \frac{6^{7/9} \mathcal{A} E_o^{5/3}}{480\pi^{11/9} \lambda_s a_o^{1/3} Q^{5/3} Z^{8/9} N_A \rho} \tag{14.5}$$

For expediency in calculation, and in [cgs] units where $[\mathcal{A}] =$ grams, $[\rho] =$ grams/cm$^3$, $[R] =$ cm, $[E_o] =$ eV, and $[\rho R] =$ grams/cm$^2$, then

$$R(E_o) = \frac{5.03024 \times 10^{-12} \mathcal{A}\, E_o^{5/3}}{\lambda_s\, Z^{8/9} \rho} \tag{14.6}$$

which (apart from small differences in values used for the fundamental constants) is as given by Kanaya and Okayama. When relativistic corrections are incorporated, then it becomes

$$R(E_o) = \frac{5.03024 \times 10^{-12} \mathcal{A}\, E_o^{5/3}}{\lambda_s\, Z^{8/9} \rho} \left( \frac{2mc^2 + E_o}{2mc^2 + 2E_o} \right) \tag{14.7}$$

If the primary energy $E_o$ is in the MeV range rather than the keV range, then relativistic corrections become important. For lower energy primaries on the order of 10–20 keV, though, the non-relativistic approximation $R(E_o) \propto E_o^{5/3}$ can be used in Eq. (2.27) to find the yield $\delta(E_o)$. Flöttmann [121] used an empirical fit to the yield according to which

$$\frac{\delta(E_o)}{\delta_{max}} = \frac{s\xi}{s - 1 + \xi^2} \tag{14.8}$$

where $\delta(E_o)$ is the yield when the energy of the primaries is $E_o$, $\delta_{max} = \delta(E_m)$ is the maximum secondary yield occurring at $E_o = E_m$, $\xi = E_o/E_m$, and $s > 1$ is a fit parameter (not the same as the $s$ of Kanaya and Okayama).

## 14.3 Bethe Model

To move beyond semi-empirical models and to account for energy loss by secondaries, a more refined estimate of $dE/dx$ is required. The Bethe model [24] is often used [56, 115, 189, 262, 263, 381] for that purpose, and notable for working at lower energies than some expected [371]. The yield of a material struck by a high-energy beam of primary electrons is a measure of the number of secondary electrons generated by the primary beam as it collides with, and thereby shares, its energy via electron–electron collisions with core electrons bound to atoms. The subsequent evolution of the secondaries commonly uses a three-step model [15, 189, 244, 311, 348, 352, 367, 368]. Bethe's "stopping power" (SP) model governed how the loss of primary electron energy occurred [154], and gives a means to update the account given in Section 2.2.4. For heavier atoms, it is more complex, but in the case of carbon, it becomes quite simple because carbon atoms are light and only have a small number of shells. Because diamond is an excellent secondary emitter that motivated its consideration as the electron source in a diamond electron current amplifier [21, 189, 317, 367, 368]; considering carbon in particular becomes very useful here. Bethe's formula is

$$\frac{dE}{dx} = -\frac{192\pi N Q^2}{E} \sum_{n,l} f_{nl} \ln\left(\frac{2E}{BE_n}\right) \tag{14.9}$$

where $f_{nl}$ is the sum of the oscillator strengths for all transitions of an electron of the $n, l$ shell based on the number of energy levels of the atoms making up the material. Carbon is in the second row of the periodic table. Therefore, the sum over $l$ has only one term, and $n \le 2$. As a result, $f_{nl} = 2^n$ is simply obtained. The ionization energy of a hydrogen atom is the same as the Rydberg energy, or $R_y = 2Q/a_o = 13.6057$ eV. An electron in the $n^{th}$ level will have an energy $BE_n = (2Q/a_o)(Z/n)^2$, where $Z$ is the atomic number, as before. Let $\mathcal{N} = \rho/\mathcal{A} = (3.512\ \text{grams/cm}^3)/(12.0107\ \text{grams/mole})$ or $\mathcal{N} = 176.091$ [#/nm$^3$]. Only two terms contribute to the sum which, on using $\ln(x) + \ln(y) = \ln(xy)$, becomes

$$\frac{dE}{dx} = -\frac{1152\pi \mathcal{N} Q^2}{E} \ln\left(\frac{\nu E a_o}{Q}\right) \tag{14.10}$$

where $\nu \equiv 2^{4/3}/Z^2 = 0.0699956$. Unfortunately, the logarithm unphysically vanishes at $E = E_c \equiv Q/\nu a_o = 97.1896$ eV. To avoid that problem, let $E = uE_c$ and then insert a factor of order unity into the logarithm argument by $E \rightarrow E_c + E = (1+u)E_c$ to prevent it from vanishing (or becoming negative). The relation for $dE/dx$ can then be rewritten as [157]

$$\frac{ba_o}{E_c}\left(\frac{dE}{dx}\right) = -\frac{\ln(1+u)}{u} \tag{14.11}$$

where $b \equiv 1/(1152\pi v^2 \mathcal{N} a_o^3) = 2.16130$ for diamond. Now the right-hand side does not vanish. Letting $\ln(u) \rightarrow \ln(1+u)$ is, though, an arbitrary choice. A more flexible choice and one that allows better correspondence to experimental data is to instead let $\ln(u) \rightarrow \ln[d_1 + (u/d_2)]$ with $d_1$ and $d_2$ chosen to achieve better correspondence to other metals [139].

The previous power-law relationship of $dE/dx$ is recoverable, but the optimal choice of parameters is dictated by what the primary energy range $E_o$ is, and therefore the power-law representation is a local fit. Let the power-law approximation be given by

$$\frac{ba_o}{E_c}\left(\frac{dE}{dx}\right) = -\frac{C}{u^{n-1}} \tag{14.12}$$

If the right-hand side of Eq. (14.11) is called $f(u)$ and that of Eq. (14.12) called $f_a(u)$, then finding the local power-law about a specific point $u_o$ amounts to finding when $f(u_o) = f_a(u_o)$ as well as when $\partial_u f(u_o) = \partial_u f_a(u_o)$. It is found

$$\begin{aligned} n &= 2 - \frac{u_o}{(1+u_o)\ln(1+u_o)} \\ C &= \frac{\ln(1+u_o)}{u_o^{2-n}} \end{aligned} \tag{14.13}$$

As an example, the value of $n = 1.39$ suggested by Shih et al. [311]; for diamond[1] occurs when $u_o = 1.96214$, or $E_o = 190.7$ eV, which is on the lower side of the range of validity of the power-law representation at that energy; alternately, the value $n = 5/3$ of [196] occurs when $u_o = 15.8589$. As shown in Figure 14.6 (see Algorithm B.4.10), the power-law representation improves for larger $u_o$ for $C$ and $n$ calculated using Eq. (14.13).

The corresponding range equation can be determined from the distance $R(E)$ an electron with energy $E$ travels by using Eq. (14.13) to find

$$R(E) = ba_o \int_0^{E/E_c} \frac{u}{\ln(1+u)} du \tag{14.14}$$

where $ba_o = 0.114371$ nm, by setting $E \rightarrow E_o$ (the energy of the primary electron). This equation can also demonstrate that the majority of secondary electrons are generated near the maximum range $R(E_o)$ rather than distributed along the path of the primary. Let the primary path be a straight line into the material (a reasonable approximation). Secondaries that it generates are orthogonal to that path, and the generation of each secondary removes $\Delta E \approx 2.5E_g$ from the primary's energy, where $E_g$ is the bandgap energy [154, 189]: for diamond, $E_g = 5.5$ eV and $\Delta E = 13.75$ eV. After the generation of the $j^{th}$ secondary, the energy of the primary electron is $E_j = E_o - j\Delta E$, and so finding $R(E_j)$ shows how the primary electron comes to a stop (where it sheds most of the secondaries it generates), with

$$R(E_j) = R(E_{j+1}) - ba_o \int_{u_{j+1}}^{u_j} \frac{u}{\ln(1+u)} du \tag{14.15}$$

1 Note that the notation $n$ of Ref. [311] is $n-1$ here.

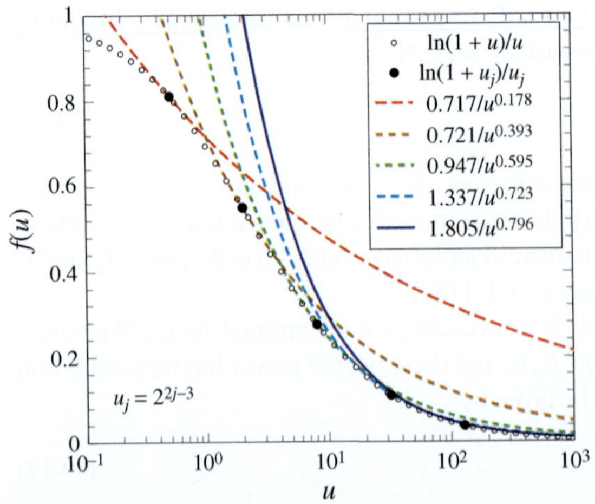


**Figure 14.6** Comparison of the Bethe-like approximation $f(u) = \ln(1 + u)/u$ to the power-law approximation $f_j(u) = C_j/u^{n_j-1}$, where $C_j$ and $n_j$ are determined at $u_j = 2^{2j-3}$ for $j \in (1, ..., 5)$ using Eq. (14.13).

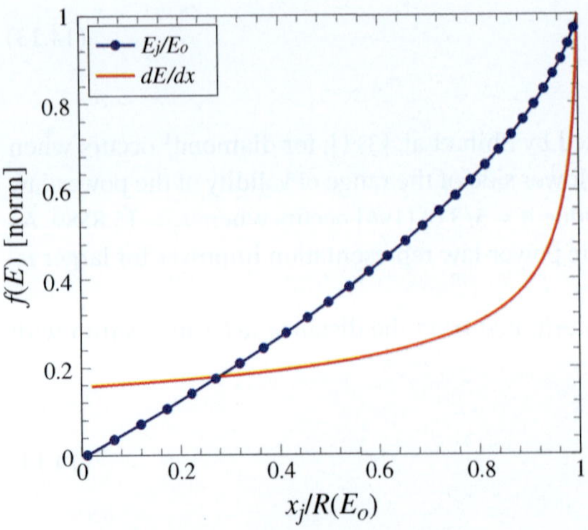


**Figure 14.7** Shedding of secondaries for $E_o = 1.9937$ keV for $E_o/\Delta E = 145$ generated secondaries. The number of secondaries generated per unit length is proportional to $dE/dx \approx (E_{j+1} - E_j)/(x_{j+1} - x_j)$. For $E_j$, a mark (•) is shown for every $5^{th}$ collision. Compare $dE/dx$ curve to Figure 1 of Ref. [311].

The factor $dE/dx$ is proportional to the number of electrons generated per unit length, and as Figure 14.7 (see Algorithm B.4.10) shows, it spikes near $x \approx R(E_o)$, that is, near where the primary comes to a stop.

The Bethe model therefore allows for the approximation of $\delta(E_o)$ using the approximation of Eq. (14.2), but a subtlety is involved: as Figures 14.6 and 14.8 (see Algorithm B.4.10) indicate, both $dE/dx \propto 1/E^{n-1}$ and $R(E_o) \propto E_o^p$, but the best fit values of $(n, p)$ are not related by $p = n + 1$ because the function $u/\ln(1 + u)$ affects each differently. In practice, then, determining each separately is the better approach. It then follows

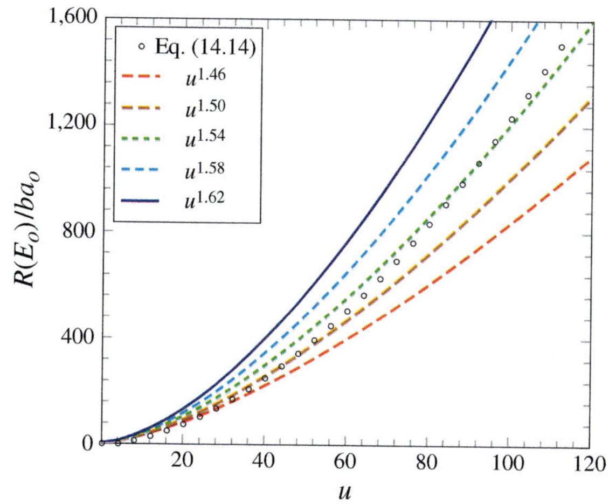


**Figure 14.8** Comparison of $R(E_o)/ba_o$ using Eq. (14.14) to various $u^p$ with $p$ shown in the legend. *Source:* After [157] / AIP Publishing LLC/CC BY 4.0.

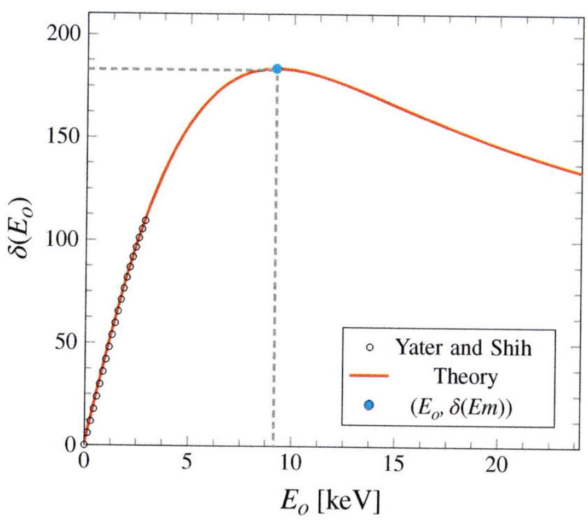


**Figure 14.9** Yield from H/C(111) Diamond from the data of Yater and Shih [189, 367]. Values used in the theoretical model (Eq. 14.16) are $(n, p) = (1.65, 1.5)$ and $(B', \gamma') = (3460, 1/800)$. It is found $\delta(E_m) = 183$ at $E_m = 9.17$ blue •. *Source:* After [157] / AIP Publishing LLC/CC BY 4.0.

$$\begin{aligned} \delta(E_o) &\approx B\frac{CE_c}{n\gamma ba_o}\left[\frac{1-\exp(-n\gamma ba_o u_o^p)}{u_o^{p-1}}\right] \\ &= \frac{B'}{n}\left[\frac{1-\exp(-n\gamma' u_o^p)}{u_o^{p-1}}\right] \end{aligned} \tag{14.16}$$

where the terms of Eq. (2.27) are carried forward here. Let $B' = BCE_c/\gamma'$ and $\gamma' = \gamma ba_o$. Choose $n$ and $p$ from the regimes of Figures 14.6 and 14.7 that match the experimental range of interest for $u_o$. If $dE/dx \approx \eta/u^{n-1}$ (after Eq. 2.25) in the energy range of $E_o$, with $\eta \sim O(1)$, then it can be used to redefine of $b$ in Eq. (14.11) if the right-hand side is replaced by $(-1/u^{n-1})$ when the power-law form is used to find $\delta(E_o)$. The decay factor $\gamma$ and the other parameters in $\delta$ can be found from experimental data or using other means like Monte Carlo (if the scattering factors are known) [58, 353]. For present purposes, consider the data of Yater and Shih [189, 369]. For matching the experimental data, values of $n$ and $p$ are needed for $u_o \sim 30$. Let $n = 1.65$ and $p = 1.5$. Adjusting $\gamma'$ until experimental data and theoretical curves sufficiently coincide gives $\gamma' \approx 1/800$. Setting $B' = 3460$ enables the last experimental data point at $E = 2.89$ keV to coincide with the theory prediction there. As shown in Figure 14.9, $\delta(E_m) = 183$ occurs at $E_m \approx 9.17$ keV demonstrating that the yield of diamond is exceptionally high, and generating interest in it as a possible electron source for a diamond amplifier [21, 317, 367, 368].

## 14.4 Exercises

**Exercise 14.1** Show that the location of the maximum of $y(x) = xG_n(x)$ for $G_n(x)$ given by Eq. (14.2) can be found by iteratively solving the equation $s = \ln\{[\ln(s+1) - 1]/(n-1)]\}$. For $n = 1.35$, what is the asymptotic solution $s_o$? What is $y[x(s_o)]$? What value of $x$ is entailed by that solution? Give answers to 5 significant digits.

**Solution** $s_o = 2.2077$. $x(s_o) = 1.7789$. $y[x(s_o)] = 0.71715$.

**Exercise 14.2** Let Eq. (14.4) be written as $dE/dx = \Delta^n / LE_o^{n-1}$. For diamond, use mass of C-12 atom, $\rho = 2.2$ g/cm$^3$, $n = 5/3$, Kanaya and Okayama's $\lambda_s$, and $\Delta = 2.5E_g$, to find $L$ in units of Bohr's radius $a_0$.

**Solution** $L = 0.2587a_0$.

**Exercise 14.3** For diamond and a primary electron energy at the $E_m$ value of Yater and Shih [367] (location of maximum yield), find $R(E_o)$ using Kanaya and Okayama model and parameters of Ex. 14.2. What is it if $E_o = 3$ keV?

**Solution** $R(E_m) = 418.13$ nm. $R(E_o) = 64.947$ nm.

# 15

# Space Charge

*... this haste was hot in question.*
*And many limits of the charge set down ...*

W. Shakespeare, *Henry 4.1* I.1

## 15.1 Child–Langmuir Law

In a one-dimensional problem, charge in the anode–cathode (AK) gap is visualized as sheets of charge, as that model is computationally simple to work with [25]. A sheet of charge gives rise to a constant field on either side of the sheet, as can be demonstrated using a Gaussian pillbox argument. It is based on the divergence theorem for a vector field $\vec{A}(\vec{r})$ which satisfies

$$\int_{\mathcal{V}} \vec{\nabla} \cdot \vec{A}\, d\mathcal{V} = \int_{\Omega} \vec{A} \cdot d\vec{S} \tag{15.1}$$

where $\mathcal{V}$ is a volume and $\Omega$ is the surface of that volume. In carefully crafted situations where the value of $\vec{A} \cdot \hat{n}$ (the normal of $\vec{A}$ to the surface) is constant over chosen sections of the surface, the latter integral is easily performed. The pillbox, a cylinder with flat ends aligned so that its axis of symmetry is normal to the plane bisecting the pillbox, is a convenient surface for a sheet of charge, but a point charge is simpler, and so consider it first.

For a point charge (like an electron, but the sign right now does not matter), $\vec{A}$ is the electric field $\vec{\mathcal{E}} \cdot \hat{r} = \mathcal{E}$ when the charge $q$ is at the origin. The differential surface area is $d\vec{S} = r^2 \sin\theta\, d\theta\, d\phi\, \hat{r}$. The potential energy between the point charge and a test charge depends only on the magnitude of their separation, and so the electric field is radial, and its magnitude is likewise dependent upon the distance of separation. Relating Poisson's equation to the surface integral when the magnitude of the electric field is constant over the sphere gives

$$\frac{q}{\varepsilon_0} = \int_{\mathcal{V}} \vec{\nabla} \cdot \vec{\mathcal{E}} d\mathcal{V} = \mathcal{E} \int_{\Omega} r^2 \sin\theta d\theta d\phi = 4\pi r^2 \mathcal{E} \tag{15.2}$$

where $\varepsilon_0$ is the permittivity of free space as usual. And so, $q\mathcal{E} = q^2/4\pi\varepsilon_0 r = 4Q/r$ for a point charge, as expected. Electric fields add by superposition, and more complex arrangements can be systematically constructed.

*Fundamentals of Electron Emission Physics*, First Edition. Kevin L. Jensen.

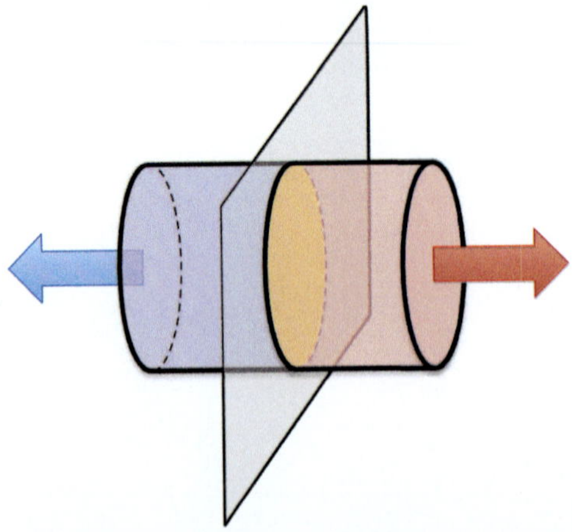

**Figure 15.1** Gaussian cylinder (pillbox) enclosing a portion of a sheet of charge (yellow oval). Only non-zero electric fields at the surface are to the right (red) and to the left (blue): no electric fields are normal to the sides of the cylinder by symmetry (moving the cylinder in the plane of the sheet of charge would not change anything). Charge enclosed is $\sigma\ dA = \rho\ dA\ dx$.

With that success, now consider a cylinder for which the axis of symmetry is along the $\hat{x}$-axis and that bounds a small area $dA$ of a sheet of charge with a charge density $\sigma$ [78, 151], as in Figure 15.1. Because the sheet of charge only has electric fields normal to its plane, there are no electric fields normal to the cylinder sides ($\vec{\mathcal{E}} \cdot \hat{n} = 0$ there), only at the cylinder ends, each one pointing along the cylinder axis $\hat{z}$. Letting $\vec{\mathcal{E}} \cdot \hat{z} = \mathcal{E}$, then $\int_\Omega \vec{\mathcal{E}} \cdot d\vec{S} = 2\pi R^2 \mathcal{E}$, where $R$ is the radius of the cylinder. The amount of charge on the sheet is $q\sigma\pi R^2 = q\rho\pi R^2 dx$. Therefore, the electric field magnitude on either side of the sheet of charge is $F = q\mathcal{E} = q^2\sigma/2\varepsilon_0 = q^2\rho\ dx/2\varepsilon_0 = 8\pi Q\rho\ dx$, which is independent of distance, where $\sigma$ and $\rho$ are the number of charges per unit area and per unit volume, respectively. Units are checked by recalling that $F$ has units of [eV/nm], $\sigma$ of [1/nm$^2$], and $Q$ of [eV nm].

By superposition, then, if more than one sheet is in the AK gap, the total electric field is the sum of the electric fields for each sheet, and so if the field due to one of the sheets gives $dF$, then

$$dF = \frac{q^2}{K_s \varepsilon_0} \rho(x) dx$$

where $K_s$ is the dielectric constant of the gap region ($K_s = 1$ for vacuum). This can be rewritten in [efnq] as

$$\frac{dF}{dx} = \frac{d^2}{dx^2} V(x) = \frac{16\pi Q}{K_s} \rho(x) \tag{15.3}$$

where $V(x) = q\varphi(x)$ is a potential energy, when $\varphi(x)$ is a potential, and the sign of the equation is a consequence of the negative charge of the electron. The solution depends on whether number density $\rho(x)$ is constant (e.g. a crystal lattice with uniform doping) or whether current density $J(x)$ is constant as for steady-state emission from an electron source.

For practice, consider a constant number density $\rho(x) = \rho_o$ first. For a gap of width $D$, a double integration yields [154]

$$\frac{V_a}{D} = F_c + \frac{8\pi Q D}{K_s} \rho_o \tag{15.4}$$

where $F_c$ is the field at the boundary between metal and the (doped) semiconductor. The depletion width $w$ is defined by the conditions $F_c = 0$, $\rho_o \to N_d$ (the doping level), and $D \to w$ to obtain

$$V_a = \frac{8\pi Q w^2}{K_s} N_d \tag{15.5}$$

where $V_a$ is the barrier preventing electron injection from a metal into a semiconductor thin film. The shape of $V(x)$ is parabolic with the origin at the metal-semiconductor interface, and the minimum occurs where $x = w$, beyond which $V(x)$ is flat. As an example for diamond parameters [368], let $V_a = 4.64$ eV, $K_s = 5.7$ and assume a boron doping density of $N_d = 4 \times 10^{13}$ atoms / cm$^3$. It then follows that $w = 8.5487$ $\mu$m. Were the doping to drop to $N_d \to 10^{16}$ atoms / cm$^3$, then $w = 540.67$ nm results [157].

The Child–Langmuir equation is associated with constant current density. To find $J_{CL}$, use the relation $J = qv(x)\rho(x)$, where $v(x)$ is the electron velocity in vacuum, where $K_s \to 1$. Solving Poisson's equation for the space charge problem is widely treated [25, 223, 87, 302, 154, 376]. It amounts to writing

$$\frac{d^2}{dx^2} V(x) = \frac{dF}{dx} = \frac{1}{2} \frac{dF^2}{dV} \tag{15.6}$$

Conservation of energy requires $(1/2)mv(x)^2 = V_m + V(x) \equiv U(x)$, where $V_m$ is the potential energy at the cathode and set to 0. Because potential energy rather than potential is under discussion here, the anode is at a lower potential energy than the cathode (electrons "fall" to the anode), or $V(D) = -q\varphi(D) = -V_a$ where by convention $V_a$ is taken as a positive quantity so as to not run afoul of sign errors later. Poisson's equation is therefore rewritten as

$$dF^2 = \frac{16\pi Q}{q} J \left[\frac{2m}{U}\right]^{1/2} dU \tag{15.7}$$

the integration of which gives

$$F^2 - F_m^2 = \frac{16\pi Q J \sqrt{2m}}{q} \sqrt{U} \tag{15.8}$$

where $F_m = F(0)$. When $F_m = 0$, then using $F \to dU/dx$, a second integration from $x = 0$ to $D$ and $U$ from 0 to $V_a$ recovers the Child–Langmuir result $J = J_{CL}$ of Eq. (2.33), or

$$J_{CL}(V_a, D) = \frac{q}{18\pi Q \sqrt{2m}} \frac{V_a^{3/2}}{D^2} \tag{15.9}$$

As an example, for $V_a = 1$ eV and $D = 1$ $\mu$m, $J_{CL} = 1.45674 \times 10^{-8}$ q/fs nm$^2$ = 233.395 Amp/cm$^2$, corresponding to about 15 electrons every picosecond from a square micron area.

Because current rather than current density is measured, it is common to see the CL equation in terms of a perveance $P$ such that $I_{CL} = J_{CL} A_o = P\varphi_a^{3/2}$, where $A_o$ is the emission area, indicating that for $I_{CL}$ measured in amperes, then

$$P = 2.333952 \times 10^{-6} \left(\frac{q}{\text{eV}}\right)^{3/2} \frac{A_o}{D^2} \tag{15.10}$$

Observe that [q/eV] = [1/volt], and so the [efnq] and [MKSA] representations are synergistic with $V_a = -q\varphi_a$, with the former measured in [eV] and the latter in [volts]. Perveance becomes a useful parameter for convergent beams in linear beam devices such as microwave amplifiers (or "tubes" as they are somewhat anachronistically called) [107]: in a Pierce gun, the focusing mimics a portion of space-charge limited flow between two concentric spheres. In cylindrical or spherical coordinates, the factor $D$ in Eq. (15.9) changes. If the radius of the cathode and anode surfaces are $R_c$ and $R_a$, respectively, then letting $\eta$ designate a function of the ratio of the radii $R_c/R_a$ gives [221, 222, 376]

$$D^2 \rightarrow \begin{cases} [R_c \ln(\eta)]^2 & \text{(cylindrical)} \\ \eta^2(R_c - R_a)^2 & \text{(spherical)} \end{cases} \tag{15.11}$$

where the function $\eta$ depends on $R_a/R_c$. The Langmuir-Blodgett form of the space charge limited current for concentric cylinders is

$$J_{LB} = \frac{4}{9}\varepsilon_0 \left(\frac{2q}{m}\right)^{1/2} \frac{\varphi^{3/2}}{R_c^2 \eta^2} = \frac{qV_a^{3/2}}{18\pi Q\sqrt{2m}R_c^2\eta^2} \tag{15.12}$$

and $\eta$ given by a table for a range of values of $R_a/R_c$. Zhu et al. [380] have found that $J_{LB}$ for both cylindrical and spherical geometries is in fact well approximated over a sizable range of values $R_a/R_c$ by

$$J_{LB} \approx \frac{q}{18\pi Q\sqrt{m}} \frac{\mathcal{E}^{3/2}}{\sqrt{D}} \tag{15.13}$$

where $D$ is the gap spacing between anode and cathode and $F = q\mathcal{E}$ is the field at the surface. Lastly, for a tip-to-plate geometry, where $D$ is the distance to the anode from the apex of the tip and $R = \eta D$ the radius of the tip [331]

$$\frac{J_{t-p,CL}}{J_{CL}} = \frac{\eta(\eta+1)}{\left[\ln\left(\sqrt{1+\eta}+\sqrt{\eta}\right)\right]^2} \tag{15.14}$$

The prior analytical results are made possible by working in appropriate orthogonal coordinate systems familiar from typical problems found in electrostatics [6, 78]. There are problems that do not fit in such a tidy picture, but are more relevant to how electron beams are actually generated and used. In the case of a flat emission strip of width $W$ and infinitely long, with $D$ again the distance to the anode, then beam optics simulations using particle-in-cell (PIC) codes are required to identify the limiting two-dimensional current $J_{2D-CL} \equiv J(2)$. Its ratio with the one-dimensional $J_{1D-CL} \equiv J(1)$ was found using the OOPIC and MAGIC particle-in-cell codes to be [236]

$$\frac{J(2)}{J(1)} = 1 + 0.3145\left(\frac{D}{W}\right) - 0.0004\left(\frac{D}{W}\right)^2 \tag{15.15}$$

A magnetic field was imposed to constrain the flow, but the behavior of $J_{2D-CL}$ was found to be virtually independent of the field's magnitude. What is more interesting is that, in the absence of a magnetic field, the current showed oscillatory behavior by expanding and contracting, so that the space charge limit was an average behavior. Subsequently, Lau [223] showed that the analytic relation for $W \gg D$ was

$$\frac{J(2)}{J(1)} = 1 + \frac{D}{\pi W} \tag{15.16}$$

and that in practice, his equation was surprisingly good for $W/D$ as small as 1/10. In the case of emission restricted to a circular patch of radius $R \ll D$, then $\pi \rightarrow 4$ in the second term, in agreement with later PIC simulations [238]. Zhang et al. [377] argued that an extension to a 3D relation for space charge limited flow from a patch could be written as

$$\frac{J(3)}{J(1)} = 1 + \mathcal{F} \times \mathcal{G} \tag{15.17}$$

where the dynamical term $\mathcal{F}$ is a measure of the mean position of the electrons in various operating regimes and $\mathcal{G}$ is a geometrical term related to the size and shape of the cathode patch emitting current.

## 15.2 Space Charge and Surface Fields

Oscillatory behavior can manifest in simulations of photoemission from a hemisphere on a plane in an AK gap, the dimensions of which are large compared to the size of the hemisphere. In particle-in-cell codes, the charge of the macro-particles is indicated by color (passing through the spectrum with red denoting high charge content down to blue denoting low charge content). Because the field is strongest on the apex of the hemisphere, and because of Schottky barrier lowering there, emission is generally greatest at the apex if space charge is not present. Oscillations in the amount of charge at the apex are therefore an indication of the oscillatory nature of the apex field caused by space charge: sheets of varying charge density cause changes in the surface field as they reach the anode and are absorbed. Moreover, the cylindrical radius of the beam changes size in an oscillatory manner as well, although that is less discernible. All these oscillations are visible in Figure 15.2 for PIC simulations (MICHELLE) [280], where periodic boundary conditions on the top and bottom are assumed, and the simulation region is $\sim 10\times$ larger than the window shown.

Although Child's law is understood as space charge-limited current and entails that $F_m$ vanishes at the surface of the cathode, space charge-affected current is when it does not. The emission equations, even in their simplest forms of Section 2.2, exhibit some dependence on $F$, and so in general the field at the surface $F_m$ and the current density $J$ emitted by the cathode find a happy medium where neither is zero. Unfortunately, the integrals become harder [14, 87, 154, 213, 302], but some elegance in the manifestation of their interdependence is possible. Start from Eq. (15.8), and introduce dimensionless coordinates $(f, j)$ defined by

$$F_m \equiv fF_o; \quad J \equiv jJ_o \tag{15.18}$$

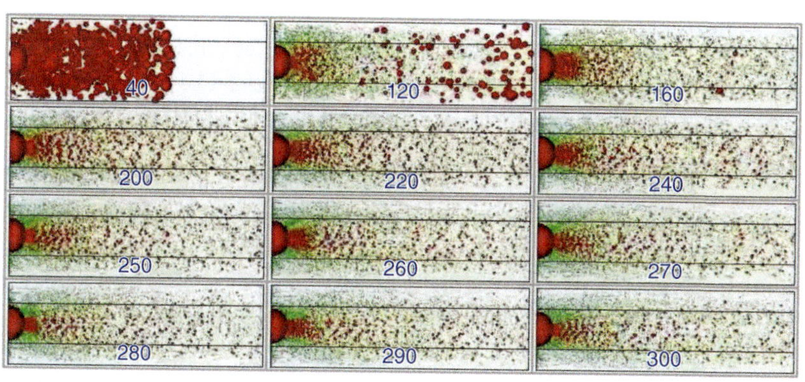


**Figure 15.2** Photoemission from a hemisphere on a plane (left) to a flat anode (right) at a distance $D$, where the length of the view region is $D/10$, for a field $F = V_a/D \sim 10^{-3}$ eV/nm, as generated using the PIC code MICHELLE. Time step is in blue on the bottom center. High charge macro particles are red, low charge are blue. After the gap is filled, the amount of charge emitted at the apex of the hemisphere oscillates as the field there oscillates. Courtesy of J. Petillo (Leidos).

where $F_o = V_a/D$ and $J_o = (16/9)J_{CL}$ (see Eq. (15.9)). Introduce the dimensionless term $y$ defined as the ratio of $F$ to the field in the AK gap when no charge is present, or $F \equiv yF_o$, and let $u$ do the same with potential energy or $U = uV_a$. Solving Eq. (15.8) for $y$ then gives

$$y = \sqrt{f^2 + j\sqrt{u}} \tag{15.19}$$

A second integration using $F = dU/dx = (V_a/D)(du/dy)$ for $0 \leq y \leq 1$ and $0 \leq u \leq 1$ gives[1]

$$1 = \frac{4}{3}\frac{\sqrt{f^2+j} + 2f}{\left(\sqrt{f^2+j}+f\right)^2} \tag{15.20}$$

Solving for $j$ now gives

$$j = \frac{4}{9}\left[2 + (2-3f)\sqrt{1+3f}\right] \tag{15.21}$$

Note the limiting cases: $j = 0$ when $f = 1$ (no charge in the gap), and $f = 0$, when $j = 16/9$ and Eq. (15.9) is recovered. An equivalent form [87] is

$$j\left(1 - \frac{9j}{16}\right) = 3f^2(1-f) \tag{15.22}$$

from which it is obvious that $f = (0, 1)$ corresponds to $j = (16/9, 0)$, but this more aesthetic form introduces unphysical solutions [302], making Eq. (15.21) preferable for computation.

There are two unknowns, $j$ and $f$ in Eq. (15.21), and therefore, to find a unique solution, a second equation is required. The equation governing electron emission is it. Consider some obvious candidates, like $j = af$ (linear) and $j = af^2$ (quadratic), which allow $f$ to be analytically found. For something more complex, like $j = a\sqrt{f}$ (root) or $j(f) = af^{2-\nu}\exp(-b/f)$ (FN), as suggested by Eq. (10.16), then numerical means are needed, but for pedagogical consistency, consider numerical methods applied to all. For linear, quadratic, and root cases, let $a = 16/9$, whereas for the FN cases, let

$$a = \frac{2Q}{\pi\hbar\Phi}\sqrt{2mV_a}\left(\frac{\Phi^2 e^6}{4QbF}\right)^{\nu}; \quad b = \frac{4}{3\hbar F}\sqrt{2m\Phi^3}; \quad \nu = \frac{8Q}{9\hbar}\sqrt{\frac{2m}{\Phi}} \tag{15.23}$$

The unique solution is found graphically by finding the intersection of the emission $j(f)$ (call it $j_e(f)$) with the space charge $j(f)$ from Eq. (15.21) (call it $j_c(f)$), as in Figure 15.3 (see Algorithm B.4.11). That intersection point is the root of $j_c(f) - j_e(f) = 0$, shown in Figure 15.4 (see Algorithm B.4.11).

Modern numerical packages [e.g. MATLAB [248]] generally contain spline interpolation functions that require only a small number of evaluations of a function to form a good fit to lines close to linear or parabolic. An efficacious way to proceed, then, is to evaluate $j_c(f) - j_e(f)$ for a small and uniformly spaced number of $f$ given by $f_i = i/N$, with the cases $\Delta j_i = j_c(f_i) - j_e(f_i)$ for $i = (0, N)$ being trivial. The spline of $f_i$ vs. $\Delta j_i$ can then quickly return that value of $f$ for which $\Delta j = 0$. Generally, a value of $N \approx 50$ returns better than six-digit accuracy. Using that method, the intersection points in Figure 15.4 are found. Alternate methods for the same result can use the method of bisection to zero in on the root, then linear (Newton's method) or quadratic interpolation to finally evaluate the root to high accuracy [157], a method that reduces the overall number of evaluations of $j_c(f)$ and $j_e(f)$ that are required.

1 Although the present treatment is motivated by Ref. [302], $j$ here differs from there because $J_o$ differs, introducing factors of 9/16 in Eq. (15.22).

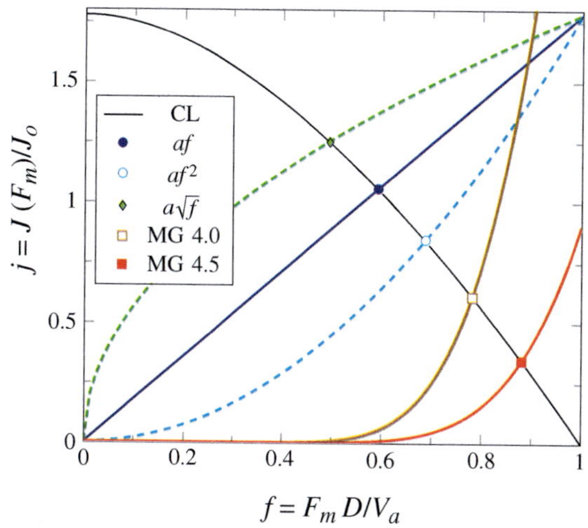


**Figure 15.3** The intersection of $j(f)$ from Eq. (15.21) with $j(f)$ from sample emission equations given by (*i*) linear ($j = af$), (*ii*) quadratic ($j = af^2$), (*iii*) root ($j = a\sqrt{f}$), and (*iv*) two cases of MG-like $j = af^{2-\nu}e^{-b/f}$ (Eq. (10.16)). Cases (*i*) – (*iii*) use $a = 16/9$. The MG cases use $\Phi = 4.0$ and 4.5 eV, $V_a = 5500$ eV and $V_a/D = 10$ eV/nm in Eq. (15.23). After Ref. [157]/AIP Publishing LLC/CC BY 4.0.

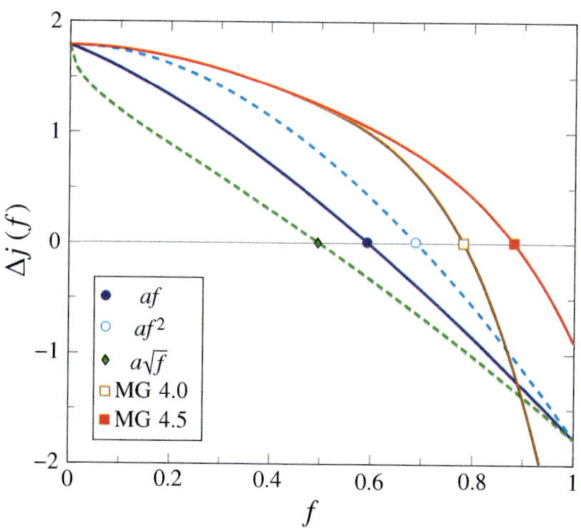


**Figure 15.4** Recasting of Figure 15.3 as the difference $\Delta j(f) = j_c(f) - j_e(f)$. Splining a number of pairs of points $(\Delta j_i, f_i)$ allows $f(\Delta j)$ to be found at $\Delta j = 0$.

Identifying the impact of space charge is now possible by comparing the current density in the absence of space charge (field at the cathode surface is $F_o = V_a/D_o$) compared to with space charge $F = fF_o$. As $f$ decreases from unity, the two curves part. This is most readily seen in the case of field emission, as that mechanism is much more strongly dependent on $F$ than the other emission mechanisms. Consider two cases for a cathode with different work functions $\Phi = 4.0$ eV and $\Phi = 4.5$ eV. For units of $V_a$ in eV, and letting $D_o = 500$ nm, then in [efnq] units where the unit of current density is [q / fs nm$^2$], the CL current density is

$$J_{CL} = 8.24056 \times 10^{-8} \frac{V_a^{3/2}}{D_o^2} \tag{15.24}$$

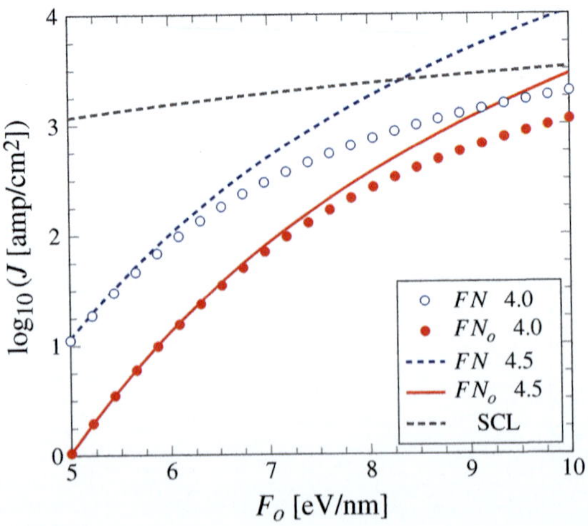


**Figure 15.5** The one-dimensional current density $J_{FN}(F)$ for two cases of work function $\Phi = (4.0, 4.5)$ eV (blue ∘ and red • symbols) compared to $J_{CL}$ (gray dashed line) for an AK gap of $D_o = 500$ nm. Lines use $F = F_o$ instead of $F = fF_o$ in $J_{FN}$. SCL is space charge limited, or $J_{CL}$.

To convert to the conventional [Amps/cm$^2$], multiply by $1.60218 \times 10^{10}$. Performing the calculation gives the results shown in Figure 15.5 (see Algorithm B.4.11). The gray line shows $J_{CL}(V_a, D_o)$: that $J_{FN}(F_o)$ crosses it shows how rapidly field emission can approach the space charge limit. When the argument is $fF_o$, though, the impact of space charge on the surface field suppresses emission and keeps $J_{FN} < J_{CL}$.

The same behavior survives in measured data, even though field emission generally requires very sharpened emitters to generate the necessary fields in the GV/m range. Such data was generated by examining tungsten wires under high anode voltages, measurements of great interest in the mid-$20^{th}$ century given strong interest field emission initiated vacuum arcs between the AK electrodes for flat surfaces that sported microprotrusions, thereby limiting the voltage stand-off that could be endured between electrodes. In a series of studies, Dyke and Trolan [72, 75] themselves and separately with Dolan et al., (1953) examined how field emission changed with the onset of large apex currents from a tungsten wire and how rapid heating of the tip preceded arc formation. Given the heightened sensitivity of field emission on surface field, these emitters provide a natural place to examine how Figure 15.4 manifests itself in practice. Consider first representative data from Figure 5 of Ref. [61], where pulsed and direct current measurements were made of current from a conical tungsten tip that had an apex radius under a micron. Direct current measurements tended to heat the apex of the emitter, so larger currents were obtained by pulsing the emitter. The measured data is shown in Figure 15.6 (see Algorithm B.4.11), and Fowler–Nordheim-like fits to curves of the form $\ln(I_a/\varphi_a^2) \approx A + B/\varphi_a$, where $\varphi_a$ is the magnitude of the anode voltage, give a signature of field emission. The pulsed current shows the onset of the impact of space charge in a manner reminiscent of Figure 15.5 after some modifications to the analysis. To account for space charge effects for a needle-like geometry, Eq. (15.13) is used in place of the one-dimensional Eq. (15.9). In order to use that equation, an estimate of the field at the apex of the tungsten emitter is required. The experimental configuration is that of a needle at a distance $D_o = 0.5$ cm to the anode. If that needle is modeled by a hyperboloid in prolate spheroidal coordinates, then Eq. (3.36) can be used to approximate the apex field, with $D_o = L \cos \upsilon_o$. The radius of the apex is estimated from the electron micrographs of Dyke et al. for which a value of $a = 1200$ nm is representative. The field at the

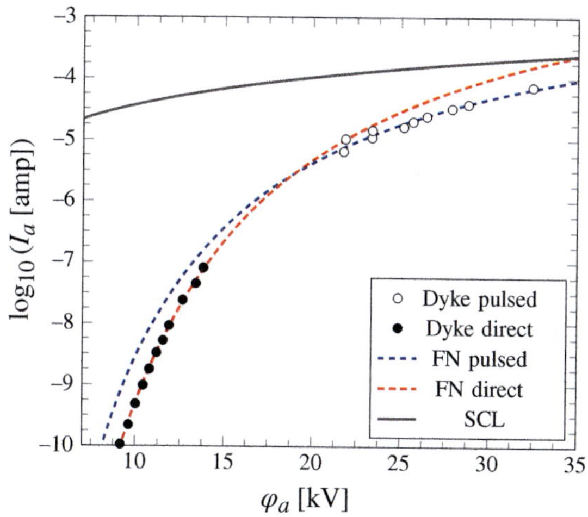


**Figure 15.6** Data from Figure 5 of Ref. [72] digitally extracted and recast as current *vs.* anode voltage $\varphi_a$. Symbols (• for direct measurement, ∘ for pulsed measurement) is contrasted with FN fits. Gray line is Eq. (15.13) adapted to the experimental configuration. SCL is space charge limited, or $J_{CL}$.

apex is estimated from the hyperbolic cone model. In terms of the tip-to-anode distance $D_o$, the tip radius $a$, and the anode $V_a$, and using the analysis of Section 3.6.1

$$F_{tip} = q\mathcal{E}_{tip} = \frac{\cos(\upsilon_o)}{(\sin \upsilon_o)^2\, Q_0(\cos \upsilon_o)} \frac{V_a}{D_o} \equiv \beta_o F_o \tag{15.25}$$

where $F_o = V_a/D_o$. The angle $\upsilon_o$ is defined by the relation $\tan \upsilon_o = \sqrt{a/D_o}$. For the parameters of Dyke and Trolan of $V_a = q\varphi_a = 35$ keV, $D_o = 0.5$ cm, and taking the apex radius to be $a = 1200$ nm, then $\upsilon_o = 0.015491$, $F_{tip} = 6.0013$ eV/nm, $\beta_o = 857.3283$ and $\beta_o/D_o = 1714.6565\ \text{cm}^{-1}$. Compare the last quantity to the estimation by Dyke et al. that $\beta_o/D_o = (9.9 \pm 1.5) \times 10^3\ \text{cm}^{-1}$: although not equal, they are comparable and reflect differences in the model of the emitter used. The entire hemisphere at the apex does not field emit because the field off-axis drops quickly as per Eq. (3.36): assume that an area $\pi r_o^2$ provides the notional emission area and let $r_o \approx 100$ nm. The area of emission is therefore crudely $A_o = \pi R_o^2 (r_o/R_o)^2 = \pi r_o^2$. As a result, for $\varphi_a$ in [volts/cm] and $D_o$ in [cm], then

$$I_{LB} = J_{LB} A_o = 3.30071 \times 10^{-6} \frac{(\beta_o V_a/D_o)^{3/2}}{\sqrt{D_o}} \pi r_o^2 \tag{15.26}$$

for $I_{LB}$ in units of amps, $V_a$ in eV, and $r_o$ in the same units as $D_o$. This equation gives the gray line marked SLC in Figure 15.6, which begins to approach the FN curves and is therefore responsible for the change in the FN fit between the pulsed and direct lines.

A similar analysis but now allowing different work functions to characterize the apex of the tungsten emitters is indicative of the impact of space charge for the emitters of Barbour al. [14], where the work function on the tips was reduced by increasing the amount of cesium on their surface. Cesiated surfaces have increasingly lower work functions as the submonolayer coverage of cesium atoms approaches unity [118, 149, 161, 336]. As more cesium is deposited on the tips, the work function is lowered to the value shown in the legend of the figure. The data of Barbour et al. is shown in Figure 15.7 in the form of a Millikan plot [89, 254], whereby the log of the current $I_a$ (or current density $J$) is compared to the inverse of the anode potential $\varphi_a$ (or inverse field $1/F$). The estimates

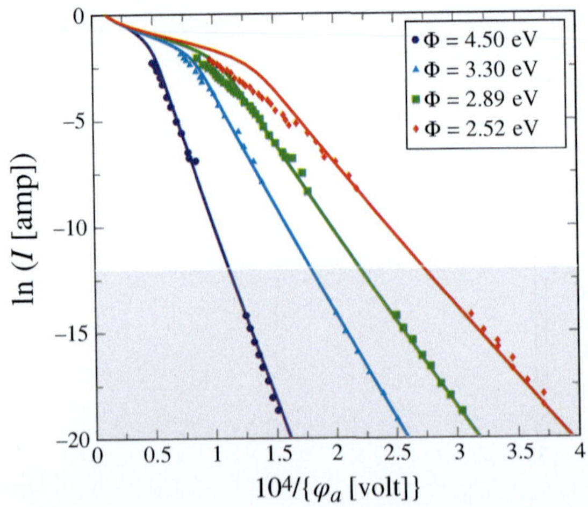


**Figure 15.7** Millikan plot of measured field emission current from a tungsten wire emitter increasingly covered with cesium to lower its work function. (symbols) Data digitally extracted from Figure 3 of Barbour et al. [14]. Points in gray region are direct current, and in white region are pulsed current. Direct current values were used to estimate the Φ shown in the legend. (lines) theoretical space charge model. After [151] (see also [170]).

of the work function Φ are based on the direct current measurements. Because the relevant space charge model is no longer one-dimensional, the theoretical space charge lines were determined using three-dimensional model of a field emitter using the point charge model of Section 3.2 and a revised space charge model capturing how often electrons are emitted from the apex and how rapidly they move away from the apex where their influence rapidly diminishes [151].

## 15.3 Transit Time Approach

Recall that a surprisingly close approximation to $J_{CL}$ was obtained in Section 2.2.5, namely Eq. (2.32), by assuming that enough charge was instantaneously injected into the AK gap to create a sheet of charge the field of which completely suppressed the field at the surface and which crossed the AK gap in the same time (the transit time) as an isolated charge would take. Estimates of the space charge limit can then be made by looking at transit time estimates [170, 345], and indeed, an impressive collection of studies makes implicit or explicit use of transit time, sheet approximation, and pulsed emission concepts [25, 46, 116, 117, 187, 201, 205, 237, 288, 345, 346, 377]. The techniques therefore merit some consideration.

Current being the amount of charge flowing in an amount of time creates an expectation that models providing reasonable estimates for each quantity can give crude estimates of the space charge limit. To begin, revise the earlier single-sheet model to consider the emission of many sheets of charge per unit area $q\sigma$ ($\sigma$ is a number density). Recall from Section 15.1 that the field $F$ from an isolated sheet of charge is constant, independent of position, and given by $F = q\mathcal{E} = q^2\sigma/2\varepsilon_0 = 8\pi\sigma Q$. The transit time argument is more than just pedagogically useful: it motivates an approach to space charge using "vacuum capacitance" [345] and is extensible to treat relativistic and quantum mechanical complications [4], or inferring how quantities scale for the useful cylindrical (2D) and spherical (3D) geometries to which the Child–Langmuir equation has been extended. In photoemission, where the shape of the beam bunch can be exquisitely controlled by the profile of the laser pulse, a pulse duration less than the transit time can create more current than allowed by $J_{CL}$, which causes a THz oscillatory behavior seen in a simulation of a train of electron bunches [277].

There are a few transit time candidates to choose from. Consider the simplest first, the ballistic transit time $\tau_o$ representing how long a single electron takes to get from one side of the AK gap to the other in the presence of a constant force $F_o$. It is given by the solution to the ballistic equation $D = v_o\tau + (F/2m)\tau^2$, or

$$\tau = \sqrt{\frac{2mD}{F_o}}\left[\sqrt{\delta} + \sqrt{1+\delta}\right]^{-1} \tag{15.27}$$

where the small dimensionless quantity $\delta = mv_o^2/2F_oD$ compares the initial kinetic energy to the energy acquired from crossing the AK gap. Let $v_o$ be approximated by the average velocity $\langle v\rangle$ of an emitted electron. Thermionic cathodes experience comparatively low fields, photoinjectors have midrange fields, and field emitters have very high fields. Therefore, consider parameters typical of the extremes. For thermal emission, the average velocity is

$$\langle v\rangle = \frac{\hbar}{m}\langle k_x\rangle = \frac{\hbar}{m}\frac{\int_0^\infty k_x e^{-\hbar^2k_x^2/2mk_BT}dk_x}{\int_0^\infty e^{-\hbar^2k_x^2/2mk_BT}dk_x} = \sqrt{\frac{2k_BT}{\pi m}} \tag{15.28}$$

For a temperature of $T = 1500$ K, an AK gap of $D = 250$ $\mu$m (or $2.5\times10^5$ nm), and an anode potential of 1000 kV (or $F_o = q\varphi_a/D = 0.004$ eV/nm), then $\delta = 4\times10^{-5}$. For field emission, many terms cancel in both the numerator and denominator of the integral ratio of current density with density (recall the methods of Section 8, leaving

$$\langle v\rangle = \frac{\hbar}{m}\langle k_x\rangle = \frac{\hbar}{m}\frac{\int_0^{k_F} k_x(k_F^2-k_x^2)e^{\beta_F\hbar^2k_x^2}dk_x}{\int_0^{k_F}(k_F^2-k_x^2)e^{\beta_F\hbar^2k_x^2}dk_x} \approx \frac{\hbar\, k_F}{m} \tag{15.29}$$

where because the integrands are so sharply peaked about $k \approx k_F$, it is adequate to let the single $k_x$ in the numerator integrand take on its value at the maximum of the integrand and be pulled out of the integral, allowing the remaining portion to cancel the denominator. A tunneling electron at the Fermi level is then $\langle v\rangle \approx \hbar k_F/m$ or 1.56919 nm/fs for $\mu = 7$ eV (copper) is an upper bound, for which the kinetic energy is comparable to $\mu$. Although this may appear perplexing (how could a tunneling electron emerge from a barrier with a velocity comparable to the Fermi level?), in fact, trajectory models of tunneling Gaussian wave packets through a rectangular or triangular barrier paradoxically suggest just that [181], whether or not it is easy to accept, and indicative of the general mysteries of quantum mechanics. To generate the field across a planar gap of $F_o = 5$ eV/nm and $D_o = 10$ $\mu$m, then $F_oD = 50$ keV and so $\delta = 0.007$. In either case, these initial kinetic energies are a very small fraction of the energy acquired in crossing representative AK gaps. Clearly, the neglect of the electron's initial velocity is extraordinarily convenient, and so define the ballistic transit time as simply the time for an electron starting at rest to cross the gap, or

$$\tau_o \equiv \sqrt{\frac{2mD}{F_o}} \tag{15.30}$$

A second definition of the transit time accounts for how velocity varies as a function of position via

$$\tau = \int_0^\tau dt = \int_0^D \left(\frac{dx}{dt}\right)^{-1} dx = \int_0^D \frac{dx}{v(x)} \tag{15.31}$$

A trivial example is a particle accelerated from rest by a constant force $F = mg$ so that $x = (g/2)t^2$ and $v = gt$. The AK gap distance is related to that time by $D = (g/2)\tau^2$. Using $x = (g/2)(v/g)^2 = v^2/2g$, then the integral in Eq. (15.31) uses $v(x) = \sqrt{2gx}$ to find $\tau = \sqrt{2D/g}$, or the same result. With that understanding, now replace $v(x)$ in the denominator of Eq. (15.31) with $v(x) = J/q\rho(x)$ where, recall, $\rho(x)$ is the number density of charged particles. When $J$ is constant then

$$\tau = \frac{q}{J}\int_0^D \rho(x)dx \tag{15.32}$$

Recalling the "slab" argument behind Eq. (2.17), it is convenient to introduce a thick slab of charge that is the sum of all the infinitesimally thin charge sheets $q\rho(x)dx$ that make it up, and so define

$$\sigma(x) \equiv \int_0^x \rho(x')dx' \tag{15.33}$$

for which $q\sigma(D)$ is the charge per unit area within the entire gap. Notice a subtle point: before, $\sigma$ was spoken of as though it were an infinitesimally thin sheet of charge $\sigma = \rho(x)dx$, but now $\sigma(x)$ is a thick sheet (a "slab") of charge where the density varies with depth into the slab. The new manner of speaking is important to keep in mind, and reflects that outside a slab of charge, the field from the slab is the same as though all the charge were in an infinitesimal plane, a consequence of superposition and the constancy of a field outside a sheet of charge (a more familiar example, in case the sheet model appears odd, is that a sphere of charge density acts for all intents and purposes as though all the charge in it is at the center [78] for purposes of calculating forces on test charges outside that sphere). As a result, $J\tau = q\sigma(D)$, which is intuitive. On the other hand, if Poisson's equation is straightforwardly integrated twice, making use of $\sigma(x)$, it is found

$$\begin{aligned} V_a &= FD + 16\pi Q \int_0^D dx \int_0^x \rho(x')dx' \\ &= FD + 16\pi Q \int_0^D \sigma(x)dx \end{aligned} \tag{15.34}$$

Introduce the dimensionless ratio

$$N \equiv \frac{\int_0^D \sigma(x)dx}{D\sigma(D)} \tag{15.35}$$

which, using $J\tau = q\sigma(D)$, renders Eq. (15.34) as

$$V_a = FD + \frac{16\pi Q}{q} NJ\tau \tag{15.36}$$

an equation which contains the essence of the transit time approach. Two questions arise. First, what kind of number is $N$? And second, if $\tau$ is replaced by the ballistic $\tau_0$, how does that compare to the result using the actual $\tau$?

Deal with $N$ first: clearly, as Figure 2.1 suggests, the sheets making up the emitted current get accelerated toward the anode, and so spread out. Therefore $\sigma(x)$ is going to decline in some manner. Suppose that $\sigma(x) = \sigma_o(D/x)^p$ where $p$ is dimensionless, $D$ is the AK gap, $x$ increases as the anode is approached, and $\sigma_o$ is a constant. When $p$ is large, then $\sigma(x)$ declines precipitously over the AK gap, whereas for $p$ small it declines weakly. For this choice, then $N = 1/(1-p)$, or, by way of example, for $p \in (1/2, 1/4, 1/9)$ then $N \in (2, 4/3, 9/8)$. The last choice, $p = 1/9$ and $N(p) = 9/8$ holds particular importance below.

Next consider the transit time $\tau$ and look for a better estimate than the ballistic approximation. Start again with Poisson's equation after its first integration from Eq. (15.8) but with the initial kinetic energy $E_o$ included in the velocity term $v(x)$, and rewritten as

$$F(x) = \left\{ \frac{32\pi Q\sqrt{2mJ}}{q} \left[\sqrt{V(x)+E_o} - \sqrt{E_o}\right] + F^2 \right\}^{1/2} \tag{15.37}$$

where now $V(x)$ directly rather than $U(x)$, and $F$ without an argument, is the field at the cathode surface. Evaluated at $x = D$ and $E_o = 0$, Eq. (15.37) gives

$$\frac{V_a}{D} = F\left\{1 + \frac{32\pi QJ}{qF^2}\sqrt{2mV_a}\right\}^{1/2} \tag{15.38}$$

Inserting that expression into $\sigma(D) = [(V_a/D) - F]/(16\pi Q)$ (obtainable from Eq. (2.30) plus the definition of $\sigma(D)$), using the trick $(a-b) = (a^2-b^2)/(a+b)$ for general $a$ and $b$, and the definition $\tau = q\sigma(D)/J$ gives the second form of the transit time [152]

$$\tau = \frac{\sqrt{8mV_a}}{F}\left[1 + \left(1 + \frac{16\pi QJ}{qF^2}\sqrt{8mV_a}\right)^{1/2}\right]^{-1} \tag{15.39}$$

showing that as $F$ on the surface of the cathode increases, then the transit time decreases, as expected. Conversely, if the field at the surface is driven to zero, or $F \to 0$, where Eq. (15.9) holds for $J = J_{CL}$, then $\tau = 3\tau_0/2$ where $\tau_o$ using Eq. (15.30). In that limit, Eq. (15.36) becomes

$$V_a = \frac{16\pi Q}{q} NDJ_{CL}\left(\frac{3\tau_o}{2}\right) \tag{15.40}$$

for which $N = 3/4$ [152]. Compare that to if the ballistic transit time $\tau \approx \tau_o$ is used, for which $N = 9/8$, a result first anticipated by Umstattd et al. [345]. The point is, the value of $N$ when the correct form of $\tau$ is used (3/4) does not differ greatly from the value (9/8) when the convenient and simple form of $\tau_0$ is used.

A final limiting case in the transit time model is now considered. What happens if the emission current density is emitted in such a quick pulse that a single infinitesimally thin sheet of charge is generated, that is, the pulsed current transit time [152] (see also Refs. [116, 117]). Let $\sigma$ be the charge in that infinitesimal sheet, so that the number density in the gap is $\rho(x) = (\sigma/D)\delta(x - \bar{x}(t))$ where $\delta(x)$ is the Dirac delta function. Such an approximation might be useful if a strong drive laser were used to generate the injected charge (charge in excess of the space charge limit would be pushed back to the cathode). If so much charge is injected that the surface field drops to zero right away, let that limiting case be characterized by $\sigma_0$. The location of the charge sheet is then given by $x(t)$, and the fields on either side of it are constant such that

$$V_a = F_+(D - \bar{x}) + F_-\bar{x} \tag{15.41}$$

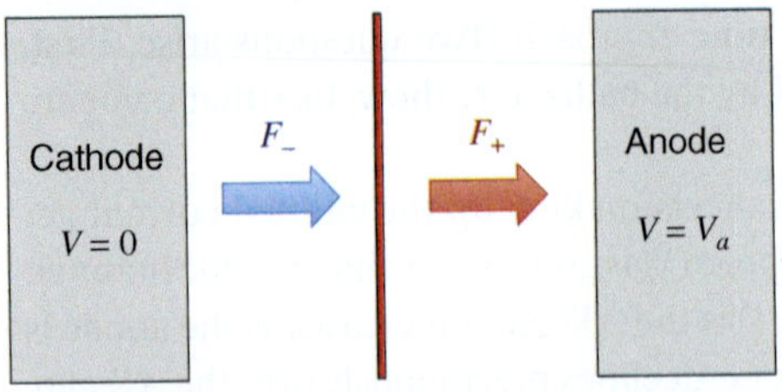


**Figure 15.8** An infinitesimal sheet of charge in an anode–cathode (AK) gap. The fields $F_\pm$ on each side of the sheet differ by an amount proportional to $\sigma$ of the sheet.

where $q^2\sigma_o = J\delta t = q^2 V_a/16\pi QD$, $\delta t \ll \tau$ is the very short duration of the laser pulse injecting the charge sheet, $F_-$ is the field between the cathode surface and the sheet, and $F_+$ the field between the sheet and anode. The fields on either side of the sheet are not necessarily equal: they are related by the Gaussian pillbox argument of Figure 15.1 which, as suggested by Figure 15.8, results in

$$16\pi Q\sigma = F_+ - F_- \tag{15.42}$$

where the sign on each term on the right-hand side is dictated by the normals on each side of the sheet of charge: one field is parallel and the other anti-parallel to the normal. With two equations and two unknowns, it follows

$$\begin{aligned} F_+ &= F_o + 16\pi Q\sigma \frac{\bar{x}(t)}{D} \\ F_- &= F_o - 16\pi Q\sigma \left(1 - \frac{\bar{x}(t)}{D}\right) \end{aligned} \tag{15.43}$$

For the special case where $\sigma = \sigma_o$, these equations simplify to

$$F_+ = F_o\left(1 + \frac{\bar{x}(t)}{D}\right); \quad F_- = F_o\frac{\bar{x}(t)}{D} \tag{15.44}$$

Use has been made of knowing that the force which accelerates the sheet of charge is the average (not the sum) of $F_+$ and $F_-$: when $\sigma$ is vanishingly small, the fields on each side are equal to $F$, and the field accelerating the sheet is $F$. As a result, the location $\bar{x}(t)$ evolves according to

$$m\frac{d^2}{dt^2}\bar{x}(t) = \frac{F_o}{2}\left(1 + 2\frac{\bar{x}(t)}{D}\right) \tag{15.45}$$

This looks like the harmonic oscillator equation, but the right-hand side is not negative, so instead of sines and cosines, sinh and cosh functions appear instead. The solution should look like $\bar{x}(t) = A\cosh(at) + B\sinh(at) + C$. Use boundary conditions, $\tau_o$ from Eq. (15.30), and the differential equation to find the $(a, A, B, C)$ terms:

- Because the initial velocity ($d\bar{x}/dt$) is zero, then $B = 0$,
- Because the initial position $\bar{x}(0) = 0$, then $C = -A$,
- From Eq. (15.45), $A = D/2$ and $a^2 = F_o/mD = \sqrt{2}/\tau_o$

Putting them together, the solution is

$$\bar{x}(t) = \frac{D}{2}\left[\cosh\left(\frac{\sqrt{2}\,t}{\tau_o}\right) - 1\right] \tag{15.46}$$

and the transit time $\tau$ specified by when $\bar{x}(\tau) = D$, or [152]

$$\tau = \frac{\tau_o}{\sqrt{2}} \cosh^{-1}(3) = 1.2465\tau_o \tag{15.47}$$

An equivalent and systematic development of the same conclusion [187] considers the field generated by the sheet and adds it to the background field of $F_o$. The field on either side of the sheet of charge can be represented as proportional to the background field. Define a factor $\xi$ such that $F_+ - F_- = \xi F_o$, and let $\sigma = \xi\sigma_o = \xi F_o/16\pi Q$. Also let $F_o = F_- s + F_+(1-s)$, where $s = \bar{x}/D$. Then simultaneous solution of these equations gives

$$F_+(1 + \xi s)F_o; \;\; F_- = [1 - \xi(1-s)]F_o \tag{15.48}$$

Now the differential equation to solve is

$$\frac{d^2}{dt^2}s = \frac{F_+ + F_-}{2mD} = \frac{F_o}{mD}[2 - \xi(1-s)] \tag{15.49}$$

which can be re-written using the usual trick of $dv/dt = (1/2)dv^2/ds$ because $v = ds/dt$. When integrated, $v(s)$ is obtained and is trivial. Integrating the inverse of $v(s)$ for $s$ ranging from 0 to 1 gives $\tau(\xi)$ and is

$$\tau(\xi) = \tau_o \int_0^1 \frac{ds}{\sqrt{2s[2 - \xi(1-s)]}} \tag{15.50}$$

For the special case $\xi = 1$ (the value of $\xi$ for which $\sigma = \sigma_o$), the integral is analytic and so the pulsed transit time is

$$\tau(1) = \sqrt{2}\tau_o \ln(1 + \sqrt{2}) = 1.2465\tau_o \tag{15.51}$$

as previously found in Eq. (15.47), and $\tau(0) = \tau_o$ as is easy to check.

Why is this important? Consider the three transit times examined: ballistic (no current), pulsed (single sheet), and constant current (innumerable sheets) given by $\tau = h\tau_o$, where $h$ takes the values 1, 1.2465, and 1.50, respectively. These values differ, but are reasonably close to each other. As a result, the generally messy condition of real, non-uniformly emitting cathodes into an AK gap, or a longer than instantaneous pulse, is in some sense a mashing together of these various limiting cases. And so, calculating the impact of space charge is far more affected by changes in $J$ than it is by changes in the transit time $\tau$. Using the ballistic transit time in Eq. (15.36) can be an easy way to find out the impact of increased current density $J$ on the surface field $F$, albeit one that is a little off but not so much as to be disagreeable. And that means paying more attention the changes in $J$ to anticipate how the surface field $F$ changes is a better use of time, as it were.

An example helps (drawn from Ref. [187]). Using the ballistic transit time $\tau_o$ and $F_o = V_a/D$, then the transit time model can be expressed as $F_o = F + (18\pi Q/q)\tau_o J$. Consider a field emission example, for which the size of $J(F)$ is strongly dependent on the surface field $F$. Let $D = 10^4$ nm, $\Phi = 4.5$ eV, and $V_a$ such that $F_o = 10$ eV/nm. If the current density is given by $J_{FN}(F)$, then Eq. (15.21) requires that $f = 0.67488$ whereas the ballistic approach would suggest $f \approx 0.68613$. The two answers are quite close, suggesting the utility of the transit time model.

A significant advantage of the simple transit time model, though, is its ability to anticipate complex time-dependent behavior that is otherwise only apparent using simulations (PIC, molecular dynamics) [80, 343]. The time dependence is most readily seen when the emission is due to field

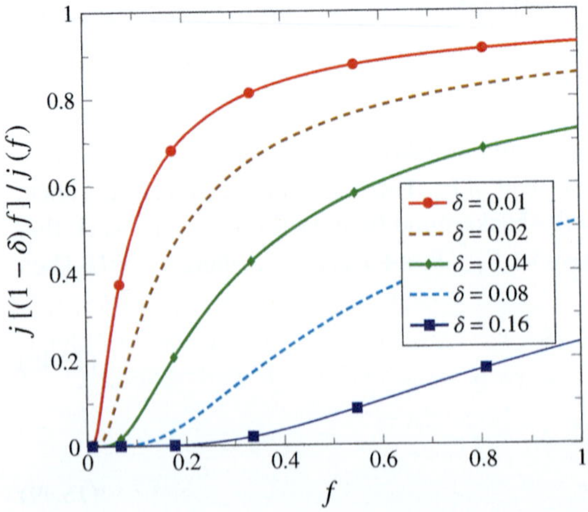


**Figure 15.9** Decrease in field emission current density $j(f) = af^{2-\nu}\exp(-b/f)$ when $f \to (1-\delta)f$ for $F_o = 10$ eV/nm and $\Phi = 4.5$ eV. Adapted from [187].

emission: any charge injected into the AK gap reduces the surface field, and $J_{FN}(F)$ drops precipitously due to the strong dependence of the FN equation on surface field. That strong dependence can be understood as follows using the dimensionless form of $j(f) = af^{2-\nu}\exp(-b/f)$, the components of which are encountered in Eq. (15.23). Remember that $f = FD/V_a$ and $j = J_{FN}/J_o$ with $J_o = (16/9)J_{CL}$. Consider when $f$ changes by a small percentage $\delta$: what is the impact on $j(f)$? It is

$$\frac{j[f(1-\delta)]}{j(f)} = (1-\delta)^{2-\nu}\exp\left[-\frac{b\delta}{f(1-\delta)}\right] \tag{15.52}$$

as seen in Figure 15.9 (see Algorithm B.4.12) and using Eq. (15.23). The factor $b$ is sizable (e.g. $b \approx 13$ for $F = 5$ eV/nm and $\Phi = 4.5$ eV), and so the ratio drops quickly as $f$ decreases. Consequently, compared to the sheet emitted an instant before, the next emitted sheet has a reduced charge density, and the more charge that is in the AK gap, the greater the reduction is.

The sheet model therefore becomes a perfect test case for ideas borne of the transit time approximation. Intuitively, what is expected to happen is straightforward. As each sheet of charge is injected into the AK gap, the field $F$ at the surface is reduced a bit in proportion to the charge density of the sheet. As more sheets are injected, $F$ drops, but depending on the size of the AK gap, it does not drop to zero: when the sheets at the start of the simulation arrive at the anode, they are absorbed, and their contribution to the reduction of the surface field ends. Each subsequently absorbed sheet that the anode absorbs contains less charge density than the sheet before it, and so the surface field rebounds, causing the injected current to rebound. The impact on the emitted sheets, however, is distributed over time, and so the emitted current density does not rebound to its initial value. An oscillatory behavior that becomes damped out over time to an equilibrium state is thereby established, with the final surface field dictated by Eq. (15.21).

Such behavior can be seen in a simple model. Assume that a sheet of charge is emitted every $\Delta t$, the only thing changing being the amount of charge in that sheet governed by $q\sigma = J_{FN}\Delta t$. Assume each sheet takes a time $\tau_o$ to cross the AK gap (an underestimation but nevertheless a reasonably close assumption). That means that at any given moment (after the first sheet crosses the gap) there are $n = \tau_o/\Delta t$ sheets in the AK gap. The force acting on each sheet that accelerates it across the AK gap, though, needs to be carefully considered. Start with the simple case $n = 1$ (equivalent to the pulsed transit time model treated above). As the sheet accelerates across the AK gap, its presence reduces the field at the surface of the cathode until such time as the sheet is absorbed by the anode. The degree to which the sheet reduces the surface field depends not only on its charge density $q\sigma$ but its position as well. The position dependence is seen as follows. For one sheet of charge, the surface $F$ is found by observing that the solution to Poisson's equation $\partial_x^2 V(x) = 16\pi Q\rho(x)$ is the sum of the particular solution $V_p$ (solution when the left-hand side is zero) and the homogeneous solution $V_h$ (solution when the right-hand side is replaced by zero), or $V(x) = V_p(x) + V_h(x)$. Both solutions must satisfy boundary conditions given by: $V_h(0) = 0$, $V_h(D) = V_a$ and $V_p(0) = V_p(D) = 0$. Thus, $V_h(x) = xV_a/D = xF_o$. $V_p(x)$ makes use of the pillbox to relate the sheet charge density to the differences of the forces on each side of the sheet and $\rho = \sigma\delta(x - \bar{x}(t))$ depends on the Dirac delta function for a sheet, to find

$$\frac{V_p(x) - V_p(0)}{x} - \frac{V_p(D) - V_p(x)}{D - x} = 16\pi Q\sigma \tag{15.53}$$

which is the same as

$$\frac{V_p(x)}{D} + \frac{V_p(x)}{D - x} = \xi F_o \tag{15.54}$$

where $\xi$ is defined by $\sigma = \xi\sigma_o$ as was introduced in Eq. (15.48). As a result,

$$V(x) = xF_o - \xi F_o x\left(1 - \frac{x}{D}\right) \tag{15.55}$$

$F$ at the surface due to the single sheet of charge in the gap at $x$ is dictated by $V(x)/x$, or or

$$F = F_o - \xi F_o\left(1 - \frac{x}{D}\right) \tag{15.56}$$

causing $F$ to be dependent upon where the sheet is. That dependence is what makes the subsequent analysis problematic, for sheets closer to the cathode have greater impact [187].

By superposition, additional sheets can be added, and an emission model set up. Let the index $j$ indicate the time $t_j$ that the sheet is emitted: then the surface field at time $t_j$ is $F(t_j) = F_j$. For $n$ sheets in the gap, the sheet charge density of the $j^{th}$ emitted sheet is $\sigma_j = \xi_j\sigma_o$, then

$$F_j = F_o - 16\pi Q\sum_{j=1}^{n}\sigma_j\left(1 - \frac{x_j}{D}\right) \tag{15.57}$$

This equation is equivalent to Eq. (A5) of Ref. [46] when the observation point is the surface of the cathode. It is rendered in a computationally better form by introducing the dimensionless position term $s_j = x_j/D$ and setting $r = 16\pi Q\sigma_o\Delta t/F_o$ to find

$$f_j = \frac{F}{F_o} = 1 - r\left(\sum_{j=1}^{n}\xi_j - \sum_{j=1}^{n}\xi_j s_j\right) \tag{15.58}$$

The problematic nature is because the positions $x_j$ (or rather $s_j$) are unknown until Eq. (15.58) is solved, but that requires knowing the trajectory of the sheet as it is accelerated under conditions that are constantly changing in the AK gap. In contrast, $\xi_j$ is known by superposition: all that is needed to calculate it is knowing how many sheets of charge are in the AK gap and their charge density, which can be sequentially solved because knowing where the sheet is not required. A measure of the difficulty of the full analytic problem (the same as calculating both sums in Eq. (15.58)) is appreciated by considering the intensive methods of Caflisch and Rosin [38] to characterize, as they name it, "unsteady emission." In contrast, solving the problem numerically faces its own difficulties: because lower surface fields accelerate a sheet more slowly, a sheet emitted at a later time but at a higher surface field has the possibility of passing a sheet emitted earlier, as can be observed in a Monte Carlo treatment where the location of the sheets are tracked [187].

Before getting into approximate solutions to Eq. (15.58), spend a moment to anticipate what the solutions should look like. The molecular dynamics (MD) simulation of Torfason et al. [343] is a distinctly accurate solution because rather than tracking sheets, individual particles were tracked after emission, and the MD simulations followed their interaction with other charges in the AK gap and the evolution of the current across the gap. A finite area planar nanodiode was simulated, with emitted charge dictated by a field emission relation equivalent to $J_{MG}(F)$ of Eq. (10.16). The largest emission area considered was a square with sides of length $L$. Each electron in the AK gap was treated as a single particle, and all coulomb interactions, along with their image charges in the electrodes, were accounted for. The data for the $L = 2500$ nm line of Figure 4 of Ref. [343] has been recast into Figure 15.10 here: being the largest of the emission areas simulated, it has the closest behavior to the sheet model behind Eq. (15.58). The MD simulation starts with an AK gap empty of charge and the field $F_o$ applied at the first-time step. Consequently, at $t = 0$, the cathode surface experiences the full field $F_o$ and so produces a current density of $J_{MG}(F_o)$. When those particles traverse the gap and impact the anode, they cease to influence $F$ at the surface. Because there are more of them at the first-time step when $F = F_o$ than at later times when $F = f_j F_o$ decreases (less charge $\sigma_j$ is injected into the gap), when the first emitted sheet of electrons is absorbed by the anode, an abrupt change in the field at the surface occurs, and so a kink in a plot of $F(t)$ is seen whereby $F(t)$ suddenly starts increasing. The location of that kink is a reasonable measure of the transit time. For purposes of comparison to the simpler sheet model, let the kink location be taken as a measure of $\tau_o$.

Now return to Eq. (15.58) and see what behavior that Torfason et al. show (see also Figure 5 of Feng and Verboncoeur [80]), can be anticipated. Consider, therefore, a simulation of Eq. (15.58). Rather than the parameters used by Torfason et al., the particular parameters of work function, anode potential, and gap spacing are altered here for pedagogical consistency to reflect earlier sections. The scaling of terms allows the oscillatory behavior of the surface field (and therefore

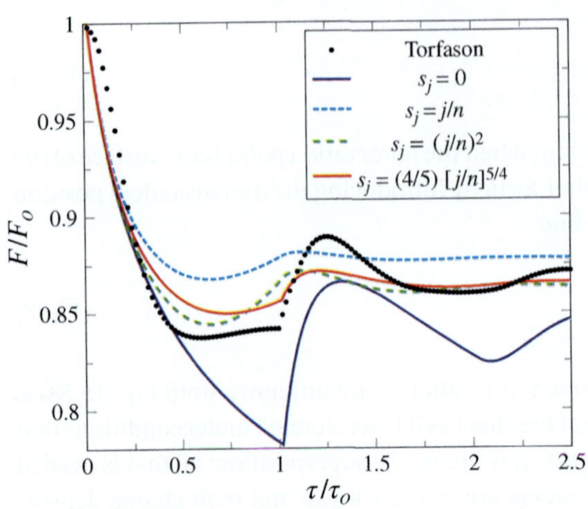


**Figure 15.10** Symbols: Oscillations in surface field of a planar nano-diode in a MD simulation using $J_{MG}(F)$ (Eq. (10.16)): after line $L = 2500$ nm of Figure 4 of Torfason et al. [343]. Lines: theory lines correspond to four cases shown for $s_j$ in Eq. (15.58) for $\Phi = 4.5$ eV, $F_o = 8$ eV/nm, $D = 4$ $\mu$m.

the emitted current density) to exhibit behavior that nevertheless hearkens to the findings of the MD simulation. Consider, therefore, an implementation of Eq. (15.58) over a series of time steps $t_j \equiv j\Delta t = j\tau_o/n$, where at $t_0 = 0$, the field on a one-dimensional cathode surface in a nanodiode is "turned on," that is, instantaneously becomes $F_o = V_a/D$. Ignore temperature effects, approximate the emitted current using $J_{MG}(F)$ with $F = fF_o$, where $J_{MG}(F)$ is as given in Eq. (10.16). Let the parameters of the simulation be: $F_o = 8$ eV/nm, $\Phi = 4.5$ eV, $D = 4000$ nm, from which it can be shown that $B_o = 65.2073$ eV/nm, $A_o = 0.00962087$ q / eV fs, $\nu = 0.7728079$, $A_{MG} = 245.363$ q / fs nm$^2$ (or $3.97838 \times 10^{12}$ A/cm$^2$). The term $\xi_j$ represents the ratio of the charge emitted at time $t_j$ compared to that emitted were no charge in the AK gap, or

$$\xi_j = \frac{J_{MG}[F(t_j)]\Delta t}{J_{MG}(F_o)\Delta t} = f_j^{2-\nu} \exp\left[-\frac{B_o(1-f_j)}{F_o f_j}\right] \tag{15.59}$$

where $f_j = F(t_j)/F_o$ and $F(t_j)$ is the space-charge lowered field at the surface at time $t_j$. For the parameters chosen, the dimensionless ratio $r$ represents the ratio of the field due to the sheet charge with that due to the anode, and is

$$r \equiv \frac{16\pi Q J_{MG}(F_o)\Delta t}{qF_o} = \frac{0.919496}{n} \tag{15.60}$$

where $n = \tau_o/\Delta t$. For sheet charge models in Figure 15.10, $n = 50$ was used.

The calculation of $\xi_j$ can be iteratively performed in the following manner. At each time increment, all the sheets emitted prior to $t_j$ are moved one step closer to the anode and the next sheet $\sigma_j = J_{MG}(F_j)\Delta t/q$ emitted based on $F_j = f_j F_o$ at the surface. In implementation, it is advantageous to visualize $n$ bins in the AK gap that can hold a sheet of charge, with $\xi_j$ being the value of $\sigma_j = \xi_j\sigma_o$ in that bin. Therefore, at $t = 0$, all the bins are empty ($\xi_j = 0$ for all $j$). At $t_1 = \Delta t$, the first bin is filled. At $t_2 = 2\Delta t$, the contents of all the bins (empty or not) are moved one step to their right (closer to the anode), and the first bin re-filled by emission conditions at that moment. By iteration, at $t_j = j\Delta t$, the contents of all the bins (empty or not) are again moved one step to the right (closer to the anode) and the first bin again re-filled by emission conditions at that moment. At every iteration, this process is repeated, and the contents of the last bin are absorbed by the anode and so are discarded from the calculation.

Such a process allows the first summation in Eq. (15.58) to be performed, as $s_j$ is not required for it. The choice to neglect the contribution of the $s_j$ term is the "$s - \tau$" approximation of Ref. [187] and corresponds to neglecting the second summation term in Eq. (15.58). The unknown values of $s_j$ impede performing the second summation because the locations of the bins must be determined by other assumptions: if they are assumed to be equally spaced in the AK gap, then $s_j = j/n$, but this implies that the sheets experience no acceleration (constant velocity) and so this approximation severely over-estimates the reducing impact of space charge on $F_j$, causing sheets to spend a longer time in the gap, and is a worst-case scenario. If the $s_j$ are spaced more densely closer to the cathode surface in a manner suggested by ballistic transport, then $x(t) \propto t^2$ implies $s_j = (j/n)^2$, but the sheets traverse the gap faster than they would in the presence of space charge. Lastly, consider $s_j = C(j/n)^p$ instead, the choice of $p < 5/4$ and $C = 4/5$ being determined to provide a reasonable match to the data of Torfason et al., motivated by discussions in the literature regarding how density and velocity vary in the nanogap: the coefficient is the approximate ratio of $\tau$ to $\tau_o$ for the pulsed (single sheet) case, and the power of 5/4 on the time $j/n$ shows some acceleration short of the ballistic case (sub-ballistic). The sheet model parameters differ from the MD simulation, but the qualitative features are evident in Figure 15.10 in that the strong oscillations at the beginning are damped out over time.

## 15.4 The Miram Curve

Although field emission was useful for discussing transit time models of space charge limited flow across an AK gap because of its strong dependence on surface field, the technological ubiquity of dispenser cathodes in radar, communications, cathode ray tubes, linear beam devices and much else [99] makes the connection between thermal emission and SCL emission important. Dispenser cathodes are an exceptionally well-established technology (see Ref. [99] and Chapters 1 and 2, in particular [99, 303]). An important driver for an understanding of the aging process of dispenser cathodes was their usage in traveling wave tubes (TWTs) [234] in space-based missions, where lifetime and power consumption are at a premium (and replacement due to failure is not an option). Lifetime is related to how fast barium was "dispensed" to replace the low work function coating that was lost due to thermal evaporation [114], and power consumption was a measure of how low the temperature of operation could be reduced: the two are intimately linked, as shown in Figure 15.11, as evaporation is a thermally regulated process, and so higher temperature led to shorter life. Consequently, even though high beam currents are desired, dispenser cathodes are typically run at a more modest current density to extend lifetime into the years range. The search has therefore been intensive for high current density variants that operate at lower temperatures, like the M-type cathode, or more recently, the scandate dispenser cathodes [103, 310, 356] that have achieved 460 A/cm$^2$ at desired temperatures.

Dispenser cathodes are run space charge limited to maximize current and enhance beam uniformity in TWT's by suppressing the noisiness of the cathode through space charge smoothing (in contrast, magnetron injection guns in gyrodevices are run temperature limited, so there is no space charge smoothing [224]). An upper limit to the current density that can be pulled from a dispenser cathode by a particular extraction grid a fixed distance away and at a particular voltage exists because of space charge limited transport. Running the cathode at a higher temperature than that needed to achieve the space charge limit is counterproductive: the same current density is achieved (the excess is reversed and sent back to the cathode), but the lifetime is compromised. Finding where the transition from temperature-limited emission governed by the RLD equation to

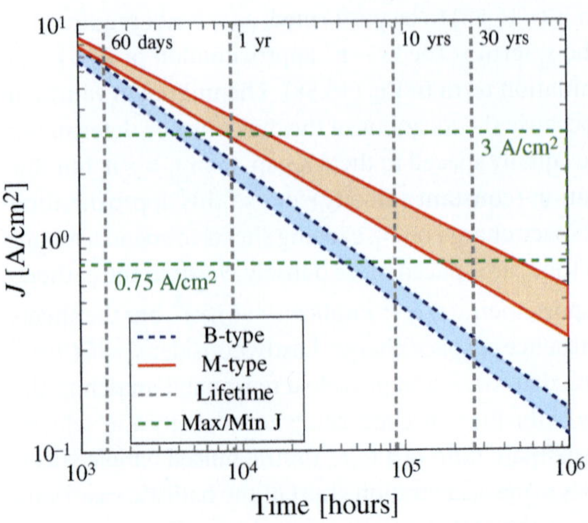


**Figure 15.11** Lifetime *vs.* operating current density for B and M-type (osmium containing, M signifying "magic") [114] dispenser cathodes (adapted from [154]). Crudely, each 40 K increase in temperature approximately halves the cathode's lifetime.

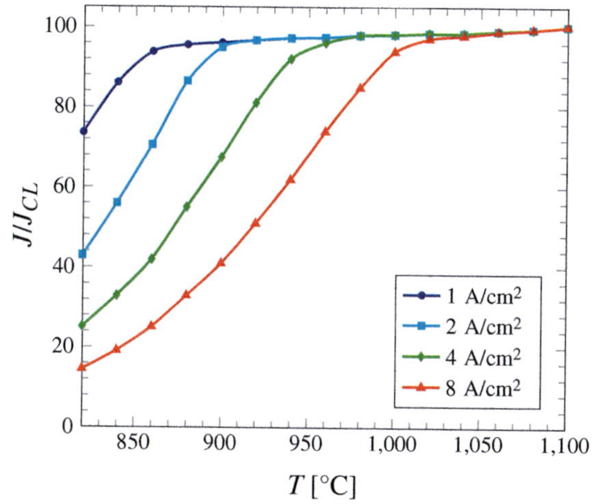


**Figure 15.12** A typical Miram plot for experimental data from a dispenser cathode (courtesy of M.C. Green) based on Ref. [79]/IEEE. The legend refers to $J_{CL}$.

the space charge limit governed by the CL equation is therefore highly desirable, particularly given that the performance of dispenser cathodes evolves over time: that need led to the development of the Miram curve [40, 79] like those shown in Figure 15.12 to characterize them.

Miram curves like Figure 15.12 normalize the emitted current with the space charge limited value and plots the relation as a function of temperature. Slow changes in emission properties are therefore revealed over time, and the curves act as a means of identifying cathodes by the shape of their "knees" on a Miram plot [79]. A good cathode exhibits the same shape for different $J$. Longo synthesized a widely used formula that characterized the Miram curve as $1/J = 1/J_{RLD} + 1/J_{CL}$. Some ambiguity is present about how to interpret $J_{RLD}$: clearly, at low temperatures, the surface of the cathode is exposed to a field $F_o = V_a/D$, and that should be reflected in a Schottky barrier lowering factor $\sqrt{4QF_o}$, but the space charge limit is characterized by the surface field vanishing. When discussing $J_{RLD}$ here, it is its behavior near the space charge limit that matters, and so $J_{RLD}$ will be evaluated using the assumption that $F = 0$ in the figures that follow. The quality of the cathode is reflected in the sharpness of the "knee," with lower quality having more rounding. Vaughan [347] observed that "well-designed and well-made" emitters were characterized by sharper knees; poorer cathodes were characterized by patchy emission, uneven heating, or other defects associated with rounded knees. He proposed a modified form of Longo's equation by raising the powers of the current density terms as shown in Figure 15.13 and offered

$$\frac{1}{J^\alpha} = \frac{1}{(J_{RLD})^\alpha} + \frac{1}{(J_{CL})^\alpha} \tag{15.61}$$

If $x = J_{RLD}/J_{CL}$, then

$$y[x(T)] = \frac{x}{(1 + x^\alpha)^{1/\alpha}} \tag{15.62}$$

from which $J = J_{CL}\, y[x(T)]$: at high temperature $(x \to \infty)$, $J \to J_{CL}$ for a thermionic cathode. "Good" cathodes were characterized by sharp knees with $\alpha \approx 6 - 10$ and "poor" ones by $\alpha \lesssim 2$. Curves change over the lifetime, acquiring more rounded knees that move to higher temperatures, and are correlated with less emission uniformity [376]. Such effects are evident in the life test data of Longo in Figure 15.13, where three different extended times of 7012, 16 199, and 25 142 hours (or

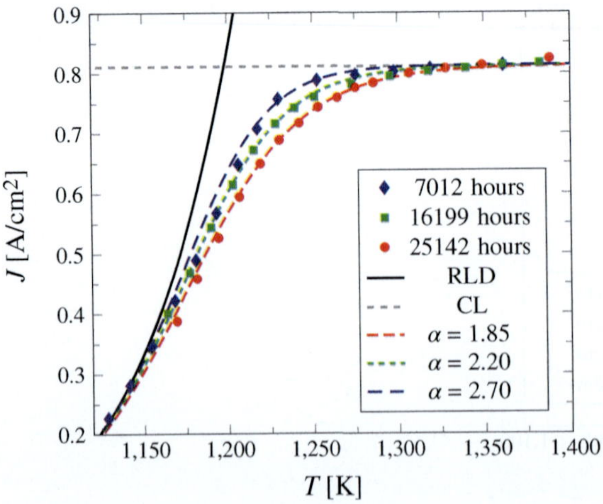


**Figure 15.13** Symbols: Evolution of dispenser cathode current density over time (data digitally extracted from Figure 4 of Ref. [234]). Observe how the knee softens as the elapsed time increases. Lines: $J_{RLD}$ with $F = 0$ (Eq. (8.10)), $J_{CL}$ (Eq. (15.9)), and $\alpha$ as shown (Eq. (15.62)).

taking a year as 8766 hours, then 0.8, 1.85, and 2.87 years, respectively) are shown compared to the Longo–Vaughan model of Eq. (15.62) with values of $\alpha$ that provide a reasonable fit: as expected, the latter in the life of the dispenser cathode, the more the quality of the cathode dropped, equivalent to a decrease in the best fit value of $\alpha$.

The choice of $x(T) = J_{RLD}(T)/J_{CL}$ has an added benefit. In terms of $T_o$ defined such that $x(T_o) = 1$, then

$$x(T) = \left(\frac{T}{T_o}\right)^2 \exp\left(B\frac{T - T_o}{T}\right) \tag{15.63}$$

where $B = (\Phi - \sqrt{4QF})/(k_B T_o) = \phi/k_B T_o$ such that $-B$ is the dimensionless exponent of the RLD equation evaluated at $T_o$, e.g. for $\Phi = 2$ eV, $F = 4 \times 10^{-4}$ eV/nm and $T_o = 1273$ K, then $\phi = 1.97600$ eV and $B = 23.5544$.

Equation (15.63) gives a nod to a useful feature of the Miram plot. Suppose a thermal cathode initially characterized by $(\Phi_i, \alpha_i)$ is tested in the temperature-limited regime (well away from the space charge limit). After extended testing, suppose those parameters shift to $(\Phi_f, \alpha_f)$. Because the work function changes by $\Phi_f - \Phi_i = \Delta\Phi$, the temperature $T$ required to produce a particular desired current density likewise changes by $\Delta T$. What is the relation between $\Delta T$ and $\Delta\Phi$? Solving $J_{RLD}(\Phi_i, T) = J_{RLD}(\Phi_f, T + \Delta T)$ results in

$$\phi_f = (1 + r)\left[\phi_i - 2k_B T \ln(1 + r)\right] \tag{15.64}$$

where $r = \Delta T/T$, $\phi_i = \Phi_i - \sqrt{4QF}$ and $\phi_f = \phi_i + \Delta\Phi$. By way of example, suppose that measurements are performed well into the thermal regime (where $J = J_{RLD}(T)$). Consider the example values of $T = 1000$ K and $\Delta T = 25$ K (so that $r = \Delta T/T = 0.025$). Then for $\phi_i = 1.97600$ eV, $\phi_f = 2.02976$ eV, and so $\Phi_f = 2.05376$ eV. Now suppose further that the measurements were done in a diode configuration where the anode was at potential of a kilovolt ($V_a = 10^3$ eV) and the AK gap was 0.25 cm ($D = 2500000$ nm) so that $F = V_a/D = 4 \times 10^{-4}$ eV/nm and (after conversion) $J_{CL} = 1.18090$ A/cm$^2$. Lastly, let $\alpha_i = 2.7$ and $\alpha_f = 2.3$. Let the value of $T_o$ be designated as $(T_{o,i}, T_{o,f})$ for the initial and final $J(T)$. A plot of $J_{RLD}(\Phi_i, T)$ *vs.* $T/T_{o,i}$ overlaps the curve $J_{RLD}(\Phi_f, T)$ *vs.* $T/T_{o,f}$ by virtue

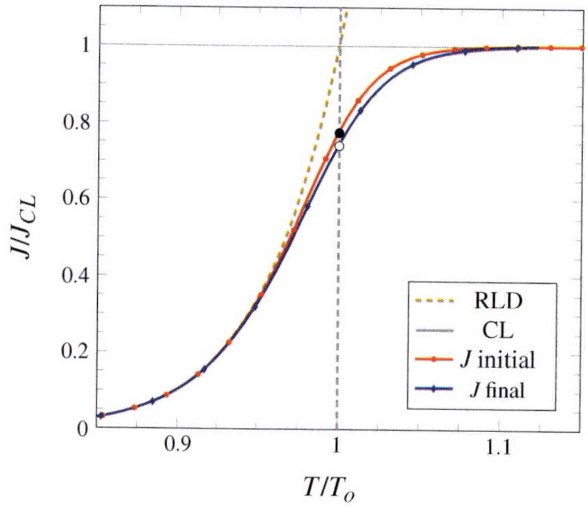


**Figure 15.14** The behavior of $J(T)$ when the $(\Phi, \alpha)$ change from $(\Phi_i, \alpha_i) = (2.0, 2.7)$ to $(\Phi_f, \alpha_f) = (2.05376, 2.3)$. The AK gap is such that $V_a = 10^3$ eV and $D = 0.25$ cm. $J_{CL} = 1.18090$ A/cm$^2$. It is found $T_{o,i} = 1217.64$ K, $T_{o,f} = 1247.55$ K. $\Delta T = 29.9132$ K, close to value based on $J_{RLD}$.

of how Eq. (15.64) is evaluated, but because $\alpha_{i,f}$ changes, the curves for $J(T)$ will change, but if the change in $\alpha$ is small, so will the change in $J$ be small as shown in Figure 15.14 (see Algorithm B.4.13), where the $(\bullet, \circ)$ dots denote $J(T_o)$ for the initial and final cases. As a result, a characterization of the Miram curve gives a close estimate of what $\Delta T$ is based on $\Delta\Phi$, and conversely, what change in work function is expected for a change in the characteristic $T_o$ even when the Miram curve evolves in shape because of changes to $\alpha$. Monitoring the knee of a cathode is therefore a diagnostic of how the cathode is changing.

The Longo–Vaughan equation, as implemented, was agnostic about the value of the field at the surface of the cathode, but clearly, as the current increases, that field is changing due to space charge. It is therefore of some importance as to whether that changing field in part causes the knee on a Miram plot. That question can be illuminated using the transit time model. If $J$ in Eq. (15.36) is replaced with $J_{RLD}$, but now explicitly including the Schottky barrier lowering factor $\sqrt{4QF}$, then using $F_o = V_a/D$, the ballistic transit time $\tau_o$, $N = 9/8$, and some rearranging, it can be recast as

$$1 = f + Ae^{B\sqrt{f}}$$
$$A = \frac{18\pi QJ}{q}\sqrt{\frac{2mD}{F_o^3}}; \quad B = \frac{4QF_o}{k_{BT}} \tag{15.65}$$

where $f \equiv F/F_o$. $A$ is so defined that when $J = J_{CL}$, then $A = 1$, which has the effect of demanding $f = 0$ for there to be a solution: the field at the cathode is driven to 0. When $A < 0$ because $J < J_{CL}$, then $f$ varies between 0 and 1, and when $J$ becomes very small so that $A \to 0$, then $f \to 1$. The evolution of $f$ from 1 to 0 therefore tracks the evolution from source-limited (thermal) to space-charge-limited flow smoothly. The question is then: does that transition explain the soft knees on experimental measurements of dispenser cathodes as observed by Longo? The answer is that it does not, as can be shown through a simple consideration of the behavior of $J_{RLD}(F, T)$ (full field at the surface) compared to $J_{RLD}(0, T)$ (no field at the surface), and both compared to a Longo–Vaughan plot using $\alpha = 2.7$. For the $V_a$ and $D$ of Figure 15.14, then $F_o = 4 \times 10^{-4}$ eV/nm, making the Schottky factor $\sqrt{4QF_o} = 0.02400$ eV. Such a small change of the effective work function $\phi$ is almost imperceptible in moving the RLD line of Figure 15.14, and so the relatively large knee even for $\alpha = 2.7$ cannot be explained by variations in $f$ at the surface of the cathode. The causes of the softness of the knee are due instead to nonuniform, *aka* patchy, emission.

Theoretical models of emission from infinite stripes characterized by different work functions [45] give agreement with numerical simulations using particle-in-cell (PIC) codes [279]: both demonstrated how the knee on a Miram curve softened depending on the distribution of randomly placed strips [140]. Subsequent simulations using molecular dynamics (wherein electrons are tracked and their Coulombic interactions are explicitly accounted for) on bounded areas with patchy work functions likewise demonstrated the softening of the knee [314, 315]. Knees are the consequence of non-uniform emission from patches of different work functions (a "spatially heterogeneous" emission surface), even if the patches are idealized in a checkerboard pattern [42].

The complexity of the theory and the computational demands that allow accurate simulations are beyond the scope here, but a simple account of the findings demonstrates the salient points. Consider, therefore, the checkerboard pattern of Jassem et al. [140] (following the prior work by Chernin et al. [45]). Using the distribution values of Table 15.1 on a $9 \times 9$ checkerboard, representative distributions of work function patches for the three random distributions in Figures 15.15 and 15.16 (the random seed was reset to the same value at the generation of each patch distribution: this ensures that if the distribution did not vary, then the checkerboard appearance would not vary, thereby showing what changes when the distribution does change, see Algorithm B.4.13). If

**Table 15.1** Work function $\Phi$ and distribution to create a chessboard pattern of patches with random placement. "EXP" = measured distribution of a cathode; "RAND" = random distributions that mimic experiments. Values rounded compared to Table 1 of Ref. [140]; highest $\Phi$ lowered to 4.5 eV to simplify $10 \times 10$ checkerboard generation in Figures 15.15 and 15.16.

| $\Phi$ | EXP | RAND 1 | RAND 2 | RAND 3 |
|---|---|---|---|---|
| 1.61 eV | 19% | 18% | 13% | 8% |
| 1.79 eV | 10% | 10% | 9% | 7% |
| 2.30 eV | 11% | 9% | 11% | 11% |
| 2.31 eV | 38% | 44% | 37% | 37% |
| 4.00 eV | 22% | 19% | 30% | 37% |

**Figure 15.15** Checkerboard pattern of three random placements of work function patches based on the distributions in Table 15.1 (left to right: Rand 1, Rand 2, Rand 3). Observe that the work function pairs (1.61,1.79) eV and (2.30, 2.31) eV will produce almost—but not quite—the same $J$ from their patches.

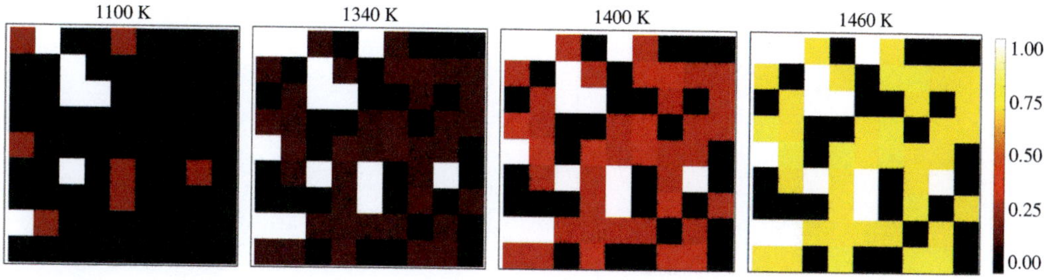


**Figure 15.16** $J/J_{CL} = \min(J_{RLD}/J_{CL}, 1)$ for each patch of Rand 3 for $T = (1100, 1340, 1400, 1460)$ K (top to right). For $V_a = 179.5$ eV and $D = 0.381$ mm (from Ref. [140]) then $J_{CL}(V_a, D) = 3.8667$ A/cm$^2$.

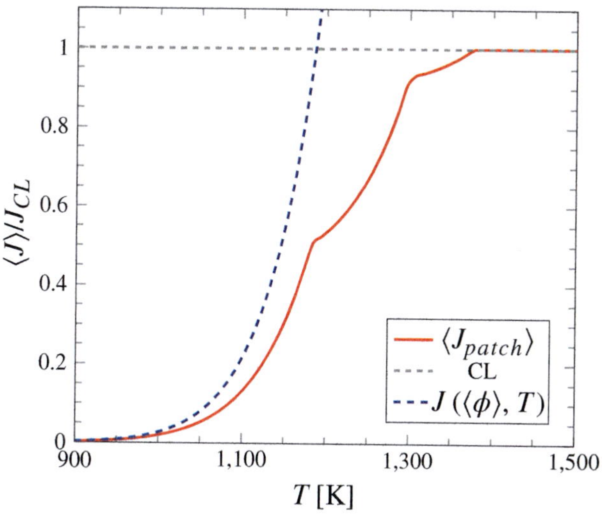


**Figure 15.17** An average of current density over the $10 \times 10$ Rand 3 distribution of Table 15.1, assuming that local (nearest neighbor) patches, if weak, can allow the center patch to emit in excess of $J_{CL}$. $J_{RLD}$ uses $\langle\Phi\rangle = 1.8$ eV. $J_{CL} = 3.8667$ A/cm$^2$.

each of the patches is constrained so that $J \le J_{CL}$, then the corresponding current density maps for the Rand 3 case are shown in Figure 15.17 (see Algorithm B.4.13), where the ratio $J/J_{CL}$ is shown for four representative temperatures. Even for such a simple representation, the behavior mimics experimental findings such as Figure 5 of Ref. [43].

Still, the local space charge limit constraint is too severe: locally, emitted charge can move over into the space above patches which are not emitting, and so the total current density emitted by a particular patch can exceed $J_{CL}$. To account for that, two approximations can be made. First, observe that most of the current density comes from the lowest two work function patches, and their neighbors typically produce well below $J_{CL}$. Therefore, assume that the local average current density is constrained, rather than each patch individually: a given patch is then constrained by $6 \times J_{CL}$ if most of its eight neighbors weakly emit. Second, evaluate $\langle J\rangle$ for the $9 \times 9$ checkerboard itself, and compare that to $J_{RLD}$ for an assumed $\langle\Phi\rangle = 1.8$ eV and $J_{CL}$ as usual. Doing so gives Figure 15.17, and it is clear that even for such a primitive model, the emergence of a knee is evident. When the PIC code MICHELLE is used to actually model emission and allow beamlets to expand over poorly emitting areas, then the agreement between PIC and theoretical models is excellent, as reproduced in Figure 15.18 based on Figure 4(d) of Ref. [140]: the knees are clearly visible, appropriately soft,

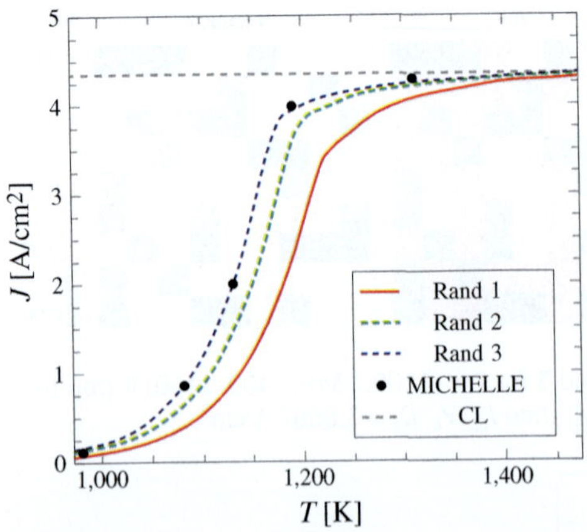


**Figure 15.18** Recreation of Figure 4d of Jassem et al. [140], showing the agreement between the PIC code MICHELLE and their theoretical model (data digitally extracted).

and representative of experimental measurements [43]. Consequently, the literature [42, 43, 45, 314, 315, 376] confirms that local variations in emission current from randomly distributed differing crystal faces of sintered dispenser cathode emitters or the random crystal faces of a metal surface is sufficient to explain why the Longo–Vaughan curve has the knees it does on a Miram plot.

## 15.5 Exercises

**Exercise 15.1** For what work function will $J_{RLD}$ equal the space charge limited current $J_{CL}$ operating at a temperature $T = 1273$ K? Assume $V_a = 1$ keV and $D = 2$ mm.

**Solution** $\Phi = 2.0535$ eV

**Exercise 15.2** Find $(a, b, \nu)$ in Eq. (15.23) for $\Phi = 4.5$ eV, $F = 6$ eV/nm, and $V_a = 600$ eV.

**Solution** $(a, b, \nu) = (201.59, 10.868, 0.77281)$

**Exercise 15.3** Find $(a, b, \nu)$ in Eq. (15.23) for $\Phi = 4.5$ eV, $F = 6$ eV/nm, and $V_a = 600$ eV.

**Solution** $(a, b, \nu) = (201.59, 10.868, 0.77281)$

**Exercise 15.4** Consider a peculiar cathode for which $j(f) = (4/3)\sqrt{1 + 3f}$. What pair $(f, j)$ satisfies $j(f)$ and the CL relation of Eq. (15.21)? An analytic solution is possible; a numerical solution is acceptable.

**Solution** $(j, f) = (1.679894733, 0.1958003506)$

**Exercise 15.5** In some simulations, electrons are injected into the vacuum with zero initial velocity. Compare the ballistic $\tau$ for $v_o = 0$ to that for $v_o = \langle v \rangle$ for thermal emission for which $T = 1373$ K, and an anode $V_a = 10$ eV.

**Solution** $\tau(0)/\tau(v_o) = 1.06325$

**Exercise 15.6** Show that $\tau(J)$ of Eq. (15.39) approaches the ballistic $\tau_o$ for $v_o = 0$ in the limit of $J \to 0$. What is their ratio for field emission-like conditions of $J = 10^8$ A/cm$^2$, $D = 50$ nm, and $V_a = 300$ eV?

**Solution** $\tau(J)/\tau(0) = 0.9221$.

**Exercise 15.7** For a nanodiode, assume a hypothetical array of molybdenum field emitters ($\Phi =$ 4.3 eV) on a square lattice with a tip-to-tip separation of $\Delta$ and governed by the $J_{MG}(F)$ of Eq. (10.16) with $t(y_o)$ replaced by $t_o = 1.06131$. Let the array be on a planar surface, and a planar anode is a distance $D_o = 6$ $\mu$m from the tips of the emitters. The shape of the emitters is such that the field at the apex of each emitter is $F = \beta_g V_a / D_o$ with $\beta_g = 75$ and $V_a = 240$ eV. Assume pulsed emission conditions (only a single sheet at a time is in the AK gap): all the field emitters emit one electron every transit time $\tau$, so that the sheet of charge in the AK gap suppresses the field at the cathode between pulses. What is $\Delta$? If instead the emitters are on an equilateral triangle lattice, now what is $\Delta$? The problem shows the importance of emission site density and its distribution.

**Solution** Square lattice: $\Delta = 734.3$ nm. Triangular lattice: $\Delta = 1115.9$ nm.

# 16

# Conclusion

*Look you now what follows...*

W. Shakespeare, *Hamlet* III.4

The scope of the present volume was constrained by its intent to examine the basic equations of electron emission and how they are calculated in a pedagogical manner. As a result, the focus was on the one-dimensional emission models, how they arise, and how they are used. The applications of those equations are taken up in the second volume. There, some of the more practical considerations, such as the behavior of contacts and resonant tunneling, can be considered. Before they are, methods beyond those to derive the current density equations for electron emission are to be examined. Some of those methods, such as the Transfer Matrix Approach for the evaluation of current density through general structures and for the treatment of non-standard complications like resonances, enable the consideration of effects that are not captured in a Gamow factor approach, specifically resonances and reflectionless transmission. A discussion of technologically fascinating notions such as the tunneling time through barriers, the behavior of multidimensional emitters that have more than one emission mechanism operating, and other applications of the canonical equations and their extensions can then be undertaken.

But there is more. Electron sources are much more than the equations used to describe their emission. There is the modeling of the sources themselves that affect how the emission equations are applied. A short but not exhaustive account is then: how low work function coatings get to the surface of dispenser cathodes; how coatings reduce work function on thermal, field, and photoemission sources; how those coatings migrate in time; how the proximity of other field emitters shields the applied field and therefore undermines emission; how charge moves through barriers and how long it takes to do so; how the concept of resonance produces unusual current–voltage relations; how statistical variation of emission sites affects current–voltage curves; how emission causes heating (or cooling), at the emission site and also how current causes heating along long fiber or nanotube emitters; how gated emitters behave; how electron beams evolve after they are emitted; and importantly, how is emittance related to cathode size, shape, and deployment.

Many of these topics have been dealt with in the archival literature behind the present volume. Indeed, some were motivated by particular applications with idiosyncratic features. And so, the present volume on foundations will be followed by a volume on the physics behind the applications. Gathering those disparate topics into an overarching narrative reveals their commonality in illuminating ways.

*Fundamentals of Electron Emission Physics*, First Edition. Kevin L. Jensen.

# A

# Methods and Functions

*Jove and my stars be prais'd! Here is yet a postscript.*

W. Shakespeare, *Twelfth Night* II.5

## A.1 Gamma Function

The defining property of the Gamma function, namely, $\Gamma(p+1) = p\Gamma(p)$, can be deduced from Euler's integral formula

$$\Gamma(p) = \int_0^\infty x^{p-1} e^{-x} dx \tag{A.1}$$

through an integration by parts. For integer $p = n$, it is equivalent to the factorial function $n! = \Gamma(n+1)$. For large $p$,

$$\Gamma(p) \approx p^{p-1} e^{-p} \sqrt{2\pi p} \left(1 + \frac{1}{12p} + \frac{1}{288p^2} + \cdots\right) \tag{A.2}$$

Stirling's approximation is recognizable in the coefficient

$$n! \approx n^n e^{-n} \sqrt{2\pi n} \tag{A.3}$$

That $\Gamma(p)$ has poles at negative integers and that $\Gamma(1/2) = \sqrt{\pi}$ are both evident from the relation

$$\Gamma(p)\Gamma(1-p) = \frac{\pi}{\sin(\pi p)} \tag{A.4}$$

## A.2 Riemann Zeta Function

The Riemann zeta function, or $\zeta(p)$, is defined by the series relation

$$\zeta(p) = \frac{1}{1 - 2^{1-p}} \sum_{j=1}^{\infty} (-1)^{p+1} j^{-p} = \sum_{j=1}^{\infty} j^{-p} \tag{A.5}$$

*Fundamentals of Electron Emission Physics*, First Edition. Kevin L. Jensen.

where the second summation is the more common definition. It is also defined by the equivalent integral relations

$$\begin{aligned}\zeta(p) &= \frac{1}{\Gamma(p)}\int_0^\infty \frac{x^{p-1}}{e^x-1}dx \\ &= \frac{1}{(1-2^{1-p})\Gamma(p)}\int_0^\infty \frac{x^{p-1}}{e^x+1}dx\end{aligned} \tag{A.6}$$

Either definition readily reproduces the common result $\zeta(1) = \ln(2)$. The asymptotic limit of $\zeta(p)$ can readily be shown from the first integral relation and the definition of $\Gamma(p)$ in Eq. (A.1): for large $p$

$$\begin{aligned}\zeta(p) - 1 &= \frac{1}{\Gamma(p)}\int_0^\infty x^{p-1}\left[\frac{1}{e^x-1} - \frac{1}{e^x}\right]dx \\ &= \frac{1}{\Gamma(p)}\int_0^\infty x^{p-1}\left[\frac{1}{e^x(e^x-1)}\right]dx \\ &\approx \frac{1}{\Gamma(p)}\int_0^\infty x^{p-1}e^{-2x}dx = 2^{-p}\end{aligned} \tag{A.7}$$

For integer $n$, $\zeta(2n)$ can be evaluated by exploiting its relation to the Fourier transform of $f(x) = x^2$ over the interval $[-\pi, \pi]$. The Fourier cosine coefficients are $a_0 = 2\pi^2/3$ and $a_n = 4(-1)^n n^{-2}$. For $x = \pi$, $\cos(nx) = (-1)^n$, and so $f(x) = a_0/2 + \sum_1^\infty a_n \cos(nx)$. It follows

$$\pi^2 = \frac{\pi^2}{3} + 4\sum_{n=1}^\infty n^{-2} = \frac{\pi^2}{3} + 4\zeta(2) \tag{A.8}$$

and so $\zeta(2) = \pi^2/6$. Similarly, $f(x) = x^{2n}$ gives $\zeta(4) = \pi^4/90$, $\zeta(6) = \pi^6/945$, and $\zeta(8) = \pi^8/9450$. The special case $\zeta(3) = 1.2020569$ defines Apéry's constant.

A comparison between the series and integral representations can reveal unexpected connections. In the integral case, the limit that $p \to 0$ seemingly runs afoul of an infinite $\Gamma(p)$, as seen from its integral definition

$$\Gamma(p) = \int_0^\infty x^{p-1}e^{-x}dx \tag{A.9}$$

but that same limit also reveals a logarithmic singularity in the integral part of $\zeta(p)$. Consider therefore the expression

$$\begin{aligned}(1-2^{1-p})\zeta(p) &= \frac{1}{\Gamma(p)}\left\{\frac{\Gamma(p)}{2} + \left[(1-2^{1-p})\Gamma(p)\zeta(p) - \frac{\Gamma(p)}{2}\right]\right\} \\ &= \frac{1}{2} + \frac{\displaystyle\int_0^\infty \frac{(1-e^{-x})}{2x^{1-p}(e^x+1)}}{\displaystyle\int_0^\infty x^{p-1}e^{-x}dx}\end{aligned} \tag{A.10}$$

The ratio of integrals term has a divergent denominator but a finite numerator as $p \to 0$, and so vanishes. Therefore, $\lim_{p\to 0} \zeta(p) = 1/2$ by analytic continuation. But that means, according to the series definition,

$$\sum_{j=0}^{\infty} j^0 = 1 + \zeta(0) = \frac{1}{2} \tag{A.11}$$

or $\zeta(0) = -1/2$. It can be shown using similar methods [154] that

$$\zeta(-1) = \sum_{j=1}^{\infty} j = -\frac{1}{12} \tag{A.12}$$

by defining the function $S(x) = \sum_{j=0}^{\infty} x^j = 1/(1-x)$ and its derivative $\partial_x S(x) = S'(x) = 1/(1-x)^2$ from which

$$S'(x) - S'(-x) = 4xS'(x^2) \tag{A.13}$$

Insert $x = 1$ and use $S'(-1) = 1/4$. Solving for $S'(1)$ gives

$$S'(1) = -\frac{1}{12} = \sum_{j=1}^{\infty} j \tag{A.14}$$

The cases of $\zeta(0)$, $\zeta(-1)$, $S(1) = 1 + \zeta(0)$ and $S'(1)$ highlight difficulties that arise with divergent series.

## A.3 Error Function

The Error function Erf($x$) is defined by

$$\text{Erf}(x) \equiv \frac{2}{\sqrt{\pi}} \int_0^x e^{-s^2} ds \tag{A.15}$$

For large argument,

$$\text{Erf}(x \gg 1) \approx 1 - \frac{2}{\sqrt{\pi}} e^{-x^2} \left( \frac{1}{2x} - \frac{1}{4x^3} \right) \tag{A.16}$$

For small argument, a series expansion is

$$\text{Erf}(x) = e^{-x^2} \sum_{j=0}^{\infty} \frac{x^{2j+1}}{\Gamma\left(j + \frac{3}{2}\right)} \tag{A.17}$$

## A.4 Legendre Polynomials

For problems with rotational symmetry (no $\phi$ dependence) then the equation $\nabla^2 \psi(r, \theta) = 0$ can be represented at $\psi(r, \theta) = \sum_l R_l(r) P_l(\theta)$. The eigenvalue equation for $P_l(\cos\theta)$ is

$$\frac{1}{\sin\theta} \frac{d}{d\theta} \left( \sin\theta \frac{dP_l}{d\theta} \right) + l(l+1) P_l(\cos\theta) = 0 \tag{A.18}$$

Letting $x = \cos\theta$, then Rodrigues' formula for the generation of $P_l(x)$ is

$$P_l(x) = \frac{(-1)^l}{2^l l!}\frac{d^l}{dx^l}(1-x^2)^l \tag{A.19}$$

for which the first few are

$$\begin{aligned}
P_0(x) &= 1\\
P_1(x) &= x\\
P_2(x) &= \frac{1}{2}(3x^2-1)\\
P_3(x) &= \frac{1}{2}(5x^3-3x)
\end{aligned} \tag{A.20}$$

A recursion relation given by

$$(l+1)P_{l+1}(x) = (2l+1)xP_l(x) - lP_{l-1}(x) \tag{A.21}$$

can be used to obtain higher $P_l(x)$. The functions satisfy the normalization relation

$$\int_{-1}^{1} P_n(x)P_l(x)dx = \frac{2}{2l+1}\delta_{nl} \tag{A.22}$$

Legendre's differential equation also admits non-polynomial solutions called Legendre Polynomials of the $2^{nd}$ kind denoted by $Q_l(x)$, for which the first few are

$$Q_0(x) = \frac{1}{2}\ln\left(\frac{1+x}{1-x}\right) \tag{A.23}$$

$$Q_1(x) = \frac{x}{2}\ln\left(\frac{1+x}{1-x}\right) - 1 \tag{A.24}$$

$$Q_2(x) = \frac{3x^2-1}{4}\ln\left(\frac{1+x}{1-x}\right) - \frac{3x}{2} \tag{A.25}$$

When the argument of $Q_l(x)$ is such that $x > 1$, then $\ln[(1+x)/(1-x)] \to \ln[(x+1)/(x-1)]$. The functions are shown in Figure A.1 (see Algorithm B.5)

## A.5 Hermite Polynomials

The Hermite polynomials naturally arise in the consideration of the quantum harmonic oscillator. They can be generated from

$$H_l(x) \equiv e^{x^2/2}\left[-\frac{d}{dx}+x\right]^l e^{-x^2/2} = e^{x^2}\left(-\frac{d}{dx}\right)^l e^{-x^2} \tag{A.26}$$

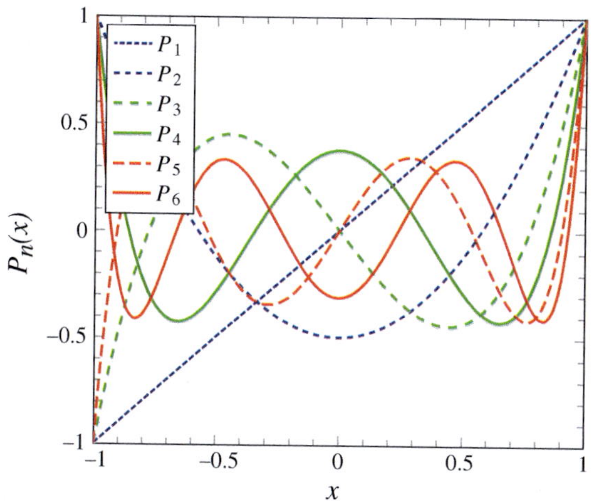


**Figure A.1** Legendre polynomials $P_n(x)$ for $n \in (1, \cdots, 6)$. Odd parity cases are dashed, even parity cases are solid lines.

for which the first few are

$$\begin{aligned} H_0(x) &= 1 \\ H_1(x) &= 2x \\ H_2(x) &= 4x^2 - 2 \\ H_3(x) &= 8x^3 - 12x \end{aligned} \tag{A.27}$$

A recursion relation is given by

$$H_{l+1}(x) = 2xH_l(x) - 2lH_{l-1}(x) \tag{A.28}$$

The functions satisfy the normalization relation

$$\int_{-\infty}^{\infty} H_n(x)H_l(x)e^{-x^2}\,dx = \frac{1}{\sqrt{\pi}2^n n!}\delta_{nl} \tag{A.29}$$

The functions are shown in Figure A.2 (see Algorithm B.5).

## A.6 Prolate Spheroidal Coordinates

Prolate spheroidal coordinates $(\eta, \upsilon, \gamma)$, which are angles to be augmented by a length scale $L$, are related to cartesian coordinates $(x, y, z)$, which are dimensioned extents, by

$$\begin{aligned} x &= L\sinh(\eta)\sin(\upsilon)\cos(\gamma) \\ y &= L\sinh(\eta)\sin(\upsilon)\sin(\gamma) \\ z &= L\cosh(\eta)\cos(\upsilon) \end{aligned} \tag{A.30}$$

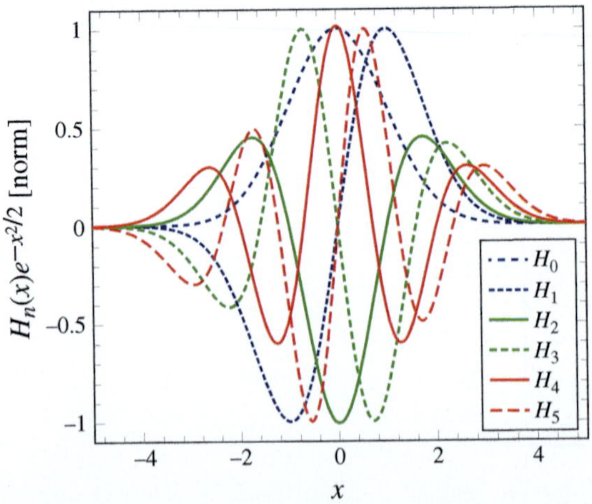


**Figure A.2** Product of Hermite polynomials with weights $H_n(x)e^{-x^2/2}$ for $n \in (0, \cdots, 5)$ and normalized. Odd parity cases are dashed, even parity cases are solid lines.

Define the $h$-factors as

$$\begin{aligned} h_\eta &= h_\upsilon = L\left(\sinh(\eta)^2 + \sin(\upsilon)^2\right) \\ h_\gamma &= L\sinh(\eta)\sin(\upsilon) \end{aligned} \tag{A.31}$$

The $\gamma$-coordinate is unneeded for rotationally symmetric geometries like cones, wires, and ellipsoids, but is retained here for generality. The components of the gradient are then

$$\vec{\nabla} = \hat{\eta}\frac{1}{h_\eta}\frac{\partial}{\partial_\eta} + \hat{\upsilon}\frac{1}{h_\upsilon}\frac{\partial}{\partial_\upsilon} + \hat{\gamma}\frac{1}{h_\gamma}\frac{\partial}{\partial_\gamma} \tag{A.32}$$

where the coordinates with hats refer to unit directions. The Laplacian is

$$\nabla^2 = \frac{1}{h_\eta h_\upsilon h_\gamma}\left[\frac{\partial}{\partial\eta}\left(\frac{h_\upsilon h_\gamma}{h_\eta}\frac{\partial}{\partial\eta}\right) + \frac{\partial}{\partial\upsilon}\left(\frac{h_\eta h_\gamma}{h_\upsilon}\frac{\partial}{\partial\upsilon}\right) + \frac{\partial}{\partial\gamma}\left(\frac{h_\eta h_\upsilon}{h_\gamma}\frac{\partial}{\partial\gamma}\right)\right] \tag{A.33}$$

where $(\eta, \upsilon, \gamma)$ can refer to orthogonal coordinates other than prolate spheroidal coordinates (that is, Eq. (A.33) is general for orthogonal coordinate systems). Substitution and collection for the prolate spheroidal case in particular results in

$$h_\eta h_\upsilon h_\gamma = L^3\left[\sinh(\eta)^2 + \sin(\upsilon)^2\right]\sinh(\eta)^2\sin(\upsilon) \tag{A.34}$$

$$\partial_\eta\left[\frac{h_\upsilon h_\gamma}{h_\eta}\partial_\eta\right] = L\sin(\upsilon)\left[\sinh(\eta)\partial_\eta^2 + \cosh(\eta)\partial_\eta\right] \tag{A.35}$$

$$\partial_\upsilon\left[\frac{h_\gamma h_\eta}{h_\upsilon}\partial_\upsilon\right] = L\sinh(\eta)\left[\sin(\upsilon)\partial_\upsilon^2 + \cos(\upsilon)\partial_\upsilon\right] \tag{A.36}$$

$$\partial_\gamma\left[\frac{h_\eta h_\upsilon}{h_\gamma}\partial_\gamma\right] = L\left[\frac{\sinh(\eta)^2 + \sin(\upsilon)^2}{\sinh(\eta)\sin(\upsilon)}\right]\partial_\gamma^2 \tag{A.37}$$

## A.7 Polynomial Fitting

A fast numerical method of finding the coefficients of a $n$-th order polynomial is often desirable. A useful one is based on the observation that a curve $y = f(x)$ that passes through $n + 1$ specified points $x_j$ with $j \in (0, 1, \cdots, n)$ is uniquely characterized by

$$f(x) \approx \sum_{j=0}^{n} C_j x^j \tag{A.38}$$

If the $C_j$ are evaluated using the following matrix method, then the curve $f$ will pass through the points $f(x_j)$.

Let the $(n+1) \times (n+1)$ matrix $\mathbf{M}$ be defined such that $[\mathbf{M}]_{j+1,k+1} \equiv f_j^k$, where $j$ is an index and $k$ is a power, and $f_j^k = f(x_j)^k$. Observe that the matrix row and column specification are non-zero numbers, but both $j$ and $k$ can be 0, which explains the indices on $\mathbf{M}$. Let the coefficient column vector $|C\rangle$ be such that its $(j+1)^{th}$ entry is $C_j$. Mathematically,

$$\begin{bmatrix} 1 & x_0 & \cdots & x_0^n \\ 1 & x_1 & \cdots & x_1^n \\ \vdots & \vdots & \ddots & \vdots \\ 1 & x_n & \cdots & x_n^n \end{bmatrix} \begin{bmatrix} C_0 \\ C_1 \\ \vdots \\ C_n \end{bmatrix} = \begin{bmatrix} f_0 \\ f_1 \\ \vdots \\ f_n \end{bmatrix} \Rightarrow \mathbf{M}|C\rangle = |f\rangle \tag{A.39}$$

Solving $\mathbf{M}|C\rangle = |f\rangle$ requires finding the inverse of the matrix $\mathbf{M}$. Then the coefficient vector satisfies

$$|C\rangle = \mathbf{M}^{-1}|f\rangle \tag{A.40}$$

where $\mathbf{M}^{-1}$ is the inverse of $\mathbf{M}$, and $|f\rangle$ is the column vector of $f_n$ values from which an interpolant is being sought. The straight line $f(x) = C_0 + C_1 x$ is an elementary example

$$\begin{bmatrix} 1 & x_0 \\ 1 & x_1 \end{bmatrix} \begin{bmatrix} C_0 \\ C_1 \end{bmatrix} = \begin{bmatrix} f_0 \\ f_1 \end{bmatrix} \Rightarrow \begin{bmatrix} C_0 \\ C_1 \end{bmatrix} = \frac{1}{x_1 - x_0} \begin{bmatrix} x_1 & -x_0 \\ -1 & 1 \end{bmatrix} \begin{bmatrix} f_0 \\ f_1 \end{bmatrix} \tag{A.41}$$

Collecting terms gives the well-known interpolation formula [292]

$$f(x) = \frac{x - x_0}{x_1 - x_0} f_0 + \frac{x_1 - x}{x_1 - x_0} f_0 \tag{A.42}$$

A second order (quadratic) polynomial results in

$$f(x) = \frac{(x_1 - x)(x - x_0)}{(x_2 - x_1)(x_2 - x_0)} f_2 + \frac{(x_2 - x)(x - x_0)}{(x_2 - x_1)(x_1 - x_0)} f_1 + \frac{(x_2 - x)(x_1 - x)}{(x_2 - x_0)(x_1 - x_0)} f_0 \tag{A.43}$$

Generalizing gives the Lagrange Interpolation formula [1, 216]

$$
\begin{aligned}
f(x) &= \sum_{j=0}^{n} L_j(x) f_j \\
L_j(x) &= \frac{(x-x_0)(x-x_1)\cdots(x-x_{j-1})(x-x_{j+1})\cdots(x-x_n)}{(x_j-x_0)(x_j-x_1)\cdots(x_j-x_{j-1})(x_j-x_{j+1})\cdots(x_j-x_n)}
\end{aligned}
\tag{A.44}
$$

where $L_j(x)$ is a ratio of products but with the $j^{th}$ entry taken out. As $n$ becomes large, Eq. (A.40) becomes sensitive and the inverse matrix difficult unless the coordinates $x_j$ are chosen and scaled with care. High order (large $n$) polynomial interpolations require careful numerical accounting to sum the series. Tricks to mitigate such problems employ uniformly spaced $x_j$ such that $x_j - x_{j-1} \equiv \delta$ and/or using better-behaved orthogonal polynomials. The Lagrange polynomials choose the points $x_j$ to be the zeros of the polynomial curve for which $y_j = 0$, which allows $y(x)$ to be simply expressed as a product of $(x - x_j)$ terms with an overall factor.

A numerical example is clearer. Consider $f(x) = x(x-7)(x+3)(x+5)$. Simple expansion shows that this is $f(x) = x^4 + x^3 - 41x^2 - 105x$, but show this using Eq. (A.40) for the points $x_j = j$ with $j = (1, 2, 3, 4, 5)$. Then

$$
\mathbf{M} = \begin{bmatrix} 1 & 1 & 1 & 1 & 1 \\ 1 & 2 & 4 & 8 & 16 \\ 1 & 3 & 9 & 27 & 81 \\ 1 & 4 & 16 & 64 & 256 \\ 1 & 5 & 25 & 125 & 625 \end{bmatrix}
$$

The product of the inverse of $\mathbf{M}$ with $|f\rangle$ is

$$
\mathbf{M}^{-1}|f\rangle = \frac{1}{24}\begin{bmatrix} 120 & -240 & 240 & -120 & 24 \\ -154 & 428 & -468 & 244 & -50 \\ 71 & -236 & 294 & -164 & 35 \\ -14 & 52 & -72 & 44 & -10 \\ 1 & -4 & 6 & -4 & 1 \end{bmatrix} \begin{bmatrix} -144 \\ -350 \\ -576 \\ -756 \\ -800 \end{bmatrix} = \begin{bmatrix} 0 \\ -105 \\ -41 \\ 1 \\ 1 \end{bmatrix}
$$

and so $\langle C| = [0, -105, -41, 1, 1]$, as expected.

## A.8 Local Maximum/Minimum

The local fitting of a function to a low order polynomial is useful for the rapid evaluation of the extremum of a smoothly varying function $f(x)$, defined as the value of $x_m$ for which

$$
\left.\frac{d}{dx} f(x)\right|_{x=x_m} = 0 \tag{A.45}
$$

For what follows, assume that the value $x_n$ is in the vicinity of an extremum of $f(x)$: programming languages and numerical packages generally have function calls that will return the value of that $x_n$ such that $f(x_n) \geq f(x_j)$ for all values of $j \neq n$ in a vector with component $f(x_j)$, and the $x_n$, so returned is by definition near the extremum.[1] Consider several approaches, all of which assume that $n$ has been found by any convenient presumably numerical means.

The first approach is useful if $x_j$ is sorted but not necessarily uniformly spaced. Define the three-component vector $x_k : k \in (-1, 0, 1)$ and $y_k = f(x_{k+n})$. Calculate the coefficients $P_j$ such that locally, $y_a(x)$ is a quadratic approximation to $y(x)$ (that is, $y_a(x_k) = y_k$) such that

$$y(x) \approx y_a(x) \equiv \sum_{j=0}^{2} P_{3-j} x^j = P_1 x^2 + P_2 x + P_3 \tag{A.46}$$

It is seen that $y_a(x)$ is the parabolic approximation to $y(x)$ near $x \approx x_n$. The $P_j$ can be determined by least squares fitting for a quadratic passing through the points $(x_k, y_k)$, or, equivalently, using Eq. (A.44). Setting the derivative of $y_a(x)$ to zero and solving for $x$ gives $x_m$ to be

$$x_m = x_n - \frac{P_2}{2P_3} \tag{A.47}$$

If the $x_j$ are equispaced, then the $P$ coefficients can be found by using finite difference approximations and the approximation (where $\Delta = x_n - x_{n-1}$)

$$\begin{aligned} y_a(x) &= y_a(x_n) + y_a'(x_n)\,(x - x_n) + \frac{1}{2} y_a''(x_n)\,(x - x_n)^2 \\ y_a'(x_n) &= \frac{y_{n+1} - y_{n-1}}{2\Delta} \\ y_a''(x_n) &= \frac{y_{n+1} - 2y_n + y_{n-1}}{\Delta^2} \end{aligned} \tag{A.48}$$

Observe that if $\tilde{x}$ is defined as $\tilde{x} \equiv x - x_n$, then the coefficients for $P_j$ are directly provided by $y_a'(\tilde{x}_0)$ and $2y_a''(\tilde{x}_0)$ already.

The second approach is of higher accuracy, but more involved and requires that $x_j$ is uniformly spaced. Again, assume that the pair $(x_n, y_n)$ is known. Now, however, calculate the coefficients for the higher order polynomial approximation, and again use $\tilde{x} = x - x_n$ as the independent variable. The higher order quartic approximation is then based on the points $\tilde{x}_k : k \in (-2, 0, 2)$ and $y_k = f(x_n + \tilde{x}_k)$

$$y_a(\tilde{x}) \equiv \sum_{j=0}^{4} P_{5-j} \tilde{x}^j = P_1 \tilde{x}^4 + P_2 \tilde{x}^3 + P_3 \tilde{x}^2 + P_4 \tilde{x} + P_5 \tag{A.49}$$

This entails that the derivative of $y_a(\tilde{x})$ is then

$$y'(\tilde{x}) \equiv \sum_{j=1}^{4} j P_{5-j} \tilde{x}^{j-1} = 4P_1 \tilde{x}^3 + 3P_2 \tilde{x}^2 + 2P_3 \tilde{x} + P_4 \tag{A.50}$$

This is a cubic equation, and so a spline fit of $\tilde{x}$ as a function of $y'$ will return $\tilde{x}_m = x_m - x_n$ when evaluated at $y' = 0$.

1 For the MATLAB package, the code `[xo,jo] = max(fj)` for a vector `fj` returns both the value of `xo` and its index `jo`, both of which are useful here.

As an example of the comparison of both methods, consider the convenient function $f(x) = xe^{-ax}$ evaluated at $x_j = x_{max}(j-1)/(N-1)$. For purposes of illustration, take $N$ to be fairly small so that the errors are readily seen: here, use $x_{max} = 7$ and $N = 15$. Lastly, let $a = 1/4$. The maximum of $f(x)$ is then trivially found from calculus to be $x_m \equiv 1/a = 4$ and $f(x_m) = 1.47152$. Numerical evaluation, on the other hand, yields $n = 10$, $x_n = 4.20000$ and $y_n = 1.46974$. The first, or quadratic, method returns $x_m \approx 4.00880$ with an error of 0.219977%. The second, or quartic, method returns $x_m \approx 3.99983$ with an error of −0.004167%.

## A.9 Finite Difference Method

Differential equations can sometimes be profitably solved numerically by using finite difference approximations to derivatives and solving a large matrix equation [154, 320]. The finite difference approximations themselves can be found by solving a matrix equation based on the Taylor expansion of a function $f(x)$ at equispaced points such that $x_j = j\Delta$. It is convenient to assume that $x_0 = 0$ because the $x$-coordinate can be shifted by a constant amount later when it is convenient to do so. Using Newton's method of having primes denote derivatives, a Taylor expansion of $f(x)$ about the origin $x = 0$ can be written as

$$f(x) = f(0) + xf'(0) + \frac{1}{2}x^2 f''(0) + \cdots = \sum_{j=0}^{\infty} \frac{x^n}{n!}\frac{d^n}{dx^n} f(x)\Big|_{x=0} \tag{A.51}$$

Introduce the compact notation $f^n \equiv \Delta^n \partial_x^n f(x)|_{x=0}$, where for notational convenience the distinction between derivatives $d/dx$ and partial derivatives $\partial/\partial x$ is suppressed. Then a Taylor expansion is easily expressed as

$$y_j \equiv f(x_j) = \sum_{n=0}^{\infty} \frac{j^n}{n!} f^n \tag{A.52}$$

where one must keep in mind that $f^n$ is not $f(x)$ raised to a power, but rather $n$ is an indication of how many times a derivative has been taken of $f(x)$. For polynomials, the series truncates because there is some maximum power of $x$ in $f(x)$. As a result, the finite difference method is reliant on the local approximation of a general function by a low order polynomial, most commonly linear and quadratic but occasionally cubic and quartic. The finite difference approximations are found by using as many values of $j$ as there are powers of $x$ in the polynomial expansion. For example, if $f(x)$ is linear, then $y_j = f^0 + jf^1$, and so two equations are required to find $(f^0, f^1)$. They are $y_0 = f^0$ and $y_{\pm 1} = f^0 \pm f^1$, where the "+" case is for points chosen after $x = 0$ (forward) and "−" for points chosen before (backward). Clearly, the $y_j$ can be put into a vector $|y\rangle$ and these two equations can be written as a matrix equation

$$|y\rangle = \begin{bmatrix} y_0 \\ y_{\pm 1} \end{bmatrix} = \begin{bmatrix} f_0 & 0 \\ f_0 & \pm f^1 \end{bmatrix} = \begin{bmatrix} 1 & 0 \\ 1 & \pm 1 \end{bmatrix} \begin{bmatrix} f^0 \\ f^1 \end{bmatrix} \equiv \mathbb{M}_2 |f\rangle \tag{A.53}$$

This matrix equation can be inverted to give the upward (+) and downward (−) finite differencing schemes, (UDS) and (DDS) respectively, for the first derivative

$$\left(\frac{df}{dx}\right)_{UDS} \approx \frac{y_1 - y_0}{\Delta}; \quad \left(\frac{df}{dx}\right)_{DDS} \approx \frac{y_0 - y_{-1}}{\Delta} \tag{A.54}$$

Approximating the second derivative requires a quadratic polynomial, and therefore the matrix $\mathbb{M}_3$ is a $3 \times 3$ matrix given by

$$\mathbb{M}_3 = \frac{1}{2}\begin{bmatrix} 2 & -2 & 1 \\ 1 & 0 & 0 \\ 2 & 2 & 1 \end{bmatrix} \tag{A.55}$$

from which $|y\rangle = \mathbb{M}_3|f\rangle$ is inverted to give

$$|f\rangle = \begin{bmatrix} f^0 \\ f^1 \\ f^2 \end{bmatrix} = \frac{1}{2}\begin{bmatrix} 2y_0 \\ y_1 - y_{-1} \\ 2y_1 - 4y_0 + 2y_{-1} \end{bmatrix} \tag{A.56}$$

from which the central difference approximation for the derivatives are

$$\begin{aligned} \left(\frac{df}{dx}\right)_{CDS} &\approx \frac{y_1 - y_{-1}}{2\Delta} \\ \left(\frac{d^2f}{dx^2}\right)_{CDS} &\approx \frac{y_1 - 2y_0 + y_{-1}}{\Delta^2} \end{aligned} \tag{A.57}$$

One need not stop there: higher order derivatives are found by using even more $y_j$ points, but one must be careful. Although the error of the approximation for the first derivative goes as $\Delta^n$, where $n$ is the order of the polynomial, in general, a polynomial fit to a function becomes "stiff" for larger values of $n$ and is reflected in polynomial fitting functions in numerical packages becoming squirrelly when $n$ is too large. That behavior is reflected in the higher derivatives $f^n$ becoming less good as approximations.

One need not use central difference schemes alone: a differential equation modeled at a boundary requires a second-order differencing scheme for the second derivative that uses points inside the simulation region, making for upwind and downwind second-order versions of $d^2y/dx^2$. They tend to be slightly less accurate than the central differencing schemes, even though the error still goes as $\Delta^2$, because points further from the origin are used, but their usage in, for example, numerically solving heat diffusion allows the numerical solution to proceed.

## A.10 Iterative Solutions

Finding the solution to some equations is open to the Method of Iteration, whereby a guessed solution can provide the means to find a more accurate solution. The equation to be solved is often rewritten in the form $x = f(x)$: if $x_j$ is the $j^{th}$ estimate of the solution $x$, then $x_{j+1} = f(x_j)$ is a concise statement of the iterative method, and is remarkably well-suited for solving using spreadsheets. The $f(x)$ and its derivatives must, however, satisfy certain constraints, one of which the gradient is not in excess of unity and $f(x_\infty)$ is finite, but for the examples considered here, those constraints are met and need not further complicate the discussion.

As a simple and historically interesting example, consider the Golden Ratio of antiquity [297] that relates a line of length $L$ cut two lines of length $a$ and $L - a$ such that the ratio of

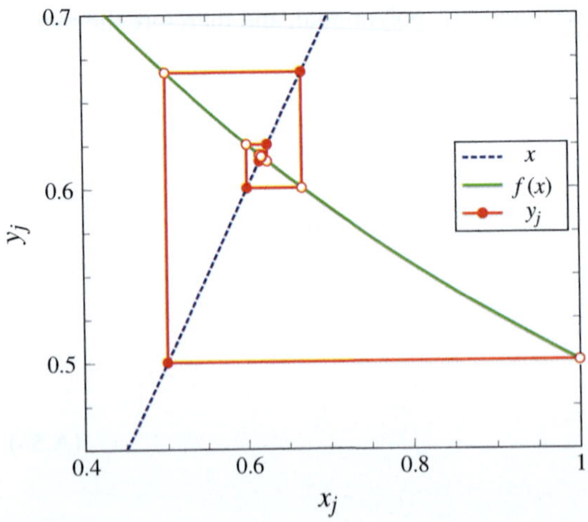


**Figure A.3** Iterative solution to Eq. (A.59). Closed circles • are the guess; open circles ∘ are the iteration $y_j = f(x_j)$ for $f(x) = 1/(1+x)$. Convergence is to where $y = x$ and $y = f(x)$ intersect. Initial guess is $(x_0, y_0) = (1, 1)$.

$a$ to $L$ is the same as that of $L - a$ is to that of $a$, or $a/L = (L-a)/a$. Letting $x = a/L$, then this is the same as

$$x^2 + x - 1 = 0 \tag{A.58}$$

or a simple quadratic equation with integer coefficients. The quadratic formula gives the two roots $x_\pm = (-1 \pm \sqrt{5})/2$, for which $x_+ = 0.618034$ is the Golden Ratio solution sought. To use iteration, rewrite the equation in the form $x = f(x)$ as

$$x = f(x) \equiv \frac{1}{x+1} \tag{A.59}$$

Graphically, Eq. (A.59) says the solution is the intersection of the two lines $y_1 = x$ and $y_2 = 1/(1+x)$. Begin the iteration with any convenient choice $x_0$ for $j = 0$ and keep solving (iterating) until $x_{j+1}$ and $x_j$ are equivalent to the level of precision desired. If $x_0 = 1$, then for $j \in (1, \cdots 10)$, $\{x_j\} = [1, 1/2, 2/3, 3/5, \ldots 89/144]$, and so to six significant digits, $x \approx x_{10} = 0.618056$ is the solution. What is happening is graphically represented in Figure A.3 (see Algorithm B.5).

# B

# Algorithms

*But in these cases*
*We still have judgement here, that we but teach*
*Bloody instructions, which being taught return*
*To plague the inventor.*

W. Shakespeare, *Macbeth* I.7

## B.1 Basic Methods

### B.1.1 Matrices and Vectors

For consistency, the algorithms have been rendered in the MATLAB programming language as it allows for direct representations of matrices and vectors, provides an immediate graphics capability that is advantageous, and has polynomial and spline-fitting functions that render certain calculations elegantly fast. Here, some standard notations in physics are re-expressed as equivalent formulations in MATLAB.

If $|v\rangle$ is a column vector and v its MATLAB equivalent, then $\langle v|$ is equivalent to v'. MATLAB scripts often work faster if the size of the vector v is declared at the outset: a column vector of $n$ entries can be initialized by a column vector $|v\rangle$ of zeros or $|w\rangle$ of ones by

```
% create a column vector v of zeros or a vector w of ones
v = zeros(n,1); w = ones(n,1);
%  v' and w' are therefore equivalent to zeros(1,n) and ones(1,n), respectively
```

If $|j\rangle$ is a column vector of zeros except for the $j^{th}$ entry which is 1, then the physics representation and its relation to the MATLAB representation of the Kronecker delta function $\delta_{ij}$ by the correspondence between $\langle i|j\rangle = \delta_{ij}$ and

```
v = zeros(n,1); v(i) = 1;
w = zeros(n,1); w(j) = 1;
w' * v = 1 % only if i = j
```

For a matrix $\mathbb{M}$ with the entry for the $i^{th}$ row and $j^{th}$ column $M_{ij}$, then $M_{ij} = \langle i|\mathbb{M}|j\rangle$ and corresponds to $M(i, j)$.

*Fundamentals of Electron Emission Physics*, First Edition. Kevin L. Jensen.

The MATLAB notation `w.*v` is equivalent to the physics notation

$$\langle w|v\rangle = \sum_{j=1}^{n} w_j v_j$$

where the "`.*`" means the element by element product and then summation. A fast method to create an $m \times n$ ($m$ rows and $n$ columns) matrix is to multiply a column vector $|v\rangle$ by a row vector $\langle w|$ via $|v\rangle\langle w|$. In MATLAB, this becomes `w*v` using "`*`", not "`.*`". For example, let what is manually entered into the MATLAB command window be preceded by "»" below. The output is then (MATLAB includes empty lines not shown here, and a computation such as `v*w` is returned on a line in the form: `ans = ...`) and is seen to be such that the $j^{th}$ column of the resulting matrix is the product of $|v\rangle$ multiplied by the $j^{th}$ element of the row vector $\langle w|$:

```
>> w = [1,2,3,4], v = [5;6;7], v*w
w =  1     2     3     4
v =  5
     6
     7
ans =5    10    15    20
     6    12    18    24
     7    14    21    28
```

### B.1.2 Structure for Fundamental Constants

The algorithms of subsequent sections make use of structure arrays. The notation herein is generally to use the notation `Px` to denote a structure array, where `x` may be any other letter, e.g. `Pn`. If `fld` is a field in `Pn`, then typing `Pn.fld` will return the contents of that field, whereas typing `Pn` will return all the entries of the structure `Pn`. Numerical values for `fld` can be treated as computational quantities. Strings in `Pn` can be used in plotting to label axes or legends. Vectors and matrices can be stored in `Pn` and used to create plots. Examples of all of these are in the scripts that follow.

The most common structure used herein is the structure array for the fundamental constants and conversion factors denoted `P` and created by calling the script `FUNdamental.m`. Explicitly,

```
function Pn = FUNdamental(P)
%% FUNCTION FUNdamental.m
%-----------------------------------------------------
% adds constants to structure if P exists:      FUNdamental(P)
% or creates structure Pn if P does not exist:  FUNdamental()
%-----------------------------------------------------
if exist('P','var')  == 1; else; P = struct; end
Pn = P;
%% fundamental Constants in units of eV, fs, nm, q
Pn.kb    = 1.0/11604.50635;  % Boltzmann's const in eV/K
Pn.hbar = 0.6582119571;     % Planck's const * 2 pi
Pn.c     = 299.7924580;      % speed of light nm/fs
Pn.mo    = 5.685630103;      % e mass in eV (fs/nm)^2
Pn.afs   = 1.0/137.035999084; % fine structure const
Pn.ao    = 0.05291772110;     % Bohr radius in nm
Pn.Qo    = Pn.afs*Pn.hbar*Pn.c/4;  % in eV nm
%% units:
% B = base unit, essentially Coulomb / q
B = 6.241509074e18;
Pn.meter = 1e9;
```

```
Pn.cm = 1e7;
Pn.Coulomb = B;
Pn.Joule = B;
Pn.second = 1e15;
Pn.Volt = 1;
Pn.Amp = Pn.Coulomb/Pn.second;
Pn.Ohm = Pn.Volt/Pn.Amp;
Pn.mole = 6.02214076e23;
Pn.kg = Pn.Joule*(Pn.second/Pn.meter)^2;
```

As an example

```
>> Pn = FUNdamental(); Pn.kb, Pn.mo*Pn.c^2, Pn.Qo
ans = 8.6173e-05
ans = 5.1100e+05
ans = 0.3600
```

where the first `and` is $k_B$, the second is the rest mass of the electron $mc^2$ in eV, and the third is the numerical value of the image charge constant $Q$. Structure files are the preferential method to send needed data into a function or subscript, either to calculate quantities of interest or to send results of a calculation to a plotting script.

### B.1.3 Structure for Line Plots

Let a matrix `Mxy` be a rectangular matrix of $N_x$ rows and $N_c + 1$ columns. Let the first column be a vector of $x$-coordinate values $x_j$, and the remaining $N_c$ columns be $y_k(x_j) \to y_{jk}$ for $k \in (1, \cdots, N_c)$. A commonly used structure file herein is `QPlot.m, which is` used for making line plots of $x$ *vs.* $y$ using `P.Mxy`:

- `P.plot` can be: `linear, inversex, inversey, loglog, logx, logy, sqrtx.`
- `P.eps` can be `true` or `false`; when true, then the generation of an output graphic using `saveas(gcf,Pn.filn,'xxxx')` can use `epsc, png, pdf, jpg` for various types of figure files.
- `P.cred` is an $N \times 3$ matrix of color values, with all of the entries between 0 and 1, e.g. `P.cred = [1,0,0;0,1,0;0,0,1]` assumes three lines, the first of which is red, second is green, and last is blue.
- `P.xyrange` is a four component row vector $[x_{min}, x_{max}, y_{min}, y_{max}]$ for the range of the axes. If it is not specified, then it is determined by the maximum and minimum entries of `P.Mxy`.
- `P.filn` is the output file name.
- `P.Lgt` is a character array for the legend.
- `P.xlabl, P.ylabl` are the character labels of the $x$- and $y$-axes.
- `P.titltx` is the title of the graphic.

The file `QPlot.m` is

```
function tfln = QPlot(Pn,ifig)
%% Function QPlot(Pn,ifig)
%      Rapid plotting of P structure data P.Mxy
%      ifig is the figure number (integer, should not be 0)
%      Types of plots set by Pn.plot:
%           linear, inversex, inversey, loglog, logx, logy, sqrtx
Mxy = Pn.Mxy;
N = size(Mxy,2);
sx = Mxy(:,1); sy = Mxy(:,2:N);
if isfield(Pn,'plot');else;Pn.plot = 'linear';end
```

```
if isfield(Pn,'xyrange');else
   Pn.xyrange = [min(sx),max(sx),min(sy(:)),max(sy(:))];
end
if isfield(Pn,'eps');else;Pn.eps = false;end
xyrange = Pn.xyrange;
if strcmp(Pn.plot,'linear')
   xlabl = join(['$',Pn.xlabl,'$']);
   ylabl = join(['$',Pn.ylabl,'$']);
elseif strcmp(Pn.plot,'loglog')
   sx = log10(sx);
   sy = log10(sy);
   xlabl = join(['$\log_{10}(',Pn.xlabl,')$']);
   ylabl = join(['$\log_{10}(',Pn.ylabl,')$']);
   xyrange = log10(xyrange);
elseif strcmp(Pn.plot,'logx')
   sx = log10(sx);
   xlabl = join(['$\log_{10}(',Pn.xlabl,')$']);
   ylabl = join(['$',Pn.ylabl,'$']);
   xyrange(1:2) = log10(xyrange(1:2));
elseif strcmp(Pn.plot,'logy')
   sy = log10(sy);
   xlabl = join(['$',Pn.xlabl,'$']);
   ylabl = join(['$\log_{10}(',Pn.ylabl,')$']);
   xyrange(3:4) = log10(xyrange(3:4));
elseif strcmp(Pn.plot,'inversex')
   sx = 1./sx;
   xlabl = join(['$1/',Pn.xlabl,'$']);
   ylabl = join(['$',Pn.ylabl,'$']);
   xyrange(1:2) = 1./xyrange(2:-1:1);
elseif strcmp(Pn.plot,'inversey')
   sy = 1./sy;
   xlabl = join(['$',Pn.xlabl,'$']);
   ylabl = join(['$1/',Pn.ylabl,'$']);
   xyrange(3:4) = 1./xyrange(4:-1:3);
elseif strcmp(Pn.plot,'sqrtx')
   sx = sqrt(sx);
   xlabl = join(['$(',Pn.xlabl,')^{1/2}$']);
   ylabl = join(['$',Pn.ylabl,'$']);
   xyrange(1:2) = sqrt(xyrange(1:2));
end
% disp(['SUB QPlot: graph type set to ',Pn.plot])
figure(ifig)
colororder(Pn.cred)
plot(sx,sy)
set(gca,'fontsize',14)
xlabel(xlabl,'FontSize',24,'Interpreter','latex')
ylabel(ylabl,'FontSize',24,'Interpreter','latex')
legend(Pn.Lgt,  'FontSize',16,'Interpreter','latex',
'Location','best')
title(Pn.titltx,'FontSize',18,'Interpreter','latex')
axis(xyrange)
% Naming convention:  filn + figure number
tfln = [Pn.filn,num2str(ifig)];
if Pn.eps; saveas(gcf,Pn.filn,'epsc'); end
```

As an example, running the script `SamplePlot.m` gives an encapsulated postscript (eps) file in Figure B.1:

```
%% Figure SamplePlot
P = struct;
P.xlabl = 'x'; P.ylabl = 'y(x)';
P.titltx = '$x \exp(- a x^2)$'; P.filn = 'SampleP';
```

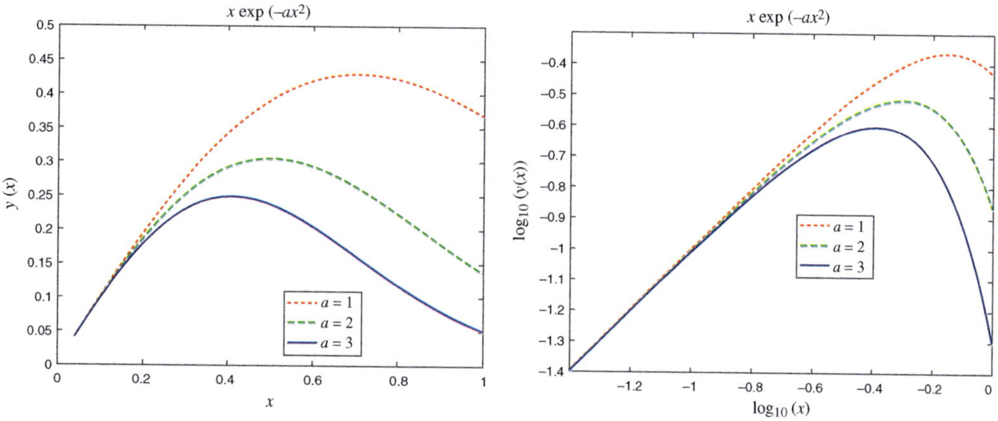


**Figure B.1** Graphical output of `Qplot(P,j)` for `P.plot = 'linear'` ($j = 1$, left) and `P.plot = 'loglog'` ($j = 2$, right), where $j$ specifies output figure.

```
P.cred = [1,0,0;0,0.7,0;0,0,1];
xj = ((10:50)'/50).^2;
a  = [1,2,3];
P.Mxy = [xj,xj.*exp(-(xj.^2)*a)];
P.Lgt = {'$a = 1$','$a = 2$','$a = 3$'};
P.xyrange = ([0,1,0,0.5]);
P.eps = true;
P.plot = 'linear'; P.filn = 'SampleA'; QPlot(P,1);
P.plot = 'loglog'; P.filn = 'SampleB'; QPlot(P,2);
```

## B.2 Review of Methods Algorithms

Select algorithms for the generation of figures in Part II. The primary script is listed first. Subordinate scripts, if they have not been listed in prior algorithm sections, are subsequently (and only) listed in the section in which they first appear. The scripts `QPlot.m` and `FUNdamental.m`, which many of the following scripts call upon, are in Sections B.1.3 and B.1.2, respectively. If the scripts were used to generate a figure, that figure is identified first.

### B.2.1 Even/Odd Eigenstates

- Figures generated by this script: Figure 6.2
- Primary script: `FIGsqwl.m`

```
%%Program FIGFriedelAnalytic
%   A plot of the analytic approximation of Friedel
%   oscillations approximation
%   The box is of width L, so its boundaries are at +- L/2
%   and the wave function of either parity vanishes there
%   Even parity:  psi = sqrt(2/L) cos(k_n x):  k_n = pi*(2n - 1)/L
%   Odd  parity:  psi = sqrt(2/L) sin(k_n x):  k_n = pi*(2n - 0)/L
%   let s = x/L, k = pi n / L, so kx = pi n x / L = pi s
```

```
clear;clc
% Create the Structure file and populate it with fundamental constants
Po = FUNdamental();
% Control parameters
Po.Ns = 160;
Po.N = 2*[2,4];
Po.cred = [1,0,0;0,0,1];   % red, green, blue, black
Po.plot = 'linear';
% Function:  Friedel
Po.rhofun = @(x) 1.0 + 3*(x.*cos(x) - sin(x))./x.^3;
 %transmission probability
% Plot type
Po.plot = 'linear';
for j = 1:2
   N = Po.N(j);
   s = 0.5*pi*(2*(1:Po.Ns)' - Po.Ns - 1)/(Po.Ns - 1);
   coss = cos(s*(2*(1:N)-1));
   sins = sin(s*(2*(1:N)-0));
   Cn   = (3/(2*N^3))*(N^2 - (1:N).^2);
   % Cn = ones(1,N)/N;
   ys   = 2*sum(Cn.*sins.^2,2);
   yc   = 2*sum(Cn.*coss.^2,2);
   Po.cred = [0,0,1;0,0.7,0;1,0,0;0.6,0.6,0.6];
   Po.Mxy = [s/pi,ys,yc,(ys + yc)/2];
   Po.xlabl = 'x / L';
   Po.ylabl = '\xi(s)';
   Po.Lgt = {'Odd','Even','Sum'};
   Po.titltx = ['N = ',num2str(N)];
   Po.filn = ['SumBasis',num2str(N)];
   Po.xyrange = [s(1)/pi,s(Po.Ns)/pi,0,2];
   % plot(sa,ysum,'bo',sa,ya,'r')
   QPlot(Po,j); hold on; plot([-2,2,2,0,0],[1,1,2,2,0]); hold off
   Po.Lgt = {'Odd','Even','Sum','unity'};
   legend(Po.Lgt)
   saveas(gcf,Po.filn,'png')
   writematrix(Po.Mxy,Po.filn,'FileType','text','Delimiter','tab')
end
```

### B.2.2 Friedel Oscillations

- Figures generated by this script: Figure 6.3
- Primary script: `FIGFriedelAnalytic.m`

```
%% Figure FIGFriedelAnalytic
%   A plot of the analytic approximation of Friedel
%   oscillations approximation
clear;clc
% Create the Structure file and populate it with fundamental constants
Po = FUNdamental();
% Control parameters
Po.Ns    = 100;   % number of s = 2 * kf * x points
Po.N     = 8;     % number of curves
Po.cred = [1,0,0;0,0,1];   % red, green, blue, black
Po.plot = 'linear';
% Function:  Friedel
Po.rhofun = @(x) 1.0 + 3*(x.*cos(x) - sin(x))./x.^3;
Po.N = [3,4,8]; Po.Ns = 125;
Nc = length(Po.N);
Xi = zeros(Po.Ns+2,2*Nc);
```

```
sa = -0.5*(0:Po.Ns+1)'/Po.Ns;
for j = 1:3
   N = Po.N(j);
   sins = sin(2*pi*sa*(1:N)).^2;
   Cn   = (3/(2*N^3))*(N^2 - (1:N).^2);
   ysum = 2*sum(Cn.*sins,2);
   ya   = Po.rhofun(2*pi*(2*N)*sa); ya(1) = 0;
   Xi(:,2*j-1:2*j) = [ya,ysum];
end
Xi(Po.Ns+2,:) = 0;
Po.Mxy = [10*sa,Xi];
Po.xlabl = '10 \times x/L';
Po.ylabl = '\xi(2 k_N x)';
Po.Lgt = {'$\xi(3)$','Sum(3)','$\xi(4)$','Sum4)','$\xi(8)$','Sum(8)'};
Po.titltx = 'Friedel oscillations';
Po.filn = 'Friedel';
Po.xyrange = ([-5 0 0 1.2]);
QPlot(Po,1);
saveas(gcf,Po.filn,'png')
writematrix([Po.Mxy,ysum],Po.filn,'FileType','text','Delimiter','tab')
```

### B.2.3 Image Charge Models

- Figures generated by this script: Figure 3.1
- Primary script: `FIGimage.m`

```
%% Function FIGImage
%   figure of image charge in the x-y plane
% function
Vq   = @(x,y,a) 1.0./sqrt(x.^2 + (y - a).^2);
s = ((0:32)'/32); s(1) = 0;
cred = [s,s,s].^0.5;
a = 1; Vo = 10; n = 2;
N = 160; Nv = length(s);
vv = linspace(-1,1,Nv+1);
vv = 2*Vo*vv.^5;
x = n*a*linspace(-1,1,N);
[x,y] = meshgrid(x,x);
Vxy = Vq(x,y,-a) - Vq(x,y,a);
Vxy = min(max(Vxy,-1.1*Vo),1.1*Vo);
figure(1)
contourf(x,-y,Vxy,vv)
axis square tight
colormap(cred)
axis off
hold on; plot([-n*a,n*a],[0,0],'r','LineWidth',2); hold off
saveas(gcf,'DipoleC','epsc');
figure(2)
surf(x,y,Vxy); axis tight square; shading interp; axis off; hold on
plot(n*[-a,a],[0,0],'r','LineWidth',2); hold off
colormap(cred)
view(135,25)
saveas(gcf,'DipoleS','epsc');
```

- Figures generated by this script: Figure 3.2
- Primary script: `FIGcapas.m`

```
%% Figure FIGcapas
%  Multiple Image charges
s = ((0:32)'/32); s(1) = 0;
cred = [s,s,s].^0.5;
% function
FVxy  = @(x,y,a,b) 1.0./sqrt((x-b).^2 + (y - a).^2);
% parameters
a = 1; Vo = 1; n = 2.5;
N = 160; Nv = length(s);
Nj = [3,5,15];
vv = linspace(-1,1,Nv+1);
vv = 2*Vo*sign(vv).*abs(vv).^2;
x = n*a*linspace(-3,3,N);
[x,y] = meshgrid(x,x);
Vxy = zeros(N);
for k = 1:3
   Nq = Nj(k);
   Vxy = 0;
   bj = 2*n*a*linspace(-1,1,Nq);
   for j = 1:Nq
       Vxy = Vxy + FVxy(x,y,-a,bj(j)) - FVxy(x,y,a,bj(j));
   end
   Vxy = min(max(Vxy/Nq,-1.1*Vo),1.1*Vo);
   figure(k)
   contourf(x,-y,Vxy,vv); hold on; plot([-n*a*3,n*a*3],[0,0],'r','LineWidth',2); hold
     off
   axis square tight
   colormap(cred)
   axis off
   saveas(gcf,['MpoleC',num2str(Nq)],'epsc');
end
```

- Figures generated by this script: Figure 3.4
- Primary script: `FIGdipole.m`

```
%% Figure FIGdipole
%   contour plot of dipole arrangement of charge and demonstration that it
%   acquires the image of a hemisphere on a plane
% Graphing terms
s = 0.5 + 0.5*((0:15)'/15);
cred = [s,s,s]; cred(1,:) = 0; cred(16,:) = 1;
% function
Vq  = @(x,z,a) 1./sqrt(x.^2 + (z + a).^2);
Nx = 2^7;
sj = linspace(-1,1,Nx);
a = 1;
x = 1.5*sj;y = x';
vv = 20*linspace(-1,1,16).^5;
del = 0.04; C = -(a/(2*del))*(a^2 - del^2);
Vxy = y - C*Vq(x,y,del) + C*Vq(x,y,-del);
Vxy = max(min(Vxy,19.9),-19.9);
contourf(x,y,Vxy,vv)
axis square tight
colormap(cred)
set(gca,'fontsize',14)
xlabel('$x/a$','FontSize',24,'Interpreter','latex')
ylabel('$z/a$','FontSize',24,'Interpreter','latex')
title(['$\delta / a = ',num2str(del),'$'],'FontSize',18,'Interpreter','latex')
hold on
plot(x,0*y,'c','LineWidth',2)
```

```
plot(cos(pi*sj),sin(pi*sj),'r','LineWidth',2);
plot(0,-del,'.b','MarkerSize',16)
plot(0,del,'.y','MarkerSize',16)
hold off
saveas(gcf,['Dipole',num2str(100*del)],'epsc')
saveas(gcf,['Dipole',num2str(100*del)],'png')
```

### B.2.4 Equipotential Models

- Figures generated by this script: Figure 3.3
- Primary script: `FIGhemisphere.m`

```
%% Figure FIGhemis[phere
% Contour plot of V(rho,z) = -Fo z (1 - (a/r)^3)
% for r^2 = rho^2 + z^2
clear; clc;
N = 200; Nc = 11;
xmax = 2;
Vo=0.3;
r = linspace(-xmax,xmax,N);
z = linspace(0,2*xmax,N)';
vv = -0.25*(0:Nc-1);
Vxy  = -z.*(1 - ((r.^2 + z.^2).^(-3)));
Vxyo = -z + 0*r;
s = (0:Nc-1)'/(Nc-1);
cred = 1-0.5*s*[1,1,1];cred(Nc,:) = 0;
contourf(r, z, Vxy, vv,'r','Linewidth',1); axis tight square; hold on;
contour(r, z, Vxyo, vv, '--b','Linewidth',0.5); hold off; axis off
colormap(cred)
axis off
saveas(gcf,'HemiC','epsc');
```

- Figures generated by this script: Figure 3.8
- Primary script: `FIGprolate.m`

```
%% Figure FIGprolate
%   contour plot of prolate spheroidal geometries
%   (hyperbolic and ellipsoidal)
Nx = 51; Nc =12;
% graphing parameters
s = 0.5 + 0.5*((0:15)'/15);
cred = [s,s,s]; cred(1,:) = 0; cred(16,:) = 1;
etah = linspace(-3,3,Nx)';
upsh = (pi/2)*(0:Nc)/Nc;
rhoh = sinh(etah)*sin(upsh);
zh   = cosh(etah)*cos(upsh);
etae = 3*(1:Nc)/Nc;
upse = (pi/2)*linspace(-1,1,Nx)';
rhoe = sin(upse)*sinh(etae);
ze   = cos(upse)*cosh(etae);
figure(1); plot(rhoh,zh,'--b',rhoe,ze,'r');
axis([-2, 2, 0, 4]); axis square; hold on;
area(rhoe(:,1),ze(:,1),'FaceColor','black'); hold off; axis off
title('Ellipsoidal','fontsize',24,'Interpreter','latex')
saveas(gcf,'Ellipsoidal','epsc');
figure(2); plot(rhoh,4 - zh,'r',rhoe,4 - ze,'--b');
axis([-2, 2, 0, 4]); axis square; hold on;
plot([0,0],[0,4],'--b');
```

```
area(rhoh(:,3),4-zh(:,3),'FaceColor','black');
plot([-2,2],[4,4],'k','LineWidth',6); hold off; axis off
title('Hyperbolic','fontsize',24,'Interpreter','latex')
saveas(gcf,'Hyperbolic','epsc');
```

### B.2.5 Field Enhancement

- Figures generated by this script: Figure 3.6
- Primary script: `FIGbetaa.m`

```
%% Program Beta - a
%  Field enhancement factor and tip radius calculation
clear; clc
% Legendre Polynomial of 2nd Kind
Qox   = @(x) 0.5*log((1+x)./(1-x));
P.xlabl = '\delta';
P.ylabl = 'y(\delta)';
P.titltx = 'Field enhancement and apex radius';
P.xyrange = [0, 1, 0, 2];
P.Lgt = {'$\beta \times \delta$','$a(\delta)/L$'};
P.filn = 'LineCbetaa';
P.xyrange = [0,1,0,3];
P.cred = [1,0,0;0,0,1];
P.plot = 'linear';
Nx = 20; x = ((1:Nx)'/Nx).^2;
betdel   = (1 + ((1+x)./(x.*(2+x).*Qox(1./(1+x)))));
adel     = (x.*(2+x)./(1+x)) + ((x.*(2+x)./(1+x)).^2).*Qox(1./(1+x));
P.Mxy = [x,x.*betdel,adel];
QPlot(P,1);
saveas(gcf,'BetaAs','png');
writematrix(P.Mxy,P.filn,'FileType','text','Delimiter','tab')
```

- Figures generated by this script: Figure 3.5
- Primary script: `FIGMakeTip.m`

```
%% Figure FIGMakeTip
%  Line Charge Model
%  Model of an ellipsoid in a constant electric field
%  using a constant linear density line charge
clear; clc
Nc = 21;
xmax   = 1; Vmax = 2*xmax;
dx     = 2*xmax/128;
rlamc = 0.05*(1:4);
rlamt = 0.10*(1:4);
%formula
urz   = @(x,y,a) 0.5*(sqrt(y.^2 + (x-a).^2) + sqrt(y.^2 + (x+a).^2));
wrz   = @(x,y,a) 0.5*(sqrt(y.^2 + (x+a).^2) - sqrt(y.^2 + (x-a).^2));
Vurz = @(x,y,a) log((urz(x,y,a) + a)./(urz(x,y,a) - a));
Qorz = @(x)  0.5*log((1.0 + x)./(1.0 - x));
% Make a gray colormap
s = (0:Nc-1)'/(Nc-1); cred = 1-0.5*[s,s,s];cred(Nc,:) = 0;
% Constant Lambda-j
Vmin = - 2*xmax;
[y, x] = meshgrid(-xmax:dx:xmax, 2*xmax:-dx:0);
for j = 1:4
    zc = -x + rlamc(j)*Vurz(x,y,1);
    zt = -x + rlamt(j)*(2*x.*Qorz(1.0./urz(x,y,1)) - 2*wrz(x,y,1));
```

```
    figure(j);contourf(y,x,zc,-2:0.2:0); axis square tight;
    colormap(cred); hold on
    plot([0,0],[0,1],'y','linewidth',4); hold off
    axis([ -xmax xmax 0 2*xmax]); axis off;
    title(['$\lambda = ',num2str(rlamc(j)),'$'],'Fontsize',14,'interpreter','latex')
    saveas(gcf,['LCo',num2str(100*rlamc(j))],'epsc')
    figure(j+4);contourf(y,x,zt,-2:0.2:0); axis square tight;
    colormap(cred); hold on
    plot([0,0],[0,1],'y','linewidth',4); hold off
    axis([ -xmax xmax 0 2*xmax]); axis off;
    title(['$\lambda_o = ',num2str(rlamt(j)),'$'],'Fontsize',14,'interpreter','latex')
    saveas(gcf,['LCt',num2str(100*rlamt(j))],'epsc')
end
```

### B.2.6 Diffusion

- Figures generated by this script: Figure 4.3
- Primary script: `FIGdiffuse.m`

```
%% Program FIGdiffuse
%  Simulation of N particles jumping left and right with proability p
clear; clc
Nb = 16; Np = 2^9;
bin = 3.6*linspace(-1,1,Nb)';
rng(5);     % iniitalize random seed
Nj = 4.^(4:9); Nc = length(Nj);
Mxy = zeros(Nb-1,Nc);
for j = 1:Nc
    x = sum(round(2*rand(Nj(j),Np) - 1))';
    sig = sqrt(var(x));
    Mxy(:,j) = histcounts(x/sig,bin);
end
bx = 0.001*round(1000*(bin(1:Nb-1) + 0.5*mean(diff(bin))));
% x = linspace(-4,4,100)';
y = mean(max(Mxy))*exp(-0.5*bx.^2);
Ns = size(Mxy,2);
s = (0:Ns)'/Ns;
P.xlabl = 'x/\sigma';
P.ylabl = 'N_{bin}';
P.titltx = 'Diffusion';
P.xyrange = [-4, 4, 0, max(Mxy(:))];
P.cred = [s.^2,4*s.*(1-s),1-s.^2];
P.Lgt = [num2str(log(Nj')/log(4));num2str(0)];
P.filn = 'Diffusion';
P.plot = 'linear';
P.Mxy = [bx,Mxy,y];
P.Lgt = 'fit';
QPlot(P,1);
saveas(gcf,P.filn,'png');
writematrix([bx,Mxy],P.filn,'FileType','text','Delimiter','tab')
```

### B.2.7 Brightness Temperature

- Figures generated by this script: Figure 4.4
- Primary script: `FIGTbright.m`

```
%% Figure FIGTbright
%   Evaluation of Brightness temperature
%   For (t, r, epsilon) = (0.93, 0.76, 0.52)
clear;clc
% Get fundamental constants
P = FUNdamental();
E = exp(linspace(log(0.05),log(20),41)');
P.Nx = length(E);
lambda = 2*pi*P.hbar*P.c./E;
C = (0.93*0.76*0.52);
Tb = 800*(1:0.5:2.5); Nt = length(Tb);
x = E./(P.kb*Tb);
y = Tb.*x./(x + log((1 - C)*exp(-x) + C));
s = (0:Nt-1)'/(Nt-1);
P.cred = [s.^2,2*s.*(1-s),1 - s.^2];
% graphics terms
P.xlabl = '\ln (\hbar \omega ~\textrm{[eV]})';
P.ylabl = 'T/T_b';
P.titltx = ['$t \times r \times \epsilon = ',num2str(C),'$'];
P.filn = 'Tbright';
P.Lgt = num2str(Tb');
P.xyrange = [-3, 3, 1, 2.5];
P.Mxy = [log(E),y./Tb];
QPlot(P,1);
saveas(gcf,P.filn,'png')
writematrix(P.Mxy,P.filn, 'FileType', 'text','Delimiter','tab')
```

### B.2.8 Wave Packet Evolution

- Figures generated by this script: Figures 6.4, 6.5, 6.6
- Primary script: `FIGDeltaWavepacket.m`
- Subordinate scripts: `FUNwavdelta.m`, `FUNsetkj.m`, `MOVrecbar.m`

```
%% Program FIGDeltaWavepacket
%   time evolution of gaussian wave packet through a dirac-delta barrier
clear; clc;
Pd = FUNdamental();
Pd.Nt = 120;
Pd.Nx = 200;
Pd.Nk = 10;
M = (0:15)/15; Pd.cred = [M',4*(1-M').*M',1-M']; clear M;
Pd.cred(1,:) = 0;
Pd.icase = 0;
Pd.Ck = 1;  % default Ck
Pd.dmin = 1e-3;
Pd.Nk = 48;
Pd.Eo = 1;
Pd.PltNam = 'DeltaVx';
Pd.L = 1;
Pd.W = 8;
Pd.gamma = 6;
Pd.iplt = round([0.1,0.5,0.9]*Pd.Nt);
Pdo = Pd; Pdo.gamma = 0;
Pd  = FUNwavdelta(Pd) ;
Pdo = FUNwavdelta(Pdo);
MOVrecbar(Pd,3)
QPlot(Pd,1);hold on; plot(Pdo.x,Pdo.rho(:,Pdo.iplt),'--g'); hold off
legend(Pdo.Lgt);
Nc = length(Pd.iplt);
```

```
x = Pd.x; rho = Pd.rho; rhoo = Pdo.rho;
Mxy = zeros(Pd.Nx+2,2*Nc + 1);
Mxy(:,1) = [-1;x;1];
Mxy(2:Pd.Nx+1,   2:  Nc+1) = rho(:,Pd.iplt);
Mxy(2:Pd.Nx+1,Nc+2:2*Nc+1) = rhoo(:,Pd.iplt);
saveas(gcf,Pd.filn,'png')
writematrix(Mxy,Pd.filn,'FileType','text','Delimiter','tab')
figure(4)
waterfall(x,Pd.tj(30:90),rho(:,30:90)');
colormap([0,0,0])
axis tight; axis off
view(0,35)
saveas(gcf,'WavePacWF','epsc')
saveas(gcf,'WavePacWF','png')
figure(5)
rho = 2*rho/max(rho(:));
vv = (0:12)'/12; cred = [vv.^2,4*vv.*(1-vv),1 - vv.^2]; cred(1,:) = 0;
contourf(x,Pd.tj,rho',vv)
set(gca,'fontsize',14)
axis([-0.5,0.5,-0.5,0.5]);hold on
colormap(cred);
plot([-1,1],[-1,1],'--y','LineWidth',3)
plot([0,0],[-1,1],'y','LineWidth',4)
hold off
colorbar
axis off
saveas(gcf,'WavePacC','epsc')
saveas(gcf,'WavePacC','png')
%% snapshots
io = [40,45,50,55,60,65,70,75,80];
s = (0:8)'/8; cred = [1-s.^2,4*s.*(1-s),s.^2];
for j = 1:9
   figure(5+j)
   area(x,rho(:,io(j)),'FaceColor',cred(j,:))
   axis([-0.6,0.6,0,1.5])
   set(gca,'XTick',[], 'YTick', [])
   % xlabel(['$t = ',num2str(io(j)),' \Delta t$'],'Interpreter','latex','FontSize',16)
   saveas(gcf,['Wavetj',num2str(io(j))],'epsc')
end
```

```
function Pnew = FUNwavdelta(Pn)
%% Propagation of a wave packet striking a delta function barrier
%  Function called by FIGDeltaWavepacket.m
Pnew = Pn;
if isfield(Pnew,'iplt');else;
   Pnew.iplt = round([1,Pnew.Nt/4,2*Pnew.Nt/4,3*Pnew.Nt/4,Pnew.Nt]);
end
alfk = Pn.hbar/(2*Pn.mo);
Nt = Pn.Nt; Nx = Pn.Nx;
ko   = sqrt(2*Pn.mo*Pn.Eo)/Pn.hbar;     Pnew.ko = ko;
tmax = Pn.W*Pn.mo/(Pn.hbar*ko);         Pnew.tmax = tmax;
kj = FUNsetkj(Pn);
tj = tmax*(2*(1:Nt) - Nt - 1)/(Nt-1);
tk = 2*kj./(2*kj + 1i*Pn.gamma);
rk = -1i*Pn.gamma./(2*kj+1i*Pn.gamma);
x = (2*(1:Nx)' - Nx - 1)*Pn.W/(Nx-1);
rho = zeros(Nx,Nt);
Ck  = exp(-(Pn.L*(kj-ko)).^2);
Nh = Nx/2;
```

```
for jt = 1:Nt
   for ix = 1:Nh
       Phir = sum(Ck.*(exp(1i*x(ix)*kj) + rk.*exp(-1i*x(ix)*kj)).*...
             exp(-1i*(kj.^2).*alfk*tj(jt)));
       rho(ix,jt) = abs(Phir)^2;
       Phit = sum(Ck.*(tk.*exp(1i*x(Nh + ix)*kj)).*...
             exp(-1i*(kj.^2).*alfk*tj(jt)));
       rho(Nh+ix,jt) = abs(Phit)^2;
   end
end
rho = rho/max(rho(:,1));
Pnew.x   = x/Pnew.W;
Pnew.tj  = tj/Pnew.tmax;
Pnew.rho = rho;
Pnew.Mxy = [Pnew.x,Pnew.rho(:,Pnew.iplt)];
Pnew.Lgt = {num2str(Pnew.iplt')};
Ns = length(Pnew.iplt); s = (0:Ns-1)'/(Ns-1);
Pnew.cred = [s.^2,4*s.*(1 - s),1-s.^2];
Pnew.xlabl = 'x / W';
Pnew.ylabl = '\rho(x,t) / \rho(x_o,0)';
Pnew.filn = 'DeltaWP';
% Pnew.tparam = ['$(E_o,L,\gamma,N_k) = $',num2str([Pnew.Eo,Pnew.L,Pnew.gamma,Pnew.Nk
    ])];
Pnew.titltx = ['$(E_o,W,\gamma,N_k) = $',...
   num2str([Pnew.Eo,Pnew.W,Pnew.gamma,Pnew.Nk])];
end
```

```
function kj = FUNsetkj(Px)
%% specification of k(j) - k(j-1) = Deltak
%  dmin is called "epsilon" in the manuscript
ko  = sqrt(2*Px.mo*Px.Eo)/Px.hbar;
Del = sqrt(-log(Px.dmin))/Px.L;
dk  = 2*Del/(Px.Nk - 1);
jo = round((ko - Del - dk)/dk);
kj = (jo + (1:Px.Nk) + 0.5)*dk;
end
```

```
function tfln = MOVrecbar(Pd,ifig)
%% SUBroutine MOVrecbar
%  To create movie files of wave packet
fln = ['DeltaVX',num2str(ifig)];
figure(ifig);
% load MTRX
sx = Pd.x;st = Pd.tj; rho = Pd.rho/max(Pd.rho(:));
pmax = max(rho(:)); smax = max(sx); smin = min(sx);
% Prepare the new file.
vidObj = VideoWriter(fln,'MPEG-4');
open(vidObj);
% Create an animation.
area(sx,rho(:,1),'FaceColor','r');
xlabel('$k_o x / \pi$','FontSize',24,'Interpreter','latex');
ylabel('$\rho(x,t)/\rho_{max}$','FontSize',24,'Interpreter','latex');
axis([smin smax 0 pmax])
% axis tight manual
set(gca,'nextplot','replacechildren');
Nt = size(rho,2);
for k = 1:Nt
```

```
    area(sx,rho(:,k),'FaceColor','r')
    title(['$k = ',num2str(k),'$'],'FontSize',18,'Interpreter','latex');
    % Write each frame to the file.
    currFrame = getframe(gcf);
    writeVideo(vidObj,currFrame);
end
% Close the file.
close(vidObj);
end
```

### B.2.9 Gamow Factor Models

- Figures generated by this script: Figure 6.8
- Primary script: `FIGthetaL.m`

```
%% Figure FIGthetaL
%  Simple Rectangular barrier
%  Comparing theta_rec to theta_step for k > k_o
clear; clc;
Prec = struct;
Prec.Nk = 160; Prec.Nl = 4;
Prec.L = 3*(2.^(0:Prec.Nl-1));
Prec.ko = 10;Prec.sjmax = 1;
uj = Prec.sjmax*((0:Prec.Nk)'/Prec.Nk).^1.5;  sj = sqrt(uj.^2 + 1);
Prec.thets = 4*sj.*uj./((sj + uj).^2);
Prec.thetg = 1./(1 + (sin(uj*Prec.L)./(2*sj.*uj)).^2);
Prec.Mxy = [uj,Prec.thets,Prec.thetg];
s = (0:Prec.Nl)'/Prec.Nl;
Prec.cred = [s.^2,4*s.*(1-s),1-s.^2]; Prec.cred(1,:) = [0.6,0.6,0.6];
Prec.xlabl = '[(k/k_o)^2 - 1]^{1/2}';
Prec.ylabl = '\theta(k) / 2 \kappa(k) L';
Prec.titltx = 'Rthetak';
Prec.xyrange = [0, 1, 0, 1];
Prec.Lgt = {num2str([0,Prec.L]')};
Prec.filn = 'Rthetarec';
QPlot(Prec,2);
saveas(gcf,Prec.filn,'png')
writematrix(Prec.Mxy,Prec.filn,'FileType','text','Delimiter','tab')
```

### B.2.10 Chemical Potential

- Figures generated by this script: Figure 7.5
- Primary script: `FIGChempot.m`

```
%% Figure FIGChempot
%   Evaluation of the temperature dependent chemical potential
Pn = FUNdamental();
Pn.chemo = 0.5;     % chemical potential in eV
Pn.Tmax  = 1800;
rmax = (pi*Pn.kb*Pn.Tmax/Pn.chemo)^2;
% function
fx = @(x,r) 1 + (1/8)*r*x.^2 + (7/640)*(r*x.^2).^2 + (31/3072)*(r*x.^2).^3;
fc = @(x,r) (3*sqrt(2*r)/(2*pi))*x.*log(1 + exp(-pi./(x*sqrt(r))));
Nc = 8;
Tj = (0:60:Pn.Tmax)'; Nt = length(Tj);
rj = (pi*Pn.kb*Tj/Pn.chemo).^2;
yj = zeros(Nt,4); yj(1,:) = 1;
```

```
y = 1; yo = 1;
for i = 2:Nt
   for j = 1:Nc
       y  = min((fx(y,rj(i)) + fc(y,rj(i))).^(2/3),1000);
       yo = min(fx(yo,rj(i)).^(2/3),1000);
   end
   yj(i,1) = y;
   yj(i,2) = (fx(y,rj(i)) + fc(y,rj(i))).^(2/3);
   yj(i,3) = yo;
   yj(i,4) = (fx(yo,rj(i))).^(2/3);
end
uj = 1./yj;
Pkubo = [-247/25920,-1/80,-1/12,1];
Pfrx  = polyfit(rj,1./yj(:,1),3); Pfrx(4) = 1;
Pfrxo = polyfit(rj,1./yj(:,3),3); Pfrxo(4) = 1;
Pn.xlabl = 'T \textrm{~[K]}';
Pn.ylabl = '\mu(T)/\mu_o';
Pn.xyrange = [0, Tj(Nt), 0.01*floor(100*uj(Nt,1)), 1];
Pn.Lgt = {'$\delta f > 0$','$\delta f = 0$','Kubo'};
Pn.filn = 'ChemPot';
Pn.cred = [1,0,0;0,0,1;0.7,0.7,0.7];
Pn.titltx = 'Chemical Potential';
Pn.Mxy = [Tj,polyval(Pfrx,rj),polyval(Pfrxo,rj),polyval(Pkubo,rj)];
QPlot(Pn,1); hold on;
plot(Tj,uj(:,1),'or',Tj,uj(:,3),'.b'); hold off
Pn.Lgt = {'fit $\delta f > 0$','fit $\delta f = 0$','Kubo',...
         '$\delta f > 0$','$\delta f = 0$'};
legend(Pn.Lgt)
saveas(gcf,Pn.filn,'png')
Mxy = [Tj/1000,Pn.Mxy(:,2:4),uj(:,1),uj(:,3)];
writematrix(Mxy,Pn.filn,'FileType','text','Delimiter','tab')
```

### B.2.11 Scattering

- Figures generated by this script: Figure 7.9
- Primary script: `FIGBlochG.m`
- Subordinate scripts: `FUNbg.m`

```
%% Program FIGBlochG
%  Evaluation of the BG function and comparison
%  with its approximations
clear;clc
% formulae
wx  = @(x) ((((x + 5).*x + 20).*x + 60).*x + 120).*x + 120;
fx = @(x) (x.^5)./((exp(x) - 1.0).*(1.0 - exp(-x)));
Wlx = @(x) 6*x.^2 - 72*log(1 + (x.^2)/12);
Whx = @(x) 124.4313306172 - wx(x).*exp(-x) - wx(2*x).*exp(-2*x)/2^5;
% axes
xj = exp(linspace(-1,3,41)');
Wx = FUNbg(xj); Wx(1) = Wlx(0);
Pn = struct();
Pn.Mxy = log([xj,Wlx(xj),Whx(xj),min(Wlx(xj),Whx(xj))]);
% graphics
Pn.xlabl = '\ln(x)';
Pn.ylabl = '\ln[W_-(5,x)]';
Pn.titltx = 'Bloch-Gr\"un';
Pn.filn = 'BlochGrun';
Pn.plot = 'linear';
Pn.cred = [1,0,0;0,0,1];
```

```
Pn.xyrange = [-1,3,-5,7];
Pn.Lgt = {'Low','High','Min'};
QPlot(Pn,1);
hold on; plot(log(xj),log(Wx),'ok'); hold off
Mxy = Pn.Mxy; Mxy(:,4) = log(Wx);
saveas(gcf,Pn.filn,'png')
writematrix(Mxy,Pn.filn, 'FileType', 'text','Delimiter','tab')
```

```
function yx = FUNbg(y)
%% Bloch-Gruneisen Function
%  Used in FIGBlochG.m
% functions
fx = @(x) (x.^5)./((exp(x) - 1.0).*(1.0 - exp(-x)));
gx = @(x) -(log(x).^5)./((1.0 - x).^2);
zo = 124.4313306172022;
Nx = length(y);
z = exp(-y);
s = zeros(Nx,1);
for j = 1:Nx
   if y(j) < 1.e-10;
       y(j) = 0;
   elseif y(j) < 20.0;
       s(j) = integral(@(x) fx(x),0.0,y(j));
   else
       s(j) = zo - integral(@(x) gx(x),0.0,z(j));
   end
end
yx = s;
```

### B.2.12 Optical Terms

- Figures generated by this script: Figures 7.10, 7.11, 7.12
- Primary script: `FIGLDRtheory.m`
- Subordinate scripts: `FUNnkcalc.m`

```
%% Figure FIGLDR
%  Generate the LD and LDR plots of optical data
clear;clc
Pth = FUNdamental();
% Functions
Pth.fres = @(x,p,g,w) ((p^2)*x.*g)./((x.^2-w.^2).^2+(x.*g).^2);
Pth.fchi = @(x,f,p,g,w) -(f*p^2)./(x.^2 - w.^2 + 1i*x.*g);
Pth.chem = 7;     % Fermi level in eV
Pth.hwp  = 10.7690682;
Pth.fdru = 0.735981;      % Drude parameters
Pth.gdru = 0.09124136;
Pth.wlor = 54.17564887;    % Lorentz parameters
Pth.flor = 8.972118;
Pth.glor = 87.8489;
Pth.titltx = 'Theoretical Optical parameters';
Pth.Nx = 160;
Pth.wmax = 12.5;
Pth.wj = [2.60, 3.60, 5.10, 8.36, 14.21];
Pth.gj = [0.90, 1.36, 2.00, 4.90,  9.00];
Pth.fj = [0.08, 0.11, 0.36, 0.35,  1.28];
Np = length(Pth.wj);
Pth.xyrange = [0, Pth.wmax, 0, 6];
```

```
xfit = linspace(0,Pth.wmax,Pth.Nx)';
yfit = Pth.fj.*Pth.fres(xfit,Pth.hwp,Pth.gj,Pth.wj);
s = (0:Np)'/Np;
Pth.cred    = [s.^2,4*s.*(1-s),1 - s.^2];
Pth.Mxy     = [xfit,yfit,sum(yfit,2)];
clear xfit yfit s
% Plot quantities: (change as needed in later routines)
Pth.filn    = 'EXPn';
Pth.plot    = 'linear';
Pth.cred = [1,0,0;0,0,1];
Pth.xlabl   = '\hbar \omega';
Pth.ylabl   = '\Delta K';
Pth.Lgt     = char({num2str(Pth.wj'); 'Sum $K_i$'});
Pth.titltx  = '$K_i(\omega_j)$ Components';
QPlot(Pth,1);
%% R and delta
Pthn = Pth;
Pthn.wmax = 20;
Pthn = FUNnkcalc(Pthn);
Pthn.Mxy(:,2:3) = [Pthn.R,Pthn.del/20];
Pthn.ylabl = 'R(\omega), \delta(\omega)/20';
Pthn.titltx = 'Optical Parameters:  $(R,\delta)$';
Pthn.xyrange = [0,20,0.01,1];
Pthn.cred = [1,0,0;0,0,1];
Pthn.Lgt = {'$R$','$\delta$'};
QPlot(Pthn,3);
saveas(gcf,[Pthn.filn,'Rdel'],'png')
writematrix(Pthn.Mxy,[Pthn.filn,'Rdel'], 'FileType', 'text','Delimiter','tab')
```

```
function Pn = FUNnkcalc(P)
%=================================================
% FUNCTION FUNnkcalc
%=================================================
% returns optical constants n, k, R, \delta
Pn = P;
w = Pn.wmax*(1:Pn.Nx)'/Pn.Nx;
chidru = - P.fdru*(Pn.hwp^2)./(w.*(w + 1i*Pn.gdru));
chilor = - (Pn.hwp.^2)*Pn.flor./(w.^2 - Pn.wlor.^2 + 1i*w*Pn.glor);
chires = - (Pn.hwp.^2)*  Pn.fj./(w.^2 -   P.wj.^2 + 1i*w*Pn.gj);
Pn.Kro = real(1 + chidru + chilor);
Pn.Kio = imag(1 + chidru + chilor);
Pn.Kr = real(1 + chidru + chilor + sum(chires,2));
Pn.Ki = imag(1 + chidru + chilor + sum(chires,2));
Kr = Pn.Kr;
Ki = Pn.Ki;
Pn.n   = sqrt(0.5*(sqrt(Kr.^2 + Ki.^2) + Kr));
Pn.k   = sqrt(0.5*(sqrt(Kr.^2 + Ki.^2) - Kr));
Pn.R   = ((Pn.n-1).^2 + Pn.k.^2)./ ((Pn.n+1).^2 + Pn.k.^2);
Pn.del = Pn.hbar*Pn.c./(2*Pn.k.*w);
Pn.Mxy = [w,abs(Pn.Ki),abs(Pn.Kr)];
Pn.filn = 'Optical';
Pn.xlabl = '\hbar \omega ~\textrm{[eV]}';
Pn.ylabl = '|(K_i,K_r)|';
Pn.cred = [1,0,0;0,0,1];
Pn.Lgt = {'$K_i$','$K_r$'};
Pn.titltx = 'Optical Parameters';
end
```

## B.3 Current and Barriers Algorithms

Select algorithms for the generation of figures in Part III.

### B.3.1 Shape Factor

- Figures generated by this script: Figure 9.1
- Primary script: `FIGsigmaexp.m`

```
%% Program FIGsigmaexp
%  Creation of shape factor for V(x) = Vo exp(-x/a)
%  The long  form is simgal (after Maple integration)
%  The short form is sigmas (after simplification)
P = FUNdamental();
Nu = 51;
sig = @(u) (u.*cot(u) - 1)./log(cos(u));
sigh = @(u) 4*(15 + u.^2)./(15*(6 + u.^2));
sigl = @(s) (-1 + (pi/2 - s).*s)./log(sin(s));
p = (0:Nu-1)'/(Nu-1); p = p.^4;
u = acos(sqrt(p));
s = (pi/2) - u;
x = 2*u/pi;
Ms = sig(u); Ms(Nu) = 2/3;
n  = 2.^((1:6)');
sn = (gamma(1.5)*(gamma(0.5*n+1))./gamma((n+3)/2))*[1,1];
P.Mxy = [p,sigh(u),sigl(s)];
P.xlabl = 'p';
P.ylabl = '\sigma_e(p V_o)';
P.titltx = 'Exponential V(x) shape factor';
P.xyrange = [0, 1, 0, 0.7];
P.Lgt = ['hi apx','lo apx'];
P.filn = 'SigmaVexp';
P.cred = [1,0,0;0,0,1];
QPlot(P,1); hold on; plot(p,Ms,'ok',[0,1],sn,'--'); hold off
legend('hi apx','lo apx','exact','$\sigma_n$')
disp(['n = ',num2str(n')])
saveas(gcf,P.filn,'png')
writematrix([p,Ms,sigh(u),sigl(s)],P.filn,'FileType','text','Delimiter','tab')
writematrix([n,sn],[P.filn,'sn'],'FileType','text','Delimiter','tab')
```

### B.3.2 Schottky–Nordheim

- Figures generated by this script: Figure 10.3
- Primary script: `FIGvyty.m`
- Subordinate scripts: `FUNvyint.m`

```
%% Figure FIGvyty:
%  Comparison of algorithms to calculate v(y) and t(y)
%  Exact, Forbes-Deane, Spindt-Brodie, cosine, polynomial
clear;clc
y = (0:0.05:1)'; Ny = length(y); y(1) = 1e-12; y(Ny) = 1 - 1e-12;
% Exact values
Py = FUNvyint(y); Py.vy(Ny) = 0;
Mex = [Py.vy,Py.ty];
Py.Mxy = [y,Mex];
Py.xlabl = 'y';
```

```
Py.ylabl = 'Error ~~[100\%]';
Py.Lgt = {'fd','sb','cs','pl'};
Py.Lgt = ['$v(y)$';'$t(y)$'];
Py.filn = 'MvytyE';
Py.xyrange = [0,1,-10,10];
Py.cred = [1,0,0;0.9,0.5,0.5;0.5,0.5,0.9;0,0,1];
% Forbes-Deane values
Mfd = [1 - (y.^2).*(3 - log(y))/3,1 + (y.^2).*(1 - log(y))/9];
Efd = 100*(1 - (Mfd./Mex));
% Spindt-Brodie values
Msb = [0.95 - (y.^2),0.95 + (y.^2)/3];
Esb = 100*(1 - (Msb./Mex)); Esb = max(min(Esb,50),-50);
% cosine
Mcs = [cos(pi*y/2),cos(pi*y/2) + (pi*y/3).*sin(pi*y/2)];
Ecs = 100*(1 - (Mcs./Mex));
% Polynomial
yj = (0:0.2:1); Pj = FUNvyint(yj);
Cv = polyfit(yj,Pj.vy,5);
Ct = polyfit(yj,Pj.ty,5);
Mpl = [polyval(Cv,y),polyval(Ct,y)];
Epl = 100*(1 - (Mpl./Mex));
Mvt = [Mfd,Msb,Mcs,Mpl];
Py.Mxy = [y,Efd,Esb,Ecs,Epl];
Py.titltx = 'Error';
Py.Lgt = {'fd','sb','cs','pl'};
Lgt = Py.Lgt; Lgt(5) = {'ex'};
Pv = Py; Pv.Cv = Cv;
Pv.Mxy = [y,Mvt(:,1:2:8)];
Pv.titltx = '$v(y)$ approximations';
Pv.ylabl = 'v(y)';
Pv.xyrange = [0,1,0,1.05];
Pt = Pv; Pt.Ct = Ct;
Pt.Mxy = [y,Mvt(:,2:2:8)];
Pt.titltx = '$t(y)$ approximations';
Pt.ylabl = 't(y)';
Pt.xyrange = [0,1,0.95,1.15];
QPlot(Pv,1);hold on;plot(y,Py.vy,'.k','MarkerSize',9);hold off
legend(Lgt);
saveas(gcf,'vy','png')
QPlot(Pt,2);hold on;plot(y,Py.ty,'.k','MarkerSize',9);hold off
legend(Lgt);
saveas(gcf,'ty','png')
QPlot(Py,3);
saveas(gcf,'Evyty','png')
writematrix(Py.Mxy,Py.filn,'FileType','text','Delimiter','tab')
Mxy = round([Py.y,Py.vy,Py.ty],6); Nm = floor(0.5*size(Mxy,1));
CMxy = NaN(Nm+1,6);
CMxy(1:Nm+1,1:3)  = Mxy(1:Nm+1,:);
CMxy(1:Nm  ,4:6) =  Mxy(Nm+2:2*Nm+1,:);
writematrix(CMxy,'TABvyvy','FileType','text','Delimiter','tab')
clear M* E* y ans Ny C*
```

```
function Py = FUNvyint(y)
%% Function FUNvyint
% Numerical integration for evaluation of v(y) and t(y) using formulae from
% K.L. Jensen, and A.K. Ganguly, Journal of Applied Physics 73, 4409 (1993).
% function
Rx = @(x,p) ((sin(x).*cos(x)).^2)./sqrt(p + (sin(x)).^2);
Py = struct;
```

```
Av = @(x,y) (1 - y).*(y + (sin(x)).^2)./sqrt(2*y + (1 - y).*(sin(x)).^2);
Bv = @(x,y) ((1-y).*(3 - 2*y).*(sin(x)).^4 + y.*(9 - 5*y).*(sin(x)).^2 +...
         4*y.^2)./(3*(2*y + (1-y).*(sin(x)).^2).^1.5);
Py.dvxy = Av;
Py.dtxy = Bv;
Ny = length(y);
vy = zeros(Ny,1); ty = vy;
for j = 1:Ny
   vy(j) = integral(@(x) Av(x,y(j)),0,pi/2);
   ty(j) = integral(@(x) Bv(x,y(j)),0,pi/2);
end
Py.Ny = Ny;
Py.y  = y;
Py.vy = vy;
Py.ty = ty;
if Ny ==1
   disp('creating integrands dv and dt')
   x = linspace(0,1,101)'; Py.x = x;
   Py.dv = Av(x,y);
   Py.dt = Bv(x,y);
end
end
```

- Figures generated by this script: Figures 10.1 and 10.2
- Primary script: `FIGSNvx.m`
- Subordinate scripts: `FIGSNvx.m`

```
%% Figure FIGSNvx
% Creation of image charge potential and Gamow factor figure
P = FUNdamental();
% Function
Vx = @(x,a,b,c) a - b*x - c./x;
chem = 7; Phio = 4.5; Fo = 5; Qo = P.Qo;
Nx = 51;
y = [0.25,0.5,0.75,0.95]; Ny = length(y);
E = chem + Phio - sqrt(4*Fo*Qo)./y;
s = -1 + 2*(linspace(0,1,Nx)').^1.5;
xo = sqrt(Qo/Fo);
vj = zeros(Nx,Ny);
for j = 1:Ny
   x = xo*(1 + s*sqrt(1 - y(j)^2))/y(j);
   vj(:,j) = Vx(x,chem + Phio - E(j),Fo,Qo)/Vx(xo,chem + Phio - E(j),Fo,Qo);
end
vj(1,:) = 0; vj(Nx,:) = 0;
P.Mxy = [s,vj];
P.xlabl = 's';
P.ylabl = '[V(x) - E] / [V(x_o) - E]';
P.titltx = 'SN:  $x = (x_o/y)[1 + s \sqrt{1 - y^2}]$';
P.xyrange = [0.2, 2.2, 0, 5];
P.Lgt = num2str(y');
P.filn = 'SNVx';
P.xyrange = [-1,1,0,1];
sc = (0:Ny-1)'/(Ny-1);
P.cred = [sc.^2,4*sc.*(1-sc),1 - sc.^2]; clear sc
QPlot(P,1);
saveas(gcf,[P.filn,'s'],'png')
writematrix(P.Mxy,[P.filn,'s'],'FileType','text','Delimiter','tab')
%% regular plot
Pn = P;
Pn.cred = [1,0,0;0,0,1];
```

```
Pn.xlabl = 'x(s)'; Pn.ylabl = 'V(x)';Pn.titltx = 'Zeros of $V(x)-E$';
xpm = (xo./y).*(1 + [-sqrt(1 - y.^2); sqrt(1 - y.^2)]);
vpm = Vx(xpm(1,:),chem + Phio,Fo,Qo); % V(xp) = V(xm), so just find one
xpm = [0,xpm(1,:);(chem+Phio)/Fo,xpm(2,:)];
vpm = [(chem+Phio),vpm;0,vpm];
ymin = sqrt(4*Qo*Fo)/(chem + Phio);
xj = xo*(1 + s*sqrt(1 - ymin^2))/ymin;
Vj = Vx(xj,chem + Phio,Fo,Qo); Vj(1) = 0; Vj(Nx) = 0;
Pn.Mxy = [xj,Vj];
Vmax = chem + Phio - sqrt(4*Qo*Fo);
Pn.xyrange = [0, xj(Nx), 0, chem+Phio];
Pn.Lgt = '$V_{SN}(x)$';
Pn.filn = 'SNvx';
QPlot(Pn,2); hold on
plot(xpm,vpm,'-',xo,Vmax,'ok'); hold off
Pn.Lgt = {'$V_{SN}(x)$','triangle'};
legend(Pn.Lgt)
saveas(gcf,[Pn.filn,'x'],'png')
writematrix(Pn.Mxy,[Pn.filn,'x'],'FileType','text','Delimiter','tab')
disp('zeros')
disp([xpm',vpm'])
disp('maximum [xo,Vmax]')
disp([xo,Vmax])
%% Standard plot of V(x) - E vs x
%   Conventional graph methods
%   $d(\theta^2)$ is the integrand^2 of $\theta(\mu)$
Po = P;
Po.Mxy = [xj,Vj-chem];
Po.Lgt = {'$V(x) - \mu$'};
Po.xyrange = [0,1,0,2];
Po.xlabl = 'x \textrm{~[nm]}';
Po.ylabl = 'V(x) - \mu \textrm{~[eV]}';
Po.titltx = 'Standard SN plot';
xo = sqrt(Qo/Fo);
Vo = Vx(xo,chem + Phio,Fo,Qo);
xpm = (Phio + sqrt(Phio^2 - 4*Qo*Fo)*[-1,1])/(2*Fo);
Vpm = Vx(xpm,chem + Phio,Fo,Qo) - chem;
QPlot(Po,3); hold on;
area(xj,Vj-chem);
plot(xo,Vo - chem,'ok','MarkerFaceColor','w','MarkerSize',12);
plot(xpm,Vpm,'ok','MarkerFaceColor','k','MarkerSize',12);
hold off
Po.Lgt = {'$V(x)-\mu$','$d(\theta(\mu)^2)$','$V(x_o) - \mu$','$V(x_\pm) - \mu$'};
legend(Po.Lgt);
saveas(gcf,Po.filn,'png')
```

- Figures generated by this script: Figure 10.5
- Primary script: `FIGshapesig.m`
- Subordinate scripts: `FIGshapesig.m`

```
%% Figure FIGshapesig:
% Creation of image charge potential and Gamow factor figure
P = FUNdamental();
P.chem = 7; P.Phio = 4.5; P.Fo = 4; P.E = P.chem - 0.5;
P.Nx = 100;
% Schottky Barrier Lowering
Pn = FUNimage(P,1); clear P
% Shape factors for standard barriers
Nx = 101;
Vo = Pn.chem + Pn.Phio - Pn.E; Fo = Pn.Fo; Qo = Pn.Qo;
```

```
Lo = sqrt(Vo^2 - 4*Qo*Fo)/Fo;
sj = (0:Nx-1)'/(Nx-1);sj = [0;sj;sj(Nx)];
yj = zeros(Nx+2,4);
yj(:,1) = 1;
yj(:,2) = 1 - sj; yj(1,2) = 0;
yj(:,3) = 1 - (2*sj - 1).^2;
xm = (Vo/(2*Fo)) - 0.5*Lo + Lo*sj;
yj(:,4) = (Vo - Fo*xm - (Qo./xm))/(Vo - sqrt(4*Qo*Fo));
yj(1,:) = 0; yj(Nx+2,:) = 0;
Ps = Pn;
Ps.Mxy = [sj,sqrt(max(yj,0))];
Ps.xlabl = '(x-x_1)/(x_2 - x_1)';
Ps.ylabl = '[(V(x) - E)/(V(x_o) - E)]^{1/2}';
Ps.xyrange = [0,1,0,1];
Ps.Lgt = {'rectangular','triangular','parabolic','Schottky'};
Ps.cred = [0.6,0.6,0.6;0,0,1;0,0.8,0;1,0,0];
Ps.filn = 'ShapeSN';
QPlot(Ps,2); axis padded;
writematrix(Ps.Mxy,Ps.filn,'Delimiter','tab')
disp(['Output written to ',Ps.filn,'.txt'])
disp('Pn.xys = [xo,yo,xs,ys]')
disp(Pn.xys)
%% image charge barrier
function P = FUNimage(Po,ifig)
P = Po;
Vx = @(x,a,b,c) a - b*x - c./x;
% theta factor
L = sqrt((P.chem + P.Phio - P.E)^2 - 4*P.Qo*P.Fo)/P.Fo;
xmin = ((P.chem + P.Phio - P.E)/(2*P.Fo)) - 0.5*L;
xmax = ((P.chem + P.Phio - P.E)/(2*P.Fo)) + 0.5*L;
x = linspace(xmin,xmax,P.Nx)';
y = Vx(x,P.chem+P.Phio,P.Fo,P.Qo)-P.E;
xo = ((P.chem + P.Phio - P.E)/P.Fo)*[0;1];
yo = (P.chem + P.Phio - P.E)*[1;0];
yu = P.chem*[1;1];
xs = sqrt(P.Qo/P.Fo)*[1;1];
ys = [P.chem + P.Phio - P.E;P.chem + P.Phio - P.E - sqrt(4*P.Qo*P.Fo)];
xz = [0,0;10,10]; yz = [ys(1),ys(2);ys(1),ys(2)];
P.xys = [xo,yo,xs,ys];
P.xlabl = 'x ~\textrm{[nm]}';
P.ylabl = 'V(x) - E \textrm{~[eV]}';
P.titltx = ['$F = ',num2str(P.Fo),'$ eV/nm; $E = ',num2str(P.E),'$ eV'];
P.xyrange = [0, (P.chem + P.Phio - P.E)/P.Fo, 0, P.chem + P.Phio - P.E+0.25];
P.cred = [1,0,0;0,0,1];
Lgt = {'Triangular','Image','$\mu$','$\sqrt{4QF}$'};
P.filn = 'ImageVx';
P.Mxy = [x,y];
P.Lgt = Lgt(1);
QPlot(P,ifig); hold on;
plot(xo,yo,'b',xo,yu-P.E,'g',xs,ys,'-oc')
plot(xz,yz,'--','color',0.8*[1,1,1]); hold off
legend(Lgt, 'FontSize',16,'Interpreter','latex','Location','northeast')
saveas(gcf,P.filn,'epsc');
writematrix(P.Mxy,P.filn,'Delimiter','tab')
disp(['Output written to ',P.filn,'.txt'])
end
```

- Figures generated by this script: Figure 10.6
- Primary script: `FIGsigsn.m`

```
%% Program FIGsigsn:
%  comparison of
%    F(y) = int[0,pi] [(sin(x))^2 / sqrt(1 + p(y) cos(x))] dx
%  as a function of y to its polynomial expansion
clear;clc
P = struct;
% function
fv = @(x,p) (sin(x).^2)./sqrt(1 + p*cos(x));
P.xlabl = '(1 - y)/(1 + y)';
P.ylabl = 'F[p(y)]';
P.titltx = 'SN:  $p(y) = \sqrt{1 - y^2}$';
P.xyrange = [0, 1, 1.5, 1.9];
P.Lgt = 'Poly';
P.filn = 'SNFpy';
P.cred = [1,0,0];
nj = (3:2:7)'; Err = zeros(length(nj),1);
for k = 1:length(nj)
   n = nj(k);
   Fj = zeros(n,1); Fj(1) = pi/2; Fj(n) = 4*sqrt(2)/3;
   xj = (0:n-1)'/(n-1);
   yj = (1-xj)./(1+xj); pj = sqrt(1 - yj.^2);
   for j = 1:n; Fj(j) = integral(@(x) fv(x,pj(j)),0,pi); end
   Cj = polyfit(xj,Fj,n-1);
   Nx = 41;
   xy = (0:Nx-1)'/(Nx-1); y = (1 - xy)./(1+xy); p = sqrt(1 - y.^2);
   Fp = zeros(Nx,1);
   for j = 1:Nx; Fp(j) = integral(@(x) fv(x,p(j)),0,pi); end
   P.titltx = ['SN:  $p(y) = \sqrt{1 - y^2}$, N = ',num2str(nj(k))];
   P.Mxy = [xy,polyval(Cj,xy)];
   QPlot(P,k); hold on; plot(xy,Fp,'ok'); hold off
   legend('Poly','Exact')
   Err(k) = max(abs(100*(1 - (P.Mxy(:,2)./Fp))));
   Mj = [yj,xj,pj,Fj,(0:n-1)',Cj(n:-1:1)'];
   saveas(gcf,[P.filn,num2str(nj(k))],'png')
   writematrix(round(Mj, 6),[P.filn,num2str(nj(k))],'FileType','text','Delimiter','tab
     ')
end
P.n = n;
P.Cj = Cj;
disp('n    Err(%)')
disp(num2str([nj,Err]))
disp('Column headers in txt file:  y, x, p, F(p), j, Cj')
%% Evaluation of $\sigma_{sn}(y)$
%   and its comparison to its asymptotic limits
%   of triangular and parabolic shapefactors
Psig = P;
y = (0:40)'/40; xy = (1-y)./(1+y);
cy = sqrt(1+y)/(2*sqrt(2));
sigy = cy.*polyval(Cj,xy);
Psig.Mxy = [y,sigy]; P.Lgt = '$\sigma(y)$';
Psig.xyrange = [0,1,0.66,0.795];
Psig.xlabl = 'y';
Psig.ylabl = '\sigma_{sn}(y)';
QPlot(Psig,11); hold on; plot([0,0;1,1],[pi/4,2/3;pi/4,2/3],'--c'); hold off
Psig.Lgt = {'$\sigma_{sn}(y)$','$\sigma_\triangleright = 2/3$',...
           '$\sigma_\cap = \pi/4$'};
legend(Psig.Lgt)
Psig.filn = 'SNsig.txt';
saveas(gcf,Psig.filn,'png')
writematrix(Psig.Mxy,Psig.filn,'FileType','text','Delimiter','tab')
```

- Figures generated by this script: Figure 11.3
- Primary script: FIGpolyNVx.m

```
%% Figure FIGpolyNVx
%  Creation of image charge potential and Gamow factor figure
%  for the cubic potential (addition of gamma x^2) to
%  standard image charge potential, so named because
%    x * V(x) is a cubic equation
Pn = FUNdamental();
% cubic image charge function
Vx = @(x,a,b,c,g) a - b*x - c./x + g*x.^2;
if exist('g','var')  == 1; else; g = 0.85;end
Pn.g = g; Pn.F = 5; Pn.H = 7;
Nx = 51;
sj = ((0:Nx-1)'/(Nx-1)).^1.5;
del = (54*g^2)*Pn.Qo/(Pn.F^3);
omg = sqrt(1 - (3*Pn.H*g / (Pn.F^2)));
phi = acos((2*(3*omg^2 - 1) + del)/(4*omg^3));
chi = acos(1 - del)/3;
xj = (Pn.F/(3*g))*(1 + 2*omg*cos((phi/3) - (2*pi*(1:3)/3)));
xj = abs(xj);
xo = (Pn.F/(6*g))*(1 + 2*cos(chi - (2*pi/3)));
xp = (Pn.F/(6*g))*(1 + 2*cos(chi));
Pn.xlabl = '(x - x_2)/(x_1 - x_2)';
Pn.ylabl = 'V_N(x) \textrm{[eV]}';
Pn.titltx = ['Quadratic Barrier for $\gamma = ',num2str(g),'$ eV/nm$^2$'];
Pn.filn = 'PolyNVx';
Pn.xyrange = [-1,1,0,1];
Pn.cred = [1,0,0;0,0,1;0.8,0.8,0.8];
x  = xj(2) + (xj(1) - xj(2))*sj;
xz = xj(1) + (4 - xj(1))*sj;
Vpoly = Vx(x,Pn.H,Pn.F,Pn.Qo,Pn.g);
Vo   = Vx([xo;xp],Pn.H,Pn.F,Pn.Qo,Pn.g);
Vxz = Vx(xz,Pn.H,Pn.F,Pn.Qo,Pn.g);
Vsn = Vx(x,Pn.H,Pn.F,Pn.Qo,0);
Pn.Mxy = [x,Vpoly,Vsn];
Pn.xyrange = [0, 4, -1, 5];
Pn.Lgt = {'Vpoly = $V_2(x)$','Vsn = $V_1(x)$'};
QPlot(Pn,1);hold on; plot([0,4],[0,0])
plot(xz,Vxz,'--r');
plot(xj,0*xj,'ok',[xo;xp],Vo,'*k')
hold off
legend(Pn.Lgt)
Mxv = [xo;xp]; Mxv = [Mxv,Vo];
disp(['Zeros = ',num2str(xj)])
disp('Inflections: (x,V(x))');
disp(Mxv)
saveas(gcf,Pn.filn,'png')
writematrix([Pn.Mxy,xz,Vxz],Pn.filn,'FileType','text','Delimiter','tab')
```

- Figures generated by this script: Figure 11.4, 11.5
- Primary script: FIGdepleteVx.m

```
%% Figure FIGdepleteVx
%   Depletion Barrier:  Plotting of \sigmia_\cup(u)
%   \sigma_\cup[u(E)] = \frac{1}{2(1-u)} \left[1 - \frac{u^2}{\sqrt{1 - u^2}} *
%                         \ln\left(\frac{1+\sqrt{1 - u^2}}{u}\right)\right]
%  First Figure:  Q = 0
clear; clc
```

```
% function
Pn.su = @(x) (1./(2*(1-x))).*(1 - ((x.^2)./sqrt(1-x.^2)).*log((1 + sqrt(1-x.^2))./x));
Pn.xlabl = 'u = \sqrt{E/\chi}';
Pn.ylabl = '\sigma_\cup(u)';
Pn.titltx = 'Depletion Barrier';
Pn.filn = 'DeplVx';
Pn.xyrange = [0,1,0.48,0.68];
Pn.cred = [1,0,0;0,0,1;0.7,0.7,0.7];
Pn.Nx = 21;
u = linspace(0,1,Pn.Nx)';
sigu = Pn.su(u);
sigu(1) = 0.5; sigu(Pn.Nx) = 2/3;
uo = (1:3)/4; siguo = Pn.su(uo);
uo = [0,uo,1]; siguo = [1/2,siguo,2/3];
Pn.Ps = polyfit(uo,siguo,4);
Pn.Lx = [0;1]; Pn.Ly = [1/2,2/3;1/2,2/3];
Pn.Mxy = [u,sigu,polyval(Pn.Ps,u)]; clear u sigu
Pn.Lgt = {'$\sigma(u)$','poly'};
QPlot(Pn,1); hold on; plot(Pn.Lx,Pn.Ly,'--b',uo,siguo,'ok'); hold off
Pn.Lgt = char([Pn.Lgt,'$\sigma_\cup(0)$','$\sigma_\triangleright$']);
legend(Pn.Lgt)
saveas(gcf,Pn.filn,'png')
writematrix(Pn.Mxy,Pn.filn,'FileType','text','Delimiter','tab')
%% Second figure:  V_2(x) with Q nonzero
Pv = FUNdamental();
Fo = 2.5; g = 0.25; Vo = 4; Qo = Pv.Qo;
Pv.Fo  = Fo;
Pv.gam = g;
Pv.Vo  = Vo;
Pv.plot = 'linear';
N = 50;
Pv = FUNDepVxj(Pv,N);
QPlot(Pv,2); hold on;
plot(Pv.Ej(N)*[0;1],[1;1]*[Pv.xo,Pv.xp],'--','Color',0.6*[1,1,1]); hold off
Lgt = [Pv.Lgt,'$x_o$','$x_p$'];
legend(Lgt)
saveas(gcf,Pv.filn,'png')
writematrix(Pv.Mxy,Pv.filn,'FileType','text','Delimiter','tab')
Lj = max(Pv.xj(:,1)-Pv.xj(:,2),0.01);
eps = Pv.xj(:,2)./Lj;
rho = Lj./Pv.xj(:,3);
hs = (pi/4)*sqrt((1 + 2*eps)./(2 - (1 + 2*eps).*rho));
Ej = Pv.Ej;
Pe = Pv;
Pe.ylabl = '\epsilon, \rho';
Pe.xlabl = 'E/V(x_o)';
Pe.filn = 'hsepsrho';
Pe.Mxy = [Ej/Ej(N),eps,rho,hs];
Pe.Lgt = {'$\epsilon$','$\rho$','$\int h(s)$'}; Pe.cred = [1,0,0;0,0,1;0,0.5,0];
Pe.xyrange = [0,1,0,1.5];
QPlot(Pe,3);
saveas(gcf,Pe.filn,'png')
writematrix(Pe.Mxy,Pe.filn,'FileType','text','Delimiter','tab')
```

- Figures generated by this script: Figure 11.8
- Primary script: `FIGSheltonB.m`

```
%% Figure FIGSheltonB
%   Thermal Emission and Shelton Data
%   H. Shelton, Phys. Rev. 107, 1553 (1957).
```

```
%   https://doi.org/10.1103/PhysRev.107.1553
%   Figure 6
clear;clc
Pn = FUNdamental();
% measured temperature
Trs  = [1503; 1552; 1605; 1651; 1706; 1765; 1833];
% measured retarded current log(Jr / T^2)
Jtr = [-30.4350; -29.3871; -28.2615; -27.2804; ...
      -26.2810; -25.2006; -24.1260];
% measured saturated current log(Js / T^2)
Jts = [-29.7539; -28.6617; -27.5093; -26.5185; ...
      -25.4317; -24.3666; -23.1776];
Nr  = length(Trs);
Pret = polyfit(1./Trs,Jtr,1);
Psat = polyfit(1./Trs,Jts,1);
N = 1e4;
Tj = linspace(1400,1900,Nr)';
Pn.Mxy = zeros(7,3);
Pn.Mxy(:,1) = N./Tj;
Pn.Mxy(:,2) = polyval(Pret,1./Tj);
Pn.Mxy(:,3) = polyval(Psat,1./Tj);
% Graphics tersm
Pn.Lgt = {'Fit Ret','Fit Sat'};
Pn.titltx = 'H. Shelton, PR107, 1553 (1957)';
Pn.filn = 'ThermalTa';
Pn.ylabl = '\log(J/T^2)';
Pn.xlabl = '10^4 / T';
Pn.cred = [1,0,0;0,0,1];
Pn.plot = 'linear';
Pn.xyrange = [5.4,6.7,-31,-23];
QPlot(Pn,1); hold on;
plot(N./Trs,Jtr,'ok',N./Trs,Jts,'*k'); hold off
Pn.Lgt = char([Pn.Lgt,'Exp Ret','Exp Sat']);
legend(Pn.Lgt)
Aret = exp(Pret(2));
Asat = exp(Psat(2));
ac   = log(Asat/Aret); % this is alpha/k_B
disp(['Aret = ',num2str(Aret),' A/cm2'])
disp(['Asat = ',num2str(Asat),' A/cm2'])
disp(['ac   = ',num2str(ac),' k_B'])
Mxy = [Pn.Mxy,N./Trs,Jtr,Jts];
saveas(gcf,Pn.filn,'png')
writematrix(Mxy,Pn.filn, 'FileType', 'text','Delimiter','tab')
```

### B.3.3 Metal–Insulator–Metal

- Figures generated by this script: Figures 11.6 and 11.5
- Primary script: `FIGmim.m`

```
%% Plot of the MIM barrier
%   comparison of the exact (Hx) to the approximate (Ha) to
%   the simmons (Hs)
% Formula
us = @(x,p,a,b) 0.5*(1 - x) + ((x.^3)./(1 + x)) + x.*(1 - x).*(p + a*x.^2 + b*x.^4);
Pn = FUNdamental();
% Plotting Parameters
n = 3; N = 200; Ns = 64;
p = FUNzetap(3);
a = FUNzetap(5) - 1;
```

```
b = FUNzetap(7) - 1;
s = 0.5*(1 - cos(pi*(1:Ns-1)'/Ns));
% Barrier parameters
Vo = 2; Fo = 0.4; W = 4; Ks = 8;
Qs = (Ks - 1)*Pn.Qo/(Ks + 1);
% This section returns Pa, the coefficients for a cubic fit
np = 4; so = (0:np-1)/(np-1);
uo = us(so,p,a,b); Pa = polyfit(so,uo,np-1);
j = (1:N);
sj = (1:9)'/10;
hj = 1./(j.*(j.^2 - s.^2));
Hx = ((1./(2*s)) + sum(hj,2)).*(s.*(1-s));
Ha = us(s,p,a,b);
% Hs = 0.637297*ones(Ns-1,1);
Hs = 0.6931*ones(Ns-1,1);
Mh = [Hx,Ha,Hs].*(Qs./(W*s.*(1-s)));
Pn.Mxy = [s,Vo - Fo*(W*s) - Mh];
Pn.xyrange = [-0.05, 1.05, 0, 2.1];
Pn.Lgt = {'Exact','Analytic','Simmons'};
Pn.xlabl  = 's = x/W';
Pn.ylabl  = 'V(x)';
Pn.cred   = [1,0,0;0,0,1;0,0.8,0];
Pn.titltx = 'MIM Barrier';
Pn.filn = 'MIMvx';
QPlot(Pn,1); hold on;
plot(s,Pn.Mxy(:,2),'.r');
plot([0,0,1,1],[0,Vo,Vo-Fo*W,0],'--','Color',[0.6,0.6,0.6])
hold off
legend(Pn.Lgt)
saveas(gcf,Pn.filn,'png')
writematrix(Pn.Mxy,Pn.filn,'FileType','text','Delimiter','tab')
%% asymptotic limit based on zeta(p) - 1 = 1/2^p
%  It leads to
%      sum(k=1,infty) [zeta(2k+3)-1] s^(2k) approx 1/8[1-(s/2)^2]
%  This is just to show that the approximation is not as good as us(x)
%  above
Sx = sum(hj,2);
Sa = ((s.^2)./(1 - (s.^2))) + FUNzetap(3) + (FUNzetap(5) - 1)*s.^2 + ...
    (FUNzetap(7) - 1)*s.^4;
Sb = ((1.^2)./(1 - (s.^2))) + (1./(8*(1 - (s/2).^2)));
figure(2)
plot(s,1./Sx,'ok',s,1./Sa,'.r',s,1./Sb,'b')
```

## B.4 Emission Algorithms

Select algorithms for the generation of figures in Part IV.

### B.4.1 FN-RLD

- Figures generated by this script: Figure 12.1
- Primary script: `FIGfnrld.m`
- Subordinate scripts: `FUNJmg.m, FUNJrld.m`

```
%% Figure FIGfnrld
%   Plot of field vs thermal emission for various Phi, F, and T
%   Field = FN-Murphy Good formulation
%   Thermal = RLD equation
```

```
clear;clc; close all
P = FUNdamental();
P.Phi = 4.5; Tj = (600:200:2000);
P.F = (0.5:0.5:10)'; Nf = length(P.F); Nt = length(Tj);
Pmg = FUNJmg(P);
Mt = zeros(Nf,Nt);
for j = 1:Nt
   Prld = P; Prld.To = Tj(j);
   Prld = FUNJrld(Prld);
   Mt(:,j) = Prld.Mxy(:,2);
end
Psum = Pmg;
s = 0.5 + 0.5*(0:Nt)'/Nt;Psum.cred = [s.^2,3*s.*(1-s),1-s.^2]; clear s
Psum.cred(1,:) = [0,0,1];
Psum.xlabl = 'F~\textrm{ [eV/nm] }';
Psum.ylabl = '\ln\{J(F,T) ~\textrm{ [A/cm}^2\textrm{] }\}';
Psum.Mxy = [P.F,Pmg.Mxy(:,2),Mt];
Psum.Lgt = num2str([0;Tj']);Psum.Lgt(1,:) = ' MG ';
Psum.titltx = ['$\Phi = ',num2str(Psum.Phi),'$ eV'];
Psum.xyrange = [min(P.F),max(P.F),min(Mt(:)),max(max(Psum.Mxy(:,2:Nt+1)))];
QPlot(Psum,3)
saveas(gcf,Psum.filn,'png')
writematrix(Psum.Mxy,Psum.filn, 'FileType', 'text','Delimiter','tab')
```

```
function Pn= FUNJrld(P)
%% Thermal Emission
%  Richardson-Laue-Dushman equation
%  Returns Jrld in units of Amp/cm^2
%  Either F can be a vector or T can, but NOT BOTH
Pn = P;
Nt = length(Pn.To); Nf = length(Pn.F);
Pn.Arld = Pn.mo*(Pn.kb^2)/(2*(pi^2)*(Pn.hbar^3)) * (Pn.cm^2)/Pn.Amp;
x = (Pn.Phi - sqrt(4*Pn.Qo*Pn.F))./(Pn.kb*Pn.To);
Jrld = log(Pn.Arld) + 2*log(Pn.To) - x;
Pn.Mxy = [x,Jrld];
Pn.cred = [1,0,0];
Pn.xlabl = '\beta_T \phi';
Pn.ylabl = '\ln\{J_{RLD}(F,T) ~\textrm{ [A/cm}^2\textrm{] }\}';
Pn.Lgt = 'RLD';
if Nf > 1
   Pn.titltx = ['$T = ',num2str(Pn.To),'$ Kelvin'];
else
   Pn.titltx = ['$F = ',num2str(Pn.F),'$ eV/nm'];
end
Pn.xyrange = [min(x),max(x),min(Jrld),max(Jrld)];
Pn.filn = 'Jrld';
end
```

```
function Pn= FUNJmg(P)
%% Fowler-Nordheim
%  with Murphy-Good image mods and Forbes-Deane
%  approximations to v(y), with t_o = t[y(mu)]
%  Amg on output has units of Amps/cm^2
%  Bo on output has units of eV/nm
Pn = P;
yo = sqrt(4*P.Qo*P.F)/P.Phi;
to = 1 + ((yo/3).^2).*(1 - log(yo));
```

```
Bo = (4/(3*P.hbar))*sqrt(2*P.mo*P.Phi^3);
nu = (8*P.Qo/(9*P.hbar))*sqrt(2*P.mo/P.Phi);
Amg = ((Bo^2)/(16*P.Phi*P.hbar*pi^2))*((P.Phi^2)*exp(6)/(4*P.Qo*Bo))^nu;
Pn.nu = nu;
Pn.Amg = Amg*(P.cm^2)./(P.Amp * to.^2);
Pn.Bo = Bo;
x = Pn.F/Pn.Bo;
Pn.Jmg = (2 - nu)*log(x) - (1./x) + log(Pn.Amg);
Pn.Mxy = [x,Pn.Jmg];
Pn.cred = [1,0,0];
Pn.xlabl = 'F/B_o';
Pn.ylabl = '\ln\{J_{MG}(F) ~\textrm{[A/cm}^2\textrm{]}\}';
Pn.Lgt = 'MG';
Pn.titltx = 'FN using MG and FD approx';
Pn.xyrange = [min(x),max(x),min(Pn.Jmg),max(Pn.Jmg)];
Pn.filn = 'Jfnmgfd';
end
```

### B.4.2 GTF Integrand

- Figures generated by this script: Figures 12.3, 12.4, 12.5, and 12.6
- Primary script: `FIGtfstudyC.m`
- Subordinate scripts: `FUNgtfEm.m`, `FUNdefe.m`, `FUNfz.m`, and `CPlot.m`

```
%% Figure FIGtfstudyC
%  Contour plot of dN(n,s):  F vs E/mu
% First contour plot
clear;clc
Pm = FUNdamental();
Pm.chem = 7; Pm.Phi = 4; Pm.T = 1500;
Pm.Nf = 81; Pm.Nx = Pm.Nf;
Pm.Fmax = 5; Pm.Fmin = 0.3;
Pm.Emin = Pm.chem - 1; Pm.Emax = Pm.chem + Pm.Phi+1;
Pm.plot = 'linear';
Pm.eps = true;
Pm = FUNgtfEm(Pm,1);
% line plot
Tj = 1000 + 200*(1:5);
Mxn = zeros(Pm.Nx,5); Myn = Mxn;
for j = 1:5
   Pm.T = Tj(j);
   Pm = FUNgtfEm(Pm,0);
   Pnft = Pm; Pnft.Mxy = Pm.Mnft;
   Mxn(:,j) = Pnft.Mxy(:,1); Myn(:,j) = Pnft.Mxy(:,2);
   Pnft.xlabl = 'F \textrm{[eV/nm]}';
   Pnft.ylabl = 'n(F,T)';
   Pnft.Lgt = '$n$';
   Pnft.filn = [Pm.filn,'Em'];
   % Pnft.xyrange = [Pnft.Fmin,Pnft.Fmax,min(Pnft.Mxy(:,2)),max(Pnft.Mxy(:,2))];
   % QPlot(Pnft,3+j); hold on; plot([Pnft.Fmin,Pnft.Fmax],[1,1],'r'); hold off
   % legend('$n$','$1$','FontSize',16,'axInterpreter','latex','Location','best')
   % writematrix(Pnft.Mxy,[Pnft.filn,num2str(Pnft.T)],'FileType','text','Delimiter','
     tab')
end
Fp = [Mxn(:,1),Mxn(:,5)]; npl = Myn(1,1)*sqrt(Fp(:,1)/Fp(1,1));
nph = Myn(Pnft.Nx,5)*(Fp(:,2)/Fp(Pnft.Nx,2));
Pnft.plot = 'linear';
s = (-1:4)'/4;
Pnft.cred = [s.^2,4*s.*(1-s),1-s.^2]; Pnft.cred(1,:) = 0.7;
```

```
Pnft.Mxy(:,2) = 1;
Pnft.xyrange = [0.3,5,0.4,2.2];
Pnft.titltx = 'Path of $dJ(E_m)$';
QPlot(Pnft,4); hold on
plot(Mxn,Myn,Fp,[npl,nph],'--k'); hold off
legend(num2str([1;Tj']))
writematrix([Mxn,Myn],[Pnft.filn,'A'],'FileType','text','Delimiter','tab')
writematrix([Fp,npl,nph],[Pnft.filn,'B'],'FileType','text','Delimiter','tab')
```

```
%% Figure FIGtfstudy
%   triangluar barrier gamow factor and supply function
%   Case A:  holding temperature fixed and varying field
%   Case B:  holding field fixed and varying temperature
%% Single Case
clear;clc
Pn = FUNdamental();
Pn.chem = 7; Pn.Phi = 4;Pn.T = 1500;
Pn.Emin = 0.6*Pn.chem; Pn.Emax = 1.8*Pn.chem;
Pn.Nx = 101;
Pn.plot = 'linear'; Pn.eps = false;
Pn.F = 1;
Pn = FUNdefeN(Pn,1);
Pn.filn = ['fDfxD',num2str(Pn.F)];
PDf = Pn;
PDf.Mxy(:,2:4) = exp(PDf.Mxy(:,2:4));
PDf.Mxy(:,4) = PDf.Mxy(:,4)./max(PDf.Mxy(:,4));
PDf.ylabl = 'D(E); f(E); D(E) f(E)';
PDf.xyrange = [Pn.Emin/PDf.chem,Pn.Emax/PDf.chem,0,1.05];
QPlot(PDf,2)
writematrix([Pn.Mxy,PDf.Mxy(:,2:4)],Pn.filn,'FileType','text','Delimiter','tab')
%% Case A
clear;clc;
Fj = [1, 2, 5, 8, 10, 14, 18]/5;
Nf = length(Fj);
Pf = FUNdamental();
Pf.chem = 7; Pf.Phi = 4; Pf.Nx = 101;
Pf.T = 1500;
Pf.Emin = 0.8*Pf.chem; Pf.Emax = 1.7*Pf.chem;
Pf.plot = 'linear';
Pf.eps = false;
Mxy = zeros(Pf.Nx,2*(1+Nf));
Pf.F = Fj(1);Pf = FUNdefeN(Pf,0);
Mxy(:,1:2) = [Pf.Mxy(:,1),Pf.Mxy(:,3)];
vxf = zeros(Nf,1); vyf = vxf; vbf = vxf;
for j = 1:Nf
   Pf.F = Fj(j);
   Pf = FUNdefeN(Pf,0);
   Mxy(:,2*j+1:2*j+2) = [Pf.Mxy(:,2),Pf.Mxy(:,4)];
   vxf(j) = Pf.Em/Pf.chem; vyf(j) = Pf.Dfm;
   vbf(j) = Pf.ntf;
end
Pf.Mxy = Mxy;
BCx = [Pf.chem,Pf.chem,Pf.chem+Pf.Phi,Pf.chem+Pf.Phi]/Pf.chem;
BCy = [-70,70,70,-70];
Pf.xyrange(3) = -40;
QPlot(Pf,2); hold on;
plot(BCx,BCy,'--','Color',0.6*[1,1,1]);
plot(vxf,vyf,'.k','MarkerSize',18)
hold off
```

```
legend(Pf.Lgt)
writematrix(Pf.Mxy,[Pf.filn,'fhold'],'FileType','text','Delimiter','tab')
writematrix([vxf,vyf],[Pf.filn,'fvxvy'],'FileType','text','Delimiter','tab')
%% Case B
Tj = [3,5,6,9,12,16,28]*100; Nt = length(Tj);
Pt = FUNdamental();
Pt.chem = 7; Pt.Phi = 2; Pt.Nx = 101;
Pt.F = 0.5;
Pt.Emin = 0.8*Pt.chem; Pt.Emax = 1.5*Pt.chem;
Pt.plot = 'linear';
Pt.eps = false;
Mxy = zeros(Pt.Nx,2*(1+Nt));
Pt.T = Tj(1);Pt = FUNdefeN(Pt,0);
Mxy(:,1:2) = [Pt.Mxy(:,1),Pt.Mxy(:,2)];
vxt = zeros(Nt,1); vyt = vxt; vbt = vxt;
for j = 1:Nt
   Pt.T = Tj(j);
   Pt = FUNdefeN(Pt,0);
   Mxy(:,2*j+1:2*j+2) = [Pt.Mxy(:,3),Pt.Mxy(:,4)];
   vxt(j) = Pt.Em/Pt.chem; vyt(j) = Pt.Dfm;
   vbt(j) = Pt.ntf;
end
Pt.Mxy = Mxy;
BCx = [Pt.chem,Pt.chem,Pt.chem+Pt.Phi,Pt.chem+Pt.Phi]/Pt.chem;
BCy = [-70,70,70,-70];
Pt.xyrange(3) = -50;
QPlot(Pt,3); hold on;
plot(BCx,BCy,'--','Color',0.6*[1,1,1]);
plot(vxt,vyt,'.k','MarkerSize',18)
hold off
legend(Pt.Lgt)
writematrix(Pt.Mxy,[Pt.filn,'thold'],'FileType','text','Delimiter','tab')
writematrix([vxt,vyt],[Pt.filn,'tvxvy'],'FileType','text','Delimiter','tab')
```

```
%%  FIGfDfxD: plot of f(E), D(E), and overlap
%   Based don Figure FIGtfstudy from JAP Tutorial 2024
%   triangluar barrier gamow factor and supply function
%   Case A:  holding temperature fixed and varying field
%   Case B:  holding field fixed and varying temperature
clear;clc
Pn = FUNdamental();
Pn.chem = 7; Pn.Phi = 4.5;Pn.T = 1500;
Pn.Emin = 0.5*Pn.chem; Pn.Emax = 2.1*Pn.chem;
Pn.Nx = 101; Nx = Pn.Nx;
Pn.plot = 'linear'; Pn.eps = false;
Fj = [1,2,6];
for j = 1:3
   Pn.F = Fj(j);
   Pn = FUNdefeN(Pn,0);
   Pn.filn = ['fDfxD',num2str(Pn.F)];
   Pn.ylabl = 'D(E); f(E); D(E) f(E)';
   xc = [1;1]*[Pn.chem,Pn.chem + Pn.Phi]/Pn.chem;
   yc = 50*[-1;0]*[1,1];
   QPlot(Pn,j); hold on; plot(xc,yc,'--','Color',0.5*[1,1,1]); hold off
   legend(Pn.Lgt)
   Mxy = Pn.Mxy;Mxy = [Mxy(1,1),-100,-100,-110;Mxy;Mxy(Nx,1),-100,-100,-100];
   writematrix(Mxy,Pn.filn,'FileType','text','Delimiter','tab')
end
```

```
function Pn = FUNgtfEm(P,iopt)
%% Function FUNgtfEm:  general thermal field integrand
Pn = P;
Nx = Pn.Nx; Nf = Pn.Nf;
Fj = linspace(Pn.Fmin,Pn.Fmax,Nf); Pn.Fj = Fj;
Mxy = zeros(Nx,Nf);
Em = zeros(1,Nf);
Ej = Pn.Emin + (Pn.Emax - Pn.Emin)*linspace(0,1,Nx)';
for j = 1:Nf
   Pn.F = Fj(j);
   Pn = FUNdefeN(Pn,0);
   Mxy(:,j) = Pn.Mxy(:,4);
   [~,io] = max(Mxy(:,j));
   Mxy(:,j) = exp(Mxy(:,j) - Mxy(io,j));
   % x = Ej(io-2:io+2); y = Mxy(io-2:io+2,j); P = polyfit(x,y,2);
   Em(j) = Pn.Em;
end
Ea = Pn.chem + 0.98*Pn.Phi*linspace(0,1,Nf);
Fa = 2*Pn.kb*Pn.T*sqrt(2*Pn.mo*(Pn.chem + Pn.Phi - Ea))./Pn.hbar;
nft = Pn.hbar*Fj./(2*Pn.kb*Pn.T*sqrt(2*Pn.mo*abs(Pn.chem - Em + Pn.Phi)));
Pn.Mnft = [Fj',nft'];
Ex = Pn.chem + Pn.Phi*[0,0,1,1]'; Fy = 1.05*Pn.Fmax*[0,1,1,0]';
Pn.Cx = Ej/Pn.chem; Pn.Cy = Fj; Pn.Mxy = Mxy';
Pn.xlabl = 'E/\mu';
Pn.ylabl = 'F ~~\textrm{[eV/nm]}';
Pn.filn = 'GTFn';
s = (0:10)'/10; Pn.cred = [s.^2,4*s.*(1-s),1-s.^2]; Pn.cred(1,:) = 0;
Pn.vv = s;
Pn.titltx = ['$\Phi = ',num2str(Pn.Phi),'$ eV; $T = ',num2str(Pn.T),'$ K'];
if iopt
   Em = Em/Pn.chem; Ea = Ea/Pn.chem; Ex = Ex/Pn.chem;
   CPlot(Pn,1); hold on;
   plot(Ea,Fa,'y','LineWidth',6)
   plot(Em,Fj,'.k','MarkerSize',12)
   plot(Ex,Fy,'--w','LineWidth',2)
   hold off
   axis([Pn.Emin/Pn.chem,Pn.Emax/Pn.chem,Pn.Fmin,Pn.Fmax])
   saveas(gcf,Pn.filn,'png');
   saveas(gcf,Pn.filn,'epsc');
end
end
```

```
function Pn = FUNdefeN(P,iopt)
%% Triangular Barrier theta and beta_F used for D(E) f(E)
Pn = P;
theta = @(x,b,V) b*sign(V-x).*(abs(V - x)).^1.5;
betaf = @(x,c,V) c*sqrt(V - x);
betat = 1.0./(Pn.kb*Pn.T);
Nx = Pn.Nx;
B = (4./(3*Pn.hbar*Pn.F))*sqrt(2*Pn.mo);
C = (2./(Pn.hbar*Pn.F))*sqrt(2*Pn.mo);
Vo = Pn.chem + Pn.Phi;
Ej = Pn.Emin + (Pn.Emax - Pn.Emin)*linspace(0,1,Nx)';
% theta calculation
th = theta(Ej,B,Vo);
[~,ia] = min(abs(th - 30)); ia = max(ia,1);
[~,ib] = min(abs(th + 30)); ib = min(ib,Nx);
```

```
DEj = zeros(Nx,1);
DEj(1:ia-1)   = -theta(Ej(1:ia-1),B,Vo);
DEj(ia:ib)    = - log(1 + exp(theta(Ej(ia:ib),B,Vo)));
DEj(ib+1:Nx) = 0;
% supply calculation
fEj = log(FUNfz(betat*(Pn.chem - Ej))) - log(betat*Pn.chem);
% Local maximum
[Pn.Em,Pn.Dfm,~,~] = FUNemdJ(Ej,DEj + fEj,DEj);
Pn.thetam = theta(Pn.Em,B,Vo);
Pn.betaf = betaf(Pn.Em,C,Vo);
Pn.ntf = betat/Pn.betaf;
Pn.Mxy = [Ej/Pn.chem,DEj,fEj,DEj + fEj];
Pn.xlabl = 'E/\mu';
Pn.xyrange = [Ej(1)/Pn.chem,Ej(Nx)/Pn.chem,max(fEj(Nx),-40),1];
Pn.ylabl = '\ln[D; f; D*f]';
Pn.titltx = ['$F = ',num2str(Pn.F),'$ eV/nm, $T = ',num2str(Pn.T),'$ [K]'];
Pn.cred = [0,0.8,0;0,0,1;1,0,0];
Pn.Lgt = {'$D(E)$','$f(E)$','$D(E)f(E)$'};
Pn.filn = 'GTF';
if iopt>0; QPlot(Pn,iopt);end
end
```

```
function y = FUNfz(x)
%% Function FUNfz
%   calculation of log(1 + z) for z = exp(x)
%   CAUTION:  x is assumed to be monotically DECREASING
% function
ln1x = @(x) x.*(x.*(x.*(3*x.*(4*x-5)+20)-30)+60)/60;
xmax = 8;
Nx = length(x); y = zeros(Nx,1); z = exp(x);
[~,ib] = min(abs(x + xmax));
[~,ia] = min(abs(x - xmax));
y(ia:ib) = log(1 + z(ia:ib));
if x(ia) > -xmax; y(1:ia)  = x(1:ia) + ln1x(1./z(1:ia)); end
if x(ib) <  xmax; y(ib:Nx) = ln1x(z(ib:Nx)); end
end
```

```
function [xmo,ymo,zmo,beta] = FUNemdJ(xj,yj,zj)
%% Function FUNxoyo
%   find location xo and local maximum of yo = y(xo) of y(x)
%   Letting zj = dy/dx, then also returns beta = -z(xo)
[~,io] = max(yj); N = length(yj);
io = max(3,min(io,N-2));
xm = xj(io-2:io+2); ym = yj(io-2:io+2); zm = zj(io-2:io+2);
Cj = polyfit(xm,ym,4);
Dj = Cj.*(4:-1:0); Dj = Dj(1:4); dy = polyval(Dj,xm);
xmo = spline(dy,xm,0); ymo = polyval(Cj,xmo);
Pj = polyfit(xm,zm,4);
Qj = Pj.*(4:-1:0); Qj = Qj(1:4);
zmo   =   polyval(Pj,xmo);
beta = - polyval(Qj,xmo);
end
```

```
function tfln = CPlot(Pn,ifig)
%% Function Plot(Pn,ifig)
```

```
%       Rapid contour plotting of P structure data P.Mxy
%       ifig is the figure number
%         (integer, should not be 0)
xlabl = join(['$',Pn.xlabl,'$']);
ylabl = join(['$',Pn.ylabl,'$']);
if ifig
   figure(ifig)
   contourf(Pn.Cx,Pn.Cy,Pn.Mxy,Pn.vv)
   colormap(Pn.cred)
   set(gca,'fontsize',16)
   title(Pn.titltx,'FontSize',18,'Interpreter','latex')
   ylabel(ylabl,'FontSize',24,'Interpreter','latex')
   xlabel(xlabl,'FontSize',24,'Interpreter','latex')
   axis tight square
   colorbar
   saveas(gcf,Pn.filn,'epsc');
end
tfln = [Pn.filn,num2str(ifig)];
end
```

### B.4.3 *N*(*n*,*s*) Function

- Figures generated by this script: Figures 12.7 and 12.8
- Primary script: `FIGNns.m`
- Subordinate scripts: `FIGnft.m`, `FUNgtfo.m`

```
%% Figure FIGNns
%   Contour plot of N(n,s) vs F and z(E)
%   polyfit approach finds the local maximum of function
%   P(2) and P(1) amount to Taylor expansion terms
clear;clc
Pn = struct;
Pn.Nx = 100; Pn.Nf = 80; Pn.s = 25;
n = linspace(0.001,1.5,Pn.Nf);
b = ceil(32 + log(max(n)));
x = linspace(-b/2,b,Pn.Nx)';
y = zeros(Pn.Nx,Pn.Nf);
xm = zeros(1,Pn.Nf);
for j = 1:Pn.Nf
   % x(:,j) = linspace(a(j),b(j),Nx)';
   y(:,j) = log(1 + exp(n(j)*(x - Pn.s)))./(1 + exp(x));
   [~,io] = max(y(:,j));
   xj = x(io-2:io+2); yj = y(io-2:io+2,j);
   P = polyfit(xj,yj,2);
   xm(j) = -P(2)/(2*P(1));
end
y = y./max(y);
Pn.vv = (0:10)/10;
s = (0:9)'/9; Pn.cred = [s.^2,4*s.*(1-s),1 - s.^2];
Pn.cred(1,:) = 0;
Pn.xlabl = 'z';
Pn.ylabl = 'n';
Pn.Cx = x; Pn.Cy = n; Pn.Mxy = y';
Pn.titltx = ['N(n,s) integrand for $s = ',num2str(Pn.s),'$'];
Pn.filn = 'Nnsz';
xw = Pn.s*[0,0,1,1]; yw = max(n)*[0,1,1,0];
CPlot(Pn,1); hold on;
plot(xm,n,'.g','MarkerSize',9)
plot(xw,yw,'w','LineWidth',2)
```

```
plot([-20,40],[1,1],'--w','LineWidth',0.7)
hold off
axis([-16,33,0,1.5])
saveas(gcf,Pn.filn,'epsc');
```

```
%% Figure FIGnft
% contour plots of dN for (F,E) and (n,E)
Nf = 100;
Fj = linspace(0.2,4.0,Nf);
nj = 0*Fj;
Pfn = struct;
Pfn = FUNdamental(Pfn);
Pfn.chem = 7; Pfn.Phi = 4;
Pfn.T = 1500; Pfn.N = 160;
Pfn.Emin = 6; Pfn.Emax = 12;
Pfn.F = Fj(1); Pfn = FUNgtfo(Pfn);
Mx = zeros(Pfn.N,Nf);
for j = 1:Nf
   Pfn.F = Fj(j);
   Pfn = FUNgtfo(Pfn);
   Mx(:,j) = Pfn.dJ;
   nj(j) = Pfn.nft;
end
E = Pfn.E/Pfn.chem;
xo = [1;1]; xp = (1 + (Pfn.Phi/Pfn.chem))*[1;1];
yo = [Fj(1);Fj(Nf)]; yn = [nj(1);nj(Nf)];
Nv = 12;
vv = (0:Nv)'/Nv;
cred = [vv,4*vv.*(1-vv),1-vv]; cred(1,:) = 0;
fln = 'FNnft';
titltx = ['$(T,\Phi,\mu) = (',num2str(Pfn.T),'; ',num2str(Pfn.Phi),'; ',...
   num2str(Pfn.chem),')$'];
xm = 1 + (Pfn.Phi/Pfn.chem)*sqrt(linspace(0,1,32)');
ym = (2*Pfn.kb*Pfn.T/Pfn.hbar)*sqrt(2*Pfn.mo*(Pfn.chem + Pfn.Phi - Pfn.chem*xm));
figure(1)
contourf(E,Fj,Mx',vv);
colormap(cred)
axis square; colorbar; hold on
plot(xm,ym,'w','LineWidth',2);
plot([xo,xp],yo,'w','LineWidth',1); hold off
axis([min(E), max(E), min(Fj), max(Fj)]);
set(gca,'fontsize',14)
xlabel('$E/\mu$','FontSize',24,'Interpreter','latex');
ylabel('$F$ [eV/nm]','FontSize',24,'Interpreter','latex');
title(titltx,'FontSize',18,'Interpreter','latex')
saveas(gcf,[fln,'F'],'epsc');
figure(2)
contourf(E,nj,Mx',vv);
colormap(cred)
axis square; colorbar; hold on
plot([xo,xp],yn,'w','LineWidth',1); hold off
set(gca,'fontsize',14)
xlabel('$E/\mu$','FontSize',24,'Interpreter','latex');
ylabel('$n = \beta_T  / \beta_F(E_m)$','FontSize',24,'Interpreter','latex');
title(titltx,'FontSize',18,'Interpreter','latex')
saveas(gcf,[fln,'n'],'epsc');
```

```
function Pnew = FUNgtfo(Pfn)
%% Function FUNgtfo
%   Triangular barrier dJ(E) integrand using analytic $\theta$ and
%   $\beta_F$
Pnew = Pfn;
chem = Pfn.chem;
Phi = Pfn.Phi;
betaT = 1/(Pfn.kb*Pfn.T);
Coft = 4*sqrt(2*Pfn.mo)/(3*Pfn.hbar*Pfn.F);
Cofb = 2*sqrt(2*Pfn.mo)/(Pfn.hbar*Pfn.F);
E = linspace(Pfn.Emin,Pfn.Emax,Pfn.N)';
theta = Coft*sign(Pfn.chem + Pfn.Phi - E).*abs(Pfn.chem + Pfn.Phi - E).^1.5;
DE  = 1.0./(1 + exp(theta));
fE = log(1 + exp((chem - E)/(Pfn.kb*Pfn.T))); fE = fE/max(fE);
dJ  = DE.*fE;
[~,io] = max(dJ);    % find index of maximum
x = E(io-2:io+2); y = dJ(io-2:io+2); P = polyfit(x,y,2);
Eo = -P(2)/(2*P(1));dJo = polyval(P,Eo);
theto = Coft*sign(chem + Phi - Eo)*abs(chem + Phi - Eo)^1.5;
betafo = Cofb*sqrt(abs(chem + Phi - Eo));
DEo = 1.0./(1 + exp(theto - betafo*(E - Eo)));
Pnew.E = E; Pnew.Eo = Eo;
Pnew.dJ = dJ/dJo;
Pnew.dJo = DEo.*fE/max(dJ);
Pnew.nft = betaT/betafo;
end
```

- Figures generated by this script: Figures 12.9 and 12.11
- Primary script: `FIGthetabeta.m`
- Subordinate scripts: `FUNgtfns.m`

```
%% Figure FIGthetabeta
%   plot of $\theta(E)$ and the FN and parabolic approximations to it
clear;clc
Psn = FUNdamental();
Nx = 32;
Psn.Fld = 2.5; Psn.Phi = 4; Psn.chem = 7;
% Generate the Theta coefficient
Psn.xo = sqrt(Psn.Qo./Psn.Fld);
Psn.Cof = 2*sqrt(2*Psn.mo*Psn.Qo*Psn.xo)/Psn.hbar;
%  Generate the Cj coefficients
Psn.fv = @(x,p) (sin(x).^2)./sqrt(1 + p*cos(x));
Psn.vy = @(x) 1 - (x.^2).*(1 - log(x)/3);
Psn.ty = @(x) 1 + (x.^2).*(1 - log(x))/9;
n = 7;Fj = zeros(n,1); Fj(1) = pi/2; Fj(n) = 4*sqrt(2)/3;
xj = (0:n-1)'/(n-1);yj = (1-xj)./(1+xj); pj = sqrt(1 - yj.^2);
for j = 2:n-1; Fj(j) = integral(@(x) Psn.fv(x,pj(j)),0,pi); end
Psn.Cj = polyfit(xj,Fj,n-1); Psn.Cj = Psn.Cj(n:-1:1);
% y vector
Psn.Ej = linspace(Psn.chem,Psn.chem + Psn.Phi - sqrt(4*Psn.Qo*Psn.Fld),Nx)';
yj = sqrt(4*Psn.Qo*Psn.Fld)./(Psn.chem + Psn.Phi - Psn.Ej);
xj = (1 - yj)./(1 + yj);
Hyj = (1 - yj.^2)./sqrt(yj.^3);
Sxj = sum(Psn.Cj.*(xj.^(0:n-1)),2);
Psn.theta = Psn.Cof*Hyj.*Sxj;
Psn.xlabl = 'E \textrm{~[eV]}';
Psn.ylabl = '\theta(E)';
Psn.xyrange = [Psn.Ej(1), Psn.Ej(Nx), 0, Psn.theta(1)];
Psn.Lgt = 'Exact';
Psn.filn = 'thetasn';
```

```
Psn.cred = [0,0,0;0,0,1;1,0,0];
Psn.titltx = 'Gamow Factor';
yo = sqrt(4*Psn.Qo*Psn.Fld)/Psn.Phi;
thetaFNo = (4/(3*Psn.hbar*Psn.Fld))*sqrt(2*Psn.mo*Psn.Phi^3)*Psn.vy(yo);
betaFN   = (2/(Psn.hbar*Psn.Fld))*sqrt(2*Psn.mo*Psn.Phi)*Psn.ty(yo);
thetaFN  = thetaFNo - betaFN*(Psn.Ej - Psn.chem);
betap    = (pi/(Psn.hbar*Psn.Fld))*sqrt(Psn.mo*Psn.Phi*yo);
thetap   =  - betap*(Psn.Ej - Psn.chem - Psn.Phi + sqrt(4*Psn.Qo*Psn.Fld));
%% Theta Plot
Psn.Mxy = [Psn.Ej,Psn.theta,thetaFN,thetap];
QPlot(Psn,1); hold on;
plot(Psn.Ej,Psn.theta,'ok');
% plot([7,7],[0,20],'--','Color',[0.6,0.6,0.6])
hold off
Psn.Lgt = {'Exact','FN','para'};
legend(Psn.Lgt)
saveas(gcf,Psn.filn,'png')
writematrix(Psn.Mxy,Psn.filn,'FileType','text','Delimiter','tab')
%% dJ(E) Plot
Psn.yo = sqrt(4*Psn.Qo*Psn.Fld)/Psn.Phi;
Tmin = (Psn.hbar*Psn.Fld/(4*Psn.kb*Psn.ty(Psn.yo)))*sqrt(2/(Psn.mo*Psn.Phi));
Tmax = Psn.hbar*Psn.Fld/(Psn.kb* pi * sqrt(Psn.mo*Psn.Phi*Psn.yo));
PdJ = Psn;
PdJ.ylabl = 'dJ(E)';
PdJ.To = round([Tmin,0.5*(Tmax + Tmin),Tmax]);
PdJ.filn = 'dJEsn';
PdJ.Ej = linspace(PdJ.chem-1.5,PdJ.chem + PdJ.Phi - sqrt(4*PdJ.Qo*PdJ.Fld)+1.5,100)';
yj = sqrt(4*PdJ.Qo*PdJ.Fld)./(PdJ.chem + PdJ.Phi - PdJ.Ej);
xj = (1 - yj)./(1 + yj);
Hyj = (1 - yj.^2)./sqrt(yj.^3);
Sxj = sum(PdJ.Cj.*(xj.^(0:n-1)),2);
PdJ.theta = PdJ.Cof*Hyj.*Sxj;
thetaFN  = thetaFNo - betaFN*(PdJ.Ej - PdJ.chem);
thetap   =  - betap*(PdJ.Ej - PdJ.chem - PdJ.Phi + sqrt(4*PdJ.Qo*PdJ.Fld));
for j = 1:3
   dJE   = log(1 + exp((PdJ.chem - PdJ.Ej)/(PdJ.kb*PdJ.To(j))))./(1 + exp(PdJ.theta));
   dJEFN = log(1 + exp((PdJ.chem - PdJ.Ej)/(PdJ.kb*PdJ.To(j))))./(1 + exp(thetaFN));
   dJEp  = log(1 + exp((PdJ.chem - PdJ.Ej)/(PdJ.kb*PdJ.To(j))))./(1 + exp(thetap));
   % PdJ.Mxy = [PdJ.Ej,dJE,dJEFN,dJEp];
   dJEo = max(dJE);
   PdJ.Mxy = [PdJ.Ej,dJE/dJEo,dJEFN/dJEo,dJEp/dJEo];
   QPlot(PdJ,j+1); axis tight
   title(['$T = ',num2str(PdJ.To(j)),'$ K'])
   saveas(gcf,[PdJ.filn,num2str(PdJ.To(j))],'png')
   writematrix(PdJ.Mxy,[PdJ.filn,num2str(PdJ.To(j))],'FileType','text','Delimiter','
     tab')
end
```

```
function Pn = FUNgtfns(P)
%% oGTF formalism
%  for calculating n, s and Jgtf, Jrld, and Jfn
Pn = P;
% functions
Pn.ty = @(x) 1 + (x.^2).*(1 - log(x))/9;
Pn.vy = @(x) 1 - (x.^2).*(3 - log(x))/3;
To = Pn.To; Fo = Pn.Fo;
mo = Pn.mo; hbar = Pn.hbar; kb = Pn.kb;
Phio = Pn.Phi; chem = Pn.chem;
phi = Phio - sqrt(4*Pn.Qo*Fo);
```

```
Pn.yo = sqrt(4*Pn.Qo*Fo)/Phio;
Pn.Tmin = (hbar*Fo/(4*kb*Pn.ty(Pn.yo)))*sqrt(2/(mo*Phio));
Pn.Tmax = hbar*Fo/(kb* pi * sqrt(mo*Phio*Pn.yo));
Pn.betaT = 1.0/(kb*To);
Pn.yo = sqrt(4*Pn.Qo*Fo)/Phio;
phi = Phio - sqrt(4*Pn.Qo*Fo);
Pn.betau = (2/(hbar*Fo))*sqrt(2*mo*Phio) * Pn.ty(Pn.yo);
Pn.betap = (pi/(hbar*Fo))*sqrt(mo*Phio*Pn.yo);
Pn.theto = (4*sqrt(2*mo*Phio^3)/(3*hbar*Fo)) * Pn.vy(Pn.yo);
if To < Pn.Tmin
   Pn.nft = Pn.betaT/Pn.betau;
   Pn.sft = Pn.theto;
elseif To > Pn.Tmax
   Pn.nft = Pn.betaT/Pn.betap;
   Pn.sft = Pn.betap*phi;
else
   A = 3*(Pn.betap + Pn.betau) - (6*Pn.theto/phi);
   B = -2*(Pn.betap + 2*Pn.betau) + (6*Pn.theto/phi);
   C = Pn.betau - Pn.betaT;
   % root of quadratic
   Pn.Cj = [A,B,C]; % Remember:  Co + Ca*p + Cb*p^2 = 0
   Pn.p = roots(Pn.Cj);
   po = (-B - sqrt(B^2 - 4*A*C))/(2*A);
   Em = chem + po*phi;
   Pn.theta = ((1 - po)^2)*(2*po+1)*Pn.theto - ...
       phi*po*(1-po)*((1-po)*Pn.betau - po*Pn.betap);
   Pn.betaF = Pn.betaT;
   Pn.nft = 1;
   Pn.sft = Pn.theta + Pn.betaT * (Em - Pn.chem);
end
Pn.Nns = FUNnsu(Pn.nft,Pn.sft);
Arld = ((mo*kb^2)/(2*pi^2 * hbar^3))*(Pn.cm^2)/Pn.Amp;
Pn.Jgtf = Arld*Pn.Nns*To^2;
Pn.Jrld = Arld*exp(-Pn.betaT*phi)*To^2;
Pn.Jfn  = (Arld/(Pn.kb*Pn.betau)^2)*exp(-Pn.theto);
end
```

### B.4.4 General Thermal-field Model

- Figures generated by this script: Figure 12.10
- Primary script: `FIGoGTF.m`
- Subordinate scripts: `FUNgtfns.m`

```
%%  Figure FIGoGTF
%   Example of oGTF for range of fields for constant T
%   and range of temperatures for constant F
clear;clc
P = FUNdamental();
P.chem = 7; P.Phi = 4.5; P.To = 1500;
Fj = (0.2:0.06:6.06)'; Nf = length(Fj);
Mxy = zeros(Nf,4); Mxy(:,1) = Fj;
for j = 1:Nf
   P.Fo = Fj(j);
   Pn = FUNgtfns(P);
   Mxy(j,2:6) = [Pn.Jrld,Pn.Jfn,Pn.Jgtf,Pn.nft,Pn.sft];
end
Pn.xlabl = 'F \textrm{~[eV/nm]}';
Pn.ylabl = 'J(F,T)';
Pn.Lgt = 'Exact';
```

```
Pn.filn = 'oGTF';
Pn.cred = [1,0,0;0.5,0.5,1;0,0,0];
Pn.titltx = ['$o$GTF Comparison: $T = ',num2str(Pn.To),'$ K'];
Pn.Lgt = {'RLD','FN','$o$GTF'};
Pn.plot = 'logy';
Pn.Mxy = [Mxy(:,1),Mxy(:,2:4)];
Pn.xyrange = [Fj(1), Fj(Nf), min(Mxy(:,4)), max(Mxy(:,4))];
%  find Fmax and Fmin
Fmax = Pn.Phi^2 / (4*Pn.Qo);
for j = 1:8
   yo = sqrt(4*Pn.Qo*Fmax)/Pn.Phi;
   Fmax = 2*Pn.kb*Pn.To*sqrt(2*Pn.mo*Pn.Phi)*Pn.ty(yo)/Pn.hbar;
end
Fmin = (pi*Pn.kb*Pn.To*sqrt(2*Pn.mo)*(Pn.Qo^0.25)/Pn.hbar)^(4/3);
% Plot output
QPlot(Pn,1); hold on
plot([Fmin,Fmax;Fmin,Fmax],60*[-1,-1;1,1],'--','color',[0.6,0.6,0.6]); hold off
legend(Pn.Lgt)
saveas(gcf,[Pn.filn,'F'],'png')
writematrix(Mxy,[Pn.filn,'F'],'FileType','text','Delimiter','tab')
clear j yo P ans Fj
%%  Example of oGTF:  Plot of range of fields for constant F
clear;clc
P = FUNdamental();
P.chem = 7; P.Phi = 4.5; P.Fo = 2;
Tj = (400:20:2000)'; Nt = length(Tj);
Mxy = zeros(Nt,4); Mxy(:,1) = Tj;
for j = 1:Nt
   P.To = Tj(j);
   Pn = FUNgtfns(P);
   Mxy(j,2:6) = [Pn.Jrld,Pn.Jfn,Pn.Jgtf,Pn.nft,Pn.sft];
end
Pn.xlabl = 'T \textrm{~[1000 K]}';
Pn.ylabl = 'J(F,T)';
Pn.Lgt = 'Exact';
Pn.filn = 'oGTF';
Pn.cred = [1,0,0;0.5,0.5,1;0,0,0];
Pn.titltx = ['$o$GTF Comparison: $F = ',num2str(Pn.Fo),'$ eV/nm'];
Pn.Lgt = {'RLD','FN','$o$GTF'};
Pn.plot = 'logy';
Pn.Mxy = [Mxy(:,1)/1000,Mxy(:,2:4)];
Pn.xyrange = [Tj(1)/1000, Tj(Nt)/1000, min(Mxy(1,4))/4, max(Mxy(:,4))];
%  find Fmax and Fmin
yo = sqrt(4*Pn.Qo*Pn.Fo)/Pn.Phi;
Tmin = (Pn.hbar*Pn.Fo/(4*Pn.kb*Pn.ty(yo)))*sqrt(2/(Pn.mo*Pn.Phi));
Tmax = Pn.hbar*Pn.Fo/(Pn.kb* pi * sqrt(Pn.mo*Pn.Phi*yo));
QPlot(Pn,2); hold on
plot([Tmin,Tmax;Tmin,Tmax]/1000,60*[-1,-1;1,1],'--','color',[0.6,0.6,0.6]); hold off
legend(Pn.Lgt)
Mxy(:,1) = Mxy(:,1)/1000;
saveas(gcf,[Pn.filn,'T'],'png')
writematrix(Mxy,[Pn.filn,'T'],'FileType','text','Delimiter','tab')
clear j yo P ans Tj
```

```
function Pn = FUNgtfns(P)
%% oGTF formalism
%  for calculating n, s and Jgtf, Jrld, and Jfn
Pn = P;
% functions
```

```
Pn.ty = @(x) 1 + (x.^2).*(1 - log(x))/9;
Pn.vy = @(x) 1 - (x.^2).*(3 - log(x))/3;
To = Pn.To; Fo = Pn.Fo;
mo = Pn.mo; hbar = Pn.hbar; kb = Pn.kb;
Phio = Pn.Phi; chem = Pn.chem;
phi = Phio - sqrt(4*Pn.Qo*Fo);
Pn.yo = sqrt(4*Pn.Qo*Fo)/Phio;
Pn.Tmin = (hbar*Fo/(4*kb*Pn.ty(Pn.yo)))*sqrt(2/(mo*Phio));
Pn.Tmax = hbar*Fo/(kb* pi * sqrt(mo*Phio*Pn.yo));
Pn.betaT = 1.0/(kb*To);
Pn.yo = sqrt(4*Pn.Qo*Fo)/Phio;
phi = Phio - sqrt(4*Pn.Qo*Fo);
Pn.betau = (2/(hbar*Fo))*sqrt(2*mo*Phio) * Pn.ty(Pn.yo);
Pn.betap = (pi/(hbar*Fo))*sqrt(mo*Phio*Pn.yo);
Pn.theto = (4*sqrt(2*mo*Phio^3)/(3*hbar*Fo)) * Pn.vy(Pn.yo);
if To < Pn.Tmin
   Pn.nft = Pn.betaT/Pn.betau;
   Pn.sft = Pn.theto;
elseif To > Pn.Tmax
   Pn.nft = Pn.betaT/Pn.betap;
   Pn.sft = Pn.betap*phi;
else
   A = 3*(Pn.betap + Pn.betau) - (6*Pn.theto/phi);
   B = -2*(Pn.betap + 2*Pn.betau) + (6*Pn.theto/phi);
   C = Pn.betau - Pn.betaT;
   % root of quadratic
   Pn.Cj = [A,B,C]; % Remember:  Co + Ca*p + Cb*p^2 = 0
   Pn.p = roots(Pn.Cj);
   po = (-B - sqrt(B^2 - 4*A*C))/(2*A);
   Em = chem + po*phi;
   Pn.theta = ((1 - po)^2)*(2*po+1)*Pn.theto - ...
       phi*po*(1-po)*((1-po)*Pn.betau - po*Pn.betap);
   Pn.betaF = Pn.betaT;
   Pn.nft = 1;
   Pn.sft = Pn.theta + Pn.betaT * (Em - Pn.chem);
end
Pn.Nns = FUNnsu(Pn.nft,Pn.sft);
Arld = ((mo*kb^2)/(2*pi^2 * hbar^3))*(Pn.cm^2)/Pn.Amp;
Pn.Jgtf = Arld*Pn.Nns*To^2;
Pn.Jrld = Arld*exp(-Pn.betaT*phi)*To^2;
Pn.Jfn  = (Arld/(Pn.kb*Pn.betau)^2)*exp(-Pn.theto);
end
```

- Figures generated by this script: Figure 12.14
- Primary script: `FIGdyke.m`
- Subordinate scripts: `FUNNnsTF.m`

```
%% Figure FIGdyke
%   Recreation of Dyke et al. measurements of
%   FN emission at various temperature.
%   Data of Dyke, et al., digially extractedl
clear;clc
% Create structure file
Pn = FUNdamental();
Pn.chem = 7; Pn.Phi = 4.5;
Pn.Tmin = 300; Pn.Tmax = 2000;
Pn.fln = [258,280,304,343,588,667]'; Fjo = Pn.fln/100;
Nf = length(Fjo);
Pn.Tdyke = [0.7702, 0.7592, 0.7539, 0.7735, 1.1020, 1.1161;  ...
           0.9412, 0.9449, 0.9344, 0.9489, 1.4020, 1.4235;  ...
```

```
          1.1171, 1.1040, 1.0974, 1.1212, 1.7318, 1.7452;  ...
          1.2636, 1.2502, 1.2486, 1.2619, 1.9395, 1.9445;  ...
          1.4318, 1.4143, 1.4102, 1.4165, 2.0091, 2.0278;  ...
          NaN,    NaN,    NaN, 1.5376,    NaN,    NaN];
Pn.RJdyke =  [0.1732, 0.1404, 0.1155, 0.0698, 0.0845, 0.0563; ...
          0.3036, 0.2416, 0.1943, 0.1218, 0.1422, 0.0846; ...
          0.4873, 0.4291, 0.3155, 0.2161, 0.2055, 0.1454; ...
          0.8226, 0.5932, 0.4871, 0.3026, 0.2612, 0.1574; ...
          1.1753, 0.9068, 0.6944, 0.4417, 0.2819, 0.1673; ...
             NaN,    NaN,    NaN, 0.5515,    NaN,    NaN];
Fj =  1.2559*Fjo - 0.5027;
Nt = 25; Tj = Pn.Tmin + (Pn.Tmax- Pn.Tmin)*(0:Nt-1)'/(Nt-1);
Pn.Ny = 4; Pn.Em = 0;
Psn = FUNNnsTF(Pn,Fj,Tj);
Psn.Lgt = {num2str(Fjo)};
% remember t(y)^2 = 1.1 to calculate Rmin
Tminj = Fj*Psn.hbar/(2*Psn.kb*sqrt(2*Psn.mo*Psn.Phi)*1.1);
RJmin = zeros(Nf,1);
for j = 1:Nf
   RJmin(j) = spline(Psn.Mxy(:,1),Psn.Mxy(:,j+1),Tminj(j)/1000);
end
PJT = polyfit(Tminj/1000,RJmin,2);
Psn.Mxy(:,Nf+2) = polyval(PJT,Psn.Mxy(:,1));
Psn.xyrange(4) = 1.2;
Psn.Lgt = char({num2str(Fjo),'$R_{min}$'});
QPlot(Psn,2); hold on
plot(Pn.Tdyke,Pn.RJdyke,'.','MarkerSize',32); hold off
legend(Psn.Lgt)
saveas(gcf,Psn.filn,'png')
writematrix([Pn.Tdyke,Pn.RJdyke],'DykeRAW', 'FileType', 'text','Delimiter','tab')
writematrix(Psn.Mxy,'DykeGTF', 'FileType', 'text','Delimiter','tab')
```

```
function Psn = FUNNnsTF(Pn,Fj,Tj)
%% Calculation of N(n,s) for range of F or T
Nt = length(Tj);
Nf = length(Fj);
Mxy = zeros(Nt,Nf);
Psn = Pn;
Psn.Tj = Tj;
Psn.Fj = Fj;
for k = 1:Nf
   Pa = Pn;
   Pa.F = Fj(k);
   for j = 1:Nt
       Pa.To = Tj(j);
       Pa = FUNEtbns(Pa);
       Mxy(j,k) = FUNnsu(Pa.nft,Pa.sft);
   end
   Mxy(:,k) = log10(Mxy(:,k)/Mxy(1,k)) + 2*log10(Tj/Tj(1));
end
Psn.Mxy = [Tj/1000,Mxy];
sc = (0:Nf-1)'/(Nf-1);
Psn.cred = [sc.^2,4*sc.*(1-sc),1 - sc.^2];
Psn.xlabl = 'T~\textrm{[1000 K]}';
Psn.ylabl = '\log_{10} [J(F,T)/J(F,300)]';
Psn.Lgt = {num2str(Fj)};
Psn.titltx = 'Comparison to Dyke, et al., PR99, 1192 (1955)';
Psn.xyrange = [0.6, 2, 0 1.2];
Psn.filn = 'DykeComp';
end
```

### B.4.5 Quantum Efficiency

- Figures generated by this script: Figures 13.1 and 13.2
- Primary script: `FIGQEthexD.m`
- Subordinate scripts: `FIGQEthexS.m`

```
%% FIGQEthexD:
%   Photoemission model For Cu
%   Based on:  K.L. Jensen, Tutorial: The equations of electron
%              emission and their evaluation, Figure 23
%              J. Appl. Phys. 135(11), 111101 (2024).
%              http://dx.doi.org/10.1063/5.0196481
Pd = FUNdamental();
Pd.filn = 'QEDowell';
Pd.phi = 4.36 + 0.1*(-2:2);
Pd.chem = 7;
Pd.To = 300;
Pd.Nq = 128;
Np = length(Pd.phi);
% Experimental reference values courtesty of D. Dowell (SLAC)
Pd.hw = [4.3050, 4.3351, 4.3656, 4.3966, 4.4280, 4.4599, 4.4922, ...
        4.5250, 4.5582, 4.5920, 4.6263, 4.6611, 4.6964, 4.7322, ...
        4.7686, 4.8056, 4.8431, 4.8813, 4.9200, 4.9594, 4.9994, ...
        5.0400, 5.0813, 5.1233, 5.1660, 5.2094, 5.2536, 5.2985, ...
        5.3441, 5.3906, 5.4379, 5.4860, 5.5350, 5.5849, 5.6356, ...
        5.6873, 5.7400, 5.7937, 5.8483, 5.9040, 5.9608, 6.0187, ...
        6.0777, 6.1378, 6.1992, 6.2618, 6.3257, 6.3909, 6.4575, ...
        6.5255, 6.5949]';
LQ = [-16.277, -14.857, -14.430, -13.546, -12.662, -11.964, ...
     -11.483, -11.191, -10.840, -10.452, -10.189, -9.8995, ...
     -9.6566, -9.5814, -9.3542, -9.2004, -9.0619, -8.9102, ...
     -8.7982, -8.7341, -8.6281, -8.5222, -8.4264, -8.3475, ...
     -8.2821, -8.2024, -8.1117, -8.0472, -7.9604, -7.9047, ...
     -7.8442, -7.7776, -7.7040, -7.6459, -7.5791, -7.5483, ...
     -7.4876, -7.4269, -7.3587, -7.3172, -7.2460, -7.2197, ...
     -7.1730, -7.1050, -7.0539, -6.9686, -6.9138, -6.8143, ...
     -6.6943, -6.5763, -6.4421]';
Pd.QE = exp(LQ);
Nw = length(Pd.hw);
rt = (pi^2/3)*(Pd.kb*Pd.To)^2;
My = zeros(Pd.Nq,Np);
[~,io] = min(abs(Pd.hw - 5));
hx = linspace(Pd.phi(1),Pd.hw(Nw),Pd.Nq)';
[~,jo] = min(abs(hx - Pd.phi));
for j = 1:Np
   C  = Pd.QE(io)/((Pd.hw(io) - Pd.phi(j))^2 + rt);
   My(:,j) = C*((hx - Pd.phi(j)).^2 + rt);
   My(1:jo(j)-1,j) = NaN;
end
Pd.Mxy = [hx/Pd.hw(io),My/Pd.QE(io)];
Pd.Lgt = num2str(Pd.phi');
Pd.xlabl = '\hbar \omega / \hbar \omega_o';
Pd.ylabl = 'QE/QE_o';
Pd.xyrange = [0.83, 1.32, 0.001, 20];
s = (0:Np-1)'/(Np-1);
Pd.cred = [s.^2,4*s.*(1-s),1-s.^2];
Pd.plot = 'logy';
Pd.titltx = 'Copper QE (Dowell)';
QPlot(Pd,2); hold on
plot(Pd.hw/Pd.hw(io),log10(Pd.QE/Pd.QE(io)),'ok');
```

```
plot(1,0,'.k','MarkerSize',32);
hold off
Pd.Lgt = char({Pd.Lgt,'Dowell'});
legend(Pd.Lgt)
disp(['Reference hw = ',num2str(Pd.hw(io)),' eV'])
disp(['Reference QE = ',num2str(Pd.QE(io)),' eV'])
saveas(gcf,Pd.filn,'png')
writematrix(Pd.Mxy,[Pd.filn,'Th'], 'FileType', 'text','Delimiter','tab')
writematrix([Pd.hw/Pd.hw(io),Pd.QE/Pd.QE(io)],[Pd.filn,'Exp'], ...
          'FileType', 'text','Delimiter','tab')
```

```
%% Figure FIGQEthexS
%   Photoemission model For Cs3Sb
%   Based on:  K.L. Jensen, Tutorial: The equations of electron
%              emission and their evaluation, Figure 24
%              J. Appl. Phys. 135(11), 111101 (2024).
%              http://dx.doi.org/10.1063/5.0196481
Ps = FUNdamental();
Ps.filn = 'QESpicer';
Ps.phi = 1.93 + 0.05*(-2:2);
Ps.To = 300;
Ps.Nq = 128;
Np = length(Ps.phi);
% Experimental reference values courtesty of D. Dowell (SLAC)
Ps.hw = [1.6470, 1.6940, 1.7410, 1.7870, 1.8530, 1.8860, 1.9320, ...
        2.0400, 2.0870, 2.1200, 2.1570, 2.3300, 2.4380, 2.4750, ...
        2.5260, 2.5830, 2.6290, 2.6860, 2.7280, 2.7790, 2.8260, ...
        2.8820, 2.9190, 2.9800, 3.0780, 3.1250, 3.1810, 3.2330, ...
        3.2800, 3.3220, 3.3780, 3.4430, 3.5320, 3.6350, 3.6910, ...
        3.7610, 3.8410]';
LQ = [-15.6394, -14.3563, -13.0726, -12.0288, -10.6148, -9.3533, ...
      -8.3744,  -6.9078,  -5.2983,  -4.9618,  -4.6052, -3.3814, ...
      -2.9759,  -2.8473,  -2.7181,  -2.7031,  -2.5903, -2.5639, ...
      -2.4418,  -2.4534,  -2.3228,  -2.3026,  -2.2634, -2.2443, ...
      -2.1286,  -2.1120,  -2.1542,  -2.0875,  -2.0479, -2.0479, ...
      -2.0250,  -2.0250,  -2.0025,  -1.9590,  -1.9805, -1.9590, ...
      -1.9173]';
Ps.QE = exp(LQ);
Nw = length(Ps.hw);
rt = (pi^2/3)*(Ps.kb*Ps.To)^2;
My = zeros(Ps.Nq,Np);
[~,io] = min(abs(Ps.hw - 2.5));
hx = linspace(1.8,4,Ps.Nq)';
[~,jo] = min(abs(hx - Ps.phi));
for j = 1:Np
   C  = Ps.QE(io)/((Ps.hw(io) - Ps.phi(j))^2 + rt);
   My(:,j) = C*((hx - Ps.phi(j)).^2 + rt);
   My(1:jo(j)-1,j) = NaN;
end
Ps.Mxy = [hx/Ps.hw(io),My/Ps.QE(io)];
Ps.Lgt = num2str(Ps.phi');
Ps.xlabl = '\hbar \omega / \hbar \omega_o';
Ps.ylabl = 'QE/QE_o';
Ps.xyrange = [0.75, 1.55, 0.005, 20];
s = (0:Np-1)'/(Np-1);
Ps.cred = [s.^2,4*s.*(1-s),1-s.^2];
Ps.plot = 'logy';
Ps.titltx = 'Cs$_3$Sb QE (Spicer)';
QPlot(Ps,3); hold on
```

```
plot(Ps.hw/Ps.hw(io),log10(Ps.QE/Ps.QE(io)),'ok');
plot(1,0,'.k','MarkerSize',32);
hold off
Ps.Lgt = char({Ps.Lgt,'Spicer'});
legend(Ps.Lgt)
disp(['Reference hw = ',num2str(Ps.hw(io)),' eV'])
disp(['Reference QE = ',num2str(Ps.QE(io)),' eV'])
saveas(gcf,Ps.filn,'png')
writematrix(Ps.Mxy,[Ps.filn,'Th'], 'FileType', 'text','Delimiter','tab')
writematrix([Ps.hw/Ps.hw(io),Ps.QE/Ps.QE(io)],[Ps.filn,'Exp'], ...
          'FileType', 'text','Delimiter','tab')
```

### B.4.6 Electron–Electron Scattering

- Figures generated by this script: Figure 13.9
- Primary script: `FIGrelaxtauB.m`
- Subordinate scripts: `FUNtauee.m`

```
%% Figure FIGrelaxtau
%   electron-electron scattering rates based on
%   calculation of Lugovskoy-Bray relaxation time
clear; clc
Pn = FUNdamental();
Pn.chem = 7.0;
Pn.Ks   = 2.3;
Pn.re   = 1;
Pn.iout = 0;
Pn.To = 300;
Pn.hw = 4.5*(0.06:0.02:1)';
Pn = FUNtauee(Pn);
% Data
Ma =    [0.5998, 99.6520; 0.6951, 96.9650; 0.7735, 90.8460; ...
0.7960, 75.0250; 0.8240, 90.5470; 0.8576, 55.7710; 0.8857, ...
72.9350; 0.9081, 61.7410; 0.9473, 55.3230; 1.0426, 49.5020; ...
1.0482, 42.6370; 1.0874, 39.5020; 1.1267, 36.0700; 1.1547, ...
34.4280; 1.2108, 33.5320; 1.2332, 40.2490; 1.2537, 34.1290; ...
1.3397, 35.6220; 1.3546, 29.4530; 1.3621, 33.3830; 1.3957, ...
28.0600; 1.4182, 26.9650; 1.4518, 29.6520; 1.4630, 26.0700; ...
1.5004, 21.6920; 1.5191, 27.5620; 1.5452, 22.0900; 1.5676, ...
24.1790; 1.5901, 22.0900; 1.6536, 20.5970; 1.6872, 22.6870; ...
1.7209, 21.0950; 1.7433, 20.9950; 1.8442, 18.5070; 1.8853, ...
17.8110; 1.9077, 14.4280; 1.9600, 16.5170; 2.0086, 14.7260; ...
2.0497, 13.8310; 2.0908, 12.0400; 2.1431, 13.3330; 2.1954, ...
13.5320; 2.2440, 11.0450; 2.2739, 12.4380; 2.3711, 11.4430; ...
2.4234, 10.3480; 2.4907, 9.3532; 2.5243, 10.0500; 2.5617,  ...
8.9552; 2.5953, 8.6567; 2.6626, 8.6567; 2.7111, 8.7562; ...
2.7485, 7.8607; 2.8830, 7.1642; 2.9354, 7.6617; 3.0362,  ...
6.5672; 3.1147, 5.8706; 3.1446, 6.0697; 3.2044, 5.8706; ...
3.2455, 5.5721; 3.2904, 6.1692; 3.3203, 5.5721; 3.3987,  ...
5.2736; 3.4922, 5.0746; 3.5632, 4.8756; 3.6229, 4.4776; ...
3.6977, 4.3781; 3.7351, 4.3781; 3.7986, 4.2786; 3.8434,  ...
4.0796; 3.8733, 6.0697; 3.8920, 3.8806; 3.9144, 7.1642; ...
3.9630, 4.5771; 3.9854, 3.3831; 4.0191, 5.0746; 4.0639,  ...
6.6667; 4.1162, 5.8706; 4.1648, 5.2736; 4.2059, 3.7811; ...
4.2433, 3.3831];
Mb =  [1.2170, 34.0830; 1.2975, 28.6881; 1.3780, 25.6785; ...
1.4585, 22.9880; 1.5390, 20.7747; 1.6195, 18.8308; 1.7000, ...
17.0774; 1.7805, 15.6933; 1.8610, 14.4064; 1.9415, 13.2985; ...
2.0220, 12.3020; 2.1025, 11.3638; 2.1830, 10.6012; 2.2635,  ...
```

```
9.8758; 2.3440, 9.2292; 2.4245, 8.6511; 2.5050, 8.1118; ...
2.5855, 7.6714; 2.6660, 7.2190; 2.7465, 6.8186; 2.8270,  ...
6.4399; 2.9075, 6.1187; 2.9880, 5.7962; 3.0685, 5.5061; ...
3.1490, 5.2409; 3.2295, 4.9911];
Mc = [1.1925, 55.3740; 1.3975, 40.1290; 1.7795, 25.2390; ...
2.1429, 18.0820; 2.5248, 13.1780; 2.8323, 10.6920; 3.2236, ...
8.4028];
% Theory
hwmin = max(Mb(1,1),Mc(1,1));
hwmax = min(max(Mb(:,1)),max(Mc(1,:)));
xj = hwmin + (hwmax-hwmin)*(0:25)'/25;
yb = exp(spline(Mb(:,1),log(Mb(:,2)),xj));
yc = exp(spline(Mc(:,1),log(Mc(:,2)),xj));
Mb = [xj,yb]; Mc = [xj,yc];
QPlot(Pn,2); hold on;
plot(Ma(:,1),log10(Ma(:,2)),'.k','MarkerSize',12);
plot(Mb(:,1),log10(Mb(:,2)),'--b','LineWidth',1);
plot(Mc(:,1),log10(Mc(:,2)),'--g','LineWidth',1);
hold off
Pn.Lgt = char(Pn.Lgt,'Ladstadter2004','Lugovskoy1998.1','Lugovskoy1998.5');
legend(Pn.Lgt)
axis tight
saveas(gcf,Pn.filn,'png')
writematrix(Pn.Mxy,[Pn.filn,'Th'],'FileType','text','Delimiter','tab')
writematrix(Ma,[Pn.filn,'Lad'],'FileType','text','Delimiter','tab')
writematrix([xj,yb,yc],[Pn.filn,'Lug'],'FileType','text','Delimiter','tab')
```

```
function Pn = FUNtauee(P)
%% Function tauee(T,E)
% electron-electron relaxation time tau-ee
% y must be a matrix or column vector
% w must be a single number or a row vector
Pn = P;
Pn.gx  = @(x) (0.25*x.^3).*((x./(x.^2 + 1)) + atan(x) ...
   - (atan(x.*sqrt(x.^2+2))./sqrt(x.^2 + 2)));
Pn.Rx  = @(x,a,c) ((c*x).^2)./(pi^2 + (x - a).^2);
E = Pn.hw + Pn.chem;
Nt = length(Pn.To); N  = length(E);
y = zeros(N,Nt);
Pn.kf = sqrt(2*Pn.re*Pn.mo*Pn.chem)/Pn.hbar;
Pn.qo = sqrt(4*Pn.re*Pn.kf/(pi*Pn.Ks*Pn.ao));
beta = 1./(Pn.kb*Pn.To);
for j = 1:N
   y(j,:) = Pn.Rx(beta*E(j),beta*Pn.chem,Pn.Ks);
end
k = sqrt(2*Pn.re*Pn.mo*E)/Pn.hbar;
w = 1./Pn.gx(2*k/Pn.qo);
tauo = Pn.hbar/(2*Pn.Qo/Pn.ao);
tau = tauo*(4*pi/Pn.re)*y.*w;
Pn.Ej = E;
Pn.tauee = tau;
Pn.Mxy = [E-Pn.chem,tau];
Pn.xlabl = '\hbar \omega - \mu ~\textrm{[eV]}';
Pn.ylabl = '\tau_{ee} ~\textrm{[fs]}';
Pn.plot = 'logy';
if Nt > 1
   s = (0:Nt-1)'./(Nt-1);
   Pn.cred = [s.^2,4*s.*(1-s),1 - s.^2];
   Pn.Lgt = [num2str(Pn.To')];
```

```
else
   Pn.cred = [1,0,0];
   Pn.Lgt = ['Theory',num2str(Pn.To)];
end
Pn.titltx = ['e-e Scattering $K_s = ',num2str(Pn.Ks),'$'];
Pn.filn = 'tauee';
end
```

### B.4.7 Multiple Scattering

- Figures generated by this script: Figure 13.11
- Primary script: `FIGQEscat.m`

```
%% Figure QEscat
%  calculation of QE factors for inclusion of scattered electrons
Pn = FUNdamental();
Pn.Nx = 50;
Pn.chem = 7.0;
Pn.delmax = 6;
wj = 1./(2.^(0:25));
Pn.Mxy = zeros(Pn.Nx,4);
Pn.Mxy(:,1) = Pn.delmax*(1:Pn.Nx)'/Pn.Nx;
for iopt = 1:3
   switch iopt
       case 2
           Pn.del  = 12; Pn.phi = 1.8; Pn.taua = 10;
       case 3
           Pn.del  = 100; Pn.phi = 4; Pn.taua = 10;
       otherwise
           Pn.del  = 12; Pn.phi = 4; Pn.taua = 10;
   end
   Pn.hw   = Pn.phi + Pn.delmax*(1:Pn.Nx)'/Pn.Nx;
   Sn = ones(Pn.Nx,1);
   for j = 1:Pn.Nx
       hw   = Pn.hw(j);
       n    = floor(log(hw/Pn.phi)/log(2)) + 1;
       taue = 25.*(wj(1:n).^2);
       tau  = Pn.taua*taue./(Pn.taua + taue);
       vn   = sqrt(2*Pn.mo*(Pn.chem + (hw.*wj(1:n))))/Pn.hbar;
       lo   = vn.*tau;
       pn = Pn.del./lo;
       xm = sqrt((Pn.chem + Pn.phi)./(Pn.chem + (hw.*wj(1:n))));
       if n > 1
           Cp =[1,pn./(1 + pn)];
           for k = 2:n+1
               Cp(k) = Cp(k)*Cp(k-1);
           end
           Cp = Cp(1:n);
           RFn = (0.5*(1 - xm).*(1 + xm - 2*pn) - (pn.^2).*log((xm + pn)./(1 + pn)));
           RFn = RFn./RFn(1);
           Sn(j) = sum(Cp.*RFn);
       end
   end
   Pn.Mxy(:,iopt+1) = 100*(Sn-1);
end
Pn.xlabl = '\hbar \omega - \phi \textrm{~ [eV]}';
Pn.ylabl = '100 \times [(QE/QE_o) - 1]';
Pn.cred  = [1,0,0;0,0.8,0;0,0,1];
Pn.filn  = 'QEscat';
```

```
Pn.titltx = 'Ratio QE for 3 cases';
Pn.Lgt = num2str((1:3)');
Pn.xyrange = [0, Pn.delmax, 0, max(Pn.Mxy(:,2))+0.1];
QPlot(Pn,1); axis tight
saveas(gcf,Pn.filn,'png')
writematrix(Pn.Mxy,Pn.filn,'FileType','text','Delimiter','tab')
```

### B.4.8 Shell-sphere Model

- Figures generated by this script: Figure 13.15
- Primary script: `FIGShellSphereN.m`

```
%% Figure FIGShellSphereN
% --------------------------------------------------
% Expanding Shell (unscattered) is BLUE
% Sphere (scattered) is RED
% Scattered electron algorithm:  e- travel for
%   a time tau and then scatter:  when scattered,
%   the electron's energy is reduced as though it
%   had scattered with an electron from the Fermi
%   level (that is, its energy goes to the average
%   energy of the two electrons)
% --------------------------------------------------
%   Basic constants
Pn = FUNdamental();
%   Particular terms
Pn.N = 600;
Pn.Nt = 26;
Pn.tau = 1;
Pn.zinit = 12;
Pn.zmin = 0;
Pn.zmax = 5*Pn.zinit;
Pn.lambda = 532;
Pn.chem = 7;
Pn.hw = 2*pi*Pn.hbar*Pn.c/Pn.lambda;
Pn.vmax = sqrt((2/Pn.mo)*(Pn.chem + Pn.hw));
% Set the random seed
rng(7)
% circle coordinates
cirx = cos(2*pi*(0:90)/90);
ciry = sin(2*pi*(0:90)/90);
voj = sqrt((2.0/Pn.mo)*(Pn.chem + (Pn.hw./2.^(0:Pn.Nt))));
phi = 2*pi*rand(Pn.N,1); theta = pi*rand(Pn.N,1);
vr = voj(1)*[sin(theta).*cos(phi),sin(theta).*sin(phi),cos(theta)];
va = vr;
rp = zeros(Pn.N,3); rp(:,3) = Pn.zinit;
ra = rp;
% Prepare the new file.
fln = ['ShellSphere',num2str(Pn.Nt)];
vidObj = VideoWriter(fln,'MPEG-4');
open(vidObj);
Pn.Rj = zeros(Pn.Nt+1,5);
Nt = Pn.Nt;
for j = 1:Nt
   rp = rp + Pn.tau*vr;
   phi = 2*pi*rand(Pn.N,1); theta = pi*rand(Pn.N,1);
   ra = ra + Pn.tau*va;
   va = voj(j+1)*[sin(theta).*cos(phi),sin(theta).*sin(phi),cos(theta)];
   scatter(rp(:,1),rp(:,3),8,'filled','b'); hold on
```

```
    % set(gca,'fontsize',14);hold on;
    rpm  = sqrt(sum(mean((rp - mean(rp)).^2)));
    plot(rpm*cirx,rpm*ciry+mean(rp(:,3)),'LineWidth',2,'Color',[0,0,0.6]);
    scatter(ra(:,1),ra(:,3),8,'filled','r');
    ram = sqrt(sum(mean((ra - mean(ra)).^2)));
    plot(ram*cirx,ram*ciry+mean(ra(:,3)),'LineWidth',2,'Color',[0.6,0,0]);
    plot([-Pn.zmax,Pn.zmax],[0,0],'k','LineWidth',6)
    hold off;
    axis(Pn.zmax*[-0.5,0.5,0,1]); axis square; grid off;
    title(['time = ',num2str(j*Pn.tau),' fs'],'FontSize',24,'Interpreter','latex');
    axis off;
    Pn.Rj(j+1,1:3) = [j*Pn.tau,rpm,ram];
    % output graphics
    if mod(j,5) == 0; saveas(gcf,['ShSp',num2str(j)],'png'); end
    currFrame = getframe(gcf);
    writeVideo(vidObj,currFrame);
end
% Close the file.
close(vidObj);
Pn.Rj(:,4) = Pn.Rj(Nt+1,2)*(0:Nt)'/Nt;
Pn.Rj(:,5) = Pn.Rj(Nt+1,3)*sqrt((0:Nt)'/Nt);
writematrix(Pn.Rj,'ShellSphereRj','FileType','text','Delimiter','tab')
%%
tj = Pn.Rj(:,1); MC = Pn.Rj(:,2:3);
FC = Pn.Rj(:,4:5);
figure(3)
plot(tj,MC,'o',tj,FC)
Ps = struct;
Ps.xlabl = 't / \tau';
Ps.ylabl = 'RMS Radius \textrm{~[au]}';
Ps.filn  = 'ShellSphereRj.txt';
Ps.titltx = 'Ballistic and Diffusive Expansion';
Ps.plot = 'linear';
Ps.Lgt = {'Linear','Square Root'};
Ps.xyrange = [0, 17, 0, 30];
Ps.cred = [0,0,1;1,0,0];
Ps.Mxy = [tj,FC];
QPlot(Ps,1); hold on; plot(tj,MC,'o'); hold off
Ps.Lgt = {'Linear', 'Square Root', 'No scattering', 'Scattering',};
legend(Ps.Lgt);
```

- Figures generated by this script: Figure 13.19
- Primary script: `FIGCurShlSph.m`

```
%% Figure FIGCurShlSph
%  Current in the Shell Sphere model
%  assume Qso = Qdo
% set the value of Ro before execution:  default = 1
if exist('Ro','var')  == 1; else; Ro = 1;end
%   Basic constants
Pn = FUNdamental();
%   Needed functions
% P.Flxp = @(x,p) (1 - x.^2)/2 - p.*(1 - x + p.*log((x+p)./(1 + p)));
%   Particular terms
Pn.N = 600;
Pn.Nt = 26;
Pn.tau = 1;
Pn.zinit = 12;
Pn.zmin = 0;
Pn.zmax = 5*Pn.zinit;
```

```
Pn.lambda = 532;
Pn.chem = 7;
Pn.hw = 2*pi*Pn.hbar*Pn.c/Pn.lambda;
Pn.vmax = sqrt((2/Pn.mo)*(Pn.chem + Pn.hw));
Pn.xlabl = 't / t_o';
Pn.ylabl = 'I(t) / I_o';
Pn.filn  = 'CurShlSph';
Pn.titltx = ['Shell Sphere:  $R_o = ',num2str(Ro),'$'];
Pn.plot = 'linear';
Pn.Nx = 101;
Pn.td = [1,2,3,6];Nt = length(Pn.td);
Pn.Lgt = num2str(Pn.td');
Pn.tmax = 8;
tj = Pn.tmax*((0:Pn.Nx-1)'/(Pn.Nx-1)).^1.5;
Pn.Mxy = zeros(Pn.Nx,Nt+1);
Pn.Mxy(:,1) = tj;
u = (0:Nt-1)'/(Nt-1);
Pn.cred  = [u.^2,4*u.*(1-u),1-u.^2]; clear u
Io = 0.5./(tj + 1).^2;
for j = 1:Nt
   Id = sqrt(Pn.td(j)./(4*pi*tj.^3)).*exp(-Pn.td(j)./tj);
   Id(1) = 0;
   Pn.Mxy(:,j+1) = Io + Ro*Id;
end
Pn.xyrange = [0, Pn.tmax,0,max(Pn.Mxy(:,2))];
QPlot(Pn,1);
saveas(gcf,[Pn.filn,num2str(Ro)],'png')
writematrix(Pn.Mxy,[Pn.filn,num2str(Ro)],'FileType','text','Delimiter','tab')
```

### B.4.9 Secondary Emission

- Figures generated by this script: Figure 14.1
- Primary script: `FIGtypes.m`

```
%% Figure FIGtypes
%   Generating a schematic of Secondary types
% Control parameters
Nx = 10; Np = 30; p = 0.9;
% x and y coordinates
xj = zeros(Nx,Np); yj = xj;
% set random seed
rng(5);
yj(1,:) = -1.25 + (p.^(0:Np-1)');
pj = sqrt((p.^(1:Nx)'))/p;
for j = 2:Nx
   r = rand(2,Np);
   for k = 1:Np
       xtemp = xj(j-1,k) + pj(j)*r(1,k).*cos(2*pi*r(2,k));
       ytemp = yj(j-1,k) + pj(j)*r(1,k).*sin(2*pi*r(2,k));
       if ytemp < 0
           xj(j,k) = xtemp; yj(j,k) = ytemp;
       else
           xj(j,k) = -10; yj(j,k) = 20;
       end
   end
end
yp(1) = 5; yp(2:Np) = yj(1,1:Np-1)';
ip = 4;
xs = xj(:,ip) + 2.25; ys = yj(:,ip); ys(1) = 5; xs(2) = xs(1);
```

```
xj = xj - 1; xp = -1 + zeros(Np,1);
xo = [3,3,-3.5]; yo = [5,-0.05,25];
area([-4,-4,4,4],[0,-5,-5,0],'FaceColor',[0.8,0.8,0.8]); hold on;
plot(-xj,yj,-xp,yp,'-ko',-xs,ys,'-r.','MarkerSize',4,'linewidth',1);
plot(-xo,yo,'-b.','linewidth',1);hold off
axis([-4,3,-3,2])
axis off
saveas(gcf,'SecType','png')
```

- Figures generated by this script: Figure 14.2
- Primary script: `FIGBaroodyF1.m`

```
%% Figure FIGBaroodyF1 (Figure 1)
%   Recreation of Baroody Figures 1&2 plotting
%   work function Phi and delta_m vs element
%   Based on K.L. Jensen, Tutorial: The equations of
%   electron emission and their evaluation,
%   Journal of Applied Physics 135, 111101 (2024)
%   https://doi.org/10.1063/5.0196481
clear; clc
Pa = FUNdamental();
Pa.cred = [1,0,0;0,0,1];
Pa.xlabl = 'AN';
Pa.ylabl = '\delta_m';
Pa.Lgt = {'Linear','$\delta_m$'};
Pa.xyrange = [11,91,0.7,1.6];
Pa.titltx = 'Baroody Figure';
Pa.xlabl = 'E/E_m';
Pa.xyrange = [0, 6, 0, 1.05];
Pa.Lgt = 'analytic';
Pa.titltx = [Pa.titltx,' 1'];
Pa.filn = 'SEY';
N = zeros(6,1);
% Mxx matricies of data:  xx is the element
% first column is E/Em, second column is \delta / \delta_m
MAl = [0.6828, 0.9567; 1.3654, 0.9589; 2.0702, 0.9147];
MCu = [0.1538, 0.5831; 0.4825, 0.8893; 1.4954, 0.9819; ...
      1.7973, 0.9348; 1.9884, 0.9468; 2.8052, 0.8208; ...
      3.4910, 0.7471; 3.8338, 0.7184; 4.4033, 0.6571; ...
      4.5026, 0.6518];
MMo = [0.3380, 0.7862; 0.5984, 0.9500; 1.2396, 0.9866; ...
      1.9950, 0.9004; 2.5000, 0.8498; 3.7557, 0.7375; ...
      4.3068, 0.6705];
MNi = [1.9885, 0.9174; 2.9933, 0.8093];
MPt = [0.2737, 0.6397; 0.5125, 0.8547; 0.5535, 0.9023; ...
      1.3758, 0.9822; 1.6687, 0.9622; 1.8009, 0.9480; ...
      2.7810, 0.9023; 3.8327, 0.7302; 4.5012, 0.6627];
MTi = [0.1730, 0.6448; 0.3645, 0.8408; 0.7374, 0.9724; ...
      1.4762, 0.9628; 1.8461, 0.9181; 2.0343, 0.9102];
N(1) = size(MAl,1);
N(2) = size(MCu,1);
N(3) = size(MMo,1);
N(4) = size(MNi,1);
N(5) = size(MPt,1);
N(6) = size(MTi,1);
s = (1:6)'/6; Pa.cred = [0,0,1;1,0,0;[s.^2,3*s.*(1-s),1 - s.^2]];
% Function
Gn  = @(x,n) (1 - exp(-x*n))./(x*n);
% Very simplistic method to find maximum of Gn:  calculate a
% huge number of values Nx, find the largest of value
```

```
Nx = 1001; na = [1,1.2,1.5,2,1.3625]; Nj = length(na);
E = linspace(0,3,Nx)'; Ea = zeros(1,Nj); yao = Ea;
ya = zeros(Nx,Nj);
for j = 1:Nj
   ya(:,j) = E.*Gn(E.^na(j),na(j));
end
[~,ia] = max(ya);
Ea = E(ia); yao = ya(ia);
% Now replace ya by a vector of values of a more reasonable size of Ns
Ns = 81;
s = linspace(0,5.2,Ns)';
ya = zeros(Ns,Nj);
for j = 1:Nj
   ya(:,j) = (s*Ea(j)).*Gn((s*Ea(j)).^na(j),na(j))/yao(j);
end
ya = ya./max(ya);
ya(1,:) = 0;
ya(Ns,:) = 0;      % Case for E = 0 has to be 0;
                   % set last to 0 for plotting elsewhere
                   % to make a filled graph region
Pa.Mxy = [s,ya];
QPlot(Pa,2); hold on
Lgt = char({num2str(na');'Al';'Cu';'Mo';'Ni';'Pt';'Ti'});
plot(MAl(:,1),MAl(:,2),'.','MarkerSize',18);
plot(MCu(:,1),MCu(:,2),'.','MarkerSize',18);
plot(MMo(:,1),MMo(:,2),'.','MarkerSize',18);
plot(MNi(:,1),MNi(:,2),'.','MarkerSize',18);
plot(MPt(:,1),MPt(:,2),'.','MarkerSize',18);
plot(MTi(:,1),MTi(:,2),'.','MarkerSize',18);
plot(Pa.Mxy(:,1),Pa.Mxy(:,Nj+1),'r','LineWidth',2);
hold off
legend(Lgt,'Location','best')
hold off;
axis([0,5,0,1.1])
Mxy = NaN(10,12);
Mxy(1:N(1),1:2) = MAl;
Mxy(1:N(2),3:4) = MCu;
Mxy(1:N(3),5:6) = MMo;
Mxy(1:N(4),7:8) = MNi;
Mxy(1:N(5),9:10) = MPt;
Mxy(1:N(6),11:12) = MTi;
saveas(gcf,Pa.filn,'png')
writematrix(Mxy,[Pa.filn,'El'], 'FileType', 'text','Delimiter','tab')
writematrix(Pa.Mxy,[Pa.filn,'n'], 'FileType', 'text','Delimiter','tab')
```

- Figures generated by this script: Figure 14.3
- Primary script: `FIGBaroodyF2.m`

```
%% Figure FIGBaroody2
%   Recreation of Baroody Figures 1&2 plotting
%   work function Phi and delta_m vs element
%   Based on K.L. Jensen, Tutorial: The equations of
%   electron emission and their evaluation,
%   Journal of Applied Physics 135, 111101 (2024)
%   https://doi.org/10.1063/5.0196481
clear; clc
Pb = FUNdamental();
Pb.cred = [1,0,0;0,0,1];
Pb.xlabl = 'AN';
Pb.ylabl = '\delta_m';
```

```
Pb.Lgt = {'Linear','$\delta_m$'};
Pb.xyrange = [11,91,0.7,1.6];
Pb.titltx = 'Baroody Figure';
Pb.filn = 'PhiDelm';
% Baroody Figure 2
%   Here, the elemental name is replaced with atomic number AN
%   Phi is scaled so it fits on the same axis as delta_m
%   Both Phi and delta_m data are rounded to hundreths place
AN = [37, 55, 19, 56, 12, 13, 90, 51, 73, 48, 40, ...
     29, 27, 42, 74, 47, 26, 79, 46, 28]';
Phi = [1.7942, 1.7958, 1.7958, 2.0928, 2.4028, ...
      2.9960, 3.4376, 3.9987, 4.0929, 4.0959, ...
      4.0959, 4.1489, 4.1955, 4.3004, 4.5108, ...
      4.6920, 4.6979, 4.7943, 4.9317, 4.9933]';
delm = [0.7090, 0.7969, 0.7969, 0.8234, 0.9509, ...
       0.9895, 1.1019, 1.5992, 1.2988, 1.0989, ...
       1.0989, 1.2982, 1.1995, 1.2514, 1.3974, ...
       1.4969, 1.2995, 1.4617, 1.2985, 1.2979]';
Pb.El = {'Rb' 'Cs' 'K' 'Ba' 'Mg' 'Al' 'Th' 'Sb' 'Ta' 'Cd' 'Zr' 'Cu' 'Co' ...
   'Mo' 'W' 'Ag' 'Fe' 'Au' 'Pd' 'Ni'}';
[~,io] = sort(Phi);
AN = AN(io); Phi = Phi(io); delm = delm(io); Pb.El = Pb.El(io);
Px = polyfit(Phi,delm,1);
yphi = polyval(Px,Phi);
MPhi = [Phi,delm,yphi];
[~,io] = sort(AN);
Pb.Mxy = [AN(io),yphi(io),delm(io)];
Pb.El = Pb.El(io);
Pb.titltx = [Pb.titltx,' 2 revised'];
QPlot(Pb,2);
saveas(gcf,Pb.filn,'png')
writematrix(Pb.Mxy,Pb.filn, 'FileType', 'text','Delimiter','tab')
writematrix(MPhi,[Pb.filn,'PD'], 'FileType', 'text','Delimiter','tab')
Mpd = [AN, Phi, delm];
disp('Element      AN          Phi          delm')
disp([string(Pb.El),num2str(Mpd)])
```

- Figures generated by this script: Figure 14.5
- Primary script: `FIGYoung.m`

```
%% Figure FIGYoung
%  Representation of Young's data
Pn = FUNdamental();
io = 7;    % index of comparison point of data
% Plotting details
Pn.xlabl = 'E_p \textrm{~[keV]}';
Pn.ylabl = 'R(E_o) ~[\mu\textrm{m}]';
Pn.filn  = 'YoungAl';
Pn.xyrange = [0, 12, 0, 1.2];
xj = [0.4501, 0.8911, 1.3683, 2.4564, 3.4982, 4.0982, ...
     5.6987, 7.0925, 10.9071]';
yj = [0.0154, 0.0374, 0.0643, 0.1282, 0.2232, 0.2708, ...
     0.4219, 0.5422, 1.0823]';
Nx = length(xj);ja = 3; jb = 1;
Pa = polyfit(xj(ja:Nx-jb),log(yj(ja:Nx-jb)),1);
Pn.Ro = exp(Pa(2));
Pn.Eo = xj(io);
Pn.alf = 1 + Pa(1);
Pn.n = [1,1.2,1.50,2.0,Pn.alf]; nj = length(Pn.n);
E = (0:0.4:12)';
```

```
Pn.Mxy = [E,yj(io)*exp(log(E/Pn.Eo)*Pn.n)];
Pn.Lgt = num2str(Pn.n');
s = (0:nj-1)'/(nj-1); Pn.cred = [s.^2,4*s.*(1-s),1-s.^2];clear s
Pn.titltx = ['Al: $R = ',num2str(Pn.Ro),...
   '*(E/',num2str(Pn.Eo),')^{',num2str(Pn.alf),'}$'];
QPlot(Pn,1); hold on;
plot(xj,yj,'ok'); hold off
Pn.Lgt = char({num2str(Pn.n');'Young'});
legend(Pn.Lgt)
saveas(gcf,Pn.filn,'png')
writematrix([xj,yj],Pn.filn,'FileType','text','Delimiter','tab')
writematrix(Pn.Mxy,[Pn.filn,'th'],'FileType','text','Delimiter','tab')
```

### B.4.10 Bethe Model

- Figures generated by this script: Figures 14.6, 14.7, and 14.8
- Primary script: `FIGdEdx.m`

```
%% Figure FIGdEdx
%   First plot:  plotting Bethe ln(1+u)/u compared to C/u^(n-1)
clear;clc
P = FUNdamental();
P.cred = [1,0,0;0,0,1];
P.xlabl = 'u';
P.ylabl = 'f(u)';
P.xyrange = [0,1000,0,1.2];
P.titltx = 'dE/dx';
P.plot = 'logx';
P.filn = 'dEdx';
yx =  @(x) log(1+x)./x;
ya = @(x,c,p) c./(x.^p);
x = exp((-3:0.2:9)'); Nx = length(x);
xo = [0.5,2,8,32,128]; Nc = length(xo);
P.Lgt = num2str((1:Nc)');
p = ((1 + xo).*log(1 + xo) - xo)./((1 + xo).*log(1 + xo));
C = log(1 + xo).*(xo.^p)./xo;
Mx = zeros(Nx,Nc);Mo = zeros(1,Nc);
for j = 1:Nc
   Mx(:,j) = ya(x,C(j),p(j));
   Mo(j) = ya(xo(j),C(j),p(j));
end
P.Mxy = [x,Mx];
s = (0:Nc-1)'/(Nc-1); P.cred = [s.^2,4*s.*(1-s),1-s.^2];
QPlot(P,1); hold on;
plot(log10(x),yx(x),'.k',log10(xo),Mo,'*k'); hold off
P.Lgt = char({num2str((1:Nc)');'$\ln(1+u)/u$';'$C/x^p$'});
legend(P.Lgt)
Mxy = [P.Mxy,yx(x)];
saveas(gcf,P.filn,'png')
writematrix(Mxy,P.filn, 'FileType', 'text','Delimiter','tab')
writematrix([xo',Mo'],[P.filn,'o'], 'FileType', 'text','Delimiter','tab')
%%   Second plot:  Range graph
clear; clc
Pn = FUNdamental();
s = (0:4)'/4; Pn.cred = [s.^2,4*s.*(1-s),1-s.^2];
Pn.xlabl = 'u';
Pn.ylabl = 'R(E_o) / b a_o';
Pn.xyrange = [0,124,0,2000];
Pn.titltx = 'Range';
```

```
Pn.plot = 'linear';
Pn.filn = 'Rubao';
% function
dR =  @(x) x./log(1+x);
x = (0.2:4:125)'; Nx = length(x);
Rx = zeros(Nx,1);
Rx(1) = x(1) + ((x(1)^2)/4) - ((x(1)^3)/36);
for j = 3:Nx
   Rx(j) = Rx(j-1) + integral(dR,x(j-1),x(j));
end
pj = (1.46:0.04:1.62);
Ra = x.^pj;
Pn.Mxy = [x,Ra];
Pn.Lgt = num2str(pj');
QPlot(Pn,2); hold on
plot(x,Rx,'ok'); hold off
Pn.Lgt = char({num2str(pj');'Exact'});
legend(Pn.Lgt);
saveas(gcf,Pn.filn,'png')
writematrix([x,Rx,Ra],Pn.filn, 'FileType', 'text','Delimiter','tab')
%%  Third plot:  distribution of secondaries graph for diamond
clear; clc
delE = 13.75;        % in eV (energy each secondary sheds)
bao = 0.114371;      % in nm (b * ao)
Nx  = 145;
Eo  = Nx * delE;      % in eV (energy of primary)
Ec  = 97.1896;       % in eV
Pn = FUNdamental();
s = (0:4)'/4; Pn.cred = [s.^2,4*s.*(1-s),1-s.^2];
Pn.ylabl = 'f(E) \textrm{~[norm]}';
Pn.xlabl = 'x_j / R(E_o)';
Pn.xyrange = [0,1.05,0,1];
Pn.titltx = ['$E_o = ',num2str(Eo/1000),'$ keV'];
Pn.plot = 'linear';
Pn.filn = 'Rjpath';
Pn.Lgt   = {'$E_j$','$dE/dx$'};
Pn.cred = [0,0,1;1,0,0];
dR =  @(x) x./log(1+x);
uj = (Eo - (1:Nx)*delE)'/Ec;
xj = zeros(Nx,1);
xj(Nx) = bao*integral(dR,0,Eo/Ec);
for j = Nx-1:-1:1
   xj(j) = xj(j+1) - bao*integral(dR,uj(j+1),uj(j));
end
Ej = (Eo - (Nx:-1:1)*delE)';
dEj = NaN(Nx,1);
dEj(2:Nx) = (Ej(2:Nx) - Ej(1:Nx-1))./(xj(2:Nx) - xj(1:Nx-1));
xj = xj/xj(Nx);
Ej = Ej/Eo;
dEj = dEj/max(dEj);
Pn.Mxy = [xj,Ej,dEj];
QPlot(Pn,3);
saveas(gcf,Pn.filn,'png')
writematrix(Pn.Mxy,Pn.filn, 'FileType', 'text','Delimiter','tab')
```

### B.4.11 Space Charge

- Figures generated by this script: Figures 15.3 and 15.4
- Primary script: `FIGLCjf.m`
- Subordinate scripts: `FIGspaceN.m` and `FUNfnabnu.m`

```
%% Figure FIGLCjf
% Child's Law vs J(F)
clear;clc;
yoj = zeros(5,1); xoj = zeros(5,1);
iplot = false; ilin = false;
iopt = 1; FIGspaceN;
Mxy = [xj,fxa(xj) - yj];
yoj(1,:) = yo;
xoj(1) = xo;
iopt = 2; FIGspaceN;
Mxy = [Mxy,fxa(xj) - yj];
yoj(2,:) = yo;
xoj(2) = xo;
iopt = 3; FIGspaceN;
Mxy = [Mxy,fxa(xj) - yj];
yoj(3,:) = yo;
xoj(3) = xo;
iopt = 4; Phi = 4; FIGspaceN;
Mxy = [Mxy,fxa(xj) - yj];
yoj(4,:) = yo;
xoj(4) = xo;
iopt = 5; Phi = 4.5; FIGspaceN;
Mxy = [Mxy,fxa(xj) - yj];
yoj(5,:) = yo;
xoj(5) = xo;
Mxy = [Mxy,fxa(xj)];
P.filn = 'LCjf';
P.Mxy = Mxy;
P.cred = [0,0,1; 0,0.8,0.8; 0,0.8,0; 0.75,0.75,0; 1,0,0; 0,0,0];
P.Lgt = {'lin','quad','sqrt','FN1','FN2','CL'};
P.titltx = 'CL versus $j(f)$';
P.xyrange = ([0, 1, 0, fxa(0)]);
QPlot(P,1); hold on;
plot(xoj,fxa(xoj),'ok'); hold off
legend(P.Lgt)
saveas(gcf,P.filn,'png');
writematrix(P.Mxy,P.filn,'Delimiter','tab')
disp(['Output written to ',P.filn,'.txt'])
Pn = P; Pn.Mxy = Pn.Mxy(:,1:6);
Pn.Mxy(:,2:6) = -P.Mxy(:,2:6) + P.Mxy(:,7);
Pn.Lgt = P.Lgt(1:5);
Pn.ylabl = 'j_{cl}(f) - j(f)';
Pn.xyrange = [0,1,-1.5,1.5];
Pn.filn = [Pn.filn,'Diff'];
QPlot(Pn,2); hold on; plot(xoj,yoj,'ok',[0,1],[0,0],'k'); hold off
legend(Pn.Lgt)
saveas(gcf,Pn.filn,'png');
writematrix(Pn.Mxy,Pn.filn,'Delimiter','tab')
disp(['Output written to ',Pn.filn,'.txt'])
```

```
%% Figure FIGspaceN
%    This routine called repeatedy by FIGLCjf.m (use it)
Nx = 11; dx = 0.001;
P = FUNdamental();
if exist('iopt','var')  == 1; else; iopt  = 4;end
if exist('Phi','var')   == 1; else; Phi   = 4.5;end
if exist('Fo','var')    == 1; else; Fo    = 10;end
if exist('Va','var')    == 1; else; Va    = 5500;end
if exist('iplot','var') == 1; else; iplot = true;end
if exist('ilin','var')  == 1; else; ilin  = false;end
if iopt == 1
   a = 1; b = 0; n = 0;
   fxb = @(x,a,b,n) a*16*x/9;
   P.titltx = 'Linear';
elseif iopt == 2
   a = 0; b = 1; n = 0;
   fxb = @(x,a,b,n) b*16*(x.^2)/9;
   P.titltx = 'Quadratic';
elseif iopt == 3
   a = 0; b = 0; n = 1;
   fxb = @(x,a,b,n) n*16*sqrt(x)/9;
   P.titltx = 'Square-root';
else
   P.Phi = Phi; P.Fo = Fo; P.Va = Va;
   [a,b,n] = FUNfnabnu(P);
   fxb = @(x,a,b,n) a*(x.^n).*exp(-b./x);
   P.titltx = ['FN-MG  $\Phi = ',num2str(Phi),'$ eV'];
end
fxa = @(x) (4/9)*(2 + (2 - 3*x).*sqrt(1 + 3*x));
x = (0:Nx)'/Nx;
y = fxa(x) - fxb(x,a,b,n);
xo = spline(y,x,0);
yo = fxa(xo) - fxb(xo,a,b,n);
yp = fxa(xo + dx) - fxb(xo + dx,a,b,n);
ym = fxa(xo - dx) - fxb(xo - dx,a,b,n);
dy  = (yp - ym)/(2*dx);
ddy = (yp - 2*yo + ym)/(dx^2);
if ilin
   xo = xo - yo/dy; disp('Linear approx');
else
   xo = xo - ((2*yo)/(dy - sqrt(dy^2 - 2*yo*ddy))); disp('quad approx')
end
yo = fxa(xo) - fxb(xo,a,b,n);
xj = linspace(0,1,80)';
yj = fxa(xj)-fxb(xj,a,b,n);
P.fxa = fxa; P.fxb = fxb;
P.cred = [1,0,0;0,0,1];
P.xlabl = 'f';
P.ylabl = 'j(f)';
P.filn = ['SpChrg',num2str(iopt)];
P.Lgt = {'$y$'};
P.Mxy = [xj,fxa(xj),fxb(xj,a,b,n)];
P.xyrange = [0,1,0,fxa(0)];
if iplot
   QPlot(P,iopt); hold on; plot(xo,fxa(xo),'ok',[0,1],[0,0],'k');hold off
   legend(P.Lgt,'Location','best')
   saveas(gcf,P.filn,'png')
   writematrix(P.Mxy,P.filn, 'FileType', 'text','Delimiter','tab')
end
% plot(x,y,'ok',xj,yj,xo,yo,'*r',[0,1],[0,0],'g')
```

```
function [a,b,n] = FUNfnabnu(P)
Phi = P.Phi;
Fo  = P.Fo;
Jo = (Fo^2)/(32*pi*P.Qo*sqrt(2*P.mo*P.Va));
n = (8*P.Qo/(9*P.hbar))*sqrt(2*P.mo/Phi);
B = (4/(3*P.hbar))*sqrt(2*P.mo*Phi^3); b = B/Fo;
C = ((exp(6)*Phi^2)/(4*P.Qo*B))^n;
a = ((C*Fo^2)/(16*P.hbar*Phi*pi^2))/Jo;
end
```

- Figures generated by this script: Figure 15.5
- Primary script: `FIGjfjcl.m`

```
%% Figure FIGjfjcl
%   Fowler Nordheim compared to Child
%   fxb(1,a,b,n)  is for FN with F = Fo
%   fxb(xo,a,b,n) is for FN with F = f Fo
Nx = 50;
P = FUNdamental();
if exist('iplot','var') == 1; else; iplot = false;end
P.Phi = 4.5; P.Do = 500;
P.xlabl = 'f';
P.ylabl = 'j(f)';
P.filn = ['SpChrgFN',num2str(10*P.Phi)];
P.Lgt = {'$y$'};
P.cred = [1,0,0;0,0,1];
P.titltx = ['FN-MG  $\Phi = ',num2str(P.Phi),'$ eV'];
fxa = @(x) (4/9)*(2 + (2 - 3*x).*sqrt(1 + 3*x));
fxb = @(x,a,b,n) a*(x.^n).*exp(-b./x);
P.fxa = fxa; P.fxb = fxb;
x = (0:Nx)'/Nx;
if iplot
   Nv = 2;
else
   Nv = 24;
end
Fa = linspace(5,10,Nv)';
Va = Fa*P.Do; jFN = zeros(Nv,2);
xoj = zeros(Nv,1);
for ix = 1:Nv
   P.Va = Va(ix); P.Fo = Fa(ix);
   [a,b,n] = FUNfnabnu(P);
   y = fxa(x) - fxb(x,a,b,n);
   xo = spline(y,x,0);
   yo = fxa(xo) - fxb(xo,a,b,n);
   jFN(ix,:) = [fxb(xo,a,b,n),fxb(1,a,b,n)];
   xoj(ix) = xo;
   if iplot
       P.Mxy = [x,fxa(x),fxb(x,a,b,n)];
       P.xyrange = [0,1,0,fxa(0)];
       QPlot(P,ix+1); hold on;    % save ifile = 1 for below
       plot(xo,fxa(xo),'ok',[0,1],[0,0],'k');hold off
       legend(P.Lgt,'Location','best')
       saveas(gcf,P.filn,'png')
       writematrix(P.Mxy,P.filn, 'FileType', 'text','Delimiter','tab')
   end
end
if Nv > 3
   Pn = P;
   Mxy = ones(Nv,3);
```

```
    JCL = 1320.282593*(Va.^1.5)./(P.Do.^2);
    Mxy(:,1:2) = jFN;
    Pn.Mxy = [Va/Pn.Do,log10((16/9)*Mxy.*JCL)];
    Pn.plot = 'linear';
    Pn.xlabl = 'F \textrm{~[eV/nm]}';
    Pn.ylabl = '\log_{10}\{J(V_a,D_o) \textrm{~[amps/cm$^2$]\}}';
    Pn.Lgt = {'$FN(f F_o)$','$FN(F_o)$','CL'};
    Pn.cred = [1,0,0;0,0,1;0.6,0.6,0.6];
    Pn.xyrange = [Va(1)/Pn.Do,Va(Nv)/Pn.Do,1,4];
    QPlot(Pn,1);
    saveas(gcf,Pn.filn,'png')
    writematrix(Pn.Mxy,P.filn, 'FileType', 'text','Delimiter','tab')
end
```

- Figures generated by this script: Figure 15.6
- Primary script: `FIGdykeB.m`

```
%% Figure Dyke 1953 Figure 5
%   Space charge affected field emision
%   Distance from flat anode to tip:  d = 0.5 cm
%   Beta factor:  beta_o = (9.9 \pm 1.5) x 1000  [1/cm]
%   Corrected version using hyperbolic emitter
Pn = FUNdamental();
Nx = 50;
Pn.Do = 0.5;
Pn.Ro = 1200.0e-7;
Pn.Ao = (pi*Pn.Ro^2)*(100e-7/Pn.Ro)^2;
vo = atan(sqrt(Pn.Ro/Pn.Do));
Pn.Qo = @(x) 0.5*log((1 + x)./(1 - x));
Pn.betao = cos(vo)/((sin(vo))^2 * Pn.Qo(cos(vo)));
Mpl = [0.4597, -3.2068; 0.4597, -3.2068; 0.4575, -2.7363; ...
      0.4280, -2.6786; 0.4277, -2.4106; 0.3973, -2.2825; ...
      0.3892, -2.0805; 0.3778, -1.8843; 0.3583, -1.5885; ...
      0.3469, -1.4312; 0.3079, -0.7786];
Mdr = [0.7245, -7.5966; 0.7474, -8.1831; 0.7908, -8.8052; ...
      0.8417, -9.7712; 0.8664, -10.337; 0.8943, -10.797; ...
      0.9267, -11.416; 0.9570, -12.018; 0.9972, -12.718; ...
      1.0372, -13.503; 1.0896, -14.239; 1.1580, -15.578; ...
      1.2475, -17.212; 1.3165, -18.629];
Mpl(:,1) = 10000./Mpl(:,1);          % convert 10^4 / V to V
Mpl(:,2) = exp(Mpl(:,2))/Pn.Amp;     % convert ln(Ia) to Ia [amps]
Mdr(:,1) = 10000./Mdr(:,1);          % convert 10^4 / V to V
Mdr(:,2) = exp(Mdr(:,2))./Pn.Amp;    % convert ln(Ia) to Ia [amps]
[~,io] = sort(Mpl(:,1)); Mpl = Mpl(io,:);
[~,io] = sort(Mdr(:,1)); Mdr = Mdr(io,:);
xp = Mpl(:,1); xd = Mdr(:,1);
yp = Mpl(:,2); yd = Mdr(:,2);
Ppl = polyfit(1./xp,log(yp),1);
Pdr = polyfit(1./xd,log(yd),1);
xmax = 35; xmin = 5;
ymax = 5e-4; ymin = 5e-13;
x = exp(linspace(log(xmin*1000),log(xmax*1000),Nx)');
ypa = exp(polyval(Ppl,1./x));
yda = exp(polyval(Pdr,1./x));
%% Remember approximation:  beta = #/Ro, so Ro cancels out
%   (beta*Ao)^2 here is therefore just some number times pi.
%   call that number Pn.betao
ycl = (3.300706484e-6*((Pn.betao*x).^1.5)./sqrt(Pn.Do))*Pn.Ao;
disp(['vo = ',num2str(vo)])
disp(['Apex field = ',num2str(Pn.betao*x(Nx)/(Pn.Do*1e7))])
```

```
disp(['beta_o = ',num2str(Pn.betao)])
disp(['beta_o/D_o = ',num2str(Pn.betao/Pn.Do)])
Pn.xlabl = '\varphi \textrm{~[kV]}';
Pn.ylabl = 'I_a \textrm{~[amps]}';
Pn.cred  = [1,0,0;0,0,1;0.5,0.5,0.5];
Pn.filn  = 'DykeFig5';
Pn.titltx = 'DykeFig5: CL lines = $R_o$ [nm]';
Pn.Lgt = {'theory P','theory D','C-L'};
Pn.plot = 'logy';
Pn.Mxy = [x/1000,ypa,yda,ycl];
Pn.xyrange = ([xmin,xmax,ymin,ymax]);
QPlot(Pn,1); hold on;
plot(xp/1000,log10(yp),'ok')
plot(xd/1000,log10(yd),'.k','MarkerSize',16); hold off
Pn.Lgt = char({'FN-P';'FN-D';'CL-LB';'Dyke-P';'Dyke-D'});
legend(Pn.Lgt);
Np = length(yp); Nd = length(yd); Mexp = NaN(max(Np,Nd),4);
Mexp(1:Np,1:2) = [xp/1000,log10(yp)]; Mexp(1:Nd,3:4) = [xd/1000,log10(yd)];
Mth = [x/1000,log10(Pn.Mxy(:,2:4))];
saveas(gcf,Pn.filn,'png')
writematrix(Mth, [Pn.filn,'th'],'FileType','text','Delimiter','tab')
writematrix(Mexp,[Pn.filn,'ex'],'FileType','text','Delimiter','tab')
```

### B.4.12 Transit Time

- Figures generated by this script: Figure 15.9
- Primary script: `FIGjfjdf.m`

Basis of Figure 15.9
`FIGjfjdf.m`

```
%%  Figure FIGjfjdf
%   Plot of the decrease of J_FN(F) for a small percentage decrease in F
clear;clc
% function
jf = @(x,b,n) exp((2-n)*log(x) - (b./x));
P = FUNdamental();
P.Phi = 4.5;   % in eV
P.Fo  = 10;    % in eV/nm
P.del = [0.01, 0.02, 0.04, 0.08, 0.16];  % percentage drop in f = F/Fo
Np = length(P.del);
P.bo  = 4*sqrt(2*P.mo*P.Phi^3)/(3*P.hbar*P.Fo);
P.nu  = (8*P.Qo/(9*P.hbar))*sqrt(2*P.mo/P.Phi);
f  = ((0.04:0.02:1)').^2; % trick to increase number of smaller values
Nx = length(f);
fp = f*(1 - P.del);
P.Mxy = [f,jf(fp,P.bo,P.nu)./jf(f,P.bo,P.nu)];
P.xlabl = 'f = F/F_o';
P.ylabl = 'j[(1-\delta)f] / j(f)';
P.cred  = [0,0,0;1,0,0;0,0,1;0,0.5,0];
P.filn  = 'jfjdf';
P.titltx = ['$\Phi = ',num2str(P.Phi),'$ eV; $F_o = ',num2str(P.Fo),'$ eV/nm'];
P.Lgt = {num2str(100*P.del')};
P.plot = 'linear';
P.xyrange = ([0,1,0,1]);
QPlot(P,1);
saveas(gcf,P.filn,'png')
writematrix(P.Mxy,P.filn, 'FileType', 'text','Delimiter','tab')
```

- Figures generated by this script: Figure 15.10
- Primary script: `FIGjfjdf.m`
- Subordinate scripts: `FUNstau.m`

```
%% Figure FIGjfjdf
%  Plot of the decrease of J_FN(F) for a small percentage decrease in F
clear;clc
% function
jf = @(x,b,n) exp((2-n)*log(x) - (b./x));
P = FUNdamental();
P.Phi = 4.5;    % in eV
P.Fo  = 10;     % in eV/nm
P.del = [0.01, 0.02, 0.04, 0.08, 0.16];  % percentage drop in f = F/Fo
Np = length(P.del);
P.bo  = 4*sqrt(2*P.mo*P.Phi^3)/(3*P.hbar*P.Fo);
P.nu  = (8*P.Qo/(9*P.hbar))*sqrt(2*P.mo/P.Phi);
f  = ((0.04:0.02:1)').^2; % trick to increase number of smaller values
Nx = length(f);
fp = f*(1 - P.del);
P.Mxy = [f,jf(fp,P.bo,P.nu)./jf(f,P.bo,P.nu)];
P.xlabl = 'f = F/F_o';
P.ylabl = 'j[(1-\delta)f] / j(f)';
P.cred  = [0,0,0;1,0,0;0,0,1;0,0.5,0];
P.filn  = 'jfjdf';
P.titltx = ['$\Phi = ',num2str(P.Phi),'$ eV; $F_o = ',num2str(P.Fo),'$ eV/nm'];
P.Lgt = {num2str(100*P.del')};
P.plot = 'linear';
P.xyrange = ([0,1,0,1]);
QPlot(P,1);
saveas(gcf,P.filn,'png')
writematrix(P.Mxy,P.filn, 'FileType', 'text','Delimiter','tab')
```

```
function Pn = FUNstau(P)
%% Evaluation of terms for the s-tau model of space charge
Pn = P;
% functions
Pn.yjf = @(x,b,v) exp(b*(x - 1)./x).*(x.^(2 - v));
Pn.tyf  = @(x) 1 + (1 - log(x)).*(x/3).^2;
Ao = 1/(16*Pn.hbar*pi^2);
Bo = 4*sqrt(2*Pn.mo*Pn.Phi^3)/(3*Pn.hbar);
nu = (8*Pn.Qo/(9*Pn.hbar))*sqrt(2*P.mo/Pn.Phi);
yo = sqrt(4*Pn.Qo*Pn.Fo)/Pn.Phi;
ty = Pn.tyf(yo);
Pn.tauo = sqrt(2*Pn.mo*Pn.Do/Pn.Fo);
Pn.Delt = Pn.tauo/Pn.n;
Pn.b     = Bo/Pn.Fo;
Pn.nu     = nu;
Pn.yo     = yo;
Pn.Ao     = Ao;
Pn.Bo     = Bo;
Pn.Amg = ((Ao*Bo^2)/(Pn.Phi * ty^2))*((exp(6)*Pn.Phi^2)/(4*Pn.Qo*Bo))^nu;
Pn.ro = (16*pi*Pn.Qo*Pn.Amg*Pn.tauo/Pn.Fo)*exp(-Pn.b)*(1/Pn.b)^(2 - nu);
Pn.sigo = Pn.Amg*exp(-Pn.b)*Pn.Delt/(Pn.b^(2 - Pn.nu));
end
```

### B.4.13 Miram Curve

- Figures generated by this script: Figure 15.14
- Primary script: `FIGPhishift.m`
- Subordinate scripts: `FUNJrldcl.m`, `FUNmiramxt.m`

```
%%  Figure FIGPhishift
%    Show Miram plot for a change in Phi and T, keeping J constant
clear;clc
P = FUNdamental();
P.Nt = 51; P.Tmin = 1000; P.Tmax = 1400;
P.Va = 1000; P.Do = 2.5e6; P.F = P.Va/P.Do;
P.Phi = 2.0;
%% initial state
Pa = P;
Pa.alf = 2.7; Po = Pa; Po.F = 0;
Pa = FUNmiramxt(Pa);
Po = FUNmiramxt(Po);
%% Final State (assuming Jrld(Phio,To) = Jrld(Phi,To + Delta t))
Pb = P;
r = 25/Pb.Tmin;
phio = Pb.Phi - sqrt(4*Pb.Qo*Pb.F);
phi = (1+r)*(phio + 2 * Pb.kb*Pb.Tmin*log(1+r));
Pb.Phi = phi + sqrt(4*Pb.Qo*Pb.F);
Pb.alf = 2.3;
Pb = FUNmiramxt(Pb);
%% Comparison
QPlot(Pa,3);hold on;
plot(Pb.Mxy(:,1),Pb.Mxy(:,3),[1,1],[Pa.yo,Pb.yo],'ok');
plot(Po.Mxy(:,1),Po.Mxy(:,2),'--c')
hold off
Lgt = {'RLD','$J(\Phi_o,\alpha_o)$','$J_{CL}$','$J(\Phi,\alpha)$','$J_o$'};
legend(Lgt)
saveas(gcf,Pb.filn,'png')
writematrix(Pa.Mxy,[Pa.filn,'A'], 'FileType', 'text','Delimiter','tab')
writematrix(Pb.Mxy,[Pb.filn,'B'], 'FileType', 'text','Delimiter','tab')
```

```
function Pn= FUNJrldcl(P)
%   Richardson-Laue-Dushman equation and Child-Langmuir equation
%   Returns J in units of Amp/cm^2
%   F presumed constant; T varies
Pn = P; Nt = length(Pn.To); Na = length(Pn.alf);
Pn.Arld = Pn.mo*(Pn.kb^2)/(2*(pi^2)*(Pn.hbar^3)) * (Pn.cm^2)/Pn.Amp;
Pn.Acl = (1/(18*pi*Pn.Qo*sqrt(2*Pn.mo)))* (Pn.cm^2)/Pn.Amp;
Pn.betaphi = (Pn.Phi - sqrt(4*Pn.Qo*Pn.F))./(Pn.kb*Pn.To);
Pn.Jrld = Pn.Arld*(Pn.To.^2).*exp(-Pn.betaphi);
Pn.Jcl  = Pn.Acl * (Pn.Va^1.5)/Pn.Do^2;
Pn.xj = Pn.Jrld/Pn.Jcl; yj = ones(Nt,Na+1);
for j = 1:Na
   yj(:,j+1) = Pn.xj./(1 + Pn.xj.^Pn.alf(j)).^(1./Pn.alf(j));
end
Pn.Mxy = [Pn.To,Pn.Jrld,Pn.Jcl*yj];
Pn.xlabl = 'T \textrm{~[K]}';
Pn.ylabl = 'J_{RLD}(T) \textrm{~[A/cm$^2$]}';
if Na > 1
   s = (0:Na-1)'/(Na-1);
   Pn.cred = [s.^2,4*s.*(1-s),1 - s.^2];Pn.cred = [0,0,0;0.6,0.6,0.6;Pn.cred];
else
```

```
    Pn.cred = [0,0,0;0.6,0.6,0.6;1,0,0];
end
Pn.filn  = 'JLongo';
Pn.titltx = 'JLongo';
Pn.Lgt = char({'RLD','CL',num2str(Pn.alf')});
Pn.plot = 'linear';
Pn.xyrange = ([Pn.To(1),Pn.To(Nt),0,1.1*Pn.Jcl]);
Pn.titltx = ['$F = ',num2str(Pn.F),'$ eV/nm'];
end
```

```
function Pn= FUNmiramxt(P)
%   construction of y[x(t)] version of Miram plot
Pn = P;
Pn.To = linspace(Pn.Tmin,Pn.Tmax,Pn.Nt)';
Pn = FUNJrldcl(Pn);
xa = Pn.Mxy(:,2)/Pn.Jcl;
ya = Pn.Mxy(:,4)/Pn.Jcl;
Tao = spline(xa,Pn.To,1);
Pn.yo = spline(Pn.To,ya,Tao);
Pn.Tao = Tao;
Pn.Mxy = [Pn.To/Tao,xa,ya,1+0*Pn.To];
Pn.xyrange = [Pn.Tmin/Tao,Pn.Tmax/Tao,0,1.1];
Pn.Lgt = {'RLD','$J/J_{CL}$','CL'};
Pn.xlabl = 'T / T_o';
Pn.ylabl = 'J / J_{CL}';
Pn.filn  = 'Miramxt';
Pn.titltx = ['Miram:  $\Phi_o = ',num2str(Pn.Phi),'$ eV'];
end
```

- Figures generated by this script: Figures 15.15–15.17
- Primary script: `FIGrldstat.m`
- Subordinate scripts: `FUNshuffle.m`

```
%% Figure RLDstat
%   Generation of an array of work function values and their emission
%   Based on Table 1 and Figure 4 of:
%   A. Jassem, D. Chernin, J.J. Petillo, Y.Y. Lau,
%             A. Jensen, S. Ovtchinnikov
%   Analysis of anode current from a thermionic cathode
%    with a 2-D work function distribution,
%   IEEE Trans. Plas. Sci. 49(2), 749-755 (2021).
%   \url{https://ieeexplore.ieee.org/abstract/document/9332265}
%
%   Observe that "save" is used to quickly reload data from prior sections
%   without having to recreate terms
clear;clc
n = 10;
Delj = zeros(5,3);
% Work function of faces
Phij    = [1.61; 1.79; 2.3; 2.31; 4.5];  % Phi values
Np = length(Phij);
% Crld = ((Np-1:-1:0)'/(Np-1))*[1,1,1];
Crld = [1,1,0.95;0.95,1,1;0.7,0.7,0.6;0.6,0.7,0.7;0,0,0];
for iopt = 1:3
   Delj(:,1) = [18; 10;  9; 44; 19]; % Proportion of total (Rand I)
   Delj(:,2) = [13;  9; 11; 37; 30]; % Proportion of total (Rand II)
   Delj(:,3) = [ 7;  8; 11; 37; 37]; % Proportion of total (Rand III)
   Nj = zeros(Np+1,3);
```

```
   % create start/end index of each crystal face
   for j = 1:Np
       Nj(j+1,:) = sum(Delj(1:j,:),1);
   end
   % Create a vector of n^2 entries and populate that vector with the
   % proportion of Phi values from Delj
   Vphi = (1:n^2); Vpatch = Vphi;
   for j = 2:Np+1
       Vphi(Nj(j-1,iopt)+1:Nj(j,iopt)) = Phij(j-1);
       Vpatch(Nj(j-1,iopt)+1:Nj(j,iopt)) = j;
   end
   % Shuffle the entries
   rng(6)
   io = FUNshuffle(n^2);
   Vphi = Vphi(io); Vpatch = Vpatch(io);
   Mphi = zeros(n); Mpatch = Mphi;
   for j = 1:n
       Mphi(:,j) = Vphi((j-1)*n+1:j*n);
       Mpatch(:,j) = Vpatch((j-1)*n+1:j*n);
   end
   % Plot the iopt case
   figure(iopt); surf(Mpatch); view(2); shading flat;
   axis square tight; colormap(Crld); axis off
   colorbar
   saveas(gcf,['PhiRand',num2str(iopt)],'pdf')
end
save matlab.mat Mphi Mpatch n
%% Current density for RAND3
clear; clc
load
Pn = FUNdamental();
Va = 179.5; Do = 381000;
Arld = Pn.mo*(Pn.kb^2)/(2*(pi^2)*(Pn.hbar^3)) * (Pn.cm^2)/Pn.Amp;
Jcl = (1/(18*pi*Pn.Qo*sqrt(2*Pn.mo)))*((Va^1.5)/(Do^2))* (Pn.cm^2)/Pn.Amp;
Tj = [1100,1340,1400,1460]; Nt = length(Tj);
for j = 1:Nt
   T = Tj(j);
   Jrld = Arld*(T^2)*exp(-Mphi/(Pn.kb*T));
   Jrld = min(Jrld,Jcl)/Jcl;
   figure(j+3); surf(Jrld); view(2); shading flat;
   axis square tight; axis off; colormap hot;colorbar
   saveas(gcf,['JrldRand',num2str(j)],'pdf')
end
save matlab.mat Mphi n Arld Jcl
%% Miram like plot
clear;clc
Pn = FUNdamental();
load
Phiave = 1.8;
Nt = 60;
Tj = linspace(900,1500,Nt)';
Jrld = zeros(size(Tj));
Aphi = mean(Mphi(:));
for j = 1:Nt
   Jrldj = Arld*(Tj(j)^2)*exp(-Mphi/(Pn.kb*Tj(j)));
   Jrldj = min(Jrldj,6*Jcl);
   Jrld(j) = min(mean(Jrldj,"all")/Jcl,1);
end
Jrldo = Arld*(Tj.^2).*exp(-Phiave./(Pn.kb*Tj))/Jcl;
Pn.Mxy = [Tj,Jrld,Jrldo];
Pn.xyrange = [Tj(1),Tj(Nt),0,1.1];
Pn.Lgt = {'Patch','RLD'};
```

```
Pn.xlabl = 'T \textrm{~[K]}';
Pn.ylabl = 'J / J_{CL}';
Pn.filn  = 'MiramPatch';
Pn.titltx = ['Miram:  $\Phi_a = ',num2str(Phiave),'$ eV'];
Pn.cred = [1,0,0;0,0,1;0.6,0.6,0.6];
QPlot(Pn,20); hold on; plot([Tj(1),Tj(Nt)],[1,1],'--'); hold off
Pn.Lgt = char([Pn.Lgt,'CL']);
legend(Pn.Lgt)
saveas(gcf,Pn.filn,'png')
writematrix(Pn.Mxy,Pn.filn, 'FileType', 'text','Delimiter','tab')
```

```
function ix = FUNshuffle(Nc)
%  returns array of indices that will randomly shuffle a vector (better
%  shuffling method than used in shuffling a deck of cards!).
%  Algorithm:  chose the topmost element, randomly replace it with an
%  element below it.  Choose next element below it, and repeat until the
%  bottom of the vector
%  Based on shuffle algorithm of D. Knuth, Numerical Methods
vx = (1:Nc);
for i = Nc:-1:1;
   io = floor(i*rand())+1;
   vo    = vx(i);
   vx(i)  = vx(io);
   vx(io) = vo;
end
ix = vx;
```

## B.5 Appendix Algorithms

- Figures generated by this script: Figures A.1 and A.2
- Primary script: `FIGorthopoly.m`

```
%% Figure FIGorthopoly
%  Legendre and Hermite Polynomials
clear;clc
P = struct;
Nx = 64; Ny = 7;
P.Nx = Nx;
x = (2*(1:Nx)' - Nx - 1)/(Nx - 1);
y = zeros(Nx,Ny);
y(:,1:3) = [x,0.5*(3*x.^2 - 1),0.5*(5*x.^3 - 3*x)];
for j = 2:Ny
   y(:,j+1) = ((2*j+1)*x.*y(:,j) - j*y(:,j-1))/(j+1);
end
s = (1:Ny-1)'/(Ny-1); P.cred = [s.^2,3*s.*(1-s),1 - s.^2];
%  Figure 1
P.Mxy     = [x,y];
P.xlabl= 'x';
P.ylabl= 'P_n(x)';
P.titltx  = 'Legendre Polynomials';
P.cred = [s.^2,4*s.*(1-s),1 - s.^2];
P.Lgt = num2str((1:Ny)');
P.filn = 'Legendre';
QPlot(P,1);
writematrix(P.Mxy,'Legendre.txt','FileType','text','Delimiter','tab')
x = 5*(2*(1:Nx)' - Nx - 1)/(Nx - 1);
```

```
y(:,1:4) = [1+0*x,2*x,4*x.^2 - 2,8*x.^3 - 12*x];
for j = 2:Ny-1
   y(:,j+1) = 2*x.*y(:,j) - 2*j*y(:,j-1);
end
y = y.*exp(-(x.^2) / 2); y = y./max(abs(y));
Pn = P;
Pn.Mxy      = [x,y];
Pn.ylabl   = 'H_n(x)';
Pn.titltx  = 'Hermite Polynomials';
Pn.filn = 'Hermite';
QPlot(Pn,2);
writematrix(Pn.Mxy,'Hermite.txt','FileType','text','Delimiter','tab')
```

- Figures generated by this script: Figure A.3
- Primary script: `FIGiteration.m`

```
%% Figure FIGiteration
%  Solve the Golden mean equation iteratively, and plot the convergence for
%  the trial function f(x) = 1/(1+x)
clear; clc
% function
fa = @(x) 1./(1 + x);
Pn = struct();
Pn.xlabl = 'x';
Pn.xyrange = [0.4,1,0.4,0.7];
Pn.ylabl = 'y(x)';
Pn.titltx = 'Iterative solutions';
Pn.cred = [1,0,0;0,0,1;0,0.8,0];
Pn.Lgt = {'y(j)'};
Pn.filn = 'IterateX';
Pn.Lgt = {'$x$';'$f(x)$'};
N = 32;
x = 0.4 + 0.6*linspace(0,1,N)';
Pn.Mxy = [x,x,fa(x)];
xj = zeros(N,1); yj = zeros(N,1);
xj(1) = 1; yj(1) = fa(xj(1));
for j = 2:N
   xj(j) =    yj(j-1);
   yj(j) = fa(xj(j-1));
end
QPlot(Pn,1); hold on
plot(xj,yj); hold off
Pn.Lgt = char({char(Pn.Lgt),'$x_j$'});
legend(Pn.Lgt)
saveas(gcf,Pn.filn,'png')
writematrix([Pn.Mxy,xj,yj],Pn.filn,'FileType','text','Delimiter','tab')
```

# References


1 M Abramowitz and IA Stegun. *Handbook of Mathematical Functions, with Formulas, Graphs, and Mathematical Tables*. Dover Publications, New York, 1965.

2 M Afif, JP Girardeau-Montaut, C Tomas, C Girardeau-Montaut, K Warda, J Delafond, and C Fayoux. Ultrashort electron emission improved by alkali ions implantation into metallic photocathodes. *Appl. Surf. Sci.*, 111: 241–245, 1997.

3 A Alexander. *Infinitesimal: How a Dangerous Mathematical Theory Shaped the Modern World*. Scientific American/Farrar, Straus and Giroux, 2014.

4 LK Ang and P Zhang. Ultrashort-pulse Child–Langmuir law in the quantum and relativistic regimes. *Phys. Rev. Lett.*, 98(16): 164802–164804, 2007. https://doi.org/10.1103/PhysRevLett.98.164802.

5 JA Appelbaum and DR Hamann. Variational calculation of the image potential near a metal surface. *Phys. Rev. B.*, 6(4): 1122, 1972. https://doi.org/10.1103/PhysRevB.6.1122.

6 GB Arfken. *Mathematical Methods for Physicists*, 2e. Academic Press, New York, 1970.

7 R Arianrhod. *Thomas Harriot: A Life in Science*. Oxford University Press, New York, NY, 2019.

8 RB Ash. *Basic Probability Theory*. Wiley, New York, 1970.

9 HA Atwater. *Introduction to General Relativity*. Pergamon Press Oxford, New York, 1974.

10 J Babiskin and J. Robert Anderson. 9d properties of metals. In D. E Gray editor, American *Institute of Physics Handbook*, 3e, pages 9d38–9d56 McGraw-Hill, New York, 1972.

11 C Bandis and BB Pate. Simultaneous field emission and photoemission from diamond. *Appl. Phys. Lett.*, 69(3): 366–368, 1996. https://doi.org/10.1063/1.118062.

12 S Banerjee and P Zhang. A generalized self-consistent model for quantum tunneling current in dissimilar metal-insulator-metal junction. *AIP Adv.* 9(8): 085302, 2019. https://doi.org/10.1063/1.5116204.

13 S Banerjee and P Zhang. Review of recent studies on nanoscale electrical junctions and contacts: quantum tunneling, current crowding, and interface engineering. *J. Vac. Sci. Technol. A: Vac. Surf. Films.*, 40(3): 030802, 2022. https://doi.org/10.1116/6.0001724.

14 JP Barbour, WW Dolan, JK Trolan, EE Martin, and WP Dyke. Space-charge effects in field emission. *Phys. Rev.*, 92(1): 45–51, 1953. https://doi.org/10.1103/PhysRev.92.45.

15 EM Baroody. A theory of secondary electron emission from metals. *Phys. Rev.*, 78(6): 780, 1950. https://doi.org/10.1103/PhysRev.78.780.

16 JD Barrow. *The Constants of Nature*. Vintage, New York, 2004.



*Fundamentals of Electron Emission Physics*, First Edition. Kevin L. Jensen.

**17** IV Bazarov, L Cultrera, A Bartnik, B Dunham, S Karkare, Y Li, X Liu, J Maxson, and W Roussel. Thermal emittance measurements of a cesium potassium antimonide photocathode. *Appl. Phys. Lett.*, 98(22): 224101, 2011. https://doi.org/10.1063/1.3596450.

**18** IV Bazarov, DG Ouzounov, BM Dunham, SA Belomestnykh, Y Li, X Liu, RE Meller, J Sikora, CK Sinclair, FW Wise, and T Miyajima. Efficient temporal shaping of electron distributions for high-brightness photoemission electron guns. *Phys. Rev. ST Accel. Beams.*, 11(4): 040702, 2008. https://doi.org/10.1103/PhysRevSTAB.11.040702.

**19** JH Bechtel. Heating of solid targets with laser pulses. *J. Appl. Phys.*, 46(4): 1585–1593, 1975. http://link.aip.org/link/?JAP/46/1585/1.

**20** JH Bechtel, W Lee Smith, and N Bloembergen. Two-photon photoemission from metals induced by picosecond laser pulses. *Phys. Rev B.*, 15(10): 4557–4563, 1977.

**21** I Ben-Zvi, T Rao, A Burrill, X Chang, J Grimes, J Rank, Z Segalov, and J Smedley. Diamond secondary emitter. *Int. J. Mod. Phys. A.*, 22(22): 3759–3775, 2007. https://doi.org/10.1142/S0217751X07037408.

**22** CN Berglund and WE Spicer. I - photoemission studies of copper and silver: theory. *Phys. Rev.*, 136(4A): A1030, 1964. https://doi.org/10.1103/PhysRev.136.A1030.

**23** S Bernal, RA Kishek, M Reiser, and I Haber. Observations and simulations of transverse density waves in a collimated space-charge dominated electron beam. *Phys. Rev. Lett.*, 82(20): 4002–4005, 1999. https://doi.org/10.1103/PhysRevLett.82.4002.

**24** H Bethe. Zur theorie des durchgangs schneller korpuskularstrahlen durch materie. *Ann. Phys.*, 397(3): 325–400, 1930. https://doi.org/10.1002/andp.19303970303.

**25** CK Birdsall and WB Bridges. *Electron Dynamics of Diode Regions*. Academic Press, New York, 1966.

**26** JS Blakemore. *Semiconductor Statistics*. Dover, New York, 1987.

**27** D Bohm and T Staver. Application of collective treatment of electron and ion vibrations to theories of conductivity and superconductivity. *Phys. Rev.*, 84(4): 836–837, 1951. https://doi.org/10.1103/PhysRev.84.836.2.

**28** HP Bonzel and C Kleint. On the history of photoemission. *Prog. Surf. Sci.*, 49(2): 107–153, 1995.

**29** JH Booske. Plasma physics and related challenges of millimeter-wave-to-terahertz and high power microwave generation. *Phys Plasmas.*, 15(5): 055502 / 1–16, 2008. https://doi.org/10.1063/1.2838240.

**30** CA Brau. High-brightness photoelectric field-emission cathodes for free-electron laser applications. *Nucl. Instrum. Methods Phys. Res. A*, 393 (1–3): 426–429, 1997.

**31** CA Brau. What brightness means. In J Rosenzweig, G Travish, and L Serafini editors, *Physics and Applications of High Brightness Electron Beams: Proceedings of the ICFA Workshop Chia Laguna*, 1–6 July 2002, pages 20–27 World Scientific Publishing Company, Sardinia, Italy, 2004. https://doi.org/10.1142/9789812705235_0002.

**32** RS Brayfield, AJ Fairbanks, AM Loveless, S Gao, A Dhanabal, W Li, C Darr, W Wu, and AL Garner. The impact of cathode surface roughness and multiple breakdown events on microscale gas breakdown at atmospheric pressure. *J. Appl. Phys.*, 125(20): 203302, 2019. https://doi.org/10.1063/1.5085217.

**33** KF Brennan and C Summers. Theory of resonant tunneling in a variably spaced multiquantum well structure - an airy function-approach. *J. Appl. Phys.*, 61(2): 614–623, 1987. https://doi.org/10.1063/1.338213.

**34** L Broglie. *The Wave Nature of the Electron*. Nobel Lecture, 1929. http://www.nobelprize.org/nobel_prizes/physics/laureates/1929/broglie-lecture.html.

**35** WM Bullis, FH Brewer, CD Kolstad, and LJ Swartzendruber. Temperature coefficient of resistivity of silicon and germanium near room temperature. *Solid-State Electron.* 11: 639–646, 1968. https://doi.org/10.1016/0038-1101(68)90065-8.

**36** FA Buot and KL Jensen. Lattice weyl-wigner formulation of exact many-body quantum-transport theory and applications to novel solid-state quantum-based devices. *Phys. Rev. B.*, 42(15): 9429, 1990. https://doi.org/10.1103/PhysRevB.42.9429.

**37** E Butkov. *Mathematical Physics.* Addison–Wesley Pub. Co Reading, Mass, 1968.

**38** RE Caflisch and MS Rosin. Beyond the Child–Langmuir limit. *Phys. Rev. E.*, 85(5): 056408, 2012. https://doi.org/10.1103/PhysRevE.85.056408.

**39** M Cahay and S Bandyopadhyay. *Problem Solving in Quantum Mechanics: From Basics to Real-world Applications for Materials Scientists*, Applied Physicists, and Devices Engineers. John Wiley & Sons, 2017.

**40** MJ Cattelino, GV Miram, and WR Ayers. A diagnostic technique for evaluation of cathode emission performance and defects in vehicle assembly, 1982. IEEE.

**41** FM Charbonnier, WP Dyke, Lf Garrett, JP Barbour. Basic and applied studies of field emission at microwave frequencies. *Proc. IEEE.*, 51(7): 991–1004, 1963.

**42** D Chen, R Jacobs, D Morgan, J Booske. Impact of nonuniform thermionic emission on the transition behavior between temperature-and space-charge-limited emission. *IEEE Trans. Electron Dev.*, 68(7): 3576–3581, 2021. https://doi.org/10.1109/TED.2021.3079876.

**43** D Chen, R Jacobs, J Petillo, V Vlahos, KL Jensen, D Morgan, and J Booske. Physics-based model for nonuniform thermionic electron emission from polycrystalline cathodes. *Phys. Rev. Appl.*, 18(5): 054010, 2022. https://doi.org/10.1103/PhysRevApplied.18.054010.

**44** D Chernin. Exact evaluation of K. Jensen's general thermal-field-photoemission integral. *IEEE Trans. Plasma Sci.*, 47(1): 336–338, 2019. https://doi.org/10.1109/TPS.2018.2871331.

**45** D Chernin, YY Lau, JJ Petillo, S Ovtchinnikov, D Chen, A Jassem, R Jacobs, D Morgan, and JH Booske. Effect of nonuniform emission on miram curves. *IEEE Trans. Plasma Sci.*, 48(1): 146–155, 2020. https://doi.org/10.1109/tps.2019.2959755.

**46** PJ Christenson, DP Chernin, AL Garner, and YY Lau. Resistive destabilization of cycloidal electron flow and universality of (near-) brillouin flow in a crossed-field gap. *Phys. Plasmas.*, 3(12): 4455–4462, 1996. https://doi.org/10.1063/1.872064.

**47** O Chubenko, S Karkare DA Dimitrov, JK Bae, L Cultrera, I Bazarov, and A Afanasev. Monte Carlo modeling of spin-polarized photoemission from p-doped bulk gaas. *J. Appl. Phys.*, 130(6): 063101, 2021. https://doi.org/10.1063/5.0060151.

**48** R Courths and S Hufner. Photoemission experiments on copper. *Phys. Rep.*, 112(2): 53–171, 1984.

**49** L Cultrera, IV Bazarov, A Bartnik, B Dunham, S Karkare, R Merluzzi, and M Nichols. Thermal emittance and response time of a cesium antimonide photocathode. *Appl. Phys. Lett.*, 99(15), 2011. https://doi.org/10.1063/1.3652758.

**50** L Cultrera, A Galdi, JK Bae, F Ikponmwen, J Maxson, and I Bazarov. Long lifetime polarized electron beam production from negative electron affinity gaas activated with sb-cs-o: trade-offs between efficiency, spin polarization, and lifetime. *Phys. Rev. Accel. Beams.*, 23(2): 023401, 2020. https://doi.org/10.1103/PhysRevAccelBeams.23.023401.

**51** PH Cutler, J He, J Miller, NM Miskovsky, B Weiss, and TE Sullivan. Theory of electron emission in high fields from atomically sharp emitters: validity of the Fowler-Nordheim equation. *Progr. Surf. Sci.*, 42(1–4): 169–185, 1993. https://doi.org/10.1016/0079-6816(93)90068-7.

**52** PH Cutler, NM Miskovsky, PB Lerner, and MS Chung. The use of internal field emission to inject electronic charge carriers into the conduction band of diamond films: a review. *Appl. Surf. Sci.*, 146 (1–4): 126–133, 1999. https://doi.org/10.1016/S0169-4332(99)00017-3.

**53** PH Cutler and D Nagy. Use of a new potential barrier model in Fowler-Nordheim theory of field emission. *Surf. Sci.*, 3(1): 71–94, 1965. https://doi.org/10.1016/0039-6028(65)90020-8.

**54** JH Deane and RG Forbes. The formal derivation of an exact series expansion for the principal Schottky-Nordheim barrier function v, using the Gauss hypergeometric differential equation. *J. Phys. A: Math. Theor.*, 41(39): 395301 / 1–9, 2008. https://doi.org/10.1088/1751-8113/41/39/395301.

**55** AJ Dekker. Secondary electron emission. In FSeitz and D Turnbull editors, *Solid State Physics: Advances in Research and Applications* Vol. 6. Academic Press, New York, 1958.

**56** AJ Dekker and A van der Ziel. Theory of the production of secondary electrons in solids. *Phys. Rev.* 86(5): 755–760, 1952.

**57** C Dewdney and B Hiley. A quantum potential description of one-dimensional time-dependent scattering from square barriers and square wells. *Found. Phys.*, 12(1): 27–48, 1982. https://doi.org/10.1007/BF00726873.

**58** DA Dimitrov, R Busby, JR Cary, I Ben-Zvi, T Rao, J Smedley, X Chang, JW Keister, and E Muller. Multiscale three-dimensional simulations of charge gain and transport in diamond. *J. Appl. Phys.*, 108(7): 073712, 2010. https://doi.org/10.1063/1.3491041.

**59** DA Dimitrov, D Smithe, JR Cary, I Ben-Zvi, T Rao, J Smedley, and E Wang. Modeling electron emission and surface effects from diamond cathodes. *J. Appl. Phys.* 117: 055708 / 1–18, 2015. https://doi.org/10.1063/1.4907393.

**60** WW Dolan and WP Dyke. Temperature-and-field emission of electrons from metals. *Phys. Rev.*, 95(2): 327–332, 1954. http://link.aps.org/abstract/PR/v95/p327.

**61** WW Dolan, WP Dyke, and JK Trolan. The field emission initiated vacuum arc. II. the resistively heated emitter. *Phys. Rev.*, 1953. https://doi.org/10.1103/PhysRev.91.1054.

**62** DH Dowell, I Bazarov, B Dunham, K Harkay, C Hernandez-Garcia, R Legg, H Padmore, T Rao, J Smedley, and W Wan. Cathode randd for future light sources. *Nucl. Instrum. Methods Phys. Res. Sect. A.*, 622: 685–697, 2010. https://doi.org/10.1016/j.nima.2010.03.104.

**63** DH Dowell, KJ Davis, KD Friddell, EL Tyson, CA Lancaster, LD Milliman, RE Rodenburg, T Aas, M Bemes, SZ Bethel, PE Johnson, K Murphy, C Whelen, JL Adamski, DJ Pistoresi, DR Shoffstall, GE Busch, and DK Remelius. First operation of a photoinjector at high-average power volume, 2013(1) pages 13–17 *Proceedings of the Society of Photo-Optical Instrumentation Engineers (SPIE)* San Diego, CA, USA 1993.

**64** DH Dowell, FK King, RE Kirby, JF Schmerge, and JM Smedley. In situ cleaning of metal cathodes using a hydrogen ion beam. *Phys. Rev. Accel. Beams.*, 9(6): 063502, 2006. https://doi.org/10.1103/PhysRevSTAB.9.063502.

**65** DH Dowell and JF Schmerge. Quantum efficiency and thermal emittance of metal photocathodes. *Phys. Rev. Accel. Beams.*, 12(7): 074201 / 1–10, 2009. https://doi.org/10.1103/PhysRevSTAB.12.074201.

**66** P Drude. On the electron theory of metals. *Ann. Phys.*, 1(3): 566–613, 1900.

**67** LA DuBridge. Theory of the energy distribution of photoelectrons. *Phys.Rev.*, 43(9): 0727–0741, 1933. https://doi.org/10.1103/PhysRev.43.727.

**68** LA DuBridge and RC Hergenrother. The effect of temperature on the energy distribution of photoelectrons. i. normal energies. *Phys. Rev.*, 44(11): 0861–0865, 1933.

**69** LA DuBridge and WW Roehr. The thermionic and photoelectric work - functions of molybdenum. *Phys. Rev.*, 42(1): 52–57, 1932.

**70** S Dushman. Thermionic emission. *Rev. Mod. Phys.*, 2(4): 0381–0476, 1930. https://doi.org/10.1103/RevModPhys.2.381.

**71** WP Dyke, JP Barbour, EE Martin, and JK Trolan. T-f emission: Experimental measurement of the average electron current density from tungsten. *Phys. Rev.*, 99(4): 1192–1195, 1955. https://doi.org/10.1103/physrev.99.1192

**72** WP Dyke and JK Trolan. Field emission - large current densities, space charge, and the vacuum arc. *Phys. Rev.*, 89(4): 799–808, 1953.

**73** WP Dyke, JK Trolan, WW Dolan, and G Barnes. The field emitter: fabrication, electron microscopy, and electric field calculations. *J. Appl. Phys.*, 24(5): 570–576, 1953. http://link.aip.org/link/?JAP/24/570/1.

**74** WP Dyke, JK Trolan, WW, Dolan, and FJ Grundhauser. Field emission current-density distribution. *J. Appl. Phys.*, 25(1): 106–112, 1954. https://doi.org/10.1063/1.1721491.

**75** WP Dyke, JK Trolan, EE Martin, and JP Barbour. The field emission initiated vacuum arc .1. experiments on arc initiation. *Phys. Rev.*, 91(5): 1043–1054, 1953. https://doi.org/10.1103/PhysRev.91.1043.

**76** A Einstein. Concerning an heuristic point of view toward the emission and transformation of light. *Ann. Phys.*, 176: 132–148, 1905.

**77** A Evtukh, H Hartnagel, O Yilmazoglu, H Mimura, and D Pavlidis. *Vacuum Nanoelectronic Devices*. John Wiley & Sons, Hoboken, 2015.

**78** L Eyges. *The Classical Electromagnetic Field*. Addison-Wesley Pub. Co Reading, Mass, 1972.

**79** M Feinleib and MC Green. *High Current Density Cathodes - An Update*, volume 30, San Francisco, CA, 1984.

**80** Y Feng and JP Verboncoeur. Transition from Fowler-Nordheim field emission to space charge limited current density. *Phys. Plasmas.*, 13(7): 073105, 2006. https://doi.org/10.1063/1.2226977.

**81** RP Feynman. Negative probability. In B J Hiley and F David Peat editors, *Quantum Implications: Essays in Honour of David Bohm*, page 235, Routledge and Kegan Paul in association with Methuen, New York, NY, 1987.

**82** DG Fisher, RE Enstrom, JS Escher, and BF Williams. Photoelectron surface escape probability of (ga,in)as: Cs-o in the 0.9 to $\sim$1.6 $\mu$m range. *J. Appl. Phys.*, 43(9): 3815–3823, 1972. https://doi.org/10.1063/1.1661817.

**83** K. Floettmann. *Some basic features of the beam emittance. Phys. Rev. Special Topics-Accelerators and Beams* 6(3): 034202, 2003. https://doi.org/10.1103/PhysRevSTAB.6.034202.

**84** S Flügge. *Practical Quantum Mechanics*. Springer Verlag, New York, 1974.

**85** RG Forbes. Field electron and ion emission from charged surfaces: a strategic historical review of theoretical concepts. *Ultramicroscopy.*, 95 (1–4): 1–18, 2003.

**86** RG Forbes. Simple good approximations for the special elliptic functions in standard Fowler-Nordheim tunneling theory for a Schottky-Nordheim barrier. *Appl. Phys. Lett.*, 89(11): 113122 / 1–3, 2006. https://doi.org/10.1063/1.2354582.

**87** RG Forbes. Exact analysis of surface field reduction due to field-emitted vacuum space charge, in parallel-plane geometry, using simple dimensionless equations. *J. Appl. Phys.*, 104(8): 084303, 2008. https://doi.org/10.1063/1.2996005.

**88** RG Forbes. Physics of generalized Fowler-Nordheim-type equations. *J. Vac. Sci. Technol. B.*, 26: 788, 2008. http://link.aip.org/link/?JVTBD9/26/788/1.

**89** RG Forbes. Use of Millikan–Lauritsen plots, rather than Fowler–Nordheim plots, to analyze field emission current-voltage data. *J. Appl. Phys.* 105(11): 114313–114318, 2009. https://doi.org/10.1063/1.3140602.

**90** RG Forbes. Renewing the mainstream theory of field and thermal electron emission. In G Gaertner, W Knapp, and R Forbes editors, *Modern Developments in Vacuum Electron Sources* pages 387–447, Springer Nature Cham, Switzerland, 2020.

**91** RG Forbes, and JH Deane. Reformulation of the standard theory of Fowler-Nordheim tunnelling and cold field electron emission. *Proc. Roy. Soc. London A.*, 463(2087): 2907–2927, 2007. https://doi.org/10.1098/rspa.2007.0030.

**92** RH Fowler. The photo-electric threshold frequency and the thermionic work function. *Proc. Roy. Soc. London A.*, 118(779): 229–232, 1928. https://doi.org/10.1098/rspa.1928.0047.

**93** RH Fowler. The electron theory of metals. *Nature*, 126: 611–618, 1930.

**94** RH Fowler. The analysis of photoelectric sensitivity curves for clean metals at various temperatures. *Phys. Rev.*, 38(1): 45–56, 1931. https://doi.org/10.1103/PhysRev.38.45.

**95** RH Fowler and L Nordheim. Electron emission in intense electric fields. *Proc. Roy. Soc. London A.*, 119(781): 173–181, 1928. https://doi.org/10.1098/rspa.1928.0091.

**96** WR Frensley. Boundary conditions for open quantum systems driven far from equilibrium. *Rev. Mod. Phys.*, 62(3): 745, 1990. http://link.aps.org/abstract/RMP/v62/p745.

**97** JW Gadzuk and EW Plummer. Energy distributions for thermal field emission. *Phys. Rev. B* 3(7): 2125–2129, 1971. http://link.aps.org/abstract/PRB/v3/p2125.

**98** G. Gaertner. Chapter 1: History of vacuum electronics and vacuum electron sources and future development trends. In Georg Gaertner, Wolfram Knapp, and Richard G Forbes editors, *Modern Developments in Vacuum Electron Sources*, pages 1–31, Springer, New York, 2020. https://doi.org/10.1007/978-3-030-47291-7_1.

**99** G Gaertner, W Knapp, and RG Forbes (eds). *Modern Developments in Vacuum Electron Sources.* Springer, New York, 2020. https://doi.org/10.1007/978-3-030-47291-7_1.

**100** R Ganter, R Bakker, C Gough, SC Leemann, M Paraliev, M Pedrozzi, F Le Pimpec, V Schlott, L Rivkin, and A Wrulich. Laser-photofield emission from needle cathodes for low-emittance electron beams. *Phys. Rev. Lett.*, 100(6): 064801, 2008. https://doi.org/ARTN064801.

**101** CH Garcia and CA Brau. Pulsed photoelectric field emission from needle cathodes. *Nucl. Instrum. Methods Phys. Res. A.*, 483 (1–2): 273–276, 2002.

**102** AL Garner, G Meng, Y Fu, AM Loveless, RS Brayfield, and AM Darr. Transitions between electron emission and gas breakdown mechanisms across length and pressure scales. *J. Appl. Phys.*, 128(21): 210903, 2020. https://doi.org/10.1063/5.0030220.

**103** G Gartner, P Geittner, D Raasch, and Du Wiechert. Supply and loss mechanisms of ba dispenser cathodes. *Appl. Surf. Sci.*, 146 (1–4): 22–30, 1999. https://doi.org/10.1016/S0169-4332(99)00016-1.

**104** VA Gasparov and R Huguenin. Electron-phonon, electron-electron and electron surface scattering in metals from ballistic effects. *Adv. Phys.*, 42(4): 393–521, 1993. https://doi.org/10.1080/00018739300101514.

**105** P Geittner, G Gartner, and D Raasch. Low temperature properties of ba-dispenser cathodes. *J. Vac. Sci. Technol. B.*, 18(2): 997–999, 2000. https://doi.org/10.1116/1.591315.

**106** BE Gilchrist, KL Jensen, AD Gallimore, and JG Severns. Space based applications for fea cathodes (feac). In *Space Based Applications for Fea Cathodes (feac)*, 621: R481–R487, 2000.

**107** AS Gilmour. Microwave tubes. In *The Artech House Microwave Library*. Artech House Dedham, MA, 1986.

**108** JP Girardeau-Montaut, and C Girardeau-Montaut. Theory of ultrashort nonlinear multiphoton photoelectric-emission from metals. *Phys. Rev. B.*, 51(19): 13560–13567, 1995.

**109** JP Girardeau-Montaut, C Girardeau-Montaut, Sd Moustaizis, and C Fotakis. High-current density produced by femtosecond nonlinear single-photon photoelectric-emission from gold. *Appl. Phys. Lett.*, 624: 426–428, 1993. http://link.aip.org/link/APPLAB/v62/i4/p426/s1.

**110** SH Gold and GS Nusinovich. Review of high-power microwave source research. *Rev. Sci. Instrum.*, 68(11): 3945–3974, 1997.

**111** H Goldstein. *Classical Mechanics.* Addison-Wesley Pub. Co. Reading, Mass. 1980.

**112** VL Granatstein and CM Armstrong. Special issue on new vistas for vacuum electronics. *Proc. IEEE.*, 87(5): 699–701, 1999.

**113** DE Gray. *American Institute of Physics Handbook*. Section editors: BH Billings [and others] Coordinating editor: DE Gray. McGraw-Hill, New York, 1972.

**114** MC Green. The m-type cathode–no longer magic? *Techn. Digest Int. Electr. Dev. Meet.* 26: 471–474, 1980. https://doi.org/10.1109/IEDM.1980.189869.

**115** JS Greeneich, and T Van Duzer. Approximate formula for electron-energy versus path length. *IEEE Trans. Electron Devices.*, 20(6): 598–600, 1973. https://doi.org/10.1109/T-ED.1973.17705.

**116** ME Griswold, NJ Fisch, and JS Wurtele. An upper bound to time-averaged space-charge limited diode currents. *Phys. Plasmas.*, 17(11), 2010. https://doi.org/10.1063/1.3503661.

**117** ME Griswold, NJ Fisch, and JS Wurtele. Amended conjecture on an upper bound to time-dependent space-charge limited current. *Phys. Plasmas.*, 19: 114503, 2012. https://doi.org/10.1063/1.3671961.

**118** EP Gyftopoulos and JD Levine. Work function variation of metals coated by metallic films. *J. Appl. Phys.*, 33(1): 67, 1962. https://doi.org/10.1063/1.1728530

**119** GA Haas and RE Thomas. Chapter 2: Thermionic emission and work function. In RF Bunshah editor, *Techniques of Metals Research* Vol. *VI, Part 1*, pages 91–262, Interscience Publishers, New York, 1968.

**120** GA Haas and RE Thomas. Work function and secondary-emission studies of various cu crystal faces. *J. Appl. Phys.*, 48(1): 86–93, 1977. https://doi.org/10.1063/1.323329.

**121** JH Han, M Krasilnikov, and K Flöttmann. Secondary electron emission in a photocathode rf gun. *Phys. Rev. Spec. Top.-Accel. Beams.*, 8(3): 033501, 2005. https://link.aps.org/pdf/10.1103/PhysRevSTAB.8.033501.

**122** JW Han, DI Moon, and M Meyyappan. Nanoscale vacuum channel transistor. *Nano Lett.*, 17(4): 2146–2151, 2017. https://doi.org/10.1021/acs.nanolett.6b04363.

**123** JW Han, J Sub Oh, and M Meyyappan. Vacuum nanoelectronics: back to the future?—gate insulated nanoscale vacuum channel transistor. *Appl. Phys. Lett.*, 100(21): 213505, 2012. https://doi.org/10.1063/1.4717751.

**124** P Hartmann, J Bermuth, D Von Harrach, J Hoffmann, S Kobis, E Reichert, K Aulenbacher, J Schuler, and M Steigerwald. A diffusion model for picosecond electron bunches from negative electron affinity gaAs photocathodes. *J. Appl. Phys.*, 86(4): 2245–2249, 1999. https://doi.org/10.1063/1.371037.

**125** G Hass and L Hadley. Optical properties of metals. In D. E Gray editor, *American Institute of Physics Handbook*, 3e. McGraw-Hill, New York, 1972.

**126** J He, PH Cutler, and NM Miskovsky. Generalization of Fowler–Nordheim field emission theory for nonplanar metal emitters. *Appl. Phys. Lett.*, 59(13): 1644–1646, 1991. https://doi.org/10.1063/1.106257.

**127** E Hecht and A Zajaĺ§c. *Optics*. Addison–Wesley Pub. Co. Reading, Mass, 1974.

**128** C Hernandez-Garcia, PG O'Shea, and ML Stutzman. Electron sources for accelerators. *Phy. Today.*, 61(2): 44–49, 2008. https://doi.org/10.1063/1.2883909.

**129** C Herring and MH Nichols. Thermionic emission. *Rev. Mod. Phys.*, 21(2): 185–270, 1949. https://doi.org/10.1103/RevModPhys.21.185.

**130** M Hillery, RF O'Connell, MO Scully, and EP Wigner. Distribution-functions in physics - fundamentals. *Phys. Rep.*, 106(3): 121–167, 1984. https://doi.org/10.1016/0370-1573(84)90160-1

**131** A Hoffmann, K Reuschel, and H Rupprecht. Measurement of the hall effect and conductivity of super-pure silicon. *J. Phys. Chem. Solids*, 11: 284–287, 1959. https://doi.org/10.1016/0022-3697(59)90228-8.

**132** EJ Houston. Notes on phenomena in incandescent lamps. *Trans. Am. Inst. Electr. Eng.*, 1: 11–18, 1884.

**133** K Huang. *Statistical Mechanics*. Wiley, New York, 1987.

**134** RE Hummel. *Electronic Properties of Materials*. Springer, New York, 2011.

**135** H Ibach, and H Lüth. *Solid-State Physics: An Introduction to Principles of Materials Science, Volume Advanced Texts in Physics*. Springer Berlin; New York, 2003. https://doi.org/10.1007/978-3-540-93804-0.

**136** LG Il'chenko and TV Goraychuk. Role of the image forces potential in the formation of the potential barrier between closely spaced metals. *Surf. Sci.*, 478(3): 169–179, 2001. https://doi.org/10.1016/S0039-6028(01)00844-5.

**137** LG Il'chenko and YV Kryuchenko. External-field penetration effect on current-field characteristics of metal emitters. *J. Vac. Sci. Technol. B.*, 13(2): 566–570, 1995. https://doi.org/10.1116/1.587915.

**138** T Inoue, S Miyamoto, S Amano, M Yatsuzuka, and T Mochizuki. Enhanced quantum efficiency of photocathode under high electric field. *Jpn. J. Appl. Phys Part 1-regular Papers Short Notes and Review Papers.*, 41(12): 7402–7406, 2002. https://doi.org/10.1143/JJAP.41.7402.

**139** A Jablonski and CJ Powell. A predictive formula for the electron stopping power. *J. Surf. Anal.*, 13(2): 170–177, 2006. http://www.sasj.jp/JSA/CONTENTS/vol.13_2/Vol.13No.2/Vol.13No.2170-177.pdf.

**140** A Jassem, D Chernin, JJ Petillo, YY Lau, A Jensen, and S Ovtchinnikov. Analysis of anode current from a thermionic cathode with a 2-d work function distribution. *IEEE Trans Plasma Sci.*, 49(2): 749–755, 2021. https://ieeexplore.ieee.org/abstract/document/9332265.

**141** BL Jensen. Quantum-theory of free carrier absorption in polar semiconductors. *Ann. Phys.*, 80(2): 284–360, 1973. https://doi.org/10.1016/0003-4916(73)90107-3.

**142** BL Jensen. Quantum-mechanical extension of Boltzmann transport-equation and optical-absorption in semiconductors. *Ann. Phys.*, 95(2): 229–266, 1975. https://doi.org/10.1016/0003-4916(75)90056-1.

**143** B Jensen. Quantum theory of the complex dielectric constant of free carriers in polar semiconductors. *IEEE J. Quantum Electron.*, 18(9): 1361–1370, 1982. https://doi.org/10.1109/JQE.1982.1071713.

**144** BL Jensen. Chapter 4: Review of the theory of infrared and far-infrared free-carrier behavior in semiconductors. In Kenneth J Button editor, *Reviews of Infrared and Millimeter Waves* Vol. 8, pages 127–171 Academic Press, New York, 1983.

**145** KL Jensen. Exchange-correlation, dipole, and image charge potentials for electron sources: temperature and field variation of the barrier height. *J. Appl. Phys.*, 85(5): 2667–2680, 1999. https://doi.org/10.1063/1.369584.

**146** KL Jensen. Exchange-correlation, dipole, and image charge potentials for electron sources: temperature and field variation of the barrier height. *J. Appl. Phys.*, 887: 4455–4456, 2000.

**147** KL Jensen. Chapter 3: Theory of field emission. In W Zhu editor, *Vacuum Microelectronics* pages 33–104 Wiley, New York, 2001. https://doi.org/10.1002/0471224332.

**148** KL Jensen. Electron emission theory and its application: Fowler-Nordheim equation and beyond. *J. Vac. Sci. Technol. B.*, 21(4): 1528–1544, 2003. https://doi.org/10.1116/1.1573664.

**149** KL Jensen. Electron emission physics. In P Hawkes editor, *Advances in Imaging and Electron Physics.*, Vol 149, pages 1–338, Elsevier, San Diego, CA, 2007. http://www.sciencedirect.com/science/bookseries/10765670/149.

**150** KL Jensen. General formulation of thermal, field, and photoinduced electron emission. *J. Appl. Phys.*, 102(2): 024911–024911, 2007. http://link.aip.org/link/?JAP/102/024911/1.

**151** KL Jensen. Space charge effects in field emission: three dimensional theory. *J. Appl. Phys.*, 107(1): 014905–014909, 2010. https://doi.org/10.1063/1.3272688.

**152** KL Jensen. A quantum dipole–modified work function for a simplified electron emission barrier. *J. Appl. Phys.*, 111: 054916 / 1–10, 2012. https://doi.org/10.1063/1.3692571.

**153** KL Jensen. Scattering and the relationship between quantum efficiency and emittance. *J. Appl. Phys.*, 113(5): 056101, 2013. https://doi.org/10.1063/1.4790874.

**154** KL Jensen. *Introduction to the Physics of Electron Emission*, 1e. John Wiley & Sons, Inc. Hoboken, New Jersey, 2017. https://doi.org/10.1002/9781119051794.

**155** KL Jensen. A tutorial on electron sources. *IEEE Trans. Plasma Sci.*, 46(6): 1881–1899, 2018. https://doi.org/10.1109/TPS.2017.2782485.

**156** KL Jensen. A reformulated general thermal-field emission equation. *J. Appl. Phys.*, 126(6): 065302/1–13 2019. https://doi.org/10.1063/1.5109676.

**157** KL Jensen. Tutorial: the equations of electron emission and their evaluation. *J. Appl. Phys.*, 135(11): 111101, 2024. https://doi.org/10.1063/5.0196481.

**158** KL Jensen and FA Buot. Numerical-simulation of intrinsic bistability and high-frequency current oscillations in resonant tunneling structures. *Phys. Rev. Lett.*, 66(8): 1078–1081, 1991. https://doi.org/10.1103/PhysRevLett.66.1078.

**159** KL Jensen and FA Buot. The methodology of simulating particle trajectories through tunneling structures using a Wigner distribution approach. *IEEE Trans. Electron Dev.*, 38(10): 2337–2347, 1991. https://doi.org/10.1109/16.88522.

**160** KL Jensen and M Cahay. General thermal-field emission equation. *Appl. Phys. Lett.*, 88(15): 154105, 2006. https://doi.org/10.1063/1.2193776.

**161** KL Jensen, DW Feldman, NA Moody, and PG O'Shea. A photoemission model for low work function coated metal surfaces and its experimental validation. *J. Appl. Phys.*, 99(12): 124905, 2006. https://doi.org/10.1063/1.2203720.

**162** KL Jensen, DW Feldman, PG O'Shea. Time dependent models of field-assisted photoemission. *J. Vac. Sci. Technol. B.*, 23(2): 621–631, 2005. https://doi.org/10.1116/1.1861047.

**163** KL Jensen, DW Feldman, M Virgo, and PG O'Shea. Measurement and analysis of thermal photoemission from a dispenser cathode. *Phys. Rev. ST Accel. Beams.*, 6(8): 083501, 2003. https://doi.org/10.1103/PhysRevSTAB.6.083501.

**164** KL Jensen, D Finkenstadt, A Shabaev, S G Lambrakos, NA Moody, JJ Petillo, H Yamaguchi, and F Liu. A photoemission moments model using density functional and transfer matrix methods applied to coating layers on surfaces: theory. *J. Appl. Phys.*, 123(4): 045301, 2018. https://doi.org/10.1063/1.5008600.

**165** KL Jensen, D Finkenstadt, DA Shiffler, A Shabaev, SG Lambrakos, NA Moody, and JJ Petillo. Analytical models of transmission probabilities for electron sources. *J. Appl. Phys.*, 123(6): 065301, 2018. https://doi.org/10.1063/1.5018602.

**166** KL Jensen and AK Ganguly. Numerical-simulation of field-emission and tunneling - a comparison of the Wigner function and transmission coefficient approaches. *J. Appl. Phys.* 73(9): 4409–4427, 1993. https://doi.org/10.1063/1.352779.

**167** WD Jensen, FB Gerhard, and Jr. DM Koffman. Electron beam lithography. In GD Considine and PH Kulik editors, *Van Nostrand's Scientific Encyclopedia*, pages 1771–1774, John Wiley and Sons Inc, Hoboken, 2005. http://www.mrw.interscience.wiley.com/emrw/9780471743989/vse/article/vse2878/current/abstract.

**168** BL Jensen and KL Jensen. The quantum mechanical extension of the Drude-Zener theory and the optical constants of an alpha semiconductor. *J. Comput. Theor. Nanosci.*, 6: 1770–1788, 2009. https://doi.org/10.1166/jctn.2009.1242.

**169** KL Jensen, B Jensen, EJ Montgomery, DW Feldman, PG O'Shea, and NA Moody. Theory of photoemission from cesium antimonide using an alpha-semiconductor model. *J. Appl. Phys.*, 104(4): 044907–044910, 2008. https://doi.org/10.1063/1.2967826.

**170** KL Jensen, J Lebowitz, YY. Lau, and J Luginsland. Space charge and quantum effects on electron emission. *J. Appl. Phys.*, 111: 054917–054919, 2012. https://doi.org/10.1063/1.3692577.

**171** KL Jensen, JL Lebowitz, JM Riga, DA Shiffler, and R Seviour. Wigner wave packets: transmission, reflection, and tunneling. *Phys. Rev. B.*, 103: 155427, 2021. https://doi.org/10.1103/PhysRevB.103.155427.

**172** KL Jensen, M McDonald, O Chubenko, JR Harris, DA Shiffler, NA Moody, JJ Petillo, and AJ Jensen. Thermal-field and photoemission from meso- and micro-scale features: effects of screening and roughness on characterization and simulation. *J. Appl. Phys.*, 125(23): 234303 / 1–25, 2019. https://doi.org/10.1063/1.5097149.

**173** KL Jensen, MS McDonald, MK Dhillon, D Finkenstadt, A Shabaev, and M Osofsky. Thermal-field emission from cones and wires. *J. Vac. Sci. Technol. B.*, 40(2): 022801, 2022. https://doi.org/10.1116/6.0001656.

**174** KL Jensen, M McDonald, JR Harris, DA Shiffler, M Cahay, and JJ Petillo. Analytic model of a compound thermal-field emitter and its performance. *J. Appl. Phys.*, 126(24): 245301, 2019. https://doi.org/10.1063/1.5132561.

**175** KL Jensen, EW Montgomery, DW Feldman, PG O'Shea, JR Harris, JW Lewellen, and N Moody. Multiple scattering effects on quantum efficiency and response time for cesiated metal photocathodes. *J. Appl. Phys.*, 110(3): 034504–034513, 2011. https://doi.org/10.1063/1.3610397.

**176** KL Jensen, NA Moody, DW Feldman, EJ Montgomery, and PG O'Shea. Photoemission from metals and cesiated surfaces. *J. Appl. Phys.*, 102(7): 074902–074919, 2007. https://doi.org/10.1063/1.2786028.

**177** KL Jensen, RJ Nemanich, P Holloway, T Trottier, W MacKie, D Temple, and J Itoh. *Electron-emissive Materials, Vacuum Microelectronics, and Flat-panel Displays* (Symposium Proceedings), volume 621, Materials Research Society Warrendale, PA, 2001.

**178** KL Jensen, PG O'Shea, and DW Feldman. Emittance of a photocathode: effects of temperature and field. *Phys. Rev. Accel. Beams.*, 13(8): 080704–080708, 2010. https://doi.org/10.1103/PhysRevSTAB.13.080704.

**179** KL Jensen, JJ Petillo, SO Dimitrios, N Panagos, NA Moody, and SG Lambrakos. Modeling emission lag after photoexcitation. *J. Appl. Phys.*, 122(16): 164501, 2017. https://doi.org/10.1063/1.5008366.

**180** KL Jensen, JJ Petillo, DN Panagos, SO Nathan, and A Moody. Delayed photo-emission model for beam optics codes. *J. Vac. Sci. Technol. B.*, 35 (2): 02C102, 2017. https://doi.org/10.1116/1.4968511.

**181** KL Jensen, J Riga, JL Lebowitz, R Seviour, and DA Shiffler. An exact tunneling model and its application to transmission and reflection delay times. *J. Appl. Phys.*, 132(12): 124303, 2022. https://doi.org/10.1063/5.0096568.

**182** KL Jensen, J Riga, A Shabaev, M Osofsky, and JC Prestigiacomo. A delta barrier in a well and the exact time evolution of its eigenstates. *J. Appl. Phys.*, 133(17): 174402, 2023. https://doi.org/10.1063/5.0150385.

**183** KL Jensen, A Shabaev, SG Lambrakos, D Finkenstadt, NA Moody, AJ Neukirch, S Tretiak, DA Shiffler, and JJ Petillo. Analytic model of electron transport through and over non-linear barriers. *J. Appl. Phys.*, 127(23): 235301, 2020. https://doi.org/10.1063/5.0009759.

**184** KL Jensen, A Shabaev, SG Lambrakos, D Finkenstadt, JJ Petillo, AM Alexander, J Smedley, NA Moody, H Yamaguchi, F Liu, AJ Neukirch, and S Tretiak. An extended moments model of quantum efficiency for metals and semiconductors. *J. Appl. Phys.*, 128(1): 015301, 2020. https://doi.org/10.1063/5.0011145.

**185** KL Jensen, JL Shaw, JE Yater, and BB Pate. Enhancing secondary yield of a diamond amplifier using a nitrogen layer. *J. Appl. Phys.*, 117(21): 214501, 2015. http://scitation.aip.org/content/aip/journal/jap/117/21/10.1063/1.4921804.

**186** KL Jensen, DA Shiffler, JR Harris, IM Rittersdorf, and JJ Petillo. 2D/3D image charge for modeling field emission. *J. Vac. Sci. Technol. B.*, 35(2): 02C101, 2017. https://doi.org/10.1116/1.4968007.

**187** KL Jensen, DA Shiffler, IM Rittersdorf, JL Lebowitz, JR Harris, YY Lau, JJ Petillo, W Tang, and JW, Luginsland. Discrete space charge affected field emission: flat and hemisphere emitters. *J. Appl. Phys.*, 117(19): 194902/1–17, 2015. https://doi.org/10.1063/1.4921186.

**188** KL Jensen, D Dimitrov, V Pavlenko, PG O'Shea. Model of photoemission and framework for relating quantum efficiency to stoichiometry. *J. Appl. Phys.*, 137(14):145102, 2025. https://doi.org/10.1063/5.0229617.

**189** KL Jensen, JE. Yater, JL Shaw, RE Myers, BB Pate James, E Butler, and T Feygelson. Bunch characteristics of an electron beam generated by a diamond secondary emitter amplifier. *J. Appl. Phys.*, 108(4): 044509–044512, 2010. https://doi.org/10.1063/1.3462437.

**190** KL Jensen, JE Yater, EG Zaidman, MA Kodis, and A Shih. Advanced emitters for next generation rf amplifiers. *J. Vac. Sci. Technol. B.*, 16(4): 2038–2048, 1998.

**191** KL Jensen and EG Zaidman. Field-emission from an elliptic boss - exact versus approximate treatments. *Appl. Phys. Lett.*, 635: 702–704, 1993.

**192** L Johnson, B Gilchrist, RD Estes, and E Lorenzini. Overview of future NASA tether applications. In HG James, WJ Raitt, and MP Sultzer editors, *Overview of Future NASA Tether Applications*, volume 24(8), pages 1055–1063, *Int Union Radio Sci; Comm Space Res*, 1999.

**193** W Jones and Norman H March. *Theoretical Solid State Physics*. Dover Publications, New York, 1985.

**194** MI Kaganov, IM Lifshitz, and LV Tanatarov. Relaxation between electrons and the crystalline lattice. *Sov. Phys. Jetp-ussr.*, 4(2): 173–178, 1957.

**195** L Kalarasse, B Bennecer, and F Kalarasse. Optical properties of the alkali antimonide semiconductors cs3Sb, cs2KSb, csk2Sb and k3sb. *J. Phys. Chem. Solids.*, 71(3): 314–322, 2010. https://doi.org/10.1016/j.jpcs.2009.12.083.

**196** KA Kanaya and S Okayama. Penetration and energy-loss theory of electrons in solid targets. *J. Phys. D: Appl. Phys.*, 5(1): 43–58, 1972. https://doi.org/10.1088/0022-3727/5/1/308.

**197** S Karkare, D Dimitrov, W Schaff, L Cultrera, A Bartnik, X Liu, E Sawyer, T Esposito, and IV Bazarov. Monte Carlo charge transport and photoemission from negative electron affinity gaas photocathodes. *J. Appl. Phys.*, 113(10): 104904, 2013. https://doi.org/10.1063/1.4794822.

**198** TC Katsouleas, R Alarcon, J Albertine, I Ben-Zvi, SG Biedron, C Brau, WB Colson, RC Davidson, PG Gaffney, II L Merminga, JD Miller, BE Newnam, PG O'Shea, CKN Patel, D Prosnitz, and E Zimet. *Scientific Assessment of High-power Free-electron Laser Technology*. National Academies Press, Washington, DC, 2009. http://search.nap.edu/nap-cgi/de.cgi?term=0309126894.

**199** AC Keser, TM Antonsen, GS Nusinovich, DG Kashyn, and KL Jensen. Heating of micro-protrusions in accelerating structures. *Phys. Rev. Accel. Beams.*, 16: 092001 / 1–8, 2013. https://doi.org/10.1103/PhysRevSTAB.16.092001.

**200** RA Kishek, S Bernal, CL Bohn, D Grote, I Haber, H Li, PG O'Shea, M Reiser, and M Walter. Simulations and experiments with space-charge-dominated beams. *Phys. Plasmas.*, 10(5): 2016–2021, 2003. https://doi.org/10.1063/1.1558291.

**201** RA Kishek and YY Lau. A novel phase focusing mechanism in multipactor discharge. *Phys. Plasmas.*, 3(51): 1481–1483, 1996. https://doi.org/10.1063/1.872027.

**202** C Kittel. *Elementary Statistical Physics*. Wiley, New York, 1958.

**203** C Kittel. *Introduction to Solid State Physics*. Wiley, New York, 1996.

**204** C Kleint. On the early history of field-emission including attempts of tunneling spectroscopy. *Prog. Surf. Sci.*, 42 (1–4): 101–115, 1993.

**205** WS Koh, LK Ang, and TJT Kwan. Multidimensional short-pulse space-charge-limited flow. *Phys. Plasmas.*, 13(6): 063102, 2006. https://doi.org/10.1063/1.2208086.

**206** SH Kong, J Kinross-Wright, DC Nguyen, and RL Sheffield. Photocathodes for free electron lasers. *Nucl. Instrum. Methods Phys. Res. A*. 358 (1–3):272–275, 1995. http://www.sciencedirect.com/science/article/B6TJM-497C66C-2R/2/e1a1e857c1a9b61b994ef9532b8b5a6a.

**207** HG Kosmahl. Analytic evaluation of field-emission enhancement factors for ellipsoidal cones and elliptic cross-section wedges. *IEEE Trans. Electron Devices.*, 386: 1534–1537, 1991. https://doi.org/10.1109/16.81650

**208** WF Krolikowski and WE Spicer. Photoemission studies of the noble metals. i. copper. *Phys. Rev.*, 1853: 882–900, 1969. https://doi.org/10.1103/PhysRev.185.882.

**209** R Kubo. *Statistical Mechanics, an Advanced Course With Problems and Solutions*. North-Holland Pub. Co. Interscience Publishers, Amsterdam, New York, 1965.

**210** A Kyritsakis. General form of the tunneling barrier for nanometrically sharp electron emitters. *J. Appl. Phys.*, 133: 11, 2023. https://doi.org/10.1063/5.0144608.

**211** A Kyritsakis and F Djurabekova. A general computational method for electron emission and thermal effects in field emitting nanotips. *Comput. Mater. Sci.*, 128: 15–21, 2017. https://doi.org/10.1016/j.commatsci.2016.11.010.

**212** A Kyritsakis, GC Kokkorakis, JP Xanthakis, TL Kirk, and D Pescia. Self focusing of field emitted electrons at an ellipsoidal tip. *Appl. Phys. Lett.*, 97(2): 023104 / 1–3, 2010. https://doi.org/10.1063/1.3462934.

**213** A Kyritsakis, M Veske, and F Djurabekova. General scaling laws of space charge effects in field emission. *New J. Phys.*, 23(6): 063003, 2021. https://doi.org/10.1088/1367-2630/abffa8.

**214** A Kyritsakis and JP Xanthakis. Extension of the general thermal field equation for nanosized emitters. *J. Appl. Phys.*, 119(4): 045303 / 1–6, 2016. https://doi.org/10.1063/1.4940721.

**215** F Ladstadter, PF de Pablos, U Hohenester, P Puschnig, C Ambrosch-Draxl, PL de, Andres, FJ Garcia-Vidal, and F Flores. Hot-electron lifetimes in metals: a combined ab initio calculation and ballistic electron emission spectroscopy analysis. *Phys. Rev. B.*, 68(8), 2003. https://doi.org/10.1103/PhysRevB.68.085107.

**216** P Lancaster and K Salkauskas. *Curve and Surface Fitting: An Introduction*. Academic Press, London; Orlando, 1986.

**217** SA Lang, AM Darr, and AL Garner. Incorporating photoemission into the theoretical unification of electron emission and space-charge limited current. *J. Vac. Sci. Technol. B.*, 39(6): 062808, 2021. https://doi.org/10.1116/6.0001515.

**218** ND Lang and W Kohn. Theory of metal surfaces - work function. *Phys. Rev. B.*, 3(4):1215–1223, 1971.

**219** ND Lang and W Kohn. Theory of metal-surfaces - induced surface charge and image potential. *Phys. Rev. B.*, 7(8): 3541–3550, 1973.

**220** I Langmuir. The effect of space charge and residual gases on thermionic currents in high vacuum. *Phys. Rev.*, 2(6): 450–486, 1913. https://journals.aps.org/pr/abstract/10.1103/PhysRev.2.450.

**221** I Langmuir and KB Blodgett. Currents limited by space charge coaxial cylinders. *Phys. Rev.*, 22(4): 0347–0356, 1923. https://doi.org/10.1103/ PhysRev.22.347.

**222** I Langmuir and KB Blodgett. Currents limited by space charge between concentric spheres. *Phys. Rev.*, 24(1): 49–59, 1924. https://doi.org/10.1103/PhysRev.24.49.

**223** YY Lau. Simple theory for the two-dimensional Child–Langmuir law. *Phys. Rev. Lett.* 87(27): 278301, 2001. https://doi.org/10.1103/PhysRevLett.87.278301.

**224** YY Lau, KL Jensen, and B Levush. A comparison of flicker noise and shot noise on a hot cathode. *IEEE Trans. Plasma Sci.*, 28(3): 794–797, 2000. https://doi.org/10.1109/27.887726.

**225** DF Lawden. *Principles of Thermodynamics and Statistical Mechanics*. Dover Publications, Mineola, NY 2005. http://www.loc.gov/catdir/enhancements/fy0625/2005049737-d.html

**226** JD Lawson, PM Lapostolle, and RL Gluckstern. Emittance, entropy and information. *Part. Accel.*, 5: 61–65, 1973. https://cds.cern.ch/record/1107925/files/p61.pdf.

**227** B Leblond. Short pulse photoemission from a dispenser cathode under the 2nd, 3rd and 4th harmonics of a picosecond nd-yag laser. *Nucl. Instrum. Methods Phys. Res. A.*, 317 (1–2):365–372, 1992.

**228** P Lerner, PH Cutler, and Nm Miskovsky. Theoretical analysis of field emission from a metal diamond cold cathode emitter. *J. Vac. Sci. Technol.* B. 15(2): 337–342, 1997. https://doi.org/10.1116/1.589317.

**229** JW Lewellen and J Noonan. Field-emission cathode gating for RF electron guns. *Phys. Rev. Accel. Beams.*, 8(3): 033502, 2005. http://link.aps.org/abstract/PRSTAB/v8/e033502.

**230** JW Lewellen, SV Milton, E Gluskin, ND Arnold, C Benson, W Berg, SG Biedron, M Borland, YC Chae, RJ Dejus, PK Den, HB Deriy, M Erdmann, YI Eidelman, MW Hahne, Z Huang, KJ Kim, Y Li, AH Lumpkin, O Makarov, ER Moog, A Nassiri, V Sajaev, R Soliday, BJ Tieman, EM Trakhtenberg, IB Vasserman, NA Vinokurov, G Wiemerslage, and BX Yang. Present status and recent results from the APS SASE FEL. *Nucl. Instrum. Methods Phys. Res. A.*, 483 (1–2): 40–45, 2002. https://doi.org/10.1016/S0168-9002(02)00283-8.

**231** JE Lilienfeld. The visible radiation of the focal spot of x-ray tubes. *Phys. Zeitschr.* 280–282, 1919.

**232** W Liu, M Poelker, J Smedley, and R Ganter. Photocathodes. In G Gaertner, W Knapp, and R G Forbes editors, *Modern Developments in Vacuum Electron Sources* pages 293–344 Springer International Publishing Cham, 2020. https://doi.org/10.1007/978-3-030-47291-7_7.

**233** F Liu, S Sidhik, MA Hoffbauer, S Lewis, AJ Neukirch, V Pavlenko, H Tsai, W Nie, J Even, S Tretiak, PM Ajayan, MG Kanatzidis, JJ Crochet, NA Moody, JC Blancon, and AD Mohite. Highly efficient photoelectric effect in halide perovskites for regenerative electron sources. *Nat Commun.*, 12(1) 673, 2021. https://pubmed.ncbi.nlm.nih.gov/33514723.

**234** RT Longo. Physics of thermionic dispenser cathode aging. *J. Appl. Phys.* 9410: 6966–6975, 2003. https://doi.org/10.1063/1.1621728.

**235** AM Loveless, AM Darr, and AL Garner. Linkage of electron emission and breakdown mechanism theories from quantum scales to paschen's law. *Phys. Plasmas.*, 28(4): 042110, 2021. https://aip.scitation.org/doi/full/10.1063/5.0040633.

**236** JW Luginsland, YY Lau, and RM Gilgenbach. Two-dimensional Child–Langmuir law. *Phys. Rev. Lett.*, 77(22): 4668–4670, 1996.

**237** JW Luginsland, MJ Arman, and YY Lau. High-power transit-time oscillators: onset of oscillation and saturation. *Phys. Plasmas.*, 4(12): 4404–4408, 1997. https://doi.org/10.1063/1.872603.

**238** JW Luginsland, YY Lau, RJ Umstattd, and JJ Watrous. Beyond the Child–Langmuir law: a review of recent results on multidimensional space-charge-limited flow. *Phys. Plasmas.*, 9(5): 2371–2376, 2002. https://doi.org/10.1063/1.1459453.

**239** AV Lugovskoy and I Bray. Electron relaxation and excitation processes in metals. *J. Phys. D: Appl. Phys.*, 31(23): L78–L84, 1998. https://doi.org/10.1088/0022-3727/31/23/002.

**240** AV Lugovskoy and I Bray. Electron-electron scattering rate in thin metal films. *Phys. Rev. B.*, 65(4) 045405, 2002. https://doi.org/10.1103/PhysRevB.65.045405.

**241** AV Lugovskoy, T Usmanov, and AV Zinoviev. Angular distribution of photoelectrons emitted at pulse laser irradiation. *J. Opt. Soc. Am. B.*, 15(1): 53–63, 1998. https://doi.org/10.1364/JOSAB.15.000053.

**242** WW Lui and M Fukuma. Exact solution of the Schrodinger equation across an arbitrary one-dimensional piecewise-linear potential barrier. *J. Appl. Phys.*, 60(5): 1555–1559, 1986.

**243** M Lundstrom. *Fundamentals of Carrier Transport*, 2e. Cambridge University Press, Cambridge, UK; New York, 2000. http://www.loc.gov/catdir/description/cam0210/99049980.html

**244** RG Lye and AJ Dekker. Theory of secondary emission. *Phys. Rev.*, 107(4): 977–981, 1957. https://doi.org/10.1103/PhysRev.107.977.

**245** JB Marion and MA Heald. *Classical Electromagnetic Radiation*. Academic Press, New York, 1980.

**246** CM Marrese, JE Polk, KL Jensen, AD Gallimore, CA Spindt, RL Fink, and WD Palmer. Chapter 11: Performance of field emission cathodes in xenon electric propulsion systems environments. In Michael M and Micci Andrew D Ketsdever editors, *Micropropulsion for Small Spacecraft* pages 271–302 American Institute of Aeronautics and Astronautics Reston, Va, 2000. https://doi.org/10.2514/5.9781600866586.0271.0302.

**247** EE Martin, JK Trolan, and WP Dyke. Stable, high density field emission cold cathode. *J. Appl. Phys.*, 31: 782, 1960. http://link.aip.org/link/?JAPIAU/31/782/1.

**248** MATLAB version (R2023b). *The Mathworks Inc.* Natick, Massachusetts, 2023.

**249** RA Matula. Electrical resistivity of copper, gold, palladium, and silver. *J. Phys. Chem. Ref. Data.*, 8(4): 1147–1298, 1979. http://scitation.aip.org/content/aip/journal/jpcrd/8/4/10.1063/1.555614

**250** J Maxson, P Musumeci, L Cultrera, S Karkare, and H Padmore. Ultrafast laser pulse heating of metallic photocathodes and its contribution to intrinsic emittance. *Nucl. Instrum. Meth. Phys. Res. Sect A*, 865: 99–104, 2017. https://doi.org/10.1016/j.nima.2016.08.032.

**251** A Mayer. A comparative study of the electron transmission through one-dimensional barriers relevant to field-emission problems. *J. Phys.: Condens. Mat.*, 22: 175007, 2010. https://doi.org/10.1088/0953-8984/22/17/175007.

**252** WL Mcmillan. Transition temperature of strong-coupled superconductors. *Phys. Rev.*, 167(2): 331–344, 1968. https://doi.org/10.1103/PhysRev.167.331.

**253** RA Millikan and CF Eyring. Laws governing the pulling of electrons out of metals by intense electrical fields. *Phys. Rev.*, 27(1): 51–67, 1926. http://link.aps.org/abstract/PR/v27/p51.

**254** RA Millikan and CC Lauritsen. Relations of field currents to thermionic currents. *Proc Natl. Acad. Sci. USA* 14: 45–49, 1928. http://www.pnas.org/content/14/1/45.long.

**255** RA Millikan and CC Lauritsen. Dependence of electron emission from metals upon field strengths and temperatures. *Phys. Rev.*, 33(4): 0598–0604, 1929.

**256** A Modinos. *Field, Thermionic, and Secondary Electron Emission Spectroscopy*. Plenum Press, New York, 1984. http://books.google.com/books?id=MtbvAAAAMAAJ.

**257** W Mönch. *Semiconductor Surfaces and Interfaces*, volume Springer series in surface sciences; 26. Springer-Verlag, New York, 1995.

**258** NA Moody, KL Jensen, DW Feldman, EJ Montgomery, and PG O'Shea. Factors affecting performance of dispenser photocathodes. *J. Appl. Phys.*, 102(10): 104901–104909, 2007. https://doi.org/10.1063/1.2805653.

**259** NA Moody, KL Jensen, A Shabaev, SG Lambrakos, J Smedley, D Finkenstadt, JM Pietryga, PM Anisimov, V Pavlenko, ER Batista, JW Lewellen, F Liu, G Gupta, A Mohite, H Yamaguchi, MA Hoffbauer, and I Robel. Perspectives on designer photocathodes for x-ray free-electron lasers: influencing emission properties with heterostructures and nanoengineered electronic states. *Phys. Rev. Appl.*, 10(4): 047002, 2018. https://doi.org/10.1103/PhysRevApplied.10.047002.

**260** D Morris, B Gilchrist, and A Gallimore. The effect of plasma density and emitter geometry on space charge limits for field emitter array electron charge emission into a space plasma. In Mohamed S El-Genk Mary J Bragg editors, *The Effect of Plasma Density and Emitter Geometry on Space Charge Limits for Field Emitter Array Electron Charge Emission into a Space Plasma* volume 552(1) pages 467–472 Albuquerque, New Mexico, 2001. AIP.

**261** Pd Mumford and M Cahay. Space-charge effects and current self-quenching in a metal/cds/las cold cathode. *J. Appl. Phys.*, 84(5): 2754–2767, 1998.

**262** K Murata and DF Kyser. Monte-Carlo methods and microlithography simulation for electron and x-ray-beams. In Peter W Hawkes editor, *Advances in Electronics and Electron Physics* Vol 69, pages 175–259 Academic Press, 1987.

**263** K Murata, DF Kyser, and CH Ting. Monte-Carlo simulation of fast secondary-electron production in electron-beam resists. *J. Appl. Phys.*, 52(7): 4396–4405, 1981. https://doi.org/10.1063/1.329366.

**264** EL Murphy and RH Good. Thermionic emission, field emission, and the transition region. *Phys. Rev.*, 102(6): 1464–1473, 1956. https://doi.org/10.1103/PhysRev.102.1464.

**265** P Musumeci, L Cultrera, M Ferrario, D Filippetto, G Gatti, MS Gutierrez, JT Moody, N Moore, JB Rosenzweig, CM Scoby, G Travish, and C Vicario. Multiphoton photoemission from a copper cathode illuminated by ultrashort laser pulses in an rf photoinjector. *Phys. Rev. Lett.*, 104(8) 084801, 2010. https://doi.org/10.1103/PhysRevLett.104.084801.

**266** GR Neil, C Behre, SV Benson, M Bevins, G Biallas, J Boyce, J Coleman, La Dillon-townes, D Douglas, Hf Dylla, R Evans, A Grippo, D Gruber, J Gubeli, D Hardy, C Hernandez-garcia, K Jordan, MJ Kelley, L Merminga, J Mammosser, W Moore, N Nishimori, E Pozdeyev, J Preble, R Rimmer, M Shinn, T Siggins, C Tennant, R Walker, GP Williams, and S Zhang. The JLAB high power ERL light source. *Nucl. Instrum. Methods Phys. Res. Sect. A* 557(1): 9–15, 2006.

**267** J Norem, V Wu, A Moretti, M Popovic, Z Qian, L Ducas, Y Torun, and N Solomey. Dark current, breakdown, and magnetic field effects in a multicell, 805 MHz cavity. *Phys. Rev. Accel. Beams.*, 6(7):072001 / 1–21 2003.

**268** JR Oppenheimer. Three notes on the quantum theory of aperiodic effects. *Phys. Rev.*, 31(1): 66, 1928.

**269** PG O'Shea, SC Bender, DA Byrd, JW Early, DW Feldman, CM Fortgang, JC Goldstein, BE Newnam, RL Sheffield, RW Warren, and TJ Zaugg. Ultraviolet free-electron laser driven by a high-brightness 45-mev electron beam. *Phys. Rev. Lett.*, 71(22): 3661, 1993. https://doi.org/10.1103/PhysRevLett.71.3661.

**270** PG O'Shea, SC Bender, BE Carlsten, JW Early, DW Feldman, RB Feldman, WJD Johnson, AH Lumpkin, RL Sheffield, RW Springer, WE Stein, and LM Young. Performance of the photoinjector accelerator for the Los Alamos free-electron laser. *Phys. Rev. Lett.*, 5: 2754–2756, 1991.

**271** M Osofsky, H Tardy, M LaMadrid, and JM Mochel. Strong and weak electron spin-orbit scattering near the metal-insulator transition. *Phys. Rev. B.*, 31(7): 4715–4717, 1985. https://doi.org/10.1103/PhysRevB.31.4715.

**272** NA Papadogiannis, SD Moustaizis, and JP Girardeau-Montaut. Electron relaxation phenomena on a copper surface via nonlinear ultrashort single-photon photoelectric emission. *J. Phys. D: Appl. Phys.*, 30(17): 2389–2396, 1997. https://doi.org/10.1088/0022-3727/30/17/004.

**273** NA Papadogiannis, SD Moustaizis, PA Loukakos, and C Kalpouzos. Temporal characterization of ultra short laser pulses based on multiple harmonic generation on a gold surface. *Appl. Phys. B.*, 65(3): 339–345, 1997.

**274** RK Parker, Jr RH Abrams, BG Danly, and B Levush. Vacuum electronics. *IEEE Trans. Microw. Theory Techniq.*, 50(3): 835–845, 2002.

**275** BD Patterson, R Abela, H-H Braun, U Flechsig, R Ganter, Y Kim, E Kirk, A Oppelt, M Pedrozzi, S Reiche, L Rivkin, Th Schmidt, B. Schmitt, VN Strocov, S Tsujino, and AF Wrulich. Coherent science at the SwissFEL X-ray laser. *New J. Phys.*, 12, 2010.

**276** JB Pawley. The development of field-emission scanning electron microscopy for imaging biological surfaces. *Scanning* 19: 324–336, 1997. http://www.zoology.wisc.edu/faculty/paw/pdfs/scanning_fesem.pdf.

**277** A Pedersen, A Manolescu, and A Valfells. Space-charge modulation in vacuum microdiodes at THz frequencies. *Phys. Rev. Lett.*, 104(17): 175002, 2010. https://doi.org/10.1103/PhysRevLett.104.175002.

**278** J Perrin and DL Hammick. *Atoms. D.* Van Nostrand Ccompany, New York, 1916.

**279** JJ Petillo, SJ Cooke, TM Antonsen, SG Ovtchinnikov, AJ Jensen, EM Nelson, KL Jensen, and B Levush. First principles codes and analysis environments for vacuum electronics simulation. *IEEE Trans. Electron Dev.*, 70(6): 2666–2679, 2023. https://doi.org/10.1109/TED.2023.3236910.

**280** JJ Petillo, EM Nelson, JF Deford, NJ Dionne, and B Levush. Recent developments to the michelle 2-d/3-d electron gun and collector modeling code. *IEEE Trans. Electron Dev.*, 52(5): 742–748, 2005. https://doi.org/10.1109/TED.2005.845800.

**281** HR Philipp and H Ehrenreich. Optical properties of semiconductors. *Phys. Rev.*, 129(4): 1550–1560, 1963. https://doi.org/10.1103/PhysRev.129.1550.

**282** CM Pierce, JK Bae, A Galdi, L Cultrera, I Bazarov, and J Maxson. Beam brightness from cs–te near the photoemission threshold. *Appl. Phys. Lett.*, 118(12): 124101, 2021. https://doi.org/10.1063/5.0044917.

**283** PJ Price. Monte Carlo caleulation of electron transport in solids. In RK Willardson and AC Beer editors, *Semiconductors and Semimetals* Volume 14, pages 249–308, Academic Press, New York, 1979.

**284** PJ Price. Resonant wavefunctions and exchange effect in nanostructures. *Superlatt. Microstruct.*, 26(6): 363–365, 1999.

**285** AD Rakic, AB Djurisic, JM Elazar, and ML Majewski. Optical properties of metallic films for vertical-cavity optoelectronic devices. *Appl. Opt.*, 37(22): 5271–5283, 1998. https://doi.org/10.1364/AO.37.005271.

**286** AD Rakic and ML Majewski. Modeling the optical dielectric function of GaAs and AlAs: extension of Adachi's model. *J. Appl. Phys.* 80: 5909, 1996. https://doi.org/10.1063/1.363586.

**287** T Rao, A Burrill, XY Chang, J Smedley, T Nishitani, CH Garcia, M Poelker, E Seddon, FE Hannon, CK Sinclair, J Lewellen, and D Feldman. Photocathodes for the energy recovery linacs. *Nucl. Instrum. Methods Phys. Res. A.*, 557(1): 124–130, 2006.

**288** FH Read and NJ Bowring. Simulation of thermionic cathodes. *Nucl. Instrum. Methods Phys. Res. A.*, 531(3): 407–415, 2004. https://doi.org/10.1016/j.nima.2004.05.092.

**289** LE Reichl. *A Modern Course in Statistical Physics*. University of Texas Press, Austin, 1980.

**290** LE Reichl. *A modern course in statistical physics*. Wiley Weihheim, 3rd rev. and updated ed edition, 2009. ISBN 9783527407828. http://www.loc.gov/catdir/enhancements/fy1009/2010287506-t.html.

**291** F Reif. *Fundamentals of Statistical and Thermal Physics*. McGraw-Hill, New York, 1965.

**292** I Reiner. *Introduction to Matrix Theory and Linear Algebra*. Holt, Rinehart and Winston, New York, 1971.

**293** M Reiser. *Theory and Design of Charged Particle Beams*, volume Wiley series in beam physics and accelerator technology. Wiley, New York, 1994. http://www.loc.gov/catdir/description/wiley032/93033110.html.

**294** EH Rhoderick. Physics of Schottky barriers. *J. Phys. D: Appl. Phys.*, 38: 1153, 1970. https://doi.org/10.1088/0022-3727/3/8/203.

**295** EH Rhoderick. Metal-semiconductor contacts. *IEE Proceedings I Solid-state and Electron Devices.*, 129: 1, 1982. https://doi.org/10.1049/ip-i-1.1982.0001.

**296** OW Richardson. The emission of electrons from tungsten at high temperatures - an experimental proof that the electric current in metals is carried by electrons. *Science*, 381: 57–61, 1913.

**297** DS Richeson. *Tales of Impossibility: The 2000-year Quest to Solve the Mathematical Problems of Antiquity*. Princeton University Press, Princeton, NJ, 2019.

298 BK Ridley. *Quantum Processes in Semiconductors*. Clarendon Press, Oxford, 1999.

299 J Riga and R Seviour. Electromagnetic analogs of quantum mechanical tunneling. *J. Appl. Phys.*, 132(20): 200901 2022. https://aip.scitation.org/doi/full/10.1063/5.0118308.

300 J Robertson. Band model for electron emission from diamond and diamond-like carbon. In *MRS Online Proceedings Library* 471 pages 217–229 Cambridge University Press, 1997. https://doi.org/10.1557/PROC-471-217[Opensinanewwindow].

301 J Robertson. Mechanisms of electron field emission from diamond, diamond-like carbon, and nanostructured carbon. *J Vac Sci Technol B.*, 17(2): 659–665, 1999. https://doi.org/10.1116/1.590613.

302 A Rokhlenko, KL Jensen, and JL Lebowitz. Space charge effects in field emission: one dimensional theory. *J. Appl. Phys.*, 107(1): 014904–014910, 2010. https://doi.org/10.1063/1.3272690.

303 JM Roquais, B Vancil, and M Green. Chapter 2: Review on impregnated and reservoir ba dispenser cathodes. In Georg Gaertner, Wolfram Knapp, and Richard G Forbes editors, *Modern Developments in Vacuum Electron Sources*. Springer, 2020. https://doi.org/10.1007/978-3-030-47291-7_1.

304 JB Rosenzweig, A Cahill, B Carlsten, G Castorina, M Croia, C Emma, A Fukusawa, B Spataro, D Alesini, V Dolgashev, M Ferrario, G Lawler, R Li, C Limborg, J Maxson, P Musumeci, R Pompili, S Tantawi, and O Williams. Ultra-high brightness electron beams from very-high field cryogenic radiofrequency photocathode sources. *Nucl. Instrum. Methods Phys. Res. A*, 909: 224–228, 2018. https://doi.org/10.1016/j.nima.2018.01.061.

305 P Rumbach and DB Go. Fundamental properties of field emission-driven direct current microdischarges. *J. Appl. Phys.*, 112(10): 103302, 2012. https://doi.org/10.1063/1.4764344.

306 LI Schiff. *Quantum Mechanics*. McGraw-Hill, New York, 1968.

307 W Schottky. Über kalte und warme elektronenentladungen. *Zeitschr. Phys.* 14(1): 63–106, 1923. https://doi.org/10.1007/BF01340034.

308 H Shelton. Thermionic emission from a planar tantalum crystal. *Phys. Rev.*, 107(6): 1553–1557, 1957. https://doi.org/10.1103/PhysRev.107.1553.

309 DA Shiffler, J Luginsland, M Ruebush, M Lacour, K Golby, K Cartwright, M Haworth, and T Spencer. Emission uniformity and shot-to-shot variation in cold field emission cathodes. *IEEE Trans. Plasma Sci.*, 32(3): 1262–1266, 2004. https://doi.org/10.1109/TPS.2004.827608.

310 A Shih, JE Yater, and C Hor. Ba and Bao on W and on $Sc_2O_3$ coated W. *Appl. Surf. Sci.*, 242 (1–2):35–54, 2005. https://doi.org/10.1016/j.apsusc.2004.07.067.

311 A Shih, J Yater, C Hor, and R Abrams. Secondary electron emission studies. *Appl. Surf. Sci.*, 111: 251–258, 1997. https://doi.org/10.1016/S0169-4332(96)00729-5.

312 JG Simmons. Generalized formula for the electric tunnel effect between similar electrodes separated by a thin insulating film. *J. Appl. Phys.*, 34(6): 1793–1803, 1963. http://scitation.aip.org/content/aip/journal/jap/34/6/10.1063/1.1702682.

313 CK Sinclair. DC photoemission electron guns as ERL sources. *Nucl. Instrum. Methods Phys. Res. A.*, 557(1): 69–74, 2006.

314 A Sitek, K Torfason, A Manolescu, and A Valfells. Edge effect on the current-temperature characteristic of finite-area thermionic cathodes. *Phys. Rev. Appl.*, 16(3): 034043, 2021. https://doi.org/10.1103/PhysRevApplied.16.034043.

315 A Sitek, K Torfason, A Manolescu, and A Valfells. Space-charge effects in the field-assisted thermionic emission from nonuniform cathodes. *Phys. Rev. Appl.*, 151, 2021. https://doi.org/10.1103/physrevapplied.15.014040.

316 M Skolnik. Role of radar in microwaves. *IEEE Trans. Microw. Theory Tech.*, 50(3): 625–632, 2002. https://doi.org/10.1109/22.989947.

**317** J Smedley, I Ben-Zvi, A Burrill, X Chang, J Grimes, T Rao, Z Segalov, and Q Wu. Electron amplification in diamond. *AIP Conf. Proc.*, 877(1): 672–679, 2006. https://doi.org/10.1063/1.2409200.

**318** J Smedley, T Rao, J Sekutowicz, P Kneisel, J Langner, P Strzyzewski, R Lefferts, and A Lipski. Progress on lead photocathodes for superconducting injectors. In *Proceedings of the, 2005. Particle Accelerator Conference* pages 4099–4101 Knoxville, TN 2005 IEEE.

**319** J Smedley, T Rao, and Q Zhao. Photoemission studies on niobium for superconducting photoinjectors. *J. Appl. Phys.*, 98(4): 043111, 2005. https://doi.org/10.1063/1.2008389.

**320** GD Smith. *Numerical Solution of Partial Differential Equations: Finite Difference Methods*, volume Oxford applied mathematics and computing science series. Clarendon Press Oxford University Press Oxford [Oxfordshire], New York, 1985. http://www.loc.gov/catdir/enhancements/fy0603/85019867-d.html.

**321** NV Smith and GB Fisher. Photoemission studies of the alkali metals. ii. rubidium and cesium. *Phys. Rev. B.*, 3(11): 3662, 1971. http://link.aps.org/abstract/PRB/v3/p3662.

**322** AH Sommer. N-type and p-type conduction in alkali-antimonide photoemitters. *J. Appl. Phys.*, 29: 1568–1569, 1958.

**323** AH Sommer. The element of luck in research - photocathodes 1930 to 1980. *J. Vac. Sci. Technol. A.*, 1: 119–124, 1983. https://doi.org/10.1116/1.571898.

**324** WE Spicer. Photoemissive, photoconductive, and optical absorption studies of alkali-antimony compounds. *Phys. Rev.*, 112: 114–122, 1958. https://doi.org/10.1103/PhysRev.112.114.

**325** WE Spicer. Photoemission and related properties of the alkali-antimonides. *J. Appl. Phys.*, 31: 2077–2084, 1960. https://doi.org/10.1063/1.1735505.

**326** WE Spicer. The production of pairs in semiconductors by low energy electrons. *J. Phys. Chem. Solids.*, 22: 365–370, 1961. http://www.sciencedirect.com/science/article/B6TXR-46SPF8W-1S/2/fd5dfb93d5e47685a76db03541bd70e9.

**327** WE Spicer. Welch memorial lecture - the development of photoemission spectroscopy and its application to the study of semiconductor interfaces - observations on the interplay between basic and applied-research. *J. Vac. Sci. Technol. A.*, 3: 461–470, 1985.

**328** WE Spicer and A Herrera-Gomez. Modern theory and applications of photocathodes. *Proc. SPIE Int. Soc. Opt. Eng.* 2022 1: 18–35, 1993. https://doi.org/10.1117/12.158575.

**329** CA Spindt, I Brodie, L Humphrey, and ER Westerberg. Physical properties of thin-film field emission cathodes with molybdenum cones. *J. Appl. Phys.*, 47(12): 5248–5263, 1976. https://doi.org/10.1063/1.322600.

**330** CA Spindt, CE Holland, I Brodie, JB Mooney, and ER Westerberg. Field-emitter arrays applied to vacuum fluorescent display. *IEEE Trans. Electron Devices* 361: 225–228, 1989. https://doi.org/10.1051/jphyscol:1988625.

**331** NR Sree Harsha, JM Halpern, AM Darr, and AL Garner. Space-charge-limited current density for nonplanar diodes with monoenergetic emission using lie-point symmetries. *Phys. Rev. E.*, 106(6), 2022. https://doi.org/10.1103/physreve.106.l063201.

**332** WG Stam, M Gaowei, EM Echeverria, K Evans-Lutterodt, J Jordan-Sweet, T Juffmann, S Karkare, J Maxson, SJ. van der Molen, C Pennington, P Saha, J Smedley, and RM Tromp. Growth of ultra-flat ultra-thin alkali antimonide photocathode films. *APL Mater.* 12(6): 061114, 2024. https://doi.org/10.1063/5.0213461.

**333** R Stratton. Field emission from semiconductors. *Proc. Phys. Soc. Sect. B* 68(10): 746–757, 1955. https://doi.org/10.1088/0370-1301/68/10/307.

**334** R Stratton. Theory of field emission from semiconductors. *Phys. Rev.*, 125(1): 67–82, 1962. https://doi.org/10.1103/PhysRev.125.67.

**335** R Stuart, WE Spicer, and F Wooten. Monte Carlo calculations pertaining to transport of hot electrons in metals. *Phys. Rev. A*, 135(2A):A495, 1964.

**336** LW Swanson and RW Strayer. Field-electron-microscopy studies of cesium layers on various refractory metals: work function change. *J. Chem. Phys.*, 48(6): 2421–2442, 1968. https://doi.org/10.1063/1.1669464.

**337** SM Sze and KK Ng. *Physics of Semiconductor Devices*. Wiley-Interscience Hoboken, N.J. 2007. http://www.loc.gov/catdir/enhancements/fy0826/2007280778-b.html.

**338** EA Taft and HR Philipp. Structure in the energy distribution of photoelectrons from K3sb and cs3Sb. *Phys. Rev.*, 115(6): 1583–1586, 1959. https://doi.org/10.1103/PhysRev.115.1583.

**339** D Temple. Recent progress in field emitter array development for high performance applications. *Mater. Sci. Eng. R-Rep.*, 24(5): 185–239, 1999.

**340** JJ Thomson. Cathode rays. *Philos. Magaz.*, 90(sup1):25–29, 2010. https://doi.org/10.1080/14786431003659214.

**341** BH Tiang, YS Ang, and LK Ang. Over-barrier photoelectron emission with rashba spin–orbit coupling. *Appl. Phys. Lett.*, 123(16), 2023. https://doi.org/10.1063/5.0170250.

**342** K Torfason, A Sitek, A Manolescu, A Valfells. Dynamics of a field emitted beam from a microscopic inhomogeneous cathode. *IEEE Trans. Electron Devices.*, 68(5): 2461–2466 2021. https://doi.org/10.1109/ted.2021.3063210.

**343** K Torfason, A Valfells, and A Manolescu. Molecular dynamics simulations of field emission from a planar nanodiode. *Phys. Plasmas.*, 22(3): 033109, 2015. https://doi.org/10.1063/1.4914855.

**344** S Tsujino, F le Pimpec, J Raabe, M Buess, M Dehler, E Kirk, J Gobrecht, and A Wrulich. Static and optical field enhancement in metallic nanotips studied by two-photon photoemission microscopy and spectroscopy excited by picosecond laser pulses. *Appl. Phys. Lett.*, 94(9): 093508/1–3, 2009.

**345** RJ Umstattd, CG Carr, CL Frenzen, JW Luginsland, and YY Lau. A simple physical derivation of Child–Langmuir space-charge-limited emission using vacuum capacitance. *Am. J. Phys.*, 73(2): 160–163, 2005. https://doi.org/10.1119/1.1781664.

**346** A Valfells, DW. Feldman, M Virgo, PG O'Shea, and YY Lau. Effects of pulse-length and emitter area on virtual cathode formation in electron guns. *Phys. Plasmas.*, 9 (5):2377–2382, 2002. https://doi.org/10.1063/1.1463065.

**347** R Vaughan. A synthesis of the longo and eng cathode emission models. *IEEE Trans. Electron Dev.*, 33(11): 1925–1927, 1986. https://doi.org/10.1109/T-ED.1986.22844.

**348** JRM Vaughan. A new formula for secondary emission yield. *IEEE Trans. Electron Dev.*, 36(9): 1963–1967, 1989. https://doi.org/10.1109/16.34278.

**349** JP Vigier, C Dewdney, PR Holland, and A Kyprianidis. Causal particle trajectories and the interpretation of quantum mechanics. In B J Hiley and F David Peat editors, *Quantum Implications: Essays in Honour of David Bohm* pages 169–204 Routledge. New York, NY, 1987.

**350** DK Wagner and R Bowers. Radio-frequency size effect - tool for investigation of conduction electron-scattering in metals. *Adv. Phys.*, 27(5): 651–746, 1978. https://doi.org/10.1080/00018737800101464.

**351** JS Walker. Prompt and Utter Destruction: *Truman and the use of Atomic Bombs Against Japan*. University of North Carolina Press, Chapel Hill, 1997. http://www.h-net.org/review/hrev-a0b3w6-aa.

**352** CGH Walker, MM El-Gomati, AMD Assa'd, and M ZadrazìŒil. The secondary electron emission yield for 24 solid elements excited by primary electrons in the range 250–5000 ev: a theory/experiment comparison. *Scanning* 30(5): 365–380, 2008. https://doi.org/10.1002/sca.20124.

**353** E Wang, I Ben-Zvi, T Rao, DA. Dimitrov, X Chang, Q Wu, and T Xin. Secondary-electron emission from hydrogen-terminated diamond: Experiments and model. *Phys. Rev. ST Accel. Beams.*, 14: 111301, 2011. https://doi.org/10.1103/PhysRevSTAB.14.111301.

**354** H Wang, AM Loveless, AM Darr, and AL Garner. Transitions between field emission and vacuum breakdown in nanoscale gaps. *J. Vac. Sci. Technol. B*, 406: 062805, 2022. https://avs.scitation.org/doi/full/10.1116/6.0002160.

**355** X Wang and P Musumeci. Future of electron sources: report of the basic energy sciences workshop on the future of electron sources. *US Depart. Energy Off. Sci.*, 2017. https://doi.org/10.2172/1616511.

**356** J Wang, W Liu, K Zhang, and J Li. Development of high current-density cathodes with scandia-doped tungsten powders. *IEEE Trans. Electron Dev.*, 54(5): 1061–1070, 2007. https://doi.org/10.1109/TED.2007.894602.

**357** JH Weaver and HPR Frederikse. Chapter 12: Optical properties of selected elements. In DR Lide editor, *CRC Handbook of Chemistry and Physics 88th Edition* page 12.126 CRC, 2007.

**358** DG Westlake and LCR Alfred. Determination of Debye characteristic temperature of vanadium from Bloch-Gruneisen relation. *J. Phys. Chem. Solids.*, 29(11): 1931–1934, 1968.

**359** GK White and SB Woods. Electrical and thermal resistivity of the transition elements at low temperatures. *Phil. Trans. R. Soc. Lond. A.*, 251(995): 273–302, 1959.

**360** AH Wilson. The electrical conductivity of the transition metals. *Proc. Roy. Soc. London A*, 167: 0580–0593 1938. https://doi.org/10.1098/rspa.1938.0156.

**361** PB Wilson. Gradient limitation in accelerating structures imposed by surface melting. SLAC-PUB-9953, 2003.

**362** PB Wilson. Formation of taylor cones on a molten metal surface followed by ion injection into the vacuum. In *Formation of Taylor Cones on a Molten Metal Surface Followed by Ion Injection Into the Vacuum Stanford*, CA, 2007.

**363** DE Wooldridge. Theory of secondary emission. *Phys. Rev.*, 56(6): 562–578, 1939.

**364** F Wooten. *Optical Properties of Solids*. Academic Press, New York, 1972.

**365** F Wooten, JP Hernandez, and WE Spicer. Photoemission and electron scattering in Cs3Bi and Cs3Sb. *J. Appl. Phys.*, 44(3): 1112–1117, 1973. http://link.aip.org/link/?JAP/44/1112/1.

**366** H Yanagisawa, S Schnepp, C Hafner, M Hengsberger, DE Kim, MF Kling, A Landsman, L Gallmann, and J Osterwalder. Delayed electron emission in strong-field driven tunnelling from a metallic nanotip in the multi-electron regime. *Sci. Rep.*, 6: 35877–35877, 2016. https://doi.org/10.1038/srep35877.

**367** JE Yater. Secondary electron emission and vacuum electronics. *J. Appl. Phys.*, 133(5): 050901, 2023. https://doi.org/10.1063/5.0130972.

**368** JE Yater, JL Shaw, KL Jensen, T Feygelson, RE Myers, BB Pate, and JE Butler. Secondary electron amplification using single-crystal CVD diamond film. *Diam. Relat. Mater.*, 20 (5–6): 798–802, 2011. http://www.sciencedirect.com/science/article/pii/S0925963511001257.

**369** JE Yater and A Shih. Secondary electron emission characteristics of single-crystal and polycrystalline diamond. *J. Appl. Phys.*, 87(11): 8103–8112, 2000. https://doi.org/10.1063/1.373505.

**370** JE Yater, A Shih, JE Butler, and PE Pehrsson. Electron transmission studies of diamond films. *Appl. Surf. Sci.*, 191 (1–4): 52–60, 2002. https://doi.org/10.1016/S0169-4332(02)00149-6.

**371** JR Young. Penetration of electrons in aluminum oxide films. *Phys. Rev.*, 103(2): 292–293, 1956. https://doi.org/10.1103/PhysRev.103.292.

**372** JR Young. Dissipation of energy by 2.5-10 kev electrons in $Al_2O_3$. *J. Appl. Phys.*, 28(5): 524–525, 1957.

**373** RD Young and EW Müller. Experimental measurement of the total-energy distribution of field-emitted electrons. *Phys. Rev.*, 113(1): 115–120, 1959. http://link.aps.org/abstract/PR/v113/p115.

**374** H Yukawa. Meson theory in its development. In Jefferson H Weaver, L Motz, and D McAdoo editors, *The World of Physics: A Small Library of the Literature of Physics from Antiquity to the Present* volume II, page 754. Simon and Schuster, New York, 1987.

**375** P Zhang. Scaling for quantum tunneling current in nano- and subnano-scale plasmonic junctions. *Sci Rep.*, 5: 9826, 2015. https://www.ncbi.nlm.nih.gov/pubmed/25988951.

**376** P Zhang, Y Sin, A Allen, L Garner, AÌ? Valfells, JW Luginsland, and LK Ang. Space–charge limited current in nanodiodes: ballistic, collisional, and dynamical effects. *J. Appl. Phys.*, 129(10): 100902, 2021. https://doi.org/10.1063/5.0042355.

**377** P Zhang, A Valfells, LK Ang, JW Luginsland, and YY Lau. 100 years of the physics of diodes. *Appl. Phys. Rev.*, 4(1): 011304, 2017. https://doi.org/10.1063/1.4978231.

**378** Y Zhou and P Zhang. A quantum model for photoemission from metal surfaces and its comparison with the three-step model and Fowler–DuBridge model. *J. Appl. Phys.*, 127(16): 164903, 2020. https://doi.org/10.1063/5.0004140.

**379** W Zhu. *Vacuum Microelectronics.* Wiley New York, 2001. https://doi.org/10.1002/0471224332

**380** YB Zhu, P Zhang, A Valfells, LK Ang, and YY Lau. Novel scaling laws for the Langmuir-Blodgett solutions in cylindrical and spherical diodes. *Phys. Rev. Lett.*, 110(26), 2013. https://doi.org/10.1103/PhysRevLett.110.265007.

**381** B Ziaja, RA London, and J Hajdu. Ionization by impact electrons in solids: electron mean free path fitted over a wide energy range. *J. Appl. Phys.*, 99: 033514 2006. https://doi.org/10.1063/1.2161821.

**382** JM Ziman. *Electrons and Phonons: The Theory of Transport Phenomena in Solids*, volume Oxford classic texts in the physical sciences. Clarendon Press Oxford University Press, Oxford, New York, 2001. http://www.loc.gov/catdir/enhancements/fy0611/2001268325-t.html.

**383** D Zuber, KL Jensen, and TE Sullivan. An analytical solution for microtip field emission current and effective emission area. *J. Appl. Phys.*, 91(11): 9379–9384, 2002. https://doi.org/10.1063/1.1474596.

# Index

Printed and bound by CPI Group (UK) Ltd, Croydon, CR0 4YY
21/09/2026
14965845-0001